Acoustic Absorbers and Diffusers

Theory, Design and Application

Third Edition

Acoustic Absorbers and Diffusers

Theory, Design and Application

Third Edition

Trevor J. Cox
University of Salford, UK

Peter D'Antonio
RPG Diffusor Systems, USA

CRC Press
Taylor & Francis Group
Boca Raton London New York

CRC Press is an imprint of the
Taylor & Francis Group, an **informa** business

CRC Press
Taylor & Francis Group
6000 Broken Sound Parkway NW, Suite 300
Boca Raton, FL 33487-2742

First issued in paperback 2020

ISBN-13: 978-1-4987-4099-9 (hbk)
ISBN-13: 978-0-367-65841-0 (pbk)

Library of Congress Cataloging-in-Publication Data

Names: Cox, Trevor J.. | D'Antonio, Peter.
Title: Acoustic absorbers and diffusers / Trevor Cox and Peter D'Antonio.
Description: Third edition. | Boca Raton : Taylor & Francis, 2017. | Includes bibliographical references and index.
Identifiers: LCCN 2016027694 | ISBN 9781498740999
Subjects: LCSH: Acoustical engineering. | Absorption of sound. | Diffusers. | Noise barriers. | Architectural acoustics.
Classification: LCC TA365 .C69 2017 | DDC 620.2--dc23
LC record available at https://lccn.loc.gov/2016027694

Visit the Taylor & Francis Web site at
http://www.taylorandfrancis.com

and the CRC Press Web site at
http://www.crcpress.com

To our families and Manfred Schroeder.

Contents

List of figures

List of symbols

Vectors are denoted in bold.

a	Half diffuser or reflector width (m)
	Fibre radius (m)
A	Total absorption of a room (m^2)
	Scaling constant
	Area
A_n	Amplitude for grating lobes (Pa)
b	Half diffuser length (m)
c	Speed of sound (ms^{-1}) in air unless otherwise stated (≈343 ms^{-1}). Subscript 0 denotes value in air where ambiguity might arise otherwise.
c_p	Specific heat capacity of air at constant pressure (≈1.01 JKg^{-1}K^{-1})
d	Cavity depth of resonant absorbers (m)
	Diffusion coefficient
	Thickness of materials (m)
d_ψ	Diffusion coefficient for incident angle ψ
d_n	Depth of the nth well in a Schroeder diffuser (m)
D	Cell width for Helmholtz absorbers (m)
E	Ratio of the specimen perimeter (P) to area (A) of absorption samples
f	Frequency (Hz)
f_0	Design frequency (Hz)
g	Acceleration due to gravity (ms^{-2})
	Green's function (time domain)
G	Green's function (frequency domain)
$H_0^{(1)}$	Hankel function of the first kind of order zero
H	Transfer function
f_s	Sampling frequency
F_p	Excitation function
FT()	Fourier transform
j	$\sqrt{-1}$
k	Wavenumber (m^{-1}). Subscript 0 denotes value in air where ambiguity might otherwise arise.
k_x	Wavenumber component in x direction, similar expressions for y and z (m^{-1})
K_e	Effective bulk modulus
l	Depth of materials (m)

m	Mass per area (kgm^{-2})
	Order of diffraction or grating lobes
	Energy attenuation coefficient for absorption in air (dB m^{-1})
	Constant relating finite to infinite sample absorption coefficients
M_T	Thermal shape factor
M_v	Viscous shape factor
n	Normal to surface, for BEM modelling this is pointing out of the surface
N	Number of wells per period
	Prime number generator and/or length of pseudorandom sequence
N_p	Prandtl number (≈0.77)
p	Pressure (Pa or Nm^{-2}). Subscript 0 denotes value in air where ambiguity might otherwise arise.
P	Perimeter
p_1	Pressure from a single diffuser/reflector (Pa)
p_a	Pressure from an array (Pa)
p_i	Pressure incident direct from a source (Pa)
p_m	Pressure of the *m*th order diffraction lobe (Pa)
P_0	Atmospheric pressure (≈101320 Pa)
p_s	Pressure scattered from a surface (Pa)
q	Viscous permeability
q'	Thermal permeability
q_0	Static viscous permeabilities
q'_0	Static thermal permeability
r	Distance (m)
	Primitive root of N
r'	Distance from image source (m)
r	Receiver position
r_0	Source position
$\mathbf{r}_1$	Vector, source to point on the surface
$\mathbf{r}_2$	Vector, point on the surface to the receiver
r_s	Point on diffuser surface
R	Pressure reflection coefficient
s, s_n	Number sequence
s	Courant number
	Diffuser surface
	Scattering coefficient
S	Surface area of a room (m^2)
	Area of hole in Helmholtz resonator (m^2)
sinc(x)	= sin(x)/x
S_{xxm}	Maximum energy in autocorrelation side lobes
S_{xym}	Maximum energy in cross-correlation
t	Time (s)
	Sheet thickness for Helmholtz and membrane absorbers (m)
t'	Sheet thickness for Helmholtz absorber including end corrections (m)
t_a	Resistive layer thickness for Helmholtz and membrane absorbers (m)
T	Temporal sampling period
T_{60}	Reverberation time (s)
u	Particle velocity (ms^{-1})
V	Volume (m^3)

w	Well width (m)
	Slot width (m)
W	Repeat distance or periodicity width (m)
x	Cartesian coordinate (m)
X	Dimensionless coordinate (Equation 6.11)
	Spatial sampling period
y	Cartesian coordinate (m)
z	Specific acoustic impedance (Pa s m^{-1} or rayl)
	Cartesian coordinate (m)
z_c	Characteristic impedance of a medium (rayl)
z_f	Flow impedance (rayl)
z_n	Normalised specific acoustic impedance (= $z/\rho c$)
z_s	Surface specific acoustic impedance
α	Absorption coefficient
α_s	Random incidence absorption coefficient
α_∞	Tortuosity (classical)
	Random incidence absorption coefficient for an infinite sized sample
β	Admittance (rayl^{-1})
β'	Admittance with outward pointing normal (rayl^{-1})
β_n	Normalised admittance (= $\beta\rho c$)
γ	Ratio of specific heat capacities ($\approx$1.4 in air)
δ	End correction factor (i.e., $\approx$0.85 if baffled)
	Delta function
δ_c	Correlation scattering coefficient
δ_h	Size of thermal boundary layer (m)
δ_v	Size of viscous boundary layer (m)
ε	Fractional open area
η	Viscosity of air (1.84×10^{-5} Nsm^{-2})
θ	Angle of reflection
κ	Thermal conductivity of air ($\approx 2.41 \times 10^{-2}$ WmK^{-1})
λ	Wavelength (m)
λ_0	Design wavelength (m)
ν	Kinemetric viscosity of air (15×10^{-6} m^{2}s^{-1})
ρ	Density (kgm^{-3}), subscript 0 denotes value in air where ambiguity might otherwise arise
ρ_e	Effective density of porous absorber
ρ_f	Density of porous absorber's fibres or grains (kgm^{-3})
ρ_m	Bulk density of porous absorber (kgm^{-3})
σ	Flow resistivity (rayl m^{-1})
σ_s	Flow resistance (rayl)
τ	Relaxation times (s)
ϕ	Open porosity
ψ	Angle of incidence
ω	Angular frequency (s^{-1})
ω_T	Thermal characteristic frequency
ω_v	Viscous characteristic frequency
Λ	Viscous characteristic length (m)
Λ'	Thermal characteristic length (m)
$*$	Convolution

List of acronyms

BEM Boundary element method
FDTD Finite difference time domain
GRAM Geometric room acoustic model
MLS Maximum length sequence
NFTFT Near field to far field transformation
PRD Primitive root diffuser
QRD Quadratic residue diffuser

Preface to the first edition

Every book tells a story, and there is a story behind every book. This story begins in 1980, in the conference room of the Laboratory for the Structure of Matter at the Naval Research Laboratory in Washington, DC, where Peter D'Antonio was employed as a diffraction physicist. Knowing Peter's interest in music, a colleague handed him the latest issue of *Physics Today* with a cover photo of Manfred Schroeder seated in an anechoic chamber. The article suggested using number theoretic diffusers in concert halls. While Peter's interest at the time was not in concert halls, he became fascinated with the thought of using these diffusers in a renovation of Underground Sound, a private studio he originally built in 1972 with Jerry Ressler. The acoustic renovation utilized a new concept called Live End Dead End proposed by Don and Carolyn Davis of Synergetic Audio Concepts (Syn-Aud-Con) and implemented successfully by Chips Davis.

At that time, Peter was examining the three-dimensional structure of matter in various phases, using electron and x-ray diffraction techniques. Peter shared the article with John Konnert, a colleague at Naval Research Laboratory, and it became apparent that the 'reflection phase gratings' suggested by Schroeder were in effect two-dimensional sonic crystals, which scatter sound in the same way that three-dimensional crystal lattices scatter electromagnetic waves. Since the diffraction theory employed in x-ray crystallographic studies was applicable to reflection phase gratings, it was straightforward to model and design the diffusers. At this time, Peter's only link to the field of acoustics was a love of composing, recording, and performing music. Having scientific backgrounds, John and Peter approached acoustics as they did the field of diffraction physics and began researching and publishing findings in the scientific literature. The Audio Engineering Society and Syn-Aud-Con offered a unique forum and community for discussing the research. In October 1983, at the 74th Audio Engineering Society (AES) Convention in New York, Peter met Bob Todrank following a presentation of Peter and John's first paper on Schroeder diffusers. Bob was designing a new studio for the Oak Ridge Boys in Hendersonville, TN and was interested in utilizing these new acoustical surfaces. The studio was a resounding success and turned out to be a harbinger of many exciting things to come.

In 1983, Peter and John measured quadratic residue and primitive root diffusers with a TEF 10 analyzer at a Syn-Aud-Con seminar in Dallas, TX, with the assistance of Don Eger of Techron. Here, Peter met Russ Berger, who was a pioneer in the use of new products in his firm's recording studios. In 1984, an intensive measurement programme was carried out using Richard Heyser's time delay spectrometry implementation. Farrell Becker was very helpful in the initial evaluation of these exciting new surfaces. Not having access to an anechoic chamber, a boundary measurement technique was developed. These measurements were initially carried out at full scale in large spaces like open fields and parking lots, eventually moving indoors to a sports arena, a motion picture sound stage, and a local high school gymnasium. The measurements enabled the theories to be validated.

The Oak Ridge Boy's Acorn Sound Recorders project was celebrated with a Syn-Aud-Con control room design workshop in 1984. This project led to many others, and collaborations with a growing community of new studio designers were undertaken. Neil Grant was an early staunch proponent of the research and products. Some of his milestone designs include Peter Gabriel's Real World Studios, Box, UK; Reba McEntire's Starstruck Studios, Nashville, TN; Sony Music, New York, NY; and Cinerama Theater, Seattle, WA. In 1989, John Storyk integrated diffusive technology in many of his designs, including Whitney Houston Studio, Mendham, NJ; Electronic Arts, Vancouver, BC; and Jazz at Lincoln Center, New York, highlighting the list. Today much of the recorded music you hear is created in music facilities utilizing RPG technology. These fledgling years established relationships that continue to this day and produced many acoustical landmarks.

Interest in recording facilities naturally spread to broadcast facilities, where diffuser technology is now commonplace. Facilities include British Broadcasting Corporation (BBC), National Public Radio (NPR), National Broadcasting Corporation (NBC), Canadian Broadcasting Corporation (CBC), and most of the broadcast networks due to Russ Berger's innovative designs. Being musicians and audiophiles led to significant involvement in residential high-end audio listening rooms, as well as production studios.

In 1989, Peter was introduced to Jack Renner, president of Telarc Records, the company that started the classical high-end recording industry on a digital journey. Jack was recording the Baltimore Symphony Orchestra at the Meyerhoff Symphony Hall and asked if RPG Diffusor Systems, Inc., could assist him. Following initial experimentation, Telarc graciously credited RPG as Telarc's exclusive acoustical system for control room and stage use for the Berlioz Symphonie Fantastique in 1990. The somewhat accidental stage use and overwhelming acceptance by musicians and conductor prompted an objective and subjective investigation of stage acoustics and acoustical shells both with small ensembles and with the Baltimore Symphony Orchestra. These chamber group studies were conducted with Tom Knab at the Cleveland Institute of Music, where Peter has been adjunct professor of acoustics since 1990, at the invitation of Jack Renner. In 1989, RPG was privileged to provide a custom number theoretic surface for the rear wall of Carnegie Hall, New York. This installation, along with the new diffusive acoustical shell development, launched RPG's involvement into performing arts applications, which eventually included the Fritz Philips Muziekcentrum, Eindhoven and the Corning Glass Center, Corning, NY.

Many of the acoustical consultants involved in the design of worship spaces began to include the use of diffusers for rear wall applications and acoustical shells. While RPG has collaborated with many acousticians, the relationship with Mike Garrison is noteworthy for the sheer number and size of the successful worship spaces produced using diffusers. The crown jewel of this collaboration is the 9000-seat South East Christian Church in Louisville, KY.

In 1990, RPG funded the Directional Incidence Scattering Coefficient Project in an attempt to devise a standard methodology for evaluating diffuser quality. In 1991, Peter proposed a directional diffusion coefficient and the Audio Engineering Society invited him to chair standards committee SC-04-02 to formerly develop an information document describing these procedures.

In 1993, David Quirt, associate editor of the *Journal of the Acoustical Society of America*, asked Peter to referee a paper by Trevor Cox entitled 'Optimization of Profiled Diffusers'. (Trevor's research journey had started a few years earlier in 1989 when, under the direction of Raf Orlowski and Yiu Wai Lam, he completed a PhD on Schroeder diffusers at Salford University, UK.) Trevor's paper outlined a process that combined boundary element modelling and multidimensional optimization techniques to make better diffusers. In Peter's view, this paper represented a creative milestone in diffuser development on par with Schroeder's seminal contribution. Peter and John's review of the paper consumed many months. It required the writing of boundary element codes and developing the first automated goniometer to measure these optimized surfaces. During the summer of 1994, Paul Kovitz helped to complete the measurement software. Trevor's revised paper, accompanied by a refereed paper of Peter and John's review, was published in 1995. Since this was nearly three years after Trevor submitted the paper to *Journal of the Acoustical Society of America*, this must have seemed to be the peer review from hell, especially as the referees' comments were 36 pages long.

Peter finally met Trevor in Amsterdam at an AES SC-04-02 standards committee meeting in 1994 and again in Arup Acoustics' office in London. Their strong mutual interests led to an informal collaboration. In 1995, Trevor became a research consultant to RPG. This relationship started with developing an automated program to optimize loudspeaker and listening positions in a critical listening room and blossomed to generate much of the contents of this book.

Realizing that good acoustical design results from an appropriate combination of absorptive, reflective, and diffusive surfaces, as mentioned in the Introduction, Peter (and later with Trevor) began developing absorption technologies as well, including hybrid abffusive (absorptive/diffusive) and diffsorptive (diffusing/absorbing) systems, concrete masonry units, low-frequency absorbing arena seating risers, nestable open-cell foam systems, and dedicated absorptive low-frequency membrane systems.

In 1995, Peter and Trevor became aware of the diffusion research of James Angus on amplitude gratings and modulated phase gratings. James has made significant contributions to the field of diffuser design, and Peter and Trevor both have great respect for his insight and enjoy their collaborations with him. Also in 1995, Peter and Trevor met Eckard Mommertz and Michael Vorlander at the 15th ICA in Trondheim, Norway. It was at this meeting that Peter and Trevor learned of their work developing a procedure to measure the random incidence scattering coefficient. All have maintained close collaboration to this day, especially as members of the ISO WG 25, chaired by Jens Holger Rindel.

To further the development of the diffusion coefficient, RPG cofunded a three-year grant with the Engineering and Physical Sciences Research Council of the United Kingdom, beginning in 1996. Trevor, Yiu Wai Lam, and Peter were the investigators and Tristan Hargreaves was the doctoral student. This research was very fruitful in that it produced the first three-dimensional measurement goniometer and yielded a robust diffusion coefficient, which has since been published as AES-4id-2001.

This diffusion coefficient has since been used as a metric to develop a range of new diffusing surfaces, including optimized welled diffusers, profile diffusers, one- and two-dimensional curved diffusers, baffled diffusers, genetic binary hybrid surfaces, flat and curved binary amplitude gratings, and fractal and aperiodically modulated surfaces, in effect many of the topics included in this book. These new optimized custom curved surfaces have found application in performance spaces like Kresge Auditorium, Boston, MA; Hummingbird Center, Toronto, Canada; and Edwina Palmer Hall, Hitchin, UK, and also recording facilities like Sony Music's premier mastering room M1, in New York.

Things began falling into place and all of the relevant diffusion research was collected into a special edition of *Applied Acoustics*, entitled 'Surface Diffusion in Room Acoustics', guest edited by Yiu Wai Lam and published in June 2000. Lam also organized a symposium in Liverpool that year. In September 2001, a special structured session on scattering in room acoustics was organized by Michael Vorlander at the 17th ICA in Rome. Having played a pioneering role in making Schroeder's theoretical suggestions a practical reality, it was personally very gratifying for Peter to be part of a session dedicated to a topic that started as an intellectual curiosity and has now turned into a diffuser industry and a field of research actively being studied by the leading acousticians of our time.

There have been many significant accomplishments over the past 20 years. We now know how to design, predict, optimize, measure, characterize, and standardize the performance of scattering surfaces. While there is still much to do, there is a general consensus in the architectural acoustics community that a solid theoretical and experimental foundation has been laid, that diffuser performance can now be quantified and standardized, and that diffusers can now be integrated into contemporary architecture, taking their rightful place along with absorbers and reflectors in the acoustical palette. The future holds many exciting possibilities.

It is a good time in the history of diffuser development to tell this story. This book has allowed Peter and Trevor to chronicle developments with sufficient scientific detail and to collect in one volume much of what is known about both diffusers and absorbers. You can contact the authors and tell them about technology and techniques that they may have inadvertently missed in this book. So stay tuned and 'Listen to the Music, Not the Room'.

Peter D'Antonio
Trevor Cox

Preface to the second edition

As society evolves, new problems arise, and these challenges must be met with new technology. For instance, sustainability is influencing the materials used in absorbers and diffusers. Intractable problems, such as environmental noise, continue to drive innovative new solutions. Furthermore, the general expectation of better quality design in the built environment has meant that designers have to concern themselves about the visual aesthetics of treatments alongside acoustic performance.

This second edition brings the technology of absorbers and diffusers up to date. For instance, the ubiquitous fabric-wrapped panel and acoustical ceiling tile no longer address all of the concerns of our day. Therefore, we have expanded the description of other absorber technologies, such as microperforated designs. The sound diffuser continues to evolve to improve performance and to meet new demands for artistic shapes. Each stage in the evolution of these technologies overcame a particular shortcoming and increased performance.

But it isn't just the absorbers and diffusers that are changing; there also have been new developments in measurement methods, standards, and prediction models. For instance, recent advances in three-dimensional solid prototyping printers greatly simplify the fabrication of diffuser test samples. To take another example, new time domain methods are being developed to predict how absorbers and diffusers interact with sound.

It is often said that new technology takes many years to be assimilated into the culture. Well, 2008 was the 25th anniversary of the founding of RPG Diffusor Systems, Inc., and it is fair to say that diffusion technology is fully integrated into every aspect of architectural acoustics. Acousticians are routinely including absorption and diffusion coefficients into design specifications and architects are embracing the innovative diffusive shapes into their projects.

Peter D'Antonio
Trevor Cox

Preface to the third edition

It is encouraging that many of the ideas and products described in the first two editions have been incorporated into architectural acoustic designs and practices by the acoustical community. The quality of the built environment has improved substantially. The field of acoustics is vibrant and dealing with the ever-changing challenges of our society. Therefore, we attempted to incorporate this innovation into this third edition, in hopes of continually advancing our ability to silence the noise, enhance the music, and increase speech intelligibility.

This third edition continues the evolution of the state of the art in absorber and diffuser theory, design, and application, as well as the latest status of measurement methods, standards, and prediction models. All chapters have been revised and brought up to date in this new edition, with additional case studies to cater to practitioners working in the measurement, modelling, and design of rooms, semienclosed spaces, as well as in noise control. Sustainability, portable vocal booths, new hybrid diffusive devices, and design charts for Helmholtz absorbers are just a few of the new sections.

We also mourn the 2009 passing of Prof. Manfred Schroeder, a creative acoustician who inspired us and to whom we have dedicated this book. We were honoured to contribute to a book that reflects the life, work, and legacy of one of the greatest acousticians of the 20th century entitled *Acoustics, Information, and Communication: Memorial Volume in Honor of Manfred R. Schroeder*, published by Springer and edited by Ning Xiang and Gerhard M. Sessler (2015).

Trevor Cox
Peter D'Antonio

Acknowledgements

We would like to acknowledge the worldwide architectural acoustics community for their continuing support and specification of our research, designs, and products, thereby enabling the growth of a diffusion industry and allowing diffusion to take its rightful place along with absorption and reflection. We would also like to thank acoustical and audio professional bodies for offering a peer review forum and community to share our research, and the diligence of working groups developing international standards to help unify approaches.

Many people have contributed to this book through its editions, not the least colleagues and students at the University of Salford and staff at RPG Diffusor Systems Inc., who have carried out measurements and predictions. Brian Rife updated RPG's data collection and processing software. We acknowledge via references throughout this book many acousticians. We are particularly thankful to the many researchers who responded kindly to requests for images or data to enable the latest findings to be included. Most notably, for the third edition, four people have contributed sections or provided detailed critiques of chapters: Bengt-Inge Dalenbäck, Jon Hargreaves, Jonathan Sheaffer, and Rodolfo Venegas.

We have been fortunate to have collaborated with some of the best scientists of our generation: Dr. Jerome Karle, chief scientist and Nobel Laureate 1985 Chemistry, Peter's supervisor at the Naval Research Laboratory; and Dr. John H. Konnert, cofounder of RPG Diffusor Systems, Inc., and author of the restrained macromolecular least squares program.

Authors

Trevor Cox, BSc, PhD, is professor of acoustic engineering at the University of Salford and a past president of the United Kingdom's Institute of Acoustics (IOA). One major strand of his research is room acoustics for intelligible speech and quality music production and reproduction. Dr. Cox's diffuser designs can be found in rooms around the world. He was awarded the Institute of Acoustics' Tyndall Medal in 2004. Dr. Cox was an associate editor for *Acta Acustica uw Acust* in 2000–13. He is currently working on two major research projects. *Making Sense of Sound* combines perceptual testing and blind signal processing to develop algorithms to make Big Audio Data searchable and reusable. The other project is investigating future technologies for spatial audio in the home. Dr. Cox is leading the perceptual work on listener experience in this multiuniversity grant that includes BBC R&D as a partner. Dr. Cox was given the Institute of Acoustics award for promoting acoustics to the public in 2009. He has developed and presented science shows to 15,000 pupils, including performing at the Royal Albert Hall, Purcell Rooms, and the Royal Institution. Dr. Cox has presented more than 20 documentaries for BBC radio, including *Life's* soundtrack, *Save our Sounds*, and *Science vs the Strad*. His popular science book *Sonic Wonderland* was published by Bodley Head in 2014 (in the United States: *The Sound Book*, WW Norton), and for this, he won the Acoustical Society of America (ASA) Science Writers Award for Professionals in Acoustics. He has written for the *Guardian*, *New Scientist*, and *Sound on Sound*. Dr. Cox publishes the blog acousticengineering.wordpress.com and can be followed on Twitter @trevor_cox.

Peter D'Antonio, BSc, PhD, has specialized in a wide variety of scientific disciplines including spectroscopy, x-ray and electron diffraction, and electron microscopy as a member of the internationally renowned Laboratory for the Structure of Matter, headed by Nobel Laureate Dr. Jerome Karle. As a musician and recording engineer, Dr. D'Antonio maintained a separate concurrent career in the music industry. In 1974, he developed a widely used design for modern recording studios at Underground Sound, Largo, MD, utilizing a temporal reflection free zone and reflection phase grating diffusors, which led to the founding of RPG Diffusor Systems, Inc., established in 1983, where he is currently chairman. He pioneered the sound diffusion industry and has significantly expanded the acoustical palette of products, for which he holds many trademarks and patents. He has lectured extensively, published numerous peer-reviewed scientific articles, and is the coauthor of the reference book *Acoustic Absorbers and Diffusers: Theory, Design and Application, 2nd Edition*, published by Taylor & Francis in 2009. He has served on several working groups to standardize the measurement of the absorption, scattering, and diffusion coefficients. He has served as adjunct professor of acoustics at the Cleveland Institute of Music, since 1991, and is a Fellow of the Acoustical Society of America and the Audio Engineering Society, a Corporate Affiliate of the American Institute of Architects, and 2012 Inductee to the TECnology Hall of Fame.

Chapter 1

Introduction

The sound that is heard in most environments is a combination of the direct sound straight from the source or sources and the indirect reflections from surfaces and other objects. For instance, in room acoustics, both the direct sound and the reflections from the walls, ceiling, and floor are key in determining the quality of the acoustic. To take another example, outdoors, the reflection from the ground can significantly reduce noise at certain low frequencies. Hence, one of the central topics in acoustics is how to manipulate these reflections that alter the way the sound behaves, and is ultimately perceived.

Sound striking a surface is transmitted, absorbed, or reflected; the amount of energy going into transmission, absorption, or reflection depends on the surface's acoustic properties. The reflected sound can either be redirected by large flat surfaces (specularly reflected) or scattered by a diffusing surface. When a significant portion of the reflected sound is spatially and temporally dispersed, this is a diffuse reflection, and the surface involved is often termed a *diffuser*. Figure 1.1 illustrates temporal and spatial characteristics of absorbing, specularly reflecting, and diffusing surfaces, which form the acoustical palette. In addition to the surface types shown in Figure 1.1, there are also hybrid surfaces, which can both absorb and diffuse to varying degrees.

For more than 100 years, since the founding of architectural acoustics by Sabine, there has been considerable effort devoted to studying surface absorption. Over this time, a considerable library of absorption coefficients has been tabulated based on accepted standards of measurement, and a reasonable understanding of how absorbers should be designed and applied has been achieved. This development continues, and in recent decades, many innovative absorber designs have been developed, and new ways to predict and measure absorptive materials have been found. For noise control, the focus is on removing energy, whereas in architectural acoustics, both absorbers and diffusers have a role to play in creating a good acoustic. However, significant scientific knowledge about the role of scattering (diffusely reflecting) surfaces has only been developed much more recently. Over the past four decades, significant research on methods to design, optimize, predict, measure, and quantify diffusing surfaces has resulted in a body of scientific knowledge and understanding. All these issues, and many more, are covered in this book.

Good architectural acoustic design requires the right room volume, an appropriate room shape, and surface treatments, utilizing an appropriate combination and placement of absorbers, diffusers, and flat surfaces. Architectural acoustic spaces can be loosely divided into sound production, sound reproduction, and noise control environments.

An example of a sound production room is the performing arts facility, such as concert halls for classical music or theatres for speech. The room acoustic contributes greatly to the perceived sound: whether music sounds beautiful and speech is intelligible. The arrival time, direction and temporal density, and level of the early reflections, coupled with the balance of the early to late energy, decay time, and temporal and spatial density of the late reflections,

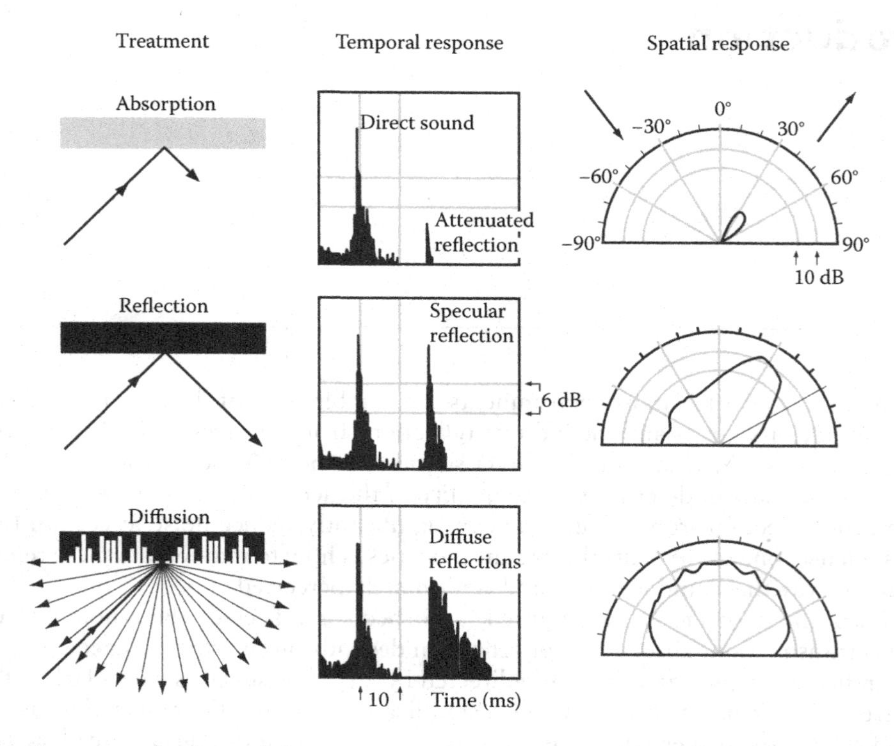

Figure 1.1 The temporal and spatial characteristics of absorbing, specularly reflecting, and diffusing surfaces.

define the quality of sound that is heard and the degree of envelopment a listener experiences. In large sound production rooms, reflection and diffuse reflection are the primary acoustic tools. This is schematically illustrated in Figure 1.2. Absorption may be used to control reverberance, but the unavoidable absorption due to paying customers must also be considered.

In contrast, the acoustics of sound reproduction rooms, like recording studios and home theatres, should be neutral. All of the spectral, timbre, and spatial information is prerecorded

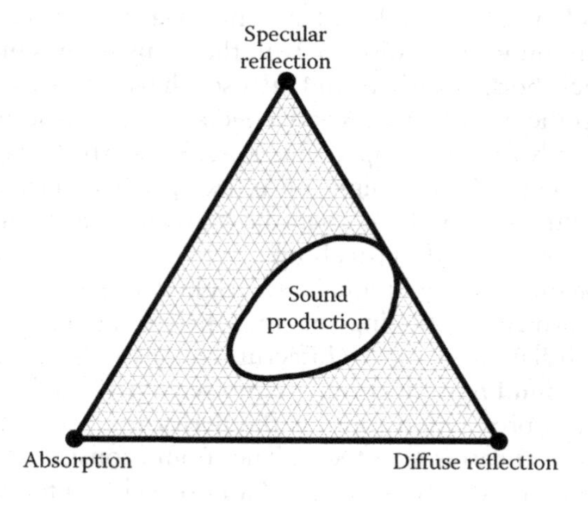

Figure 1.2 The relative importance of three acoustic treatments for sound production rooms such as concert halls, recital halls, auditoria, theatres, conference halls, courtrooms, and worship spaces.

on the playback media, and the reproduction room is only there to allow a listener to hear the audio as it was recorded. In a sound reproduction room, absorption and diffuse reflection play a key role, and specular reflection is a minor contributor. This is illustrated in Figure 1.3. Absorption and diffusion are used to control the coloration that would otherwise occur from early arriving reflections and low-frequency room modes.

In noise control situations, like gymnasiums, swimming pools, and factories, the objective is simply to reduce the reverberance and sound level. This might be done to reduce sound levels to prevent hearing damage or to improve the intelligibility of speech. Uniform distribution of absorption is the primary acoustic tool, and specular reflection and diffuse reflection have more minor roles. This is illustrated in Figure 1.4. (Diffusers can also play a useful supplementary role in disproportionate and nondiffuse spaces, but then, these figures are all generalizations of the true situation.)

Looking beyond architectural acoustics, absorbers are widely used to treat noise problems: within machine enclosures and cavity walls, on the sides of roadside barriers, within ventilation ducts and other pipework, and within hearing defender cups. In all these cases,

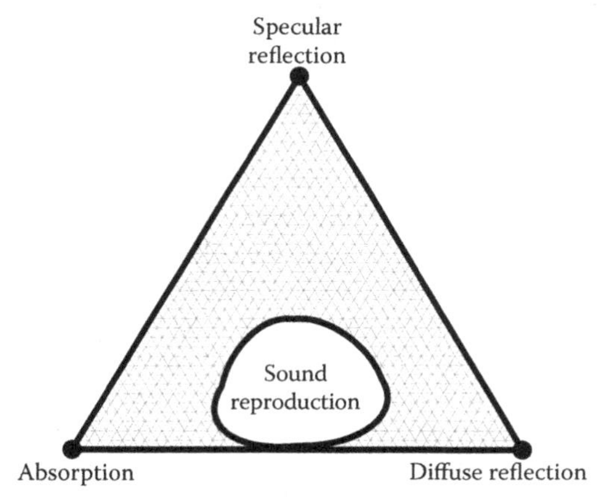

Figure 1.3 The relative importance of three acoustic treatments for sound reproduction rooms such as recording and broadcast studios, video conferencing rooms, and home theatres.

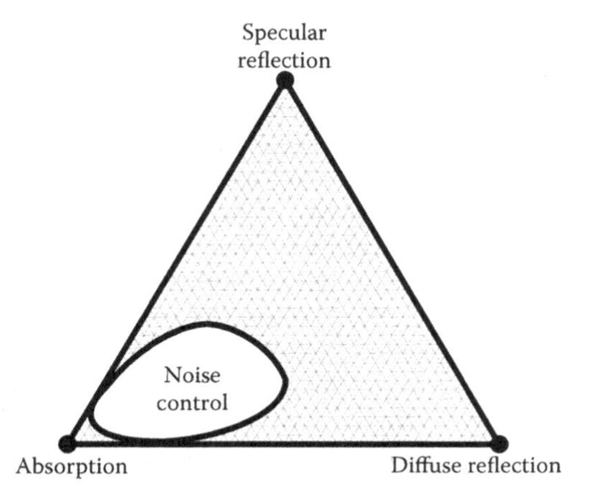

Figure 1.4 The relative importance of three acoustic treatments for noise control such as for factories, gymnasiums, swimming pools, libraries, atria, and roadside barriers.

the principles behind how the absorbers perform, and the models and measurements used in the design and characterization of the materials, are nearly always the same as for architectural acoustic applications.

Surface properties also play an important role outdoors. For instance, the absorption of the ground can have a significant impact on sound levels from ambient noise sources, such as roads and industrial premises. The treatment of noise levels might involve the use of noise barriers, and these might be treated with absorption, or less commonly, diffusers to reduce the noise levels.

This introductory description has sketched out a few of the issues concerning where and why absorbers and diffusers are applied. More detailed descriptions can be found in Chapter 2 for absorbers and Chapter 3 for diffusers. The following section, however, tries to give an overview of the relative merits of absorption and diffuse reflections.

1.1 ABSORPTION VERSUS DIFFUSE REFLECTIONS

Both absorbers and diffusers can be used to prevent acoustic distortion. For example, both can be effective in controlling echoes, coloration, and image shift, which would otherwise be caused by strong reflections. This raises the question as to which is the best treatment in which situation.

Whether absorbers or diffusers are better depends to a considerable degree on other acoustic factors, primarily on whether a decrease in reverberation and/or sound level is desirable. If a wall is causing an echo or coloration problem, and the designer wishes to conserve the reverberation time and sound energy, then a diffuser is the best solution. The diffuser is placed on the wall to disperse the reflection and to reduce the distortion without removing sound energy from the space. For this reason, in concert halls, where acoustic energy is at a premium, diffusers are to be preferred. In smaller rooms, say a lecture theatre, where intelligibility is important, absorption is the main treatment. A balance is reached with absorbers being used to adjust the reverberation time and level, and diffusers are used to ensure that early reflections, which can constructively support speech, do not produce distortion. When reflections cannot be constructively used for intelligibility, then these reflections can either be absorbed, if it doesn't make the room too dead, or diffused, thereby improving ambience and coverage.

In critical listening rooms, a mixture of absorbers and diffusers is used to control the acoustics of a space. Treatment is placed to control first-order reflections. When absorbers are used, the sonic images forming the soundstage are points in space. When diffusers are used, these images take on a more natural width and depth. Which material is correct, absorber or diffuser, is to a certain extent a matter of personal taste. If all the treatment is absorption, then the room turns out to be rather dead. While some people favour this for mixing audio, others do not, and for a listening room, a very absorbent environment is not best. Consequently, if some liveliness is to be left in the room, a combination of absorbers and diffusers must be used. Current psychoacoustical research in multichannel surround listening rooms indicates the importance of generating lateral reflections to enhance envelopment, using single plane diffusers on side/rear walls and ceiling. Broad bandwidth absorption is most effective on the front wall and in corner locations. To provide low-frequency modal control, both porous absorbers and diffusers require considerable depth to work, and the depths of treatments are often limited, because of space constraints and cost. Because of this, resonant absorbers requiring less depth are most often used. In listening rooms, resonant absorbers provide effective low-frequency control when placed in high-pressure corner locations. For example, a membrane absorber might be used. The speed of sound in a porous absorber is lower than in air, and consequently, a given thickness of absorber can work to a lower frequency than the same thickness of diffuser. For this reason, a partially absorbing diffuser, such as a hybrid structure, or a resonant absorber is usually favoured to treat low frequencies when space is a premium.

Diffusers have the advantage of generally being more robust than absorbers. Most absorber technologies involve fibrous materials, which do not stand up well to the effects of wind, rain, and toxic environments. For example, in railway stations or on streets, a large amount of particulate pollution may be generated, which over time can clog the pores of fibrous absorbents. There is a great risk with outdoor installations that fibrous absorbents will wash away over time. Consequently, if it is possible to meet the acoustic requirement using a hard diffuser, it is possible to generate a much more robust treatment than with many absorbents. Alternatively, fibreless absorbers, such as microperforated devices or washable granular absorbents, might offer a solution.

Both absorbers and diffusers have a role to play in good acoustic design. They have a complementary function, which means that when they are used appropriately, better acoustics can be achieved.

1.2 SUSTAINABLE ABSORBERS AND DIFFUSERS

The sustainability considerations when choosing treatments are similar for both absorbers and diffusers. A good acoustic enhances the usability of a building, but in most cases, sustainability is primarily driven by other environmental considerations, especially energy consumption. A British Standard gives a list of issues that might be considered during certification of a product (see Table 1.1).[1] To minimize environmental impact, acoustic treatments need to be made from recycled materials or resources that are natural, plentiful, or renewable. The

Table 1.1 Considerations to be identified and reported for sustainability

Issue	*Consideration*
Environmental	Recyclability and recycled content
	Renewability
	Harvesting or extraction impacts
	Greenhouse gas emissions
	Energy usage
	Water usage
	Transport impacts
	Biodiversity
	Eco toxicity
	Land remediation
	Waste management
Social	Workers' conditions
	Safe and healthy working conditions
	Slave and child labour
	Fair wages
	Working hours and holidays
	Freedom to join trade unions (freedom of association)
	Equality in respect of gender, ethnicity, religion, political persuasion
	Complaints and prosecutions
	Skills and training
	Community relations
Economic	Contribution to the built environment
	Ethical business practice
	Contribution to diversity and stability of the local economy
	Long-term financial viability

Source: BS 8902:2009, "Responsible sourcing sector certification schemes for construction products. Specification".

Table 1.2 Sustainability considerations for common materials used in acoustic design

Material	*Energy and resource use*	*Toxicity*	*Recycling/renewable*	*Notes*
Plywood	High in manufacture	Some health risks in manufacture, use and if burned	Limited recycling	Cannot be formed from waste wood
Plasterboard	Relatively low embodied carbon	Some poor-quality boards can leach toxins. Disposal to landfill discouraged	Gypsum mined so not renewable but can be recycled	Good alternatives include mineral particle board made from recycled gypsum and cellulose fibres such as recycled papers
Medium-density fibreboard (MDF)	Modest embodied energy	Some volatile organic compound (VOC) health risk	Older materials less recyclable	No Added Urea Formaldehyde (NAUF), Forest Stewardship Council (FSC) MDF now available
Chipboard	High energy use during manufacture. Resource depletion	Toxins during manufacture		Alternatives include a chipboard-like material made from used beverage cartons, e.g., Tectan
Hardboard	High embodied energy		Can be recycled	
Wood wool cement slabs	High energy use during manufacture			Hazardous waste
Cellulose fibre	Relatively low embodied energy	Check toxicity of fungicides and pesticides used to protect it	Good recycling potential	
Concrete	High energy use in manufacture, but new cements being developed. Provides thermal mass		Can be recycled	Durable. Consider demolition costs
Cork	Needs little energy for production. Energy costs for transportation		Renewable	Benign to the environment. Supportive of cork oak forestry and reliant habitats. Reasonably durable, although vulnerable to abrasion and indentation
Damping compounds/ sheets	May be refined from a nonrenewable source. Check plasticisers, elastomers and copolymer emulsions for embodied energy	Adhesives may carry risks in use, e.g., from VOCs	Bituminous materials can be formed from recycled asphalt. Can use recycled aluminium with viscoelastic polymer	Consider water-based compounds for vehicles. For constrained layer damping, consider impacts of using steel, zinc, or aluminium

(*Continued*)

Table 1.2 (Continued) Sustainability considerations for common materials used in acoustic design

Material	*Energy and resource use*	*Toxicity*	*Recycling/renewable*	*Notes*
Glass	High embodied energy, but glazing can contribute to a low energy building solution		Considerable opportunity for recycling, particularly where clarity of vision is not essential	
Masonry	Excluding local or reclaimed material, initial energy in manufacture may be high. Has thermal mass		Often recyclable	Check transport costs
Mineral wool and glass fibre	Moderate embodied energy. May save substantial energy in use. Thermal performance tends to dominate sustainability claims	Limit potential skin irritation and dust risks	Not renewable. Recyclable and renewable resource, but check for renewable energy used during processing. More difficult to recycle beyond production/delivery stages	Not biodegradable. Alternatives like recycled bottles, spun plastic wool, natural flax, or recycled newspaper can provide similar sound absorption performance but with lower embodied energy
Plastic foam	Typically high embodied energy. Thermal performance tends to dominate sustainability claims	Usually involves toxicity	Polyurethane (PU) recycling typically limited to powdering or rebonding fragments	Fire risk. Many options for other recycled foams. Bio based foams are developing and some from sequestered carbon
Polycarbonate	High embodied energy. Can be used in solar panels.		Postproduction waste recovery fairly good	Impressive performance e.g., light transmission, durable, high mechanical strength, for safety—impact resistant
Sealants	High embodied energy, but amount of sealant used is often small. In some applications, can have positive impact on overall energy use, e.g., through air tightness	Check for low VOC emissions. Filled reactive PU-polymer, silicone-based elastomeric and solvent-based mastic adhesives contribute to hazardous waste	Silicone sealants are difficult to recycle	Tend to use oleochemicals, polyols, PU foam. Where realistic, prefer to use water, not solvent-based systems. Lift the solid content. Aim for less material, more performance. Prefer preformed gaskets. Check durability
Steel (and other metals)	High embodied energy		Recyclable, but new uses may involve considerably more energy	Highly versatile

(Continued)

Table 1.2 (Continued) Sustainability considerations for common materials used in acoustic design

Material	*Energy and resource use*	*Toxicity*	*Recycling/renewable*	*Notes*
Straw/reeds	Low embodied energy	Treatments against fire and attack by pests may include toxins	Suitable for composting if free of toxins	May be vulnerable to moisture, fire, insect, and pest attack
Woods	Responsible sourcing needed, e.g., by reference to FSC, from a locally grown and managed forest Investigate minimum carbon release in transporting it	Consider resins used, e.g., in cross-laminated timber	Consider recycled wood or recycled plastics for noise screening material	Care by manufacturers needed to avoid encouraging illegal forestry and deforestation
Wool	Natural	Be aware of chemical treatments	Biodegradable	Wool ignites at a higher temperature than some synthetic fibres with a lower rate of flame spread and heat release, lower heat of combustion, and does not melt or drip without additional treatments. Life cycle evaluation should include methane emitted by sheep

Source: Institute of Acoustics, *Sustainable Practice Guidance Note 1* 2005.

products themselves should be recyclable, reusable, and/or biodegradable. But there is more to sustainability than just recycling materials. The materials should be sourced locally or regionally. During manufacture, waste recycling, green power, and resource-efficient manufacturing processes need to be considered. Energy consumption during production, use, and disposal must be evaluated, as energy usage is the primary driver underpinning sustainable construction. Recycled or recyclable product packaging needs to be used and the environmental impact of transportation considered. Table 1.2 summarises some common issues arising from acoustic products, as collated by the sustainability taskforce of the Institute of Acoustics.[2] *Embodied* refers to the whole life cycle of a product, e.g., during the manufacture, transport, and construction as well as end of life disposal. *Embodied carbon* refers to carbon dioxide emissions, and *embodied energy*, the total energy consumption.

Ideally, a full life cycle of a product should be assessed, but in many cases, this is not possible, so best practice is to usually check on as much of the life cycle as is practical. Table 1.3 gives some life cycle evaluations for some common acoustic materials. Some have developed single figure ratings that are formed from a weighted sum of different environmental impacts. An examples of this is the BRE Eco-Profiles[3] that features within the BRE Green Guide. The Green Guide or the Ecoinvent database[4] can give advice on selection of materials. Building products might also have green product labels such as those awarded by Natureplus.

Around the world, various competing rating systems have been developed to encourage buildings that minimize environmental impact. An example of such a scheme in the United States is the Leadership in Energy and Environmental Design (LEED) rating system developed by the US Green Building Council (USGBC).[6] This lists detailed criteria against which building projects should be rated to gain accreditation. Relevant to acoustic materials, the system deals with issues such as prerequisite in acoustic performance, recycled material content, woods from certified sources, and low-emitting material, paints, and coatings. Four levels of LEED certification for buildings are available: Platinum (80+ points), Gold (60–79 points), Silver (50–59 points), and Certified (40–49 points). To qualify under one of these four designations, buildings must satisfy all of the LEED prerequisites, such as Elimination and Control of Asbestos, Smoking Ban, etc., then go on to earn a certain number of credits. The rating system is divided into categories: Integrative Process, Location and Transportation, Material and Resources, Water Efficiency, Energy and Atmosphere, Sustainable Sites, Indoor Environmental Quality, Innovation, and Regional Priority.

Table 1.3 Comparison of natural and traditional acoustic materials based on the http://www.ecoinvent.org/ database

Material	*Density (kgm^{-3})*	*Nonrenewable energy ($MJ\ kg^{-1}$)*	*Global warming potential (kg CO_2 eq.)*	*Acidification potential (g SO_2 eq.)*
Natural rubber	6.4 (kg/m^2)	40	2.4	8.6
Coconut fibres	50	42	0	25
Flax fibres	25	4.4	0	0
Sheep wool	30	12.3	−0.3	4.6
Cellulose flocks	35–70	4.2	0.2	2.5
Expanded polystyrene	30	95	2.3	20.1
Foam glass	130	67	3.7	22.9
Glass fibre	34	43	2.1	15.5
Mineral wool	50–60	17	1.2	5.2

Source: F. Asdrubali, S. Schiavoni, and K. Horoshenkov, "A review of sustainable materials for acoustic applications", *Bldg. Acoust.*, 19(4), 283–312 (2012).

REFERENCES

1. BS 8902:2009, "Responsible sourcing sector certification schemes for construction products. Specification".
2. Institute of Acoustics, *Sustainable Practice Guidance Note 1* (2015).
3. BRE, *Green Guide to Specification*, http://www.bre.co.uk/greenguide/podpage.jsp?id=2126, accessed 10 November 2015.
4. ecoinvent Centre, The ecoinvent database, http://www.ecoinvent.org/database/database.html, accessed 10 November 2015.
5. F. Asdrubali, S. Schiavoni, and K. Horoshenkov, "A review of sustainable materials for acoustic applications", *Bldg. Acoust.*, **19**(4), 283–312 (2012).
6. U.S. Green Building Council, http://www.usgbc.org/, accessed 10 November 2015.

Chapter 2

Absorbers

Applications and basic principles

This chapter introduces the principles of absorption, including a basic explanation of the physics and some fundamental formulations that will be used in later chapters. Since this book is aimed at both practitioners and researchers, most chapters begin with an application-driven, qualitative description, followed by a quantitative description of the technology and design. Following this type of philosophy, this introduction to absorbers is written from an application or case study perspective. The style is intended to make the more theoretical sections more palatable and avoid starting with a chapter labelled *A little light mathematics.* The chapter will also introduce some of the issues concerning the design, prediction, and measurement of absorbers that will be treated in more detail in future chapters.

2.1 TYPES OF ABSORBER

There are two broad classes of absorbers: porous absorbents and resonant devices. The performances of these are most commonly evaluated using an absorption coefficient, which varies from 0 to 1, representing no and complete absorption, respectively. Figure 2.1 shows the performance of some absorbers, including both porous and resonant devices.

The most common type of porous absorbent is mineral wool, such as fibreglass, and the graph shows the absorption coefficient versus frequency for fibreglass that is 50 mm deep mounted on a rigid backing. Porous absorption is ineffective if it is too shallow compared to wavelength, and consequently, the graph looks like a high pass filter response, with high absorption being achieved from 500 Hz and above and little absorption at bass frequencies. While mineral wools are commonly used, there are many other porous materials that achieve similar performance, as outlined in Chapter 6.

The most efficient way of gaining more low-frequency absorption is to exploit resonance. One way of doing this is to place a perforated sheet in front of the porous absorbent to form a Helmholtz absorber. Another approach is to use a thin sheet in front of fibreglass that vibrates and forms a membrane or panel absorber. As these are resonant absorbers, they will have a peak of absorption and a certain bandwidth over which they operate. The graph in Figure 2.1 shows two examples: the particular Helmholtz absorber shown removes energy over about three octaves at mid-frequency, whereas this membrane design works at a lower frequency and narrower bandwidth. These types of devices are discussed in Chapter 7.

Getting broadband absorption in one device involves simultaneously exploiting both resonance and porous absorption. The example in Figure 2.1 is labelled *multilayer.* This is a device that has layers of different types of fibreglass along with an aluminium foil membrane. Another example is the microperforated wood sample that exploits the resonance of a Helmholtz absorber formed with many tiny holes to aid absorption, along with a backing layer of mineral wool to further increase absorption.

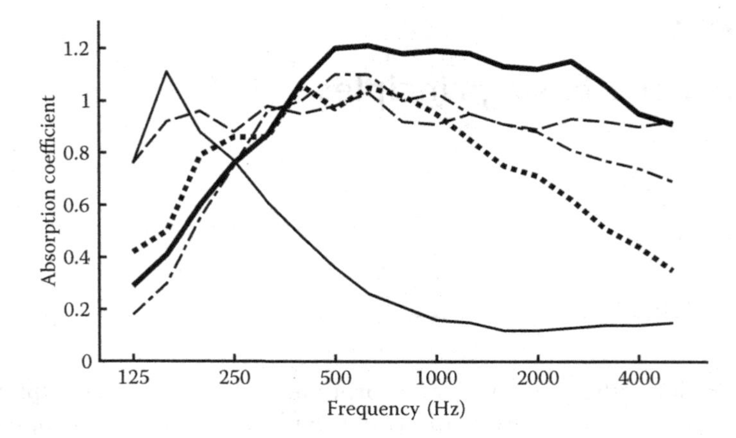

Figure 2.1 Random incidence absorption coefficient for several different absorbers: - - - - multilayer; ——— membrane; ▬▬ fibreglass; ▪▪▪▪▪▪ Helmholtz, and -·-·- microperforated wood. All 50 mm thick.

2.2 REVERBERATION CONTROL

Excessively reverberant spaces, such as atria, swimming pools, museums, railway stations, airports, and many restaurants, where the sound echoes around, making it noisy and difficult to communicate, are commonplace. In these types of spaces, people tend to slow down their speech, talk louder, and try to pronounce words more precisely in an effort to make themselves understood. For some reason, many restaurateurs seem to think that to create the right atmosphere, it is necessary to make speech communication virtually impossible! The issue here is reverberation.

Reverberation is the decay of sound after a source has stopped and it is a key feature in room acoustics.[1] Reverberation is most audible in large spaces with hard surfaces, such as cathedrals, where the acoustic waves echo around long after the source of the sound has stopped emitting. In small spaces, with plenty of soft, absorbent materials, such as a living room, the acoustic energy is quickly removed, and the sound dies away rapidly. When people talk about rooms being *live* or *dead*, this is usually about the perception of reverberance.

The amount of reverberation in a space depends on the size of the room and the amount of sound absorption. The solution to the reverberant restaurant is to add absorbers. This will reduce the reflected sound energy in the room and so decrease the reverberant energy and sound level. Problems arise in dining rooms because any surfaces close to eating or preparation areas need to be robust and washable, and many acoustic absorbers are soft and unsuitable. Consequently, the best place for absorption is the ceiling or high up on the walls out of the way.

Getting the correct amount of reverberation in a space is vital to the design of most rooms, whether the aim is to make music sound beautiful, to make speech intelligible, to reduce noise levels, or simply to make a space a pleasant place to be in.

Figure 2.2 shows a large feature wall of a library atrium, covered in an absorbing wood finish. This treatment has the advantage of being architectural yet absorptive, so it does not visually impose on the space as an add-on treatment. Figure 2.3 shows a similar treatment being used to reduce reverberance in a large board room. Microperforated wood veneers allow architects to combine absorbing and reflecting wood panelling as required. Chapter 7 discusses innovative and more conventional resonant absorber technologies.

Figure 2.2 Perforated, absorptive wood panelling covering the entire feature wall opposite an exterior glass wall (out of shot) in Morgan State University in Baltimore, MD. Acousticians: Shen, Milsom, and Wilke, Ballston, VA. (Courtesy of RPG Diffusor Systems, Inc.)

Figure 2.3 Custom microperforated (Perfecto®) angled ceiling cloud at the NFL Conference Room, New York. (Courtesy of RPG Diffusor Systems, Inc.)

A less expensive solution would be standard absorbent ceiling tiles made out of compressed mineral wool, mounted in a T-bar grid, but this is not as elegant. The visual quality of acoustic treatment is extremely important to architects, and absorbers need to fit within a design scheme rather than look like obvious add-ons. Consequently, one of the drivers for developing new acoustic materials is visual aesthetics. This is one reason why there is great interest in microperforated absorbers (see Section 7.2.4). Microperforations or microslits are barely visible at normal viewing distances and can be applied to wood and light transmitting plastics; see Figure 2.4 for an example installation. The plastic provides absorption while maintaining visibility in places with large amounts of glass. Another absorber that blends visual and sound requirements is acoustic plasters. These can provide absorbing walls and ceilings that resemble traditional plaster or painted dry-wall (see Chapter 6).

In the past, acoustic absorption might be hidden behind utilitarian plain fabrics. Nowadays, with advances in technologies to print graphics on materials, fabric-wrapped absorbent can become a stunning visual feature. Figure 2.5a shows a modern digital printing project in the atrium of Woodrow Wilson High School, Washington, DC. The architect provided an image that consisted of a superposition of solar shadows on the atrium wall, as would be cast by the glass roof structure at different times of the year—the winter and summer solstices and the spring and autumn equinoxes. The shadow patterns were printed on fabric, which was applied to a fibreglass panel, and on the summer solstice at noon, the photo illustrates that the printed and actual solar shadows aligned! In Figure 2.5b, the digital printing was done on 5-mm-thick perforated Polyethylene Terephthalate Glycol (PETg) plastic panels in the atrium of H.D. Woodson High School, which is a Science, Technology, Engineering, and Mathematics school. The architect's concept was to illustrate images of prominent scientists, student activity, and associated graphic patterns. The result was a warm light glow, through the external glazing and printed acoustical panels, as well as a pleasing acoustical ambience.

In recent years, the issue of sustainability has become increasingly important. For this reason, there is great interest in porous absorbents made from recycled materials. This is discussed in Section 6.2.3, with the more general issue of what is required from a sustainable treatment outlined in Section 1.2. The need for thermally efficient buildings also poses a challenge to acoustic consultants. More hard surfaces are being left exposed to exploit the thermal mass of ceilings

Figure 2.4 Microperforated foil window treatment (Clearsorbor Foil) at NSW Nurses and Midwives Association, Waterloo, Australia. Acoustical Contractor Acoustic Vision, LLC. (Courtesy of Acoustic Vision, LLC.)

(a)

(b)

Figure 2.5 (a) Digitally printed graphic on fabric-wrapped absorptive panels at Woodrow Wilson High School. (b) Digitally printed images on absorptive, perforated PETg panels at H.D. Woodson High School. Both schools are in Washington, DC. Acoustician: Polysonics Corporation. (Courtesy of RPG Diffusor Systems, Inc.)

and floors. These surfaces were traditionally covered with absorption to control reverberance. Therefore, new methods for controlling reverberance and sound reflections grazing across ceilings are now required. One possible approach is to suspend translucent polycarbonate microperforated foil, which transmits in the infrared, allowing radiant heating of the space below. Another approach is to use acoustic rafts with gaps between to allow the thermal mass of the ceiling to be exploited.

An extreme example that uses lots of absorption is the anechoic chamber, which is an acoustically dead space, an example of which is shown in Figure 2.6. This is a room where, above a certain cut-off frequency, there are effectively no reflections from the walls, floor, or ceiling as they are attenuated by at least 25 dB. This means that it is ideal for testing the response of diffusers because the room does not affect the measurements. Anechoic chambers are also immensely quiet—the Guinness World Record chamber has a measured background noise level of –20.35 dB(A). To remove reflections from the boundaries, every surface is covered in absorbing wedges made of open-cell foam or fibreglass. Forming the absorbent into wedges reduces reflections from impedance discontinuities at the boundaries. Some have also made chambers from multiple layers of flat absorbents, where a gradual change in impedance is used to prevent strong reflections from the flat surface (see Section 6.6.4).

Sometimes, production rooms, where sound is mixed, are made acoustically (almost) dead spaces. Cinema dubbing studios used for the latest surround-sound formats often have large amounts of wideband absorption. Another example is the nonenvironment control room.[2] Figure 2.7 shows an example midway through construction so the absorbent is revealed as well as the finished room. The room has highly absorbing, broadband absorption on the side walls, rear wall, and ceiling. Immediately against the wall boundaries is a multilayer absorbent using porous material and membranes. In front of this are angled panels that are a sandwich construction of 2–5-cm-thick porous absorbent on the outside, with chipboard on the inside. The treatment is hidden behind an acoustically transparent fabric. The front wall is hard and textured and the loudspeakers are flush mounted into this front wall. The floor is also hard and reflective. The idea is to replicate (near) free field conditions for sound radiating from the loudspeakers, to enable the monitoring of the direct sound and nothing else. But the room also allows some reflections from the front wall and floor so the space does not appear unnaturally dead when two people have a conversation in the space. Advocates of nonenvironments prefer certainty and detail over the reverberance and more natural sound that is found in more conventional control rooms, examples of which can be found in Section 3.2.

Figure 2.6 The anechoic chamber at the University of Salford, UK.

Figure 2.7 Nonenvironment production room, Neo Music Box, in Castrillo de la Vega, Spain. (Courtesy of Philip Newell.)

Returning to more general reverberation, the primary technique for control is absorption. In discussing the design, application, and measurement of absorbers, it is necessary to understand a statistical model of sound within an enclosure.[1,3] This is discussed in the next section.

2.2.1 A statistical model of reverberation

A simple model of sound propagation in a space is of particles of energy bouncing around a room in an analogous way to a snooker ball bouncing around a billiard table. The room can

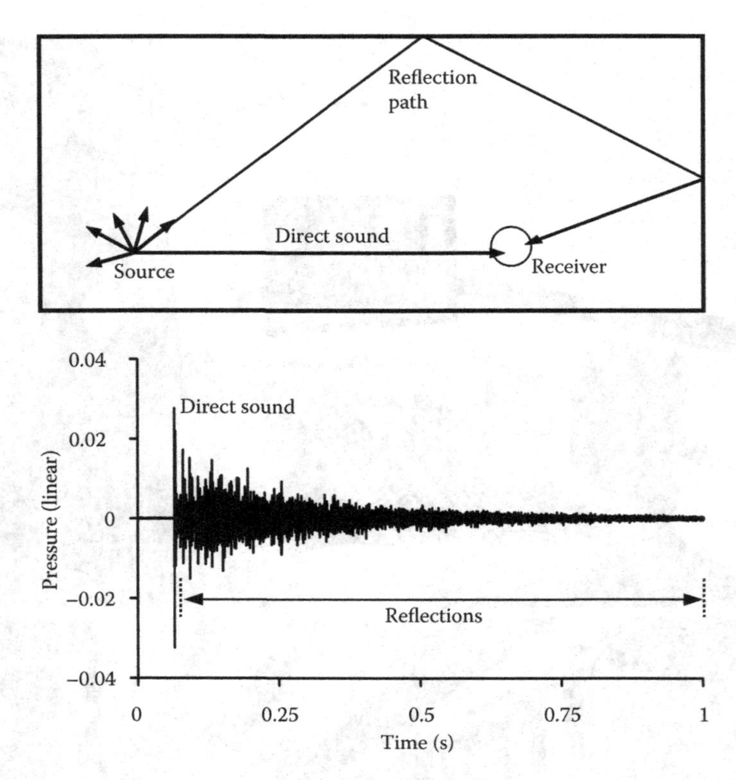

Figure 2.8 The generation of an impulse response in a room and an example impulse response from a concert hall.

be characterized by the impulse response, an example of which is shown in Figure 2.8. The *impulse response* is a pressure versus time graph showing the response at a receiver position to a short impulse created somewhere else in the room. For example, a balloon burst or a starting pistol might generate the short impulse and the response might be measured with a microphone. The first signal received is the *direct sound* from the source to receiver. Soon after, a series of *early reflections* arrive, the level of which generally decays over time, due to absorption at the room surfaces. The effects of the boundaries dominate the behaviour of sound in most rooms, and it is at the boundaries where absorption is normally found. (Only in large rooms does absorption by the air become important; see Table 4.2.) There is an increase in reflection density with time, and when the reflections become very dense this is termed the *reverberant field*. The energy of the reverberant reflections around the room is roughly constant and can be readily predicted, provided the sound field is diffuse.[4]

The reverberation time, T_{60}, measures the time taken for the sound pressure level to decay by 60 dB when a sound stops. From the impulse response, the Schroeder curve must first be calculated by backwards integration, before evaluating the reverberation time.[5] Sabine showed that the reverberation time could be calculated from the room volume and absorption by the following[6]:

$$T_{60} = \frac{55.3V}{cA}, \tag{2.1}$$

where V is the room volume, c is the speed of sound, and A is the total absorption in the room.

The total absorption can be calculated from the individual absorption coefficients of the room surfaces, using the following expression:

$$A = \sum_{i=1}^{N} S_i \alpha_i, \\ = S\bar{\alpha}, \tag{2.2}$$

where S_i is the surface area of the ith element in the room, S is the total surface area of the room, α_i is the absorption coefficient of the ith element in the room, and $\bar{\alpha}$ is the average absorption coefficient of the whole room.

The absorption coefficient of a surface is the ratio of the energy absorbed by the surface to the energy incident. It typically lies between 0 and 1, which represent non-absorbing and complete absorption, respectively. Values greater than 1 are often found in random incidence measurements, although theoretically impossible. This occurs due to weaknesses in the reverberation chamber method (see Section 4.3 for further details). The absorption coefficient can be defined for a specific angle of incidence or random incidence as required.

Equations 2.1 and 2.2 form the basis for the standard method for measuring a random incidence absorption coefficient. The reverberation time in a reverberation chamber is measured with and without the test sample. The test sample adds absorption to the room and so reduces the reverberation time. From the change in reverberation time, the absorption coefficient can be obtained. This technique is described in detail in Section 4.3. Sections 13.1.2 and 13.2 examine how to use this measurement data in room predictions and geometric room acoustic models.

For large rooms, the absorption of air should also be accounted for. The total air absorption A_{air} in a room of volume V is given by the following:

$$A_{air} = 4Vm, \tag{2.3}$$

where m is given in Table 4.2 or formulations can be found in ISO 9613-2.[7]

To allow for air absorption in the reverberation time predictions, the additional absorption calculated from Equation 2.3 is added to the denominator of Equation 2.1 to give

$$T_{60} = \frac{55.3V}{cA + 4mV}. \tag{2.4}$$

Sabine's formulation does not correctly predict the reverberation time for highly absorptive rooms. Over the years, many new formulations have been developed, the most popular of these being the Eyring equation,[8] also known as the Eyring–Norris equation:

$$T_{60} = \frac{55.3V}{-cS\ln(1-\bar{\alpha})}, \tag{2.5}$$

where ln() signifies the natural logarithm.

Alternate reverberation time equations are the topic of considerable interest. Many formulations attempt to be catch-all equations for reverberation time estimation, but it is often difficult to know a priori whether a formulation will work in a particular room. It is conceivable that better equations can be developed by analyzing rooms in more detail (such as surface size and orientation statistics and absorber and diffuser distribution), but any such attempt would require a computer model of the room to be made for the analysis. As geometric room acoustic models exist (ray tracing and variants thereof), where the impulse response of a room can be predicted, there is little need nowadays to search for ever more complex reverberation time formulations.

Using geometric room acoustic models for reverberation time estimation requires diffuse reflections to be taken into account, as discussed in Chapters 5 and 13. Despite many studies, the application of absorption coefficients in computer models is not straightforward mainly because it is difficult to know what the absorption coefficients are for surfaces, and this is a key input to the model. The accuracy of absorption coefficients is particularly important when a significant portion of the surface area of a room is very reflective, for instance if much of the room is made from concrete, glass, or wood. Furthermore, when the absorption is restricted to one plane, as is typically the case in concert halls, swimming pools, and sports halls, this means that the late decay is very dependent on the exact value of the absorption coefficient selected for the reflective surfaces. Even in a room entirely made of one material, such as a room made only of concrete, accurate absorption coefficients are critical. Changing the absorption coefficient of the concrete from 0.02 to 0.01 in such a room will double the reverberation time (except at higher frequencies in larger rooms because of air absorption). When a hard material is dominant, a very accurate estimate of the absorption coefficient is necessary for purely numerical reasons. Consequently, while there are tables of absorption coefficients in the literature and in Appendix A of this book, these cannot be blindly applied. The measured absorption coefficients can vary greatly from laboratory to laboratory, even for the same sample, as Figure 4.6 shows. Furthermore, for some products, the absorption can vary greatly from manufacturer to manufacturer—an example being carpets, as discussed in Section 6.2.5. Consequently, *in situ* methods for measuring absorption both within rooms and for outdoor applications are of interest, and these methods are discussed in Section 4.4.

The reverberation time formulations are statistical models of acoustic behaviour and are applicable only where there are a large number of reflections and the sound field is diffuse. For instance, at low frequencies, the modal behaviour of the room makes these assumptions invalid. Consequently, there is a lower frequency bound on the applicability of statistical absorption formulations. The lower bound is usually taken to be the Schroeder frequency,[9] given by the following:

$$f = 2000\sqrt{T_{60} / V}. \tag{2.6}$$

Although this formal limit has been known for many years, it does not prevent many practitioners, standards, and researchers still defining and using absorption coefficients below the Schroeder frequency, as it is convenient, even if physically incorrect. Geometric room acoustic models are also used below this limit, although they have particular difficulties predicting at frequencies where there is a low modal density, where correct modelling of phase is needed.

2.3 NOISE CONTROL IN FACTORIES AND LARGE ROOMS WITH DIFFUSE FIELDS

The noise levels within working environments must be controlled to allow safe working, as excessive levels can cause hearing loss. Consequently, there are regulations to limit the exposure of workers. There are several methods for controlling noise. The most efficient solutions control the noise at the source, but this may not always be possible. Another approach is to reduce the reverberant sound level within a space. This is only effective if the reverberant field makes a significant contribution to the noise level. For instance, the approach is ineffective if the worker is close to a noisy machine because the direct sound will dominate. The reverberant field level is reduced by the addition of absorption, and hence, the noise exposure is decreased by typically up to 3–4 dB(A). Typically, porous (or bulk) absorbers, such as mineral wool, are used as they are inexpensive, light, and effective.

The porous absorber often has to be protected from dust and so is frequently wrapped in plastic, but this decreases high-frequency absorption. There are situations where the absorbent needs to be washable, and there are a few types of porous absorber that achieve this. There are also situations where the absorbent needs to be fibreless to prevent contamination. Chapter 6 discusses the design and modelling of porous absorbents, including some innovative materials. Sections 7.2.4, 7.2.5, and 7.5.3 detail microperforation, which is one way of making fibreless absorbers. Porous absorbers are only effective at mid to high frequencies, but this is where the ear is most sensitive and consequently where noise control is most needed in the working environment.

Factories tend to be very disproportionately dimensioned, having low ceilings compared to their widths and lengths. This means that the simple diffuse field equations, such as Equation 2.1, are unlikely to work. For statistical room acoustics to hold, the space needs to be diffuse. A diffuse field is one where there is uniform reflected energy density across the whole room, and all directions of propagation are equally probable. There are many reasons why real rooms do not achieve this:

1. At low frequencies, there are standing wave modes (see Section 2.4).
2. If the room's dimensions are very dissimilar, there is a tendency to get different reverberation times in different directions. In a low room, sound will decay faster if it is propagating vertically rather than horizontally, as the vertical propagating sound will reflect more often and so get absorbed more quickly.
3. The absorption in a room should ideally be evenly distributed across all surfaces. For many cases, this is not true: in a factory or a swimming pool, the absorption might all be on the ceiling, and a reverberation chamber with a test sample has all of the absorption on the floor. In those cases, horizontally propagating sound will decay slower because it does not encounter the absorbent on the floor.
4. If the room has a distinctive shape, e.g., cylindrical, the curved surfaces can focus sound to a point, like a curved mirror does with light. The result will be an uneven sound field (see Section 11.3 for more on focussing from concave surfaces and Section 3.9 for solutions involving diffusers).

The relevance of the diffuseness of the space to absorbers is as follows. The absorption coefficient of a building element will be measured in a reverberation chamber, using Sabine's reverberation time formulation (see Section 4.3). When the absorption is applied, however, the acoustic conditions might be dramatically different, which means that the anticipated changes in noise levels might not occur. The absorption might be more or less effective than predicted; this is discussed in Section 13.1.2. A special example of the problem is considered in Sections 4.3.1 and 8.1, where auditorium and theatre seating is considered. Section 13.2 discusses the application of absorption coefficients to room acoustic models, where the issue of nondiffuseness is again important.

2.4 MODAL CONTROL IN CRITICAL LISTENING SPACES

Small rooms, like recording/broadcast studios, home theatres, and conference rooms, usually suffer from problems due to low frequency modes. At low frequencies, the standing wave modes of the room are separated in frequency. Figure 2.9 shows the frequency response for a small listening room. The frequency response is uneven, meaning that some frequencies are emphasized, where the mode(s) is strong, and some are suppressed, where the mode(s) is weak, leading to coloration of the received sound. This is most critical for music applications, particularly with the increasingly widespread use of subwoofer technology and reproduction of modern music with high bass content. Common solutions include choosing the room dimensions, loudspeaker, and listening positions correctly to flatten the frequency response of the room as much as possible and avoid degenerate modes. The dominant audible distortion created by modes is the ringing of notes caused by the slow decay of the resonances.[10] Consequently, proper treatment of room modes requires absorption to speed the decay of resonance.

Particularly prominent modes are usually treated with bass absorption, often referred to as bass traps or bins. (It is not usually possible to treat this problem with diffusion because the sizes of the diffusers become prohibitively large, although Section 3.2.3 discusses an unusual case where this has been done.) Porous absorbers are not usually used, as they would have to be extremely thick to provide significant bass absorption. Porous absorption is most effective when it is placed at a quarter wavelength from a room boundary, where the particle velocity is maximum. For a 100 Hz tone, this would be roughly 1 m from the boundary. Placing porous absorbers directly on a room boundary, while the most practical, is not efficient because the particle velocity at a boundary is 0. Too often, many people place porous absorption in corners of rooms thinking this will absorb sound, since all the modes have a *contribution* in the corners. However, while the modes have a maximum pressure in the corners, the particle velocity is very low and so the absorption is ineffective. For these reasons, resonant absorbers are preferred for treating low frequencies.

Resonant absorbers are mass–spring systems with damping to provide absorption at the resonant frequency of the system. The mass might come in the form of a membrane made of plywood or vinyl. Alternatively, the vibrating air in the neck of a hole might form the mass, as is the case for a Helmholtz absorber. The spring usually comes from an air cavity. Damping is most often provided by sound being forced through a porous material such as fibreglass or acoustic foam.

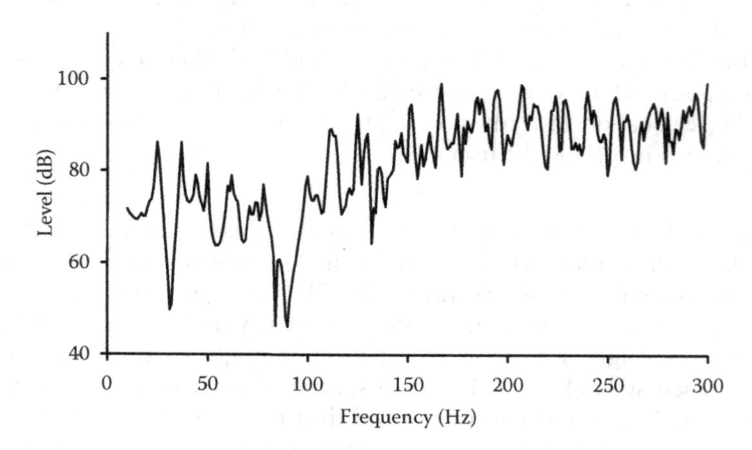

Figure 2.9 Low-frequency response of a small room.

The problem with resonant devices is that they usually provide only a narrow bandwidth of absorption. To cover a wide range of frequencies, a series of absorbers are required, each tuned to a different frequency range. Alternatively, double-layered absorbers can be used but are expensive to construct. Alternatively, multiple mechanisms can be exploited in a single device, as is done in the plate absorber described in Section 7.2.7. The vibrating mass is made from wood, metal, or vinyl and the spring is formed from foam or polyester. By placing some of the porous absorbent in front of the plate, this provides absorption over a broader bandwidth.

Resonant absorbers are discussed in Chapter 7, including microperforated absorbers, which are currently attracting considerable research interest. An alternative, but expensive, solution to bass absorption is to use active surfaces. Active absorbers have much in common with active noise control systems and are discussed in Chapter 14.

One problem with low-frequency modal control is knowing how much resonant absorption to use. Although the theories set out in Chapter 7 allow the performance of Helmholtz absorbers to be estimated, the meaning and interpretation of absorption coefficients at low frequencies are not straightforward. (Even more tricky is the lack of good prediction models for membrane absorbers, but that is another story.) At low frequency, the sound field is not diffuse, and consequently, the effect that an absorber has is not calculable through simple statistical laws. Often, practitioners pragmatically apply diffuse field theory anyway. A more complex but exact approach would use a wave-based modelling method such as finite or boundary element.

2.5 ECHO CONTROL IN AUDITORIA AND LECTURE THEATRES—BASIC SOUND PROPAGATION MODELS

A late arriving reflection appears as an echo, if its level is significantly above the general reverberation. In a large auditorium, a reflection from the rear wall is a common source of echo problems for audience members near the front of the stalls (main level seating) or performers on the stage. Echoes are very likely if the rear wall forms a concave arc that focuses the reflections at the front of the hall. The physical and subjective processes are the same as for echoes heard in mountain ranges or in cities with large building façades.

One technique for removing the echo is to apply absorption to the rear wall. The absorption attenuates the reflection, making it inaudible as a separate acoustic source. The problem with using absorption in large halls for classical music is that it removes acoustic energy, which is at a premium, and so diffusion is currently the preferred solution (see Section 3.1). For smaller spaces, where reverberation needs reducing, absorption can be used to treat echoes. The absorption needs to act at mid to high frequencies, as echoes are most notable for directional instruments. Consequently, a layer of porous material can be used.

Flutter echoes can occur in spaces with two large parallel walls. The regular pattern of reflections caused by sound bouncing back and forth between the walls causes coloration. By *coloration*, it is meant that the frequency response of the sound is detrimentally altered. If you go into many stairwells with parallel walls and clap your hands, a high-frequency ringing will be heard; this is a flutter echo. Flutter echoes are common in lecture theatres. One remedial measure is to apply absorbent to at least one of the two walls to absorb the reflections. A relatively thin layer of porous absorbent can achieve this, as it is mid to high frequency treatment that is needed. Alternatively, diffusers are sometimes preferable because they control the flutter echo, while also uniformly dispersing the sound for better coverage and intelligibility.

Porous absorbers are any material where sound propagation occurs in a network of interconnected pores, in such a way that viscous and thermal effects cause acoustic energy to be dissipated as heat. Common examples are mineral wools, fibreglass, open cell foams, acoustic tiles, carpets, and curtains. Current concerns about sustainability have also led to porous absorbers being constructed from recycled materials. Section 6.2.3 details some examples. To gain a proper theoretical understanding of porous absorbers, it is necessary to understand the theories of sound propagating in a medium. Some basic models of sound propagation, which are the basis for much of the absorber and diffuser modelling in this book, are presented in the next section.

2.5.1 Sound propagation—a wave approach

To understand and design absorbers, it is necessary to have a basic understanding of the terminology used and the fundamental mathematical constructs used for sound propagation. This section introduces some basic constructs, concepts, and terms.

A complex number representation of waves will be adopted throughout this book. The pressure of a plane wave propagating in a direction **r** is as follows:

$$p(t,\mathbf{r}) = Ae^{j(\omega t-\mathbf{k}\cdot\mathbf{r})} = Ae^{j(\omega t-k_x x-k_y y-k_z z)}, \tag{2.7}$$

where $k = \{k_x, k_y, k_z\}$ is the wavenumber, with k_x being the component in the x direction, $k^2 = |\mathbf{k}|^2 = k_x^2 + k_y^2 + k_z^2$; A is a constant related to the magnitude of the wave; $r = \{x, y, z\}$ is the location of the observation point; t is time; and $\omega = 2\pi f = kc$ is the angular frequency, where f is the frequency and c is the speed of sound.

The same conventions as used in Reference 3 are being adopted, so this is useful background reading for those who find this introduction too brief. The time dependence is $e^{+j\omega t}$. Unfortunately, there is no standard convention for the sign of this time dependence, so some of the literature uses a negative power in the exponential, leading to equations and results that are complex conjugates of those given in this book. Some texts and papers used a propagation constant, $\gamma = jk$, in their equations instead of the wavenumber.

Consider a plane wave propagating through an acoustic medium; this could be air or a porous absorber. The plane wave will be taken to propagate in the x direction for convenience. The pressure and particle velocity are given by

$$p = Ae^{j(\omega t - kx)}, \tag{2.8}$$

and

$$\mathbf{u} = \frac{A}{\rho c} e^{j(\omega t - kx)}, \tag{2.9}$$

where ρ is the density and c is the speed of sound of the acoustic medium. The ratio of pressure to velocity gives the characteristic specific acoustic impedance of the medium, z_c:

$$z_c = \rho c. \tag{2.10}$$

The characteristic acoustic impedance is a very useful property of the material when calculating the transmission of acoustic waves within and between different media.

The characteristic impedance of plane waves in air is purely real, with a value of about 415 rayls. In a porous material, it will be complex, with a characteristic resistance and reactance, which are the real and imaginary parts of the impedance, respectively.

Once the characteristic impedance and wavenumber within an acoustic medium are known, it is possible to predict the sound propagation. While it is possible to characterize a medium with the characteristic impedance and the wavenumber, it is also possible to use two other variables, the effective density, ρ_e, and bulk modulus, K_e. The word *effective* is used to signify that this is the density experienced by the acoustic waves rather than the more normal definition of mass divided by volume. The bulk modulus is the ratio of the pressure applied to a material to the resultant fractional change in volume it undergoes. It is the reciprocal of the compressibility. For a porous absorber, the effective density and bulk modulus can be related to the characteristic impedance and wavenumber by the following formulations. The characteristic impedance is given by

$$z_c = \sqrt{K_e \rho_e}, \tag{2.11}$$

and the propagation wavenumber by

$$k = \omega\sqrt{\frac{\rho_e}{K_e}}. \tag{2.12}$$

Where possible, this book will work with the impedance and wavenumber. Some porous absorbent formulations explicitly give values for the bulk modulus and the effective density, however, and so these terms are often used in the literature.

2.5.2 Surface impedance, admittance, reflection coefficient, and absorption coefficient

The effect that a surface has on an acoustic wave can be characterized by four interrelated acoustic quantities: the surface impedance, the admittance, the pressure reflection coefficient, and the absorption coefficient. The first three (surface impedance, admittance, and pressure reflection coefficient) give information about both the magnitude and phase change on reflection. The absorption coefficient does not contain phase data but only gives information about the energy change on reflection.

Knowledge of these four acoustic quantities is fundamental to understanding absorbing materials. These will now be defined mathematically by considering a wave propagating between two media. Consider a plane wave incident at an angle to a boundary between two acoustic media at $x = 0$, as illustrated in Figure 2.10. A simple model for a porous absorber assumes that it behaves as an acoustic medium like air, only with a different speed of sound c_1 and density ρ_1. The incident p_i, reflected p_r, and transmitted p_t pressures are given by the following:

$$p_i = A_i e^{j[\omega t - kx\cos(\psi) - ky\sin(\psi)]}, \tag{2.13}$$

$$p_r = A_r e^{j[\omega t + kx\cos(\theta) - ky\sin(\theta)]}, \tag{2.14}$$

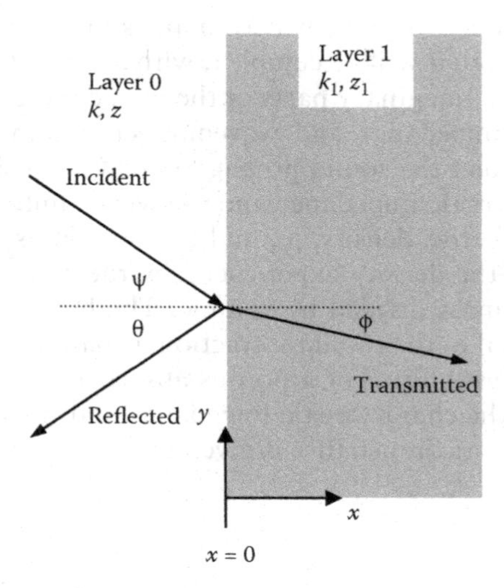

Figure 2.10 Sound incident on a surface and being reflected and transmitted.

and

$$p_t = A_t e^{j[\omega t - k_1 x \cos(\phi) - k_1 y \sin(\phi)]}, \tag{2.15}$$

where A_i, A_r, and A_t are the magnitudes of the plane waves incident, reflected, and transmitted, respectively, and the angles are defined in Figure 2.10. Applying continuity of pressure, $p_r + p_i = p_t$, gives the following relationship:

$$A_r e^{j[\omega t + kx\cos(\theta) - ky\sin(\theta)]} + A_i e^{j[\omega t - kx\cos(\psi) - ky\sin(\psi)]} = A_t e^{j[\omega t - k_1 x\cos(\phi) - k_1 y\sin(\phi)]}. \tag{2.16}$$

This must be true for all times and for all values of y as this was a plane wave. Consequently, a relationship relating the angles of propagation can be obtained from Fermat's principle:

$$\sin(\psi) = \sin(\theta). \tag{2.17}$$

For the refraction of the sound entering layer 1, the relationship between the angles comes from Snell's law:

$$\frac{\sin(\psi)}{c} = \frac{\sin(\phi)}{c_1}, \tag{2.18}$$

or

$$k\sin(\psi) = k_1 \sin(\phi). \tag{2.19}$$

The behaviour of the sound wave therefore depends on the relative values of the speeds of sound in the two media. For most absorbents, the speed of sound is much less than that in air. Consequently, the angle of propagation in the medium is smaller than in the air. In fact, for many absorbents, the angle of propagation can be assumed to be normal to the surface, i.e., $\phi \to 0$.

The pressure reflection coefficient, R (sometimes referred to as a reflection factor), gives the ratio of the reflected and incident pressure, i.e.,

$$R = \frac{p_r}{p_i}. \tag{2.20}$$

Therefore, the pressure reflection coefficient includes both magnitude and phase information about the reflection of sound. There is also an intensity reflection coefficient, but it is not used in this book.

The need for continuity of particle velocity normal to the surface enables the derivation of an expression for the specific acoustic impedance of the surface. The relationships between pressure reflection coefficient and impedance will be used repeatedly. For oblique incidence, these are as follows:

$$R = \frac{\frac{z_1}{\rho c}\cos(\psi) - 1}{\frac{z_1}{\rho c}\cos(\psi) + 1}, \tag{2.21}$$

and

$$\frac{z_1}{\rho c}\cos(\psi) = \frac{1+R}{1-R}. \tag{2.22}$$

The admittance is the reciprocal of the impedance:

$$\beta = 1/z_1. \tag{2.23}$$

Often, the surface admittance and impedance are normalized to the characteristic impedance of air, and these are denoted with a subscript of n.

The surface impedance is often split into the real term (resistance) and imaginary term (reactance). In general, the real term of surface impedance is associated with energy losses, and the imaginary term with phase changes. So a simple inspection of the surface acoustic impedance gives more insight into the absorbing properties of a material than the absorption coefficient.

As the absorption coefficient, α, is a ratio of the absorbed and incident energy, this enables the following expression to be derived:

$$\alpha = 1 - |R|^2, \tag{2.24}$$

where $|R|$ is the magnitude of the pressure reflection coefficient.

The above formulations have assumed a plane wave case; however, in certain cases, for example, for a source close to a reflecting surface, a spherical wave formulation is most

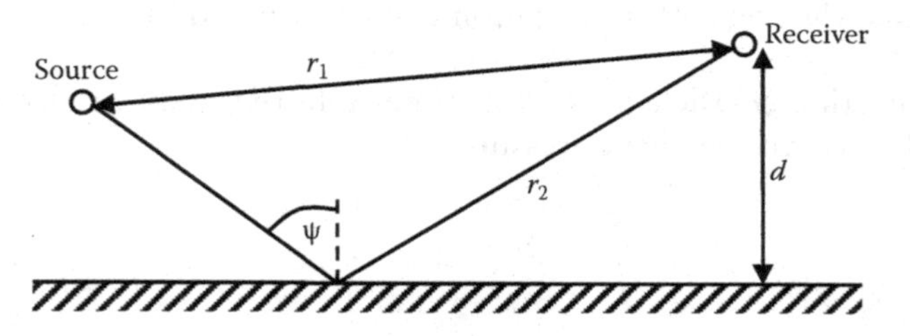

Figure 2.11 Geometry for spherical wave reflection.

appropriate. The F-term solution of Nobile and Hayek[11] gives the velocity potential at a distance d above an impedance plane as

$$\varphi(d) = \frac{e^{-jkr_1}}{r_1} + Q\frac{e^{-jkr_2}}{r_2}, \tag{2.25}$$

where r_1 and r_2 are defined in Figure 2.11. Q is the spherical wave reflection coefficient given by

$$Q = R + (1 - R)F, \tag{2.26}$$

where R is the plane wave reflection coefficient as previously used. So this asymptotic solution uses the plane wave solution plus a correction term. F is the boundary loss function given by

$$F(w) = 1 - j\pi^{1/2} w e^{-w^2} erfc(jw), \tag{2.27}$$

where $erfc()$ is the complementary error function and w is the numerical distance given by

$$w = \cos(\psi + \beta_n)\sqrt{-\frac{1}{2} jkr_2}. \tag{2.28}$$

Strictly speaking, reflections should be modelled using spherical wave reflection coefficients. In reality, however, the extra complication of the above equations is unnecessary in many cases. For instance, in the reverberant field in a room, where the number of reflections is large, a plane wave coefficient is sufficient. In contrast, when considering grazing reflections from sources close to the ground outdoors, the spherical wave reflection coefficient should be used.

This section has described a number of key terms for sound propagation fundamental to absorber and diffuser modelling. These will be used throughout the theoretical sections of this book where prediction models are developed for both absorbers and diffusers. These terms are also used in Chapters 4 and 5 when measurements are considered.

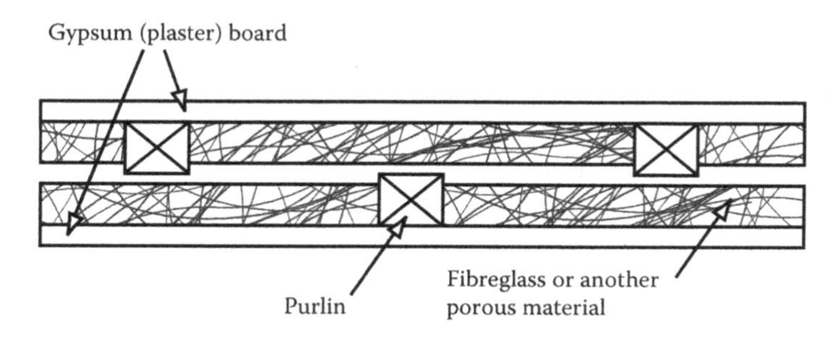

Figure 2.12 Mineral wool in a partition.

2.6 ABSORPTION IN SOUND INSULATION—TRANSFER MATRIX MODELLING

Porous materials are widely used in sound insulation. Lightweight constructions are often based on double leaf partitions with an air gap in between, as shown in Figure 2.12. It is normal for the air gap to contain a porous absorber. The absorber is used to attenuate resonances of the air cavity. Without damping, at the resonant frequencies, sound will pass easier through the partitions, and so the insulation will be degraded. It is important that the absorbing material is lightly packed; otherwise, it can form a vibration path bridging between the two partitions. This would reduce the sound insulation. Many different porous materials are effective in the partition, and this is not the most critical application for the design of absorbents.

As the issue of transmission is being discussed, it seems appropriate to discuss the transfer matrix approach to modelling transmission through, and absorption from, porous and resonant absorbers. The transfer matrix approach is the basis for many of the prediction techniques given in Chapters 6 and 7. A similar process is used in transducer modelling, where the method is called a two-port network.

2.6.1 Transfer matrix modelling

The transfer matrix approach to modelling sound propagation is a very powerful technique most often applied to porous absorption with and without membrane or perforated facings. It enables the surface impedance of single and multiple layers of absorbent to be calculated. For instance, it enables the case of a rigidly backed porous absorbent to be considered. This is the most important case because it represents the common example of an absorbent placed on a wall, floor, or ceiling.

Consider the situation shown in Figure 2.13. It is assumed that only plane waves exist within the layers. It is further assumed that the propagation is entirely contained in the x–y plane. By considering the continuity of pressure and velocity at the boundaries, it is possible to relate the surface pressure of one layer to the next.

$$\begin{Bmatrix} p_{3b} \\ u_{3b} \end{Bmatrix} = \begin{Bmatrix} p_{2t} \\ u_{2t} \end{Bmatrix} = \begin{Bmatrix} \cos(k_{x2}d_2) & j\dfrac{\omega\rho_2}{k_{x2}}\sin(k_{x2}d_2) \\ j\dfrac{k_{x2}}{\omega\rho_2}\sin(k_{x2}d_2) & \cos(k_{x2}d_2) \end{Bmatrix} \begin{Bmatrix} p_{2b} \\ u_{2b} \end{Bmatrix}, \tag{2.29}$$

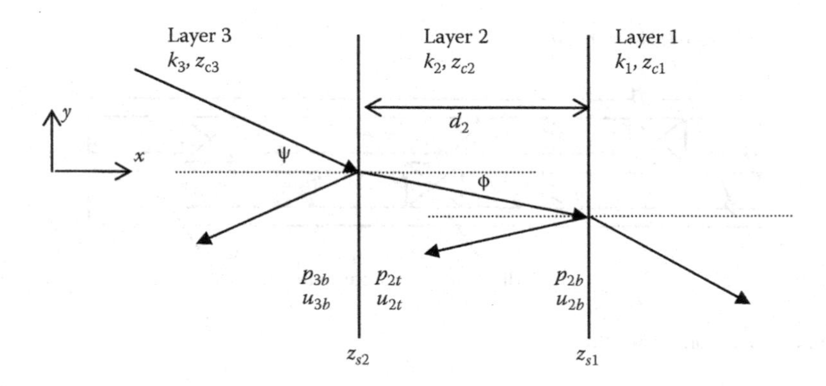

Figure 2.13 Geometry for transfer matrix modelling of a multilayer absorbent.

where p_{2b} and u_{2b} are the pressure and particle velocity, respectively, at the bottom of the second layer; for velocity, this is defined to be in the x direction; p_{3b} and u_{3b} are the pressure and particle velocity, respectively, at the bottom of the third layer; p_{2t} and u_{2t} are the pressure and particle velocity, respectively, at the top of the second layer; d_2 is the thickness of the layer 2; ρ_2 is the density of layer 2; and k_{x2} is the x direction component of the complex wavenumber for the 2nd layer and z_{c2} the corresponding characteristic impedance.

The component of the wavenumber in the x direction is calculated by considering Snell's law (Equation 2.18):

$$k_{x2} = k_2\sqrt{1-\sin(\phi)} = \sqrt{k_2^2 - k_3^2 \sin(\psi)}. \tag{2.30}$$

Many porous absorbents have a small speed of sound in comparison to air, and so often $k_{x2} \approx k_2$ as $\phi \approx 0$.

Equation 2.29 is a recursive equation from which the pressure and velocity of any layer can be determined from boundary and incident sound wave conditions. Although this process can be used to determine absolute values for the pressure and velocity, the technique is most powerful in determining surface impedance values. The surface impedance is calculated for the top of the first layer; this is then used to calculate the impedance at the top of the second layer. The process is then repeated until all layers have been evaluated. The relationship that enables this process relates the surface impedance at the top of layer 2, z_{s2}, to the impedance at the top of layer 1, z_{s1}[12]:

$$z_{s2} = \frac{-jz_{s1}z_{c2}\dfrac{k_2}{k_{x2}}\cot(k_{x2}d_2) + \left(z_{c2}\dfrac{k_2}{k_{x2}}\right)^2}{z_{s1} - jz_{c2}\dfrac{k_2}{k_{x2}}\cot(k_{x2}d_2)}. \tag{2.31}$$

To illustrate the application of Equation 2.31, consider the most common case, which is a rigidly backed single layer. In this case, $z_{s1} \rightarrow \infty$, and the equation reduces to give a formulation for the surface impedance of the rigid back absorbent as follows:

$$z_{s2} = -jz_{c2}\frac{k_2}{k_{x2}}\cot(k_{x2}d_2). \tag{2.32}$$

2.7 PIPES, DUCTS, AND SILENCERS—POROUS ABSORBER CHARACTERISTICS

Air conditioning ducts and other pipelines are a common source of noise.[13] The sound generated by fans, blowers, and internal combustion engines can propagate along the duct with little attenuation and radiate from outlets and exhausts. In addition, breakout noise can radiate through the sides of pipelines and ducts. The most effective treatment is to reduce the noise at the source, but where this is not possible, the application of absorbent material along the duct can be effective.

For pipelines, internal treatment is not often possible, and in this case, external lagging can be used to reduce breakout noise. The external lagging is often a combination of mineral wool and a heavy limp mass jacket made of metal, although the evidence is that foam is more effective. Below 300 Hz, the lagging of pipelines is not effective, and indeed, treatment around 300 Hz can often result in increased noise breakout. References 13 and 14 give design charts and equations to enable the effectiveness of pipeline and duct lagging to be calculated, although the prediction can be inaccurate unless proper manufacturer's data are known.

For ventilation ducts, internal treatment of the duct is most effective and 2.5–5-cm-thick linings are typically used. Internal duct liners are generally made of porous absorbent, with the type of material not being that important from an acoustic perspective. It is often necessary to use a protective coating for non-acoustic reasons. This might be a spray-on polyurethane coating, impervious lightweight plastic sheet, neoprene, or perforated metal. The protective coating can have a significant effect on the absorption.

Porous absorbers are also used as part of silencers and mufflers used to attenuate sound within pipework. Introducing a silencer will introduce a pressure drop, as it will inhibit the flow through the system, and this needs to be considered alongside the acoustic performance, size, and cost. It is important to know the environmental conditions that the absorbent will be subject to, both to be able to evaluate whether the material will be sufficiently robust over time and not, for instance, become clogged.

The acoustic properties of the material can also change with high temperatures and flow.[13] If sound is propagating along a lined duct against a laminar flow with a Mach number less than 0.3, then the effective length of the absorbent increases, and so does the attenuation. When the sound is propagating in the same direction of the flow, the attenuation decreases. If the flow Mach number is greater than 0.1, then shear in the flow will change the amount of attenuation also. If sound is propagating in the same direction of the flow, it will be refracted towards the lining, increasing absorption. With shear and the sound and flow in opposite directions, absorption decreases. But the flow will also change the effective impedance discontinuity between the fluid and the absorbent and so alter the attenuation. So the overall effects of flow are complex. In harsh environments, perforated or sintered metal is a

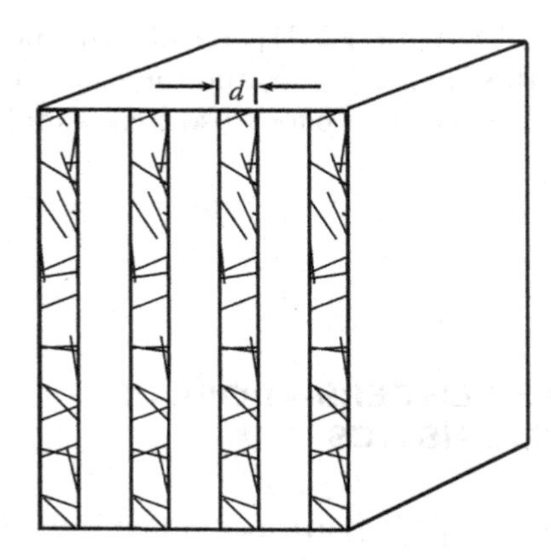

Figure 2.14 A typical parallel-baffle absorptive silencer used in a ventilation system.

good choice. Another possibility would be to use microperforated absorbers. In situations where sound pressure levels are very high, the non-linear behaviour of sound within the absorbent will need to be considered.

Silencers come in three main forms: reactive, absorptive, and a combination of reactive and absorptive. Reactive silencers change the cross-section of pipes to achieve attenuation at selective frequencies and are commonly used on outlet exhausts in harsh environments. As they do not feature acoustic absorption, they will not be discussed further. Absorptive or reactive/absorptive silencers, as the names imply, are of more interest because they remove sound energy using porous materials. The porous absorber will be most effective at mid to high frequencies. Absorptive silencers are most useful where a minimum pressure drop is required and so are commonly used in ventilation ducts, intake, and exhaust ducts of gas turbines and access openings of acoustic enclosures. Figure 2.14 shows one of the most common types of parallel-baffle silencers, where the shaded regions are made of porous absorbent. The attenuation is proportional to the perimeter area ratio, P/A, where P is the lined perimeter and A is the cross-section of the silencer. The porous material should have a low enough flow resistivity so the sound enters the absorbent, but not too low; otherwise, no dissipation occurs (flow resistivity is defined in the next section). The spacing between baffles should be less than a wavelength, as above the frequency where the wavelength equals the baffle spacing the attenuation falls off quite rapidly. The low frequency performance is determined by the thickness of the baffles, with $d \approx \lambda/8$ being optimal. Thickness rarely exceeds 200 mm, with thinner baffles usually producing more attenuation. Detailed design equations are given in Reference 14.

2.7.1 Characterizing porous absorbers

To theoretically model the sound propagation through a porous material, it is necessary first to have measurements characterizing the properties of the absorber as an acoustic medium. So far, in this chapter, the wavenumber and characteristic impedance have been discussed, but these cannot be directly measured. There are other parameters that are needed by absorber designers. Given a porous material, often, a researcher will start by measuring the flow resistivity, σ, and porosity, ϕ. The flow resistivity gives a measure of the resistance to flow that the porous absorber offers and porosity the amount of open volume in the absorber available to sound waves. Chapter 4 outlines measurement methods for obtaining these values, although for many porous absorbents, it is possible to assume a porosity of one. These are probably the two most important determining parameters for porous absorbents. Once the flow resistivity and porosity are known, it is then possible to get the characteristic impedance and wavenumber via the empirical formulations outlined in Chapter 6 and predict the absorption properties of the sample.

It is possible to go further and use a more refined model of porous absorbers, which need further measurements of properties (methods are given in Chapter 4). There are a variety of models in the literature, with two detailed in Chapter 6. The first uses three additional parameters: two characteristic lengths and tortuosity, and the second model also needs the static thermal permeability. Measuring the characteristic lengths is tricky, but there are a variety of methods for accurately getting the tortuosity.

It is possible to directly measure the surface impedance, pressure reflection coefficient, or absorption coefficient of a sample. This is often done in the development of prediction models for absorbents. As mentioned previously, the absorption coefficient can be measured through a reverberation chamber method, but the absorption coefficient is often influenced by edge effects. Furthermore, it is not possible to directly get phase information from the reverberation chamber, and this is very useful for understanding how an absorber works or why a theory succeeds or fails. To get phase information, a measurement for a particular incident angle needs to be made. The easiest approach is to measure the impedance in a tube, where only normal incidence plane waves exist. To get oblique incidence coefficients, it is necessary to use large samples in an anechoic chamber; the most common technique is a two microphone method, which is limited to homogeneous, isotropic, planar samples. All these techniques are discussed in Chapter 4.

2.8 ENCLOSURES, BARRIERS, AND ROADS

Enclosures might be used around a single machine to reduce noise, or they might be personnel enclosures within which workers can escape from excessive noise. Issues such as access and ventilation must be considered, and the need for these can compromise performance. Enclosures should be lined with porous absorbents to reduce the build-up of reverberant energy, which otherwise would compromise the sound attenuation.

Traffic is a major cause of noise problems, and although modern cars are quieter than their older ancestors, the increase in traffic levels has meant that average noise levels have not changed very much in recent decades. One partial solution is to use porous asphalt to reduce noise, the properties of which are discussed in Chapter 6. Another possibility is to use barriers to reduce noise propagation from roads to neighbouring houses and other buildings. Double reflections from high-sided vehicles enable some of the noise to bypass barriers, however, as shown in Figure 2.15a. Even more important, reflections from barriers on one side of the road can pass over barriers on the other side, as shown in Figure 2.15b. These additional reflections typically change sound levels by between 2 and 6 dB(A).[15] One solution is to apply absorption as shown in Figure 2.15c. The problem with absorption is that it tends to wear badly under the harsh conditions of high winds, salt, and water, which are common next to busy roads. Consequently, absorber performance is likely to decrease over time unless specialized and expensive durable absorbers are used.

Barriers also have a role indoors; for example, office screens are used to give acoustic isolation from noisy equipment and between workplaces in open plan offices. Barriers are often ineffective in a highly reverberant environment, however. The performance of indoor barriers can be improved by hanging absorbent or placing sound absorbing material directly on the ceiling to attenuate the first-order overhead reflection that would otherwise happen.

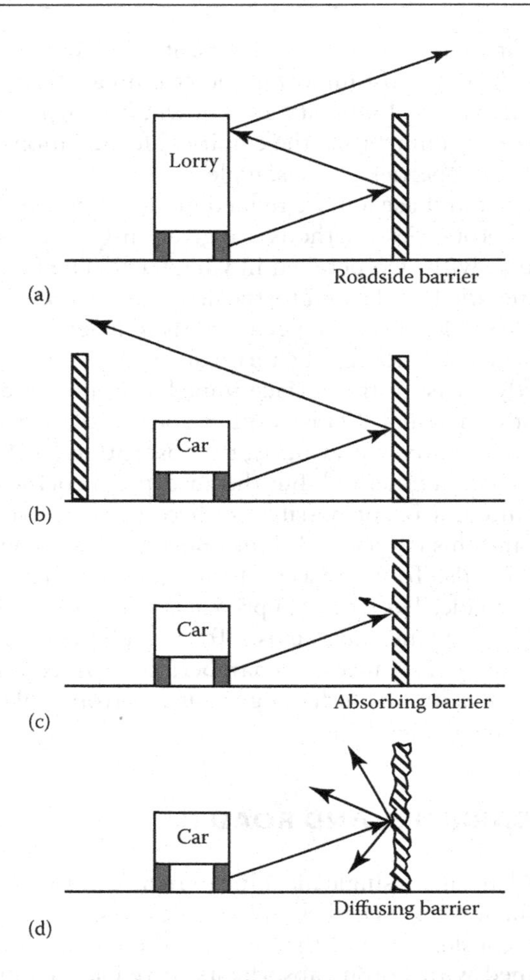

Figure 2.15 Schematics illustrating absorbers and diffusers to improve noise barriers. (a) Double reflections from high sided vehicles can pass over the barrier; (b) barrier reflections from one side of the road can pass over a barrier on the other side; (c) a sound absorbing barrier; (d) a sound diffusing barrier.

2.9 NATURAL NOISE CONTROL

It is common to find grass- or tree-covered areas around major noise sources, and optimizing natural features to maximize attenuation offers a sustainable solution to noise. Recent research findings have challenged the conventional wisdom that natural elements, such as trees, either on their own or with purpose-built barriers, vegetation, and ground, have no practical part to play in the controlled reduction of noise, in the propagation path between source and receiver. Where space allows, the use of natural means (wide tree belts, green walls, and ground), rather than artificial barriers, has the advantage of contributing to other issues in sustainability, such as reducing air pollution, generating corridors to encourage walking and cycling, generating access to local green areas, and reversing the long-term decline in wildlife habitats and populations. Within urban environments, green roofs and walls (see Figure 2.16)[16] can reduce noise in semi-enclosed spaces. Chapter 6 discusses how ground is modelled and affects noise levels. Chapter 8 examines the use of trees, shrubs, and green walls to attenuate sound. Trees can also improve the performance of conventional noise barriers by reducing downward refraction caused by wind.

Figure 2.16 A green wall at Longwood Gardens. (Courtesy of Daderot, commons.wikimedia.org/wiki/File:Green_wall_-_Longwood_Gardens_-_DSC01042.JPG.)

2.10 HEARING PROTECTION DEVICES

Ear defenders are used to reduce noise exposure, although where possible, it is better to control the noise at the source or along the path between the source and listener. They have cups that are designed to resist sound transmission, resting on cushions that should provide a comfortable and leak-free seal between the cup and the side of the head (see Figure 2.17). The resonant vibration of the cup against the combined spring formed from the air cavity and cushion compromises performance, however, typically in the region 100–200 Hz. Within the cup, porous absorption is applied to improve attenuation. Acoustic foam is most commonly used because it does not deteriorate in the warm and moist atmosphere next to the head. The addition of the porous absorbent increases

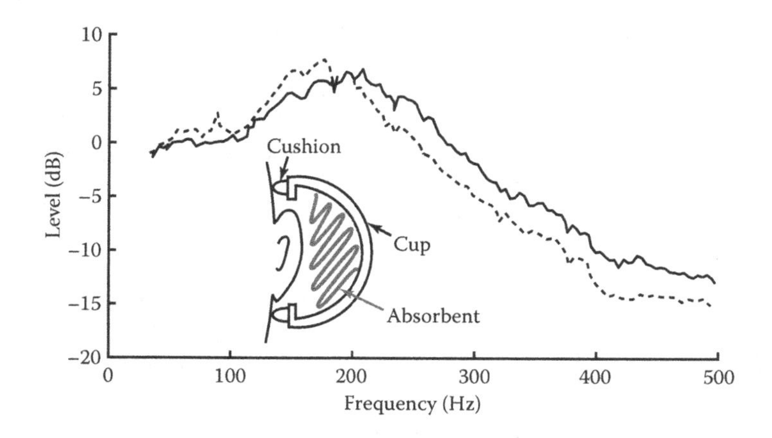

Figure 2.17 Level inside ear defender cup: ——— without and ------- with absorbent.

the compliance of the resonant system and also increases the damping.[17] Performance is less good at lower frequencies around the resonance but improved at higher frequencies away from resonance.

Bechwati et al.[18] experimentally demonstrated that activated carbon could be used to improve the performance of ear muffs by 5–10 dB below 300 Hz. The reasons for activated carbon's high absorption of low frequencies are discussed in Section 6.2.8, and its ability to change an enclosure's compliance, in Section 7.5.6. The hydrophilic nature of activated carbon poses practical problems in applying this concept.

2.11 LOUDSPEAKER CABINETS

Most conventional loudspeakers are mounted within cabinets to prevent sound generated by the rear of the driver interfering with that radiating from the front. The enclosure changes the behaviour of the driver, because the air cavity forms a compliance which alters the mechanical behaviour of the system, and this must be allowed for in the loudspeaker design. An empty enclosure has a series of resonances, which means the effect of the cavity on the driver varies with frequency. By placing absorption within the cavity, the resonant modes are damped and the sound quality improved because the effect of the enclosure is more even with respect to frequency. Usually, it is sufficient to use a porous wadding that has a relatively low resistivity because it can fill the cavity. It should be fixed in the middle of the cabinet, where the particle velocity is greatest. Some have used activated carbon within the cabinet because as well as providing absorption, the sorption of air molecules changes the compliance of the cavity; Sections 6.2.8 and 7.5.6 discuss activated carbon further.

2.12 AUTOMOTIVE ABSORBENTS AND VEHICLE REFINEMENT

Absorbents play an important role in reducing the noise created by cars and other vehicles.[19,20] Within the engine compartment, absorptive materials are applied under the hood (mainly the hoodliner) and, if possible, on other interior surfaces to reduce noise that will otherwise escape into the cabin or outside. In the exhaust system, dissipative silencers (mufflers) also include fibrous materials such as fibreglass or steel wool to reduce noise.

The noise within the cabin comes from a combination of airborne and structure-borne paths. Adding absorption, for example, via the headliner in the ceiling, can help reduce the noise. But there is a limit to how effective this can be, because for some sources, the noise is dominated by the direct rather than the reverberant sound field. An example of this would be sound from the vibration of the side window closest to a driver or passenger. Furthermore, adding too much absorption can affect the speech intelligibility between the occupants, by removing beneficial reflections.

Figure 2.18 shows the relative contribution of different surfaces in the cabin to reducing the interior noise. About half the absorption comes from the seating, so careful choice of appropriate open cell foam and fabric is needed. About a quarter of the absorption comes from the headliner. The preformed headliner is typically 6–12 mm thick and is made from a multilayer construction: glass fibre, binder, acoustic foam, and a fabric covering. As with all porous absorbents, if this can be mounted away from the ceiling with an air gap behind, and so placing the material where the particle velocity is highest, then more absorption results. Even so, there is a limited depth for the headliner, and so it does not provide much absorption below 500–1000 Hz.

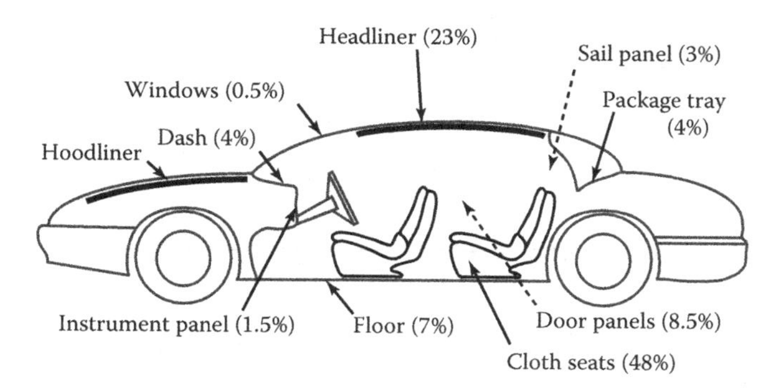

Figure 2.18 Parts of a car sound package with a typical percent absorption contribution for components in the cabin shown. (Absorption values from Harrison, M., *Vehicle Refinement: Controlling Noise and Vibration in Road Vehicles*, Butterworth-Heinemann, 2004.)

2.13 PORTABLE VOCAL BOOTHS

In 2006, a new product called a *reflection filter* was introduced to the recording studio market. They are generically referred to as *portable vocal booths*, and an example is shown in Figure 2.19. The idea is that this semicircular, portable acoustic screen, which is placed behind a microphone, allows a quasi-anechoic recording in a reverberant space by removing unwanted room reflections. Figure 2.20 shows the impulse response measured with and

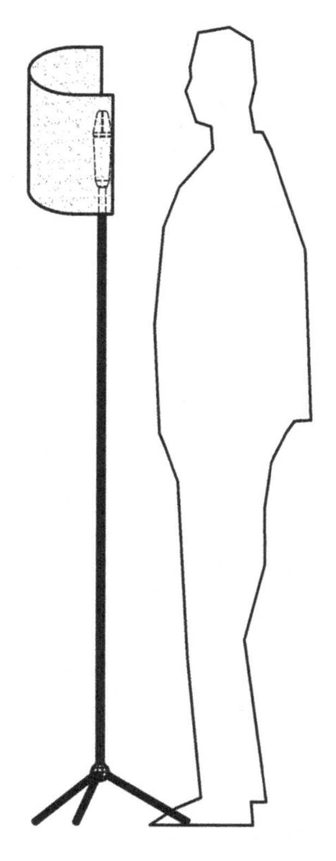

Figure 2.19 A portable microphone screen for audio recording.

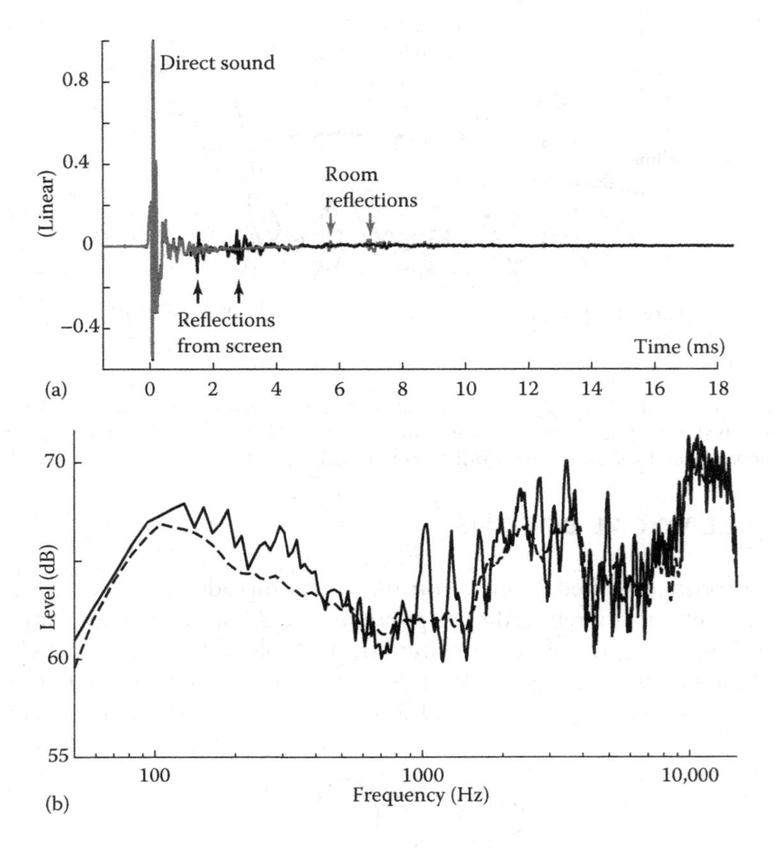

Figure 2.20 (a) Impulse response on a cardioid microphone in a small room. —— without and ▬▬ with a portable vocal booth. The order of the line plotting changes after about 5 ms to allow the room reflections to be visible. The direct sound for both measurements overlay each other. (b) Frequency responses measured: —— in a small room with the vocal booth and ------- for the loudspeaker and microphone alone in an anechoic chamber. (The anechoic loudspeaker response is uneven because it is being measured very close to a coaxial driver.)

without such a vocal booth in a small room. The grey line is the measurement without the vocal booth and shows small room reflections about 6 ms after the direct sound. The black line is with the booth present, and the room reflections have been reduced.

Reflections off the curved screen back to the microphone can cause significant comb filtering in the frequency response, however. Figure 2.20 shows that the frequency response is corrupted by wave interference. This is most prominent in the example shown from 1000 to 5000 Hz. These back reflections are probably amplified by the curved shell that is used on many screens. The curved shell is lined with porous absorbent (foam) to reduce reflections, and some designs have perforated shells that will also reduce reflections. Most commercial screens have an inner lining about 2.5 cm thick. As porous absorption is ineffective when it is thin compared to wavelength, this implies that the absorbent only removes reflections above 1.25 kHz or so. As the example in Figure 2.20 shows, even then, some commercial models are poor at minimizing coloration.

One way of evaluating the ability of the screen to protect the microphone from room reflections is to examine the direct-to-reverberant ratio (D/R ratio). In Figure 2.21, the average D/R ratio improvement for several acoustic screens on the market is shown. This is determined by subtracting the D/R ratio of an omnidirectional microphone with no screen as a reference.

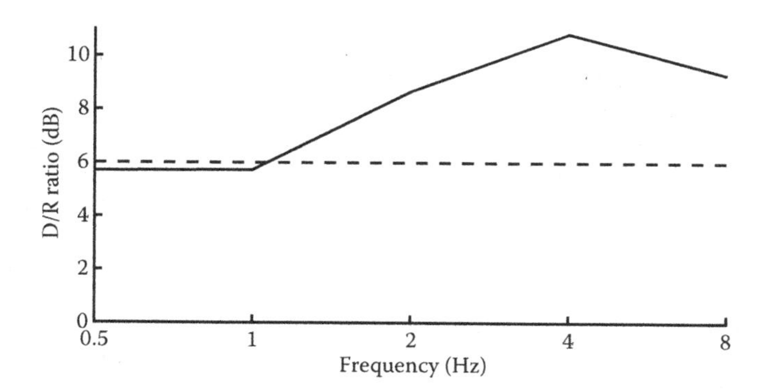

Figure 2.21 —— Average improvement in D/R ratio for some typical portable vocal booths and the ------ JND for the ratio.

This is compared to a Just Noticeable Difference (JND) for the D/R ratio of 6–8 dB.[21] It can be seen that the effect of the acoustic screen should be audible roughly above 1 kHz. In a test on 10 commercial vocal booths, Cox, Patil, and Inglis[22] came to a similar conclusion, although one larger screen achieved attenuation for an octave below that.

Cox, Patil, and Inglis also showed that the screens perform poorly at protecting the microphone from external noises such as ventilation and traffic noise. They found the protection from noise to be very uneven across frequency, with the greatest attenuation being around 800–1000 kHz. Above 2000 Hz, the attenuation was typically 2–5 dB. At lower frequencies, 200–400 Hz, most screens actually amplified unwanted noise!

Commercial acoustic screens are usually too small to reject frequencies below 1000 Hz, and the lower frequencies simply diffract around the screen. Noise and reverberant sound will continue to arrive at the microphone from above, below, and directly into the microphone, and additional improvement can be achieved only by further shielding along the sides, top, bottom, and behind the sound source. Using a cardioid microphone is also important.

2.14 SUMMARY

This chapter has outlined some absorber applications and touched on some of the issues that will be important in future chapters. It has also introduced some necessary mathematical principles. The remaining chapters concerning absorption are as follows:

- Chapter 3 introduces diffusers.
- Chapter 4, measuring absorber properties from the microscopic to the macroscopic.
- Chapter 6, application, design, and theoretical modelling of porous absorbers.
- Chapter 7, application, design, and theoretical modelling of resonant absorption, especially Helmholtz and membrane devices.
- Chapter 8, miscellaneous absorbers that did not fit into Chapters 6 and 7. Seating in auditoria turning Schroeder diffusers into absorbers, sonic crystals, and metamaterials, trees, green walls, and vegetation are considered.
- Chapter 12, hybrid treatments that are both absorbers and diffusers.
- Chapter 13, how to use predictions and laboratory measurements of absorption coefficients in room predictions, including the role of absorption coefficients in geometric room acoustic models.
- Chapter 14, active absorption.

REFERENCES

1. H. Kuttruff, *Room Acoustics*, 5th edn, Spon Press, Oxon, UK (2009).
2. P. Newell, *Recording Studio Design*, Focal Press, Oxon, UK (2011).
3. L. E. Kinsler, A. R. Frey, A. B. Coppens, and J. V. Sanders, *Fundamentals of Acoustics*, 4th edn, John Wiley & Sons, New York (2000).
4. M. Barron and L.-J. Lee, "Energy relations in concert auditoriums, I", *J. Acoust. Soc. Am.*, **84**(2), 618–28 (1988).
5. M. R. Schroeder, "New method of measuring reverberation time", *J. Acoust. Soc. Am.*, **37**, 409–12 (1965).
6. W. C. Sabine, *Collected Papers on Acoustics*, Harvard University Press, Cambridge, MA (1922).
7. ISO 9613-2:1996, "Acoustics—Attenuation of sound during propagation outdoors. Part 2: General method of calculation".
8. C. F. Eyring, "Reverberation time in 'dead' rooms", *J. Acoust. Soc. Am.*, **1**, 217–26 (1930).
9. M. R. Schroeder and H. Kuttruff, "On frequency response curves in rooms. Comparison of experimental, theoretical, and Monte Carlo results for the average frequency spacing between maxima", *J. Acoust. Soc. Am.*, **34**(1), 76–80 (1962).
10. M. Wankling, B. Fazenda, and W. J. Davies, "The assessment of low-frequency room acoustic parameters using descriptive analysis", *J. Audio Eng. Soc.*, **60**, 325–37 (2012).
11. M. A. Nobile and S. I. Hayek, "Acoustic propagation over an impedance plane", *J. Acoust. Soc. Am.*, **78**(4), 1325–36 (1985).
12. J. F. Allard and N. Atalla, *Propagation of Sound in Porous Media: Modelling Sound Absorbing Materials*, John Wiley & Sons, Chichester, UK (2009).
13. D. A. Bies and C. H. Hansen, *Engineering Noise Control: Theory and Practice*, 4th edn, CRC Press, Abingdon, UK (2009).
14. L. L. Beranek and I. L. Vér (eds), *Noise and Vibration Control Engineering*, John Wiley & Sons, Hoboken, NJ (2005).
15. C. S. Y. Lee and G. G. Fleming, "Measurement of highway-related noise", Report no. FHWA-PD-96-046 and DOT-VNTSC-FHWA-96-5, US Department of Transportation, Springfield, VA (1996).
16. Daderot, commons.wikimedia.org/wiki/File:Green_wall_-_Longwood_Gardens_-_DSC01042.JPG, accessed 27 December 2015.
17. M. R. Paurobally and J. Pan, "The mechanisms of passive ear defenders", *Appl. Acoust.*, **60**(3), 293–311 (2000).
18. F. Bechwati, T. J. Cox, M. R. Avis, and O. Umnova, "Adsorption in activated carbon and its effects on the low frequency performance of hearing defenders", *J. Acoust. Soc. Am.*, **123**(5), 3036 (2008).
19. D. Vigé, "Vehicle interior noise refinement—Cabin sound package design and development", in X. Wang (Ed), *Vehicle Noise and Vibration Refinement*, CRC Press, Boca Raton, FL (2010).
20. M. Harrison, *Vehicle Refinement: Controlling Noise and Vibration in Road Vehicles*, Butterworth-Heinemann, Oxford, UK (2004).
21. D. Griesinger, "The importance of the direct to reverberant ratio in the perception of distance, localization, clarity and envelopment", *Proc. Audio Eng. Soc. 126th convention*, preprint 7724 (2009).
22. T. J. Cox, N. Patil, and S. Inglis, "How effective are portable vocal booths?", *Sound on Sound* (October 2014).

Chapter 3

Diffusers

Applications and basic principles

In this chapter, the basic principles of diffusers will be developed. Following the same style of Chapter 2, this chapter will be driven by application. In many respects, the right and wrong places to use diffusers are still being worked out.[1,2] There are locations, such as rear walls of large auditoria used for classical music, where there is a broad consensus that diffusers are a good treatment to prevent echoes and are better than absorbers. The case for using diffusers in some other places is less clear-cut, and until further research is undertaken, some applications are going to be based more on precedence and intuition rather than scientific fact. Having said this, much has been learned in recent decades, which can help to ensure that diffusers are used where they are needed.

This chapter will outline how diffusers can be applied and the effects that their application will have on the physical acoustics and the listener response. It will also introduce some basic physics that will be needed to understand the more detailed chapters on diffuser prediction, design, measurement, and characterization later in this book.

3.1 ECHO CONTROL IN AUDITORIA

In Section 2.5, the problems of echoes and flutter echoes were discussed. To recap, echoes are caused by late-arriving reflections with a level significantly above the general reverberation. For instance, they can often be heard at the front of a badly designed theatre, with the echo being caused by a reflection from the rear wall. The echo might also come from a balcony front or other multiple-reflection paths. Flutter echoes are caused by repeated reflections from parallel walls and are often heard in lecture theatres, corridors, and meeting rooms.

In Chapter 2, absorbers were suggested as a treatment for echoes, but diffusers should be used when sound energy needs to be conserved. This would be the case in a large auditorium with an orchestra because every part of the sound energy generated by the musicians should be preserved for as long as possible and not lost to avoidable absorption. In other cases, the choice between diffusers and absorbers will normally rest on whether the energy lost to absorption will detract or improve other aspects of the acoustics, such as the reverberance, envelopment, or intelligibility.

3.1.1 Example applications

Absorbers can be used on the rear wall of auditoria to reduce echo problems. Nowadays, in large performance spaces, diffusers are usually the preferred treatment because they can solve the echo problem without decreasing the sound level for seats towards the rear of the hall. Figure 3.1 shows quadratic residue diffusers (QRDs®) applied to the rear wall of

Figure 3.1 Schroeder diffusers (QRDs) applied to the rear wall of Carnegie Hall to prevent echoes. (Courtesy of D'Antonio, P., Cox, T.J., *J. Audio Eng. Soc.*, 46, 955–976, 1998.)

Carnegie Hall, New York. QRDs are a type of Schroeder diffuser described in Chapter 10. This form of *reflection phase grating* was the starting catalyst for modern diffuser research about four decades ago. The diffusers were installed in Carnegie Hall because a long delayed reflection from the rear wall caused an echo to be heard on the stage, making it difficult for musicians to play in time with each other. Adding diffusers dispersed the reflection, reducing the reflection level returning to the stage and consequently making the echo inaudible. The diffusers also improved spaciousness on the main floor by uniformly diffusing rear wall reflections and masking echoes from the boxes.[3]

Figure 3.2 shows the application of optimized curved diffusers to the side walls of the Hummingbird Centre, Toronto. A refurbishment of the hall was going to involve adding a reverberation enhancement system. This system added additional *reflections* from loudspeakers mounted in the side walls. Unfortunately, this was a shallow splay, fan-shaped hall, and as is common with auditoria of this shape, if sources are on the side walls, echoes across the width of the audience area are heard. This was not a problem while the sound was being generated from the stage but would have been a problem when the artificial reflections were generated from the side walls by the loudspeakers. It made no sense to treat the echoes with absorption because that would just remove sound energy and defeat the point of using an artificial enhancement system. The solution was therefore to use diffusers, which break up lateral propagating reflections in the audience area.

The original concept was to use Schroeder diffusers that were to be covered in cloth to hide their visual appearance. This is a common story in diffuser design. The acoustical treatment needs to complement the visual appearance of the room to be acceptable to architects. If the visual aesthetic is not agreeable, the treatment has to be hidden behind

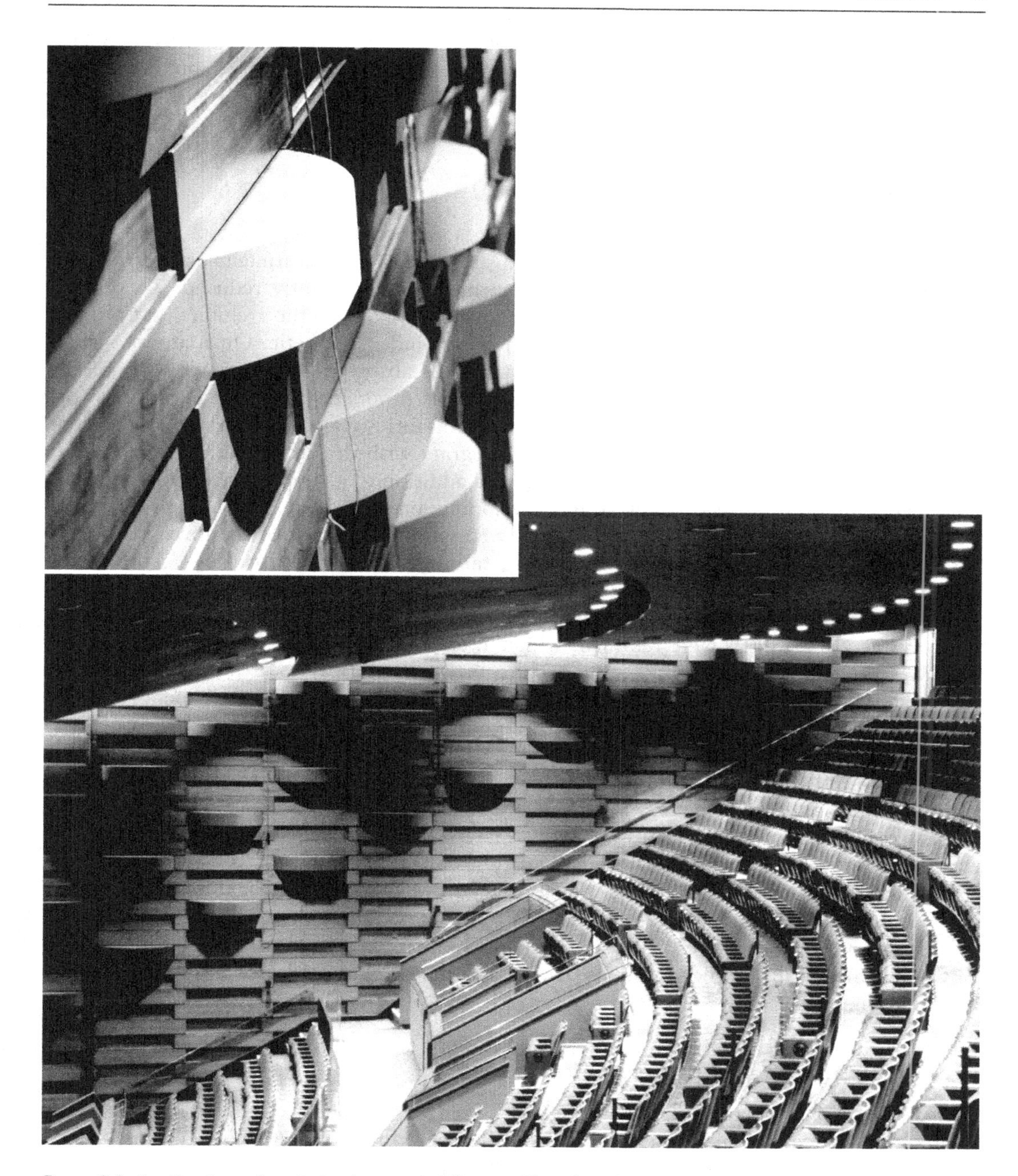

Figure 3.2 Application of optimized curved diffusers (OptiCurve) in the Hummingbird Centre, Toronto. Consultant: John O'Keefe, Aercoustics Engineering Ltd. (Courtesy of D'Antonio, P., Cox, T.J., *Appl. Acoust.*, 60, 113–142, 2000.)

fabric. But as discussed in Chapters 8 and 10, this would have turned the side wall diffusers into absorbers. (Another solution architects often apply to *ugly* acoustical treatment is to remove it completely. This solves the visual aesthetic problem and saves money!) In the Hummingbird Centre, designers tried a new method for designing optimized curved surfaces, which is outlined in Chapter 11. The shapes then complemented the appearance of the space and met the visual requirements of the architect. The concept was a basket weave that would thread in and out of the cherry wood side walls. O'Keefe comments, 'The renovated room and, in particular, the enhancement system have been well received by the owners and their tenants'.[4]

3.1.2 Aesthetics

For acoustics to be considered part of a project, it is important to be involved in the initial stages so that appropriate budgets can be established. If acoustic treatments are introduced later in the process, the chances of them being included are greatly reduced. Therefore, acoustic consultants and treatment manufacturers must explain to the architect that proper acoustical design does not have to infringe on the intended aesthetic. On many projects, architects prefer custom design, so that the diffuser or absorber has their signature on it, and not that of an acoustical manufacturer's catalogue products. There are exceptions for functional projects, where absorbing ceiling tiles and fabric-wrapped absorbers are chosen to reduce costs. How can acoustic treatments integrate with the architecture? This is very important because every project has a function, in addition to its appearance, whether that is achieving intelligible speech or clear and enveloping music.

In classical architecture, scattering surfaces were an integral part of the form, with elements such as columns, relief ornamentation, statuary, and coffered ceilings. These surfaces provided structural, aesthetic, and acoustic functions. As architecture evolved into less ornate and simpler rectangular forms, as exemplified by the International Style, very few of these architectural motifs were retained, which altered the acoustic. Postmodernism was a reaction to the austere appearance, and architecture moved away to create more complex forms. The challenge was then to find acoustic treatments whose forms complement contemporary architecture. One approach for designing diffusers is optimization—the results of which are shown in many figures in this chapter and the methodology of which is discussed in Chapter 11. This allows collaboration with the architect to shape treatment to satisfy acoustic and visual requirements.

For these diffusing surfaces, any reflective material can be used. It is possible to shape concrete, wood, glass, or other materials that the architect has in mind. (See Section 1.2 for sustainability considerations.) This is the modern equivalent of the classical relief ornamentation, columns and statuary, complementing the contemporary design both visually and acoustically. Of course, there still remains the possibility to cover treatments with acoustically transparent fabrics and perforated wood and metals, although this approach is now less common.

3.1.3 Wavefronts and diffuse reflections

How do diffusers disperse reflections? Figure 3.3 shows a cylindrical wave reflected from a flat rigid surface, calculated using a finite difference time domain (FDTD) model—see Section 9.5 for more on this model. The case shown is where the wavelength is much smaller than the surface width. An impulse was generated creating a single cylindrical wavefront as seen in frame 0, which then travels from right to left towards the surface. The wave simply changes direction on reflection, travelling back in the specular reflection direction, where the angle of incidence equals the angle of reflection (in this case back towards the source). The reflected wavefront is unaltered from the incident sound, except the direction of travel is reversed.

This could lead to the reflection being perceived as an echo, especially if the source is a directional instrument, such as a trumpet. Therefore, the role of a diffuser is to break up

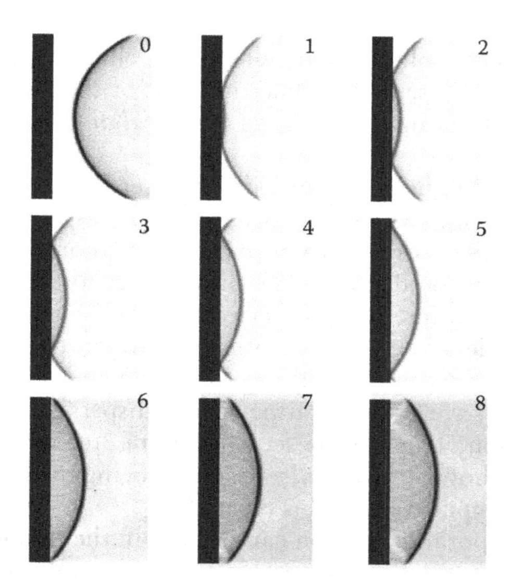

Figure 3.3 Cylindrical wave reflected from a flat surface calculated using an FDTD model. The numbers indicate the frame order of the snapshots. (Courtesy of Brian Horner.)

or *diffuse* the reflection, so that the sound energy is dispersed and sound from directional instruments such as trumpets do not remain in narrow beams.

Figure 3.4 shows the effect of changing the flat surface to help disperse the reflection. In this case, part of an ellipse is used. It can be seen that the reflected wavefront is more bowed. The change is not great because the curve of the ellipse was quite gentle; even so, in

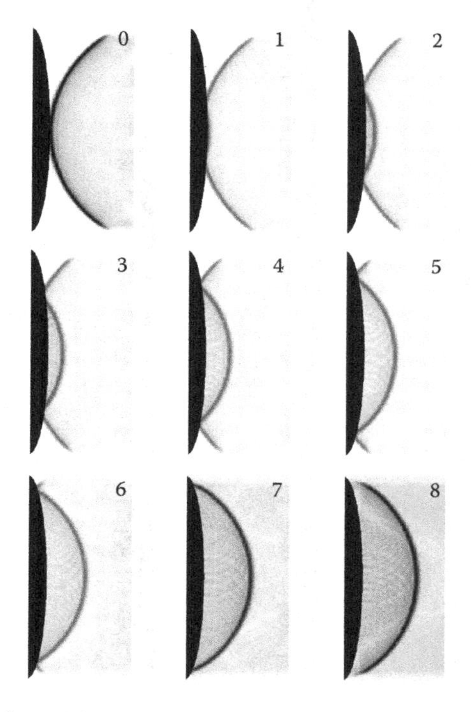

Figure 3.4 Cylindrical wave reflected from a curved surface calculated using an FDTD model.

the far field, the sound will be more spatially dispersed. The wavefront generated is still very ordered, however, so although solitary semicylinders or ellipses are good at spatial dispersion, they are not the best diffusers. A listener in front of the ellipse will hear a reflected wavefront that is not spread out in time; it lacks *temporal dispersion*. Solutions to this will be discussed in Chapter 11.

Figure 3.5 shows the effect of using a Schroeder diffuser. The reflected wavefront is much more complex than the previous examples. Inspection of the different frames shows why this complexity arises. Sound takes time to propagate in and out of the wells, causing parts of the reflected wavefronts to be delayed. The different depths of the wells cause various delay times, and the resulting interference between the reflected waves forms a complex pattern. Chapter 10 gives details of what depths are chosen to gain dispersion and why. An observer listening to this reflection will receive sound over an extended time, as these wavefronts travel past. So as well as generating spatial dispersion, Schroeder diffusers also generate temporal dispersion. This can be seen in Figure 3.6, where the impulse response for a Schroeder diffuser is shown (the *impulse response* being the response measured on the microphone when a short impulsive sound is created).

The consequences of temporal dispersion can be seen in the frequency domain. In Figure 3.7, the time and frequency responses of a single specular reflection from a flat surface (Figure 3.7a,b) and a diffuse reflection from a diffuser (Figure 3.7c,d) are shown. Figure 3.7a shows an impulse response comprising the direct sound and a specular reflection; some other room reflections are also shown. In Figure 3.7b, the frequency response of just the specular reflection is given. (The time window is illustrated by the two vertical lines in Figure 3.7c.)

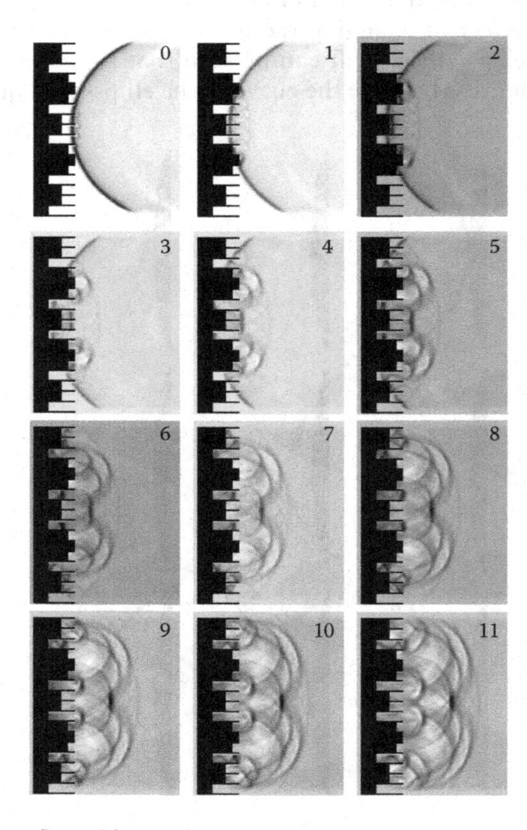

Figure 3.5 Cylindrical wave reflected from a Schroeder diffuser calculated using an FDTD model.

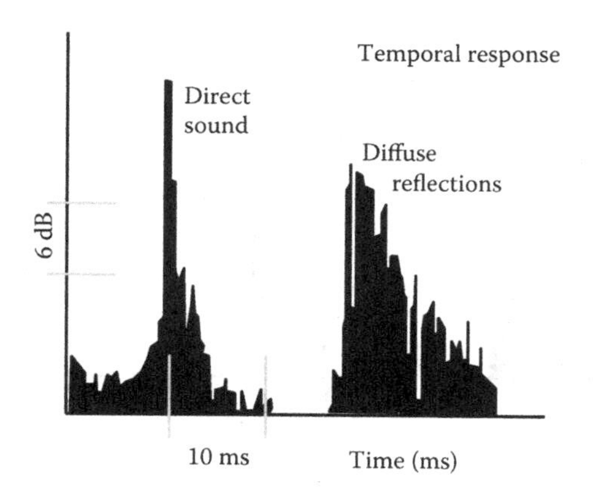

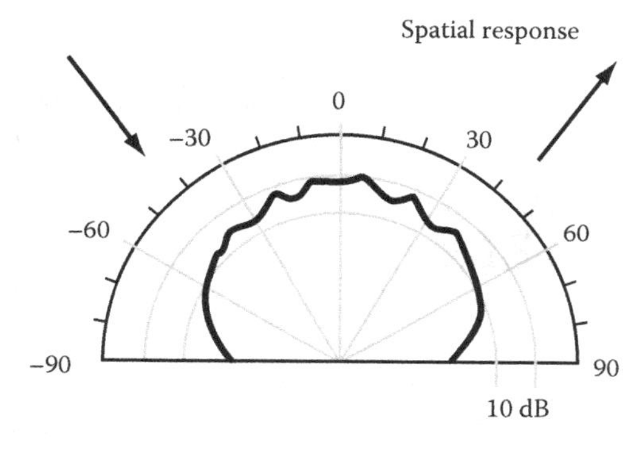

Figure 3.6 Measured spatial and temporal dispersion generated by a Schroeder diffuser.

This frequency response is characterized by a high-pass filter response, determined by the size and shape of the reflecting surface (see Section 11.1 for discussions of flat reflectors). In Figure 3.7c, where the reflection from a diffuser is shown, the specular reflection is now temporally dispersed and shown as a diffuse reflection. The frequency response of the reflection (Figure 3.7d), is characterized by a random distribution of irregularly spaced nulls and peaks. Many heuristic diffuser designs assume that any temporal distribution will provide a satisfactory frequency spectrum, when in reality, excessive coloration can be inadvertently introduced. (*Coloration* refers to an audible change in timbre of musical notes, usually caused by early reflections.)

In Figure 3.8, the time and frequency responses of the total fields, consisting of the direct sound and the reflection(s), are shown. When the direct sound and a specular reflection combine, they form a *comb filter*. The time delay between the direct sound and the reflection determines the frequency spacing of the minima and maxima, and the relative amplitudes of the sound, the levels of the minima and maxima. Comb filtering is an effect that should be avoided in critical listening rooms and performance spaces. It can happen if large flat reflectors and nearby walls are not treated with absorbers or diffusers. When the direct sound combines with a diffuse reflection, the regularity of the comb filtering is removed and the variation in levels is reduced. The spectral content of the direct sound can be more fully perceived. In addition to providing uniform spatial dispersion, reducing comb filtering is a principal reason for using diffusers in many applications.

Sometimes, absorption and diffusion are combined into a single treatment. Figure 3.9 shows an FDTD prediction for a wall formed from alternate patches of absorbing and

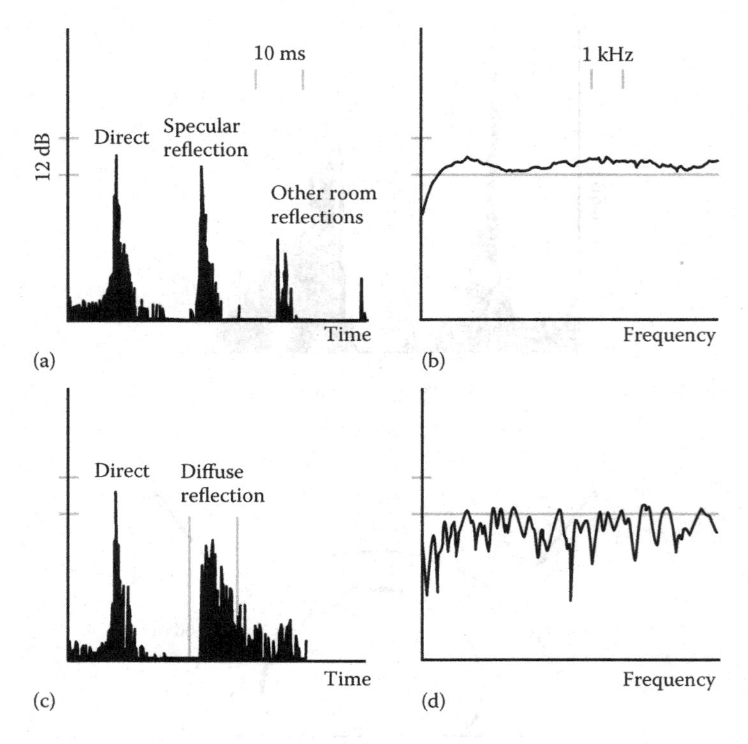

Figure 3.7 Time and frequency responses for a flat surface (a, b) and a diffuser (c, d). The frequency responses are for the reflected sound only. (Courtesy of D'Antonio, P., Cox, T.J., *Appl. Acoust.*, 60, 113–142, 2000.)

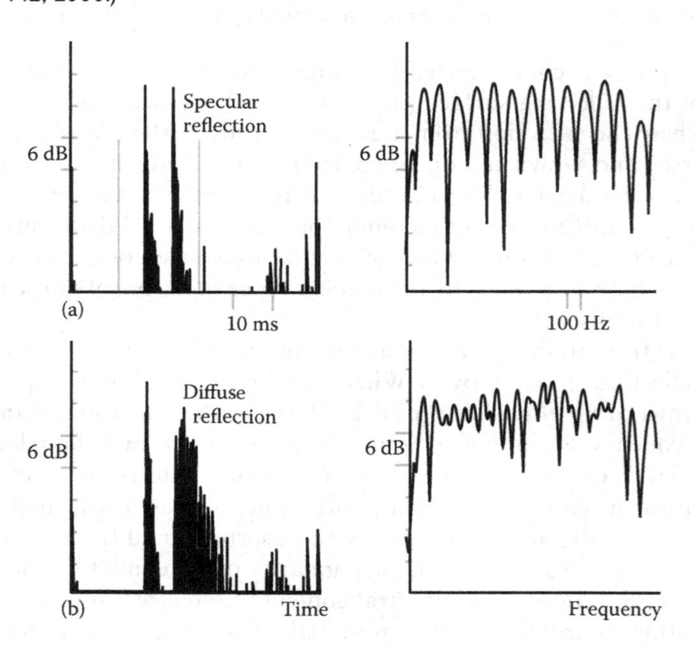

Figure 3.8 Measured temporal and frequency response of the total field, consisting of the direct and reflected sound. (a) For specular reflection and (b) for diffuse reflection. (Courtesy of D'Antonio, P., Cox, T.J., *Appl. Acoust.*, 60, 113–142, 2000.)

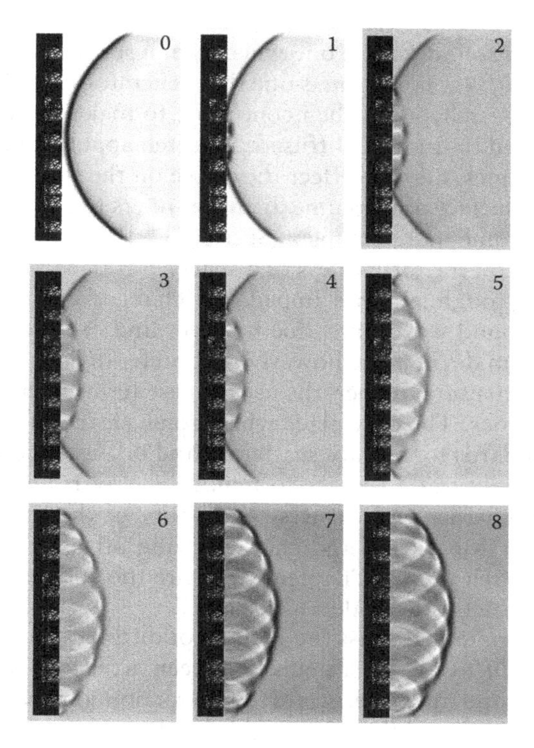

Figure 3.9 Cylindrical wave reflected from a hybrid surface using an FDTD model; the shaded wells are filled with absorbent.

reflecting material (the absorbent is placed inside a well for convenience of prediction). There is some dispersion, but it is not as large as that generated by Schroeder diffusers, however. To gain more diffusion, the surface can be curved or a ternary sequence used. This type of *hybrid surface* is discussed in more detail in Chapter 12.

3.1.4 Some terminology

Many might describe the reflection from a flat, rigid surface as being *coherent* and the reflections from diffusers as being *incoherent*. This is a misleading use of terminology, however. From a purely physical standpoint, coherence occurs when there is a fixed, time invariant phase relationship between separate parts of a wavefront. For both flat surfaces and complex diffusers, there is a fixed phase relationship. In this linear system, there is usually time invariance. The wavefront from a diffuser is just complicated, not incoherent. To achieve physical incoherence, the surface would have to move or change shape over time.

Another terminology commonly used is to say the reflection is *diffuse*. There is no formal definition of this, but it refers to the case where the reflection is dispersed both spatially and temporally. The sound should be distributed more widely, and the impulse response at a receiver should have quieter reflections spread over a longer time. Many diffusers are designed by examining the spatial dispersion, and assuming that this will be accompanied by temporal dispersion. In addition, many diffusers are designed simply assuming that any temporal variation will produce uniform spatial dispersion and an acceptable frequency response; however, this is not necessarily the case.

It is also important that diffuse reflection is not mixed up with whether a room has a diffuse sound field. A *diffuse sound field* is one where the propagation of sound is completely isotropic. For surfaces, a *diffuse reflection* is one that generates spatial and temporal dispersion. Whether these diffuse reflections then contribute to make a sound field more diffuse is not what is being referred to. Indeed, diffusers are often applied to treat first-order reflection problems, such as echoes, and the effect they have on the reverberant sound field is of secondary concern. Both surface and volume diffusion refers to cases where the sound field, or surface reflections, become more complex. Section 3.10.2 further discusses some of the effects that diffusers have on the diffuseness of a space.

The effects of surface roughness and impedance changes are to generate *diffraction*, i.e., the breaking up of sound wavefronts due to edges and other effects. Acousticians do not seem to favour the term *diffraction*, however, but prefer *diffusion*, *diffuse reflection*, or *scattering*. To complicate matters further, the use of these terms is inconsistent between different disciplines in acoustics. The only place where some clear differentiation has been set is with international standards for surface scattering and diffusion coefficients. A scattering coefficient refers to the ability of a surface to remove energy from the specular reflection direction. The diffusion coefficient measures the quality of the spatial uniformity of the reflections from a surface. Measuring and characterizing diffuse reflections are discussed in Chapter 5. This is also relevant to Chapter 13, where the application of scattering coefficients to geometric room acoustic models is discussed.

Finally, it seems appropriate to discuss the spelling of diffusers. Schroeder and D'Antonio chose to use the spelling *diffusor* to distinguish between acoustic and other diffusers, such as those used for lighting and air dispersion. However, common usage has drifted towards *diffuser*, and this will be employed throughout this book.

3.2 REDUCING COLORATION IN SMALL REPRODUCTION ROOMS

Of interest in this section are small rooms where sound is being reproduced through a loudspeaker system; examples include recreational listening rooms, recording or broadcast control rooms, teleconferencing, or distance learning rooms. It is often useful to consider extreme boundary conditions when attempting to explain a room acoustic problem. In the case of a critical listening room, one extreme is an anechoic chamber, and the other, a reverberation chamber. Anyone who has spent any time in these rooms realizes that neither is an exciting place for music listening. An ideal critical listening room will usually lie somewhere between the two extremes. It is also important to realize that since this is a sound reproduction room, the room can only corrupt what is being reproduced by the loudspeakers. The unwanted artefacts added by the room are acoustic *distortion*.

Mostly, these rooms are not used for recreational listening, but rather as acoustical *sonoscopes* to accurately perceive spectral balance and spatial imaging. Modern diffuser design has played a significant role in developing these state-of-the-art facilities. For late reflections, there has been general agreement in most designs, over many decades, that the decay time should be small, say with a reverberation time of about 0.3–0.4 s. But opinions about what should be done about strong early reflections have varied over many decades. It is interesting to map out how opinions have evolved, from the early two-channel designs to today's multichannel surround.[5]

The early two-channel designs of Tom Hidley emphasized early ceiling reflections, using a *compression* ceiling and promoted a reflective front surrounding the monitors with the rear of the room being anechoic. The introduction of time delay spectrometry by Dick Heyser fostered an era that aimed at eliminating early frontal reflections due to the measurable comb filtering that otherwise resulted. This led to the live end dead end (LEDE)[6] studio control room developed by Don and Carolyn Davis, suggesting that the front of the room should be absorptive, or dead, and the rear, live. Designers became aware of the research of Schroeder[7] describing the importance of early lateral reflections in providing envelopment in concert halls. This led Peter D'Antonio[8] to the use of Schroeder diffusers on the rear wall of LEDE control rooms to provide passive surround sound. He also proposed broadband control of early frontal reflections by creating a temporal and spatial reflection free zone (RFZ)[9] surrounding the mix position. Some of the concepts have been adapted and adopted for listening room standards.[10] Bob Walker used only sound redirection to create an RFZ in his Controlled Image Design.[11] Tom Hidley and Philip Newell promoted nonenvironment approaches, with significant broadband sound absorption, thus removing early reflections (see Figure 2.7). Peter D'Antonio, Jamie Angus, and George Massenburg then experimented with diffusing early frontal reflections, rather than absorbing them in control rooms, leading to reflection-rich zones and significant diffusion—a control room with massive diffusers will be described later in Section 3.2.3. A variant on this was developed by Davidovic and Petrovic[12] (see Figure 12.5 for an example that uses porous diffusers placed in front of absorption to achieve a similar result).

As the industry moved from two-channel to multichannel, early reflections began to be viewed constructively, rather than being seen as destructive to the critical listening process. This concept was further promoted by a review of psychoacoustics research by Floyd Toole.[13] There are still studio engineers and producers who prefer to control early reflections with absorption. An analogy to music performance may be informative. When musicians practice, they generally prefer a somewhat dead space, so that they can hear every nuance of their instrument. However, in performance, they prefer a more reverberant space to support the instrument. These practice rooms may be analogously compared to critical listening rooms, where spatial and spectral nuances must be monitored and tailored. The performance space may be compared to a recreational listening room in which enveloping reflections enhance the listening experience.

3.2.1 Reflection free zone

The debate about the best small critical listening design continues and probably will never be resolved because some choices are a matter of taste. However, for now, consider one of the most successful approaches, the RFZ. This design strives to minimize the influence of the room acoustic on the sound reproduced and so provides a neutral critical listening space. The design creates a spatial and temporal RFZ surrounding the primary mixing or listening position(s). The zone is spatial because it exists only within a certain area in the room, and it is temporal because the interfering reflections are controlled only over a certain time window, between the arrival of the direct sound and prior to reflections arriving from the rest of the room.

It is well established that early reflections affect the perceived sound, changing the timbre and causing image shift (*image shift* meaning that the sound source appears to be coming from the wrong place).[14] Acoustical treatment can be used to create a space in which spatial and spectral textures can be accurately perceived. Furthermore, the room aberrations can mask important artefacts on recordings. While the auditory system can adapt to interfering acoustic distortion, if undesirable artefacts are masked by room aberrations, they will not be perceived even after the listener has habituated to the room acoustic. Absorption can be used to control first-order reflections between the source and the listener and so remove early arriving, strong reflections, which might produce coloration and image shift. Applying large numbers of absorbers leads to a dead room, and so diffusers are used to delay and temporally diffuse reflections while preserving sound energy. This is done to minimize distortion caused by interference with the direct sound. Diffusers on the rear wall essentially produce passive surround sound that provides ambience in the room and envelopment.

Figure 3.10 shows the energy time curves measured before and after treatment in a small critical listening room. In Figure 3.10a, the direct sound and interfering side wall, floor, ceiling, and sparse reflections from the room are identified. Isolated and intense early specular reflections cause coloration, image shift, and broadening of the image width and depth. These problems can be addressed by controlling the early reflections using absorbers and diffusers. This creates an initial time delay gap before the reflections from the rear wall arrive, forming an RFZ that extends to roughly 18 ms, as shown in the lower graph. If absorption is used to remove the early reflections, psychoacoustic experiments indicate that the sonic images will be very small, as if sound comes from a point in space. If diffusers are used, the sonic images take on greater size and appear more realistic. Following the application of diffusers on the rear wall, the effects of which are shown in Figure 3.10b, the sparse

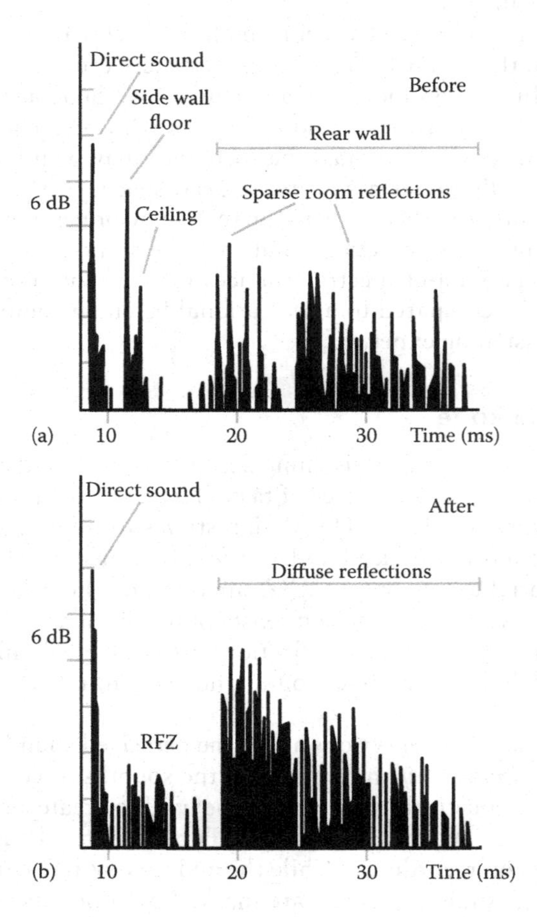

Figure 3.10 Impulse responses in a small critical listening room before (a) and after (b) treatment. (Courtesy of D'Antonio, P., Cox, T.J., *Appl. Acoust.*, 60, 113–142, 2000.)

room reflections more resemble a reverberant field of a larger room, with increased spatial and temporal reflection density. Using this technique, it is possible to create a reverberant sound field with a linear slope within a small room.

D'Antonio et al.[15] carried out a study on a recording control room; a drawing of the room plan can be seen in Figure 3.11. In this example, an RFZ is achieved by flush mounting the loudspeakers (L) with the woofers close to the front trihedral corners of the room (where two walls meet the ceiling). The massive front walls are splayed outward and treated with porous absorption to provide broadband control of first-order reflections. The rear walls are treated with broadband diffusers (QRDs) to disperse first-order reflections away from the listeners, while providing ambience and envelopment. The shaded area represents the spatial RFZ, 24 dB below the direct sound, with an initial time delay gap of approximately 17 ms, where the predominant energy is from the monitor loudspeakers. Two types of reflections are indicated in Figure 3.11: an undesirable (S) specular reflection from the glass doors forming the machine room and (D) diffuse reflections that arise from 7 m^2 of diffusers located on the rear wall.

The rear wall diffusers reduce the level of the reflections reaching the listener early and so coloration effects are reduced. Figure 3.8a shows the time and frequency response for a listener close to a large plane wall with no other surfaces present. The similarity between the incident and reflected time responses can be seen. (Some minor differences caused by diffraction are seen because the measured surface was not infinitely large.) The incident and reflected sounds interfere to cause a comb filter response (shown top right). This gives emphasis to some frequency components, while others are absent. This will change the relative magnitude of the harmonics in music and so lead to a coloured sound where the timbre is not true. Figure 3.8b shows the case for a listener close to a diffuser. The diffuser introduces temporal dispersion of the reflected sound, which leads to a more complicated frequency response. The regularity of the comb filtering is minimized, and consequently, its audibility is diminished.

Unfortunately, although the coloration can be measured, there is no formal method for evaluating the audibility of comb filtering. Chapter 5 discusses some possible approaches

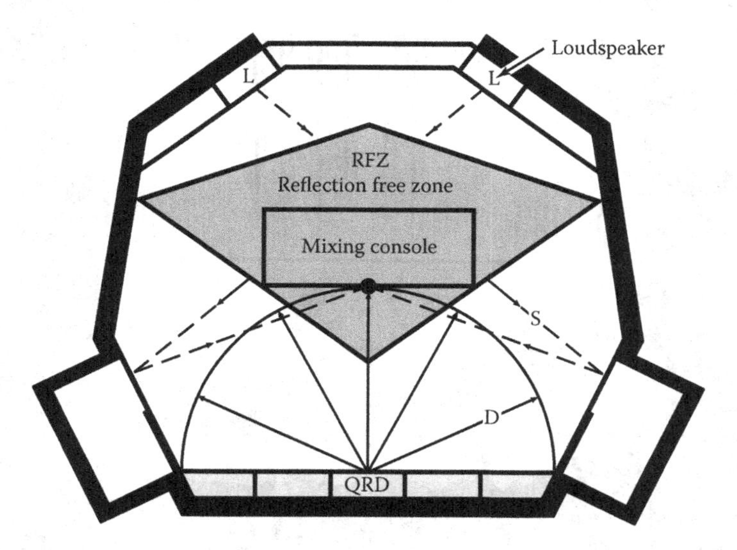

Figure 3.11 A sound reproduction room with an RFZ, shown shaded, and QRD (Schroeder diffusers) on the rear wall. (Courtesy of D'Antonio, P., Cox, T.J., *Appl. Acoust.*, 60, 113–142, 2000.)

but cannot draw any definite conclusions about a suitable method. Furthermore, a more complete model is required that allows for the masking effects of other room reflections. Comb filtering becomes less of a problem when there are either many additional reflections to mask it, or when you have multichannel loudspeakers that dominate the natural acoustics of the listening room.

The most efficient placement for a diffuser (or an absorber for that matter) is at a *geometric reflection point*, the location on a surface where the angle of reflection equals the angle of incidence and specular reflections are produced. Consider a room with a small mirror on the wall. The geometric reflection point is the location of the mirror that allows a listener to see the loudspeaker in the mirror.

The number of diffusers that should be used depends on personal preference. If interfering reflections are absorbed, then one experiences the highest resolution sonic images, which are essentially points in space. If diffusers are used to control these reflections, the apparent size of the image is broadened. If done properly, some have described these as having a more natural size, similar to what might be experienced in the presence of an actual sound source. So a balance has to be reached in which the desired apparent source width and depth are achieved, while creating an appropriate ambience. While some people favour very dead spaces for mixing audio, others do not. Some designers like to create a nonenvironment where only the direct sound is received by the sound engineer. Whether this is a desirable acoustic or not is a matter of personal preference. High levels of absorption remove most of the room effects such as coloration but lead to a very dead room that some find oppressive. It leaves the sound engineer the job of interpolating between the dead mixing room and

Figure 3.12 Gateway Mastering, Portland, ME, showing a fractal diffusing rear wall (Diffractal®). (Courtesy of Gateway Mastering + DVD.)

more normal listening environments such as living rooms but does ensure that the engineer receives a very pure sound where details can be easily detected.

If some liveliness is to be left in the room, a combination of absorbers and diffusers is better than absorption and flat walls. Consequently, many of the industry's leading mastering facilities use this combination of treatments. To take a few examples, one of the industry's most successful mastering rooms is Gateway Mastering + DVD, Portland, ME, shown in Figure 3.12, and one of the most successful recording studios is The Hit Factory, New York, shown in Figure 3.13 (unfortunately, no longer in existence). Live performance studios also usually employ a mixture of absorbers and diffusers.

In a small room, the walls, floor, and ceiling are usually close to the listener. How far away should a listener be from the diffusers? The distance between listener and diffusers can be determined by considering the scattered and total field. (The sound received by a listener can be considered to be broken down into the direct sound and the reflections. In modelling, the direct sound is often referred to as the *incident sound field*, and the reflections, the *scattered sound field*. The *total sound field* refers to everything the listener hears, the incident plus the scattered sound [see Figure 3.14]. In this context, *incident* refers to the sound field at the receivers without the diffuser, not what sound is illuminating the diffuser.)

Figure 3.13 The Hit Factory, New York. (Courtesy of The Hit Factory, New York, and Harris Grant Associates.)

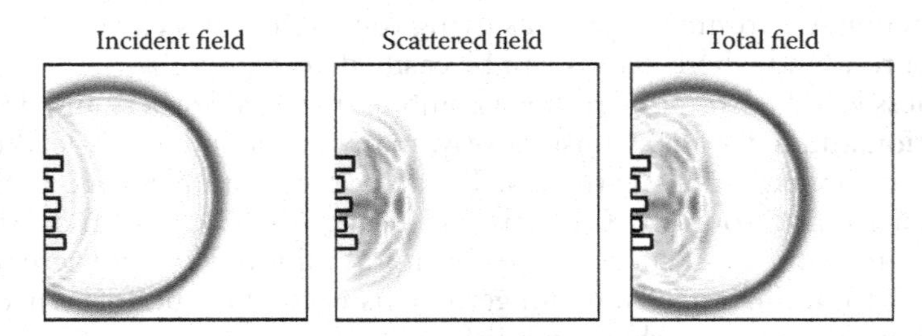

Figure 3.14 FDTD predictions of sound reflecting from a diffuser, showing the incident field (direct sound), scattered field, and total field.

First consider the scattered field, in other words just the reflections from the diffuser. A diffuser requires a certain time or distance to form a wavefront. There is an analogy to loudspeakers that can be made here. A listener would not sit 30 cm from a multiway loudspeaker because that would be in the *near field* of the device, where the sound field varies greatly with position due to constructive and destructive interference of the sound from the individual drivers. At some distance from the loudspeaker, all individual high-, mid-, and low-frequency waves from the individual drivers combine to form a coherent wavefront. The same is true for diffusers. They can also be thought of in terms of near and far field, although the situation is a bit more complex than for loudspeakers.

It is common to describe the scattered field by its spatial response. This is similar to the far field polar response of a loudspeaker; however, the polar response of a diffuser is much more difficult to measure, and this has been the subject of extensive research as reported in Chapter 5. In the far field, the polar response of an ideal diffuser is invariant to the angle of incidence, the angle of observation, and the frequency, within its operational bandwidth. Unfortunately, in most critical listening rooms, it is usual for sources and receivers to be in the near rather than the far field. Consequently, listeners should be positioned as far from scattering surfaces as possible. Precedence has shown that it is best if the listener is at least three wavelengths away from diffusers. Since diffusers usually have a lower frequency limit of roughly 300–500 Hz, this means a minimum distance of 3 m is recommended. For smaller rooms, this distance may have to be compromised.

A listener positioned near a multiway loudspeaker with their ear close to the midrange driver hears sonic anomalies, and the same is true when a listener gets too close to a diffuser. Many of the phasing anomalies reported by room designers are due to the fact that they are listening too close to the diffuser. Some listeners have even put their heads in the wells of large low-frequency diffusers and then claim something is wrong because it sounds odd! Furthermore, getting too close to a diffuser means that the temporal response is overly dominated by the surface close to the ear, which means that the temporal and spatial dispersion generated by the diffuser is not heard. The direct and reflected sounds are then rather similar and comb filtering gets worse. This naturally leads to a consideration of the total field.

When listening to music in a room, the total field is heard, which is a combination of the direct sound and reflections. If the scattered sound predominates, aberrations are heard. Just as room reflections affect the size and directionality of sonic images, they can also introduce coloration by changing the frequency response, distorting the spectral content or timbre of the direct sound. Studying the total field offers some insight into why scattering surfaces may introduce coloration.

Consider a listener approximately 1 m from a scattering surface. If the reflection comes from a flat surface, the reflected and direct sound are comparable in level and the result is a comb filter (Figure 3.8a). The total sound is not very representative of the content of the direct sound alone. While this looks bad, the comb filtering may or may not be perceived (see Section 5.4.2 for further discussions). In addition, masking by other reflections may reduce the audibility of the coloration. If the reflection comes from a diffuser, the scattered energy is dispersed in time, and the frequency response consists of an irregular spacing of nulls and peaks (Figure 3.8b). The frequency response of the total field more closely resembles the direct sound, since the diffuse reflections have randomized the constructive and destructive interference. Importantly, the listener no longer picks up the regularity of the peaks and troughs that occur for a flat surface, and so the spectral changes introduced may be less noticeable. Figure 3.8 shows the case for a diffuser that scatters in one plane. Diffusers that scatter hemispherically will direct more energy away from the listener and so will further reduce the comb filtering. The level of the scattered sound and the resulting interference in the total field decrease in the following order: flat surface, curved surface, single-plane diffuser, hemispherical diffuser, and absorber.

Hybrid surfaces provide absorption at low to mid frequencies and diffuse reflection at mid to high frequencies. These may allow the listener to get even closer to the scattering surface. The design of hybrid surfaces is described in detail in Chapter 12.

In light of these remarks, it is important to consider the temporal, spatial, and spectral response of a sound diffusing surface. Casual or arbitrary shaping of surfaces is unwise and designers should solicit theoretical or experimental proof of the performance characteristics for diffusers. Chapter 5 discusses how this might be done.

3.2.2 Surround sound

How should listening spaces be adapted for surround sound? The design of these spaces is heavily dependent on how one controls reflections from the front and surround loudspeakers, as well as how the subwoofers couple with the room. The goal is to have the room complement the additional speakers. Today, with surround-sound reproduction formats becoming increasing popular, the two-channel concepts given above are still valid but need to be employed differently. The rooms are not polarized between live and dead zones and tend to be more uniform, with diffusers being used to enhance the envelopment and immersion provided by the surround speakers and to provide the desired degree of ambience. One approach suggested by D'Antonio utilizes broadband absorption down to the modal frequencies in all the corners of the room and across the front wall behind the left, centre, and right loudspeakers (see Chapter 7 for possible designs). The side wall space between the front loudspeakers and the listeners can be controlled with broadband absorbers or single-plane diffusers. Broadband, optimized diffusers are used on the side and rear walls to disperse sound from the surround loudspeakers, which are preferably wall mounted or free standing (it is important to keep in mind that loudspeakers have better performance when they are surface mounted or far from boundary surfaces). These wall-mounted diffusers work in conjunction with the surround loudspeakers and enhance envelopment. Diffusing clouds—with or without broadband absorption down to the modal frequencies—are placed above the listeners. These ceiling diffusers scatter incident sound to the side walls to provide enveloping lateral reflections, and can also provide additional modal control and a convenient surface for lighting and heating, ventilating and air conditioning (HVAC).

The room height can be divided roughly into thirds with diffusers in the central section, to the extent that the ear is covered in both seated and standing positions. The lower and upper areas can remain untreated. If necessary, there is the possibility of using distributed absorbers or diffusers on the upper third to control flutter echoes. Wall–ceiling soffits and wall–wall intersections should be used to provide low-frequency absorption (see Section 2.4 for more discussion of modal control in small rooms).

Others favour the use of very dead spaces for surround sound (see Section 2.2), especially for more immersive surround sound mixes where the rear loudspeakers are used for distinct audio objects rather than just to provide a general aural ambience.

3.2.3 Ambechoic

An extreme example of a critical listening environment is the so-called ambechoic space, a concept devised by George Massenburg. Figure 3.15 shows a picture of the front of the

Figure 3.15 Blackbird Studios, http://www.blackbirdstudio.com. (Courtesy of George Massenburg and Blackbird Studios.)

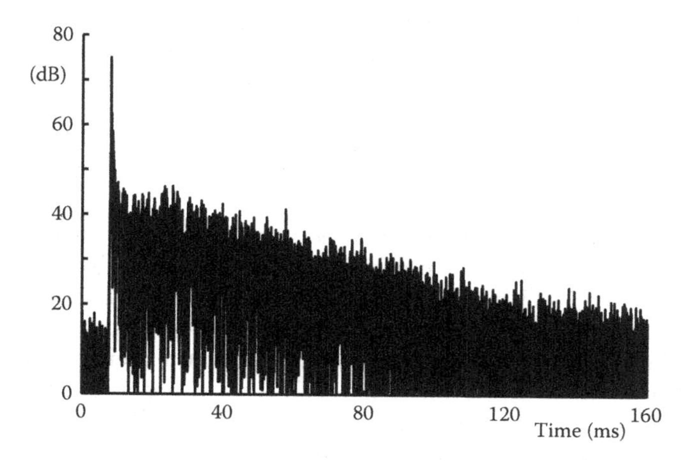

Figure 3.16 The impulse response in an ambechoic listening room.

room showing some of the extremely deep and complex diffusers used to alter the sound field. Figure 3.16 shows the beginning of the impulse response measured, showing many interesting features, including the extremely high reflection density at very early times. There is no time delay before the reverberance begins. This space can be described as an ambient anechoic or Ambechoic space, because although the individual reflections are similar in level to first order reflections in an anechoic chamber, the reflection density is very high with a decay time of about 0.3 s. Therefore, there is a definite sense of reverberance and ambience. One can comfortably hold a conversation while listening to music in the room—the room is not like an anechoic chamber. However, the low level of reverberance means that acoustic images are very precise.

Unusually, the diffusers operate down to 50–100 Hz and so are important in controlling the low-frequency modal response of the room as well. (Additional resonant absorbers are also used, as seen in the corners of the photo.) Forming rooms from very much more complex shapes by, for example, covering them in very deep diffusers can increase the number of room resonances (standing wave modes) at low frequencies.[16–19] This can reduce the unevenness of the frequency response somewhat. However, it is unusual for this to be done in a room because of the space required, and normally, resonant absorbers would be used on their own to treat low frequencies.

Another way of achieving an acoustic similar to an ambechoic, but with less deep treatment, is to use hybrid devices that partly absorb sound, while any reflections are diffused. Davidovic and Petrovic[12] did this by using diffusers with an open structure in front of porous absorption (see Section 12.1).

3.3 REDUCING COLORATION IN SMALL LIVE ROOMS

The design of the live performance rooms in a studio requires as diffuse a space as possible, along with areas within which to isolate particular instruments and vocalists. Often, an isolation booth, with views of the studio and control room, is provided for vocalists and/or drummers. The acoustics might also be variable to allow this to be manipulated according to the desired sound.

Figure 3.17 shows the live room at The Hit Factory in New York with an array of 2D QRDs in the T-bar ceiling grid, along with additional 1D Schroeder diffusers on the wall.

Figure 3.18 shows the live performance studio at XM Satellite Radio. The glass wall is a Schroeder diffuser, and on the opposite wall, there are optimized curved diffusers. Both are used to diffuse sound in the performance space (Chapters 10 and 11 discuss these designs). Other treatments not visible in the photo include hybrid devices, which are discussed in Chapter 12.

Practice rooms are even smaller spaces for music making. For all the extensive hours students spend in them, they are often uninviting and uninspiring cubicles. They are usually small rectangular rooms, 8–25 m^3, fabricated from concrete block, with traditional compressed acoustic ceiling tiles, some curtains to allow the acoustics to be varied, a concrete floor, and a full-length mirror, which students use to monitor their posture and fingering. In other words, they are a low-cost, functional, and student-proof space. Since the surfaces are usually concrete and the volume is small, the rooms typically have audible distortions caused by modes. Some designs feature non-parallel walls to minimize flutter echoes. In addition to awareness of the unwanted buzzes and squeaks, students studying articulation, tone production, and intonation are hampered by poor room acoustics. Another approach to making a music practice room is to use a prefabricated isolation cubicle. These rooms are typically small with absorbent on the room surfaces, making the space relatively dead. As with other small rooms, these can benefit from diffusers to give the musician some reverberance while minimizing coloration.

Consider a candidate room, which was 4.5 m long, 2.1 m wide on one end, 2.4 m wide on the other, and 2.7 m high. The room had a conventional compressed acoustical ceiling glued

Figure 3.17 Live room at The Hit Factory, New York, illustrating a 2D Schroeder diffuser ceiling treatment (Omniffusors). Acoustician: Neil Grant. (Courtesy of Harris Grant Associates.)

Figure 3.18 A live performance studio seen through a glass Schroeder diffuser. The opposite wall features Waveform® Spline, an optimized curved diffuser. XM Satellite Radio, Washington, DC. Acoustician: Francis Daniel Consulting Alliance. (Courtesy of Michael Moran Photography.)

to dry-wall, a concrete floor, cinder block walls, and a thin curtain to allow the acoustics to be varied somewhat. Before and after changes were made, objective measurements and subjective impressions from the musicians were gathered. The study introduced three acoustical elements: (i) an acoustical concrete masonry block, (ii) a hemispherically scattering ceiling diffuser, and (iii) a wall-mounted hybrid surface. The first two of these elements are shown in Figure 3.19.[20,21] The acoustical concrete masonry block provided low-frequency modal control via resonant absorption, structural walls, and diffuse reflections in a single plane to promote ambience, intonation, tone production, and support. In this connection, the word *ambience* is used to denote a high spatial impression and envelopment. The hemispherically scattering ceiling diffuser provided ambience, enhanced ability to hear intonation issues, and a more diffuse sound field. The hybrid surface provides mid-frequency absorption, with any high-frequency reflected energy being diffused. This provided the desired amount of articulation control for critical listening. Thus, the practice room provided detailed resolution, space, or ambience; improved feedback for intonation and tone production; provided support; reduced modal coloration; and lessened playing fatigue.

The impulse responses of the room before and after treatment are shown in Figures 3.20 and 3.21. These show that the sound decay is more even and linear after treatment is

Figure 3.19 A music practice room treated with diffusers (Skyline® on the ceiling[20] and DiffusorBlox® on rear wall[21]) and absorbers. (Courtesy of D'Antonio, P., Cox, T.J., *Appl. Acoust.*, 60, 113–142, 2000.)

added. Low-frequency modal measurements were made before and after the addition of the masonry units. The masonry units made a significant improvement in reducing the unevenness in the frequency response produced by the modes, as seen in Figure 3.22. Figure 3.23 shows the decrease in the reverberation time in the frequency range where the masonry units are tuned for maximum absorption. The experiments verified that with treatment, a more functional and enjoyable practice room was attained.

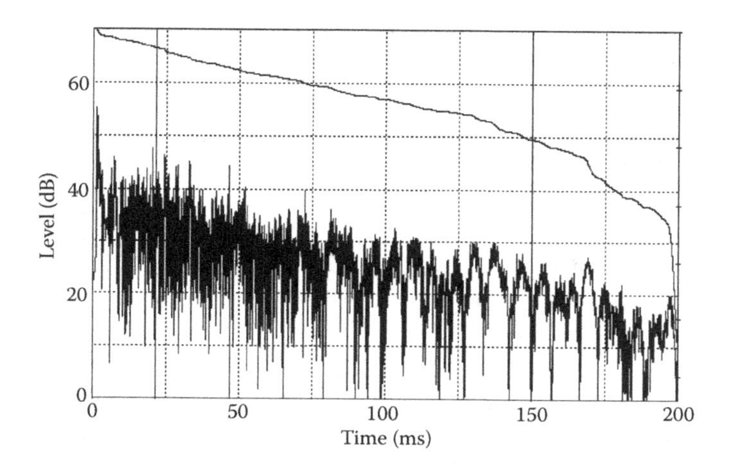

Figure 3.20 Impulse response and decay curve of music practice room shown in Figure 3.19 at early stages of treatment. (Courtesy of D'Antonio, P., Cox, T.J., *Appl. Acoust.*, 60, 113–142, 2000.)

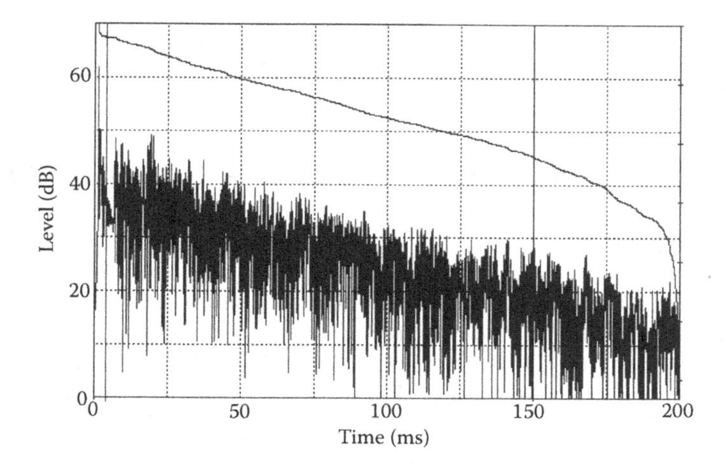

Figure 3.21 Impulse response of music practice room shown at end of treatment. (Courtesy of D'Antonio, P., Cox, T.J., *Appl. Acoust.*, 60, 113–142, 2000.)

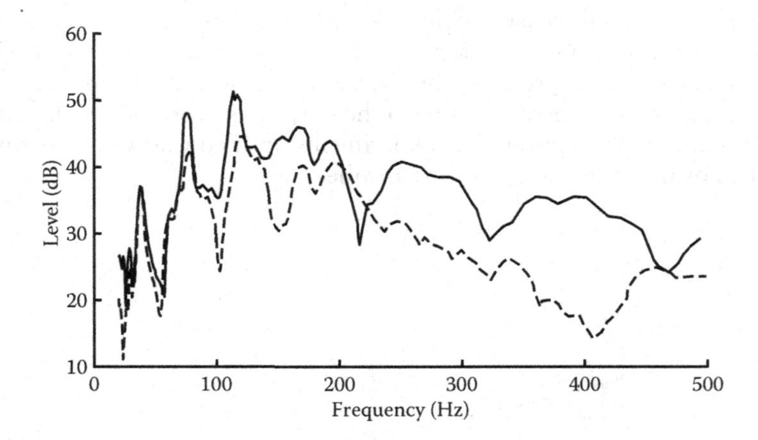

Figure 3.22 Frequency response of the music practice room: —— before, - - - - and after bass treatment. (Courtesy of D'Antonio, P., Cox, T.J., *Appl. Acoust.*, 60, 113–142, 2000.)

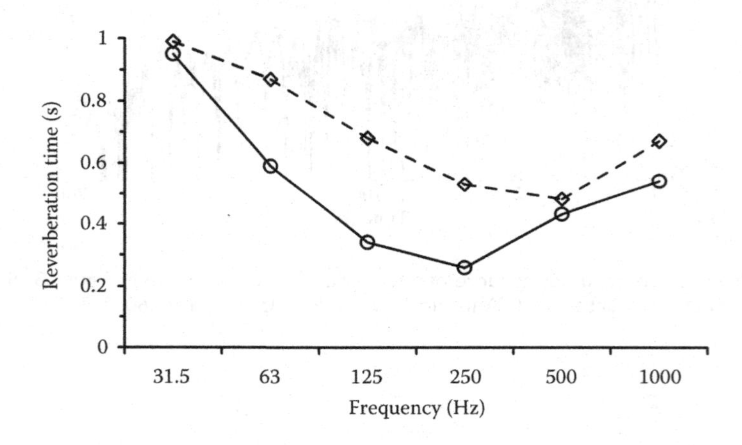

Figure 3.23 Change in reverberation time caused by adding absorbing/diffusing masonry units in the music practice room: - - - - without and —— with masonry units. (Courtesy of D'Antonio, P., Cox, T.J., *Appl. Acoust.*, 60, 113–142, 2000.)

3.4 PROMOTING DIFFUSE FIELDS IN REVERBERATION CHAMBERS

Reverberation chambers are designed to produce a diffuse sound field, one where the reflected energy density is the same throughout the room and all directions of propagation are equally probable. Reverberation chambers need to achieve this condition because they are a reference test environment providing repeatable results, which can be interpreted and matched by other laboratories. Unfortunately, a completely diffuse field is not achievable, and this is one of the reasons that round robin tests usually show significant anomalies (see Figure 4.6) and why the absorption coefficients measured in reverberation chambers need to be used with care in prediction models (see Section 13.2).

One of the methods used to achieve a more diffuse field is applying surface or volume diffusers. At low frequency, standing wave modes in the reverberation chamber cause the energy density to be uneven. Placing diffusers in the paths of modal propagation creates additional modes, which make the sound field more uniform spatially and with respect to frequency. But surface diffusers need to be of the order of half a wavelength or deeper to have a significant effect on the sound field. At mid and high frequency, problems occur when measuring absorption because the sample is usually placed on the floor, while the walls and ceiling are highly reflecting. Without diffusers, there are horizontal propagation paths that never interact with the absorbent on the floor and so decay slowly. Paths with strong vertical components decay much more rapidly when they reflect from the floor. This creates a nondiffuse field with double decays. Volume and surface diffusers deflect sound from the strongly horizontal and vertical paths and help improve the diffuseness.

Surface diffusers must be applied to at least three of the boundaries, so that opposite surface pairs have at least one surface treated (e.g., treat the ceiling). Surface diffusers can only influence sound over a hemisphere, as they only receive waves from 2π space. Volume diffusers, on the other hand, can influence a full sphere, and so it is possible to get greater diffusion from this type. Figure 3.24 shows a reverberation chamber with both types of diffusers. A final option is to use rotating elements, but this makes the acoustic time variant, and so is incompatible with modern measurement methods using maximum length sequences and swept sine waves.

Figure 3.24 The small reverberation chamber at RPG Diffusor Systems, Inc., showing installation of surface and volume diffusers. (Courtesy of RPG Diffusor Systems, Inc.)

It may be that using volumetric diffusers away from wall surfaces might offer a useful acoustic treatment in other rooms. Absorbers hung in the volume of large performance spaces are used to control reverberation, so there are places in some rooms that volumetric diffusers could be placed without interfering with sight lines. Diffusers could be hung in the propagation path of echoes. They could also influence the reverberation and distribution of sound in the room. Sonic crystals are examples of such devices where periodic arrangements of spheres and cylinders are used. However, to achieve good scattering and broadband performance, periodicity must be avoided. A fractal structure would be a good choice because it has objects on various length scales to scatter sound of different wavelengths and so can cover a broader bandwidth (Section 8.3 discusses volumetric diffusers and absorbers, including metamaterials).

3.5 IMPROVING SPEECH INTELLIGIBILITY

Many underground (subway) stations are nondiffuse spaces. The long and narrow shape results in the mean free paths along the length of the station being much longer than those for transverse propagation (where the mean free path is the average time between reflections). Consequently, sound decays are much faster for transverse propagating sound than for acoustic waves travelling up and down the length of the station. This results in a double decay and two reverberation times in the space. The long reverberation time causes problems with speech intelligibility, as the reverberation causes words to run into each other and become difficult to distinguish. Kang[22] showed that by applying diffusers to the side walls of a station, more transverse propagation can be promoted to decrease the reverberant energy and improve speech intelligibility.

A similar situation is commonly found in school and college gymnasiums. The lower portions of the walls usually have to be flat and robust, which means that absorption treatment needs to be placed high on the side walls and the ceiling. This creates a nondiffuse space that often has an excessively long reverberation time. By placing diffusers on the walls, this creates a more diffuse space where the absorption is more effective and the reverberation time can decrease dramatically. Even better would be to use a single treatment that combines absorption and diffusion in one device. For instance, the concrete masonry units in Figure 7.14 are an evolution of the more traditional slotted block, with the addition of a diffusive topology. The resonator slots provide low-frequency absorption, the surface porosity provides high-frequency absorption, and the quadratic residue topology provides sound diffusion.

3.6 PROMOTING SPACIOUSNESS IN AUDITORIA

One of the pioneering applications of Schroeder diffusers was by Marshall and Hyde in the Michael Fowler Centre, New Zealand.[23,24] Figure 3.25 illustrates the application. Large overhead reflectors provide early reflections to the audience in the balconies. This was a method whereby a hall could have good clarity and yet maintain a large volume for reverberance. The large volume partly comes from the space behind the diffusers.

Not many years before the design of the hall, Marshall hypothesised that lateral reflections were important in concert halls as these promote a sense of envelopment or spatial impression in rooms.[25] Music outdoors may be popular when accompanied with fireworks, but the quality of the sound is usually poor. Move indoors and the sound comes alive, enveloping

Figure 3.25 Schroeder diffusers in the Michael Fowler Centre, New Zealand. (Courtesy of Dr. Harold Marshall of Marshall Day Acoustics.)

and involving the listener in the music making process. Outdoors, listeners receive sound straight from the orchestra, there are no reflections from walls, and the sound appears distant. When music is played in a room, reflections from the walls, ceiling, and floor embellish the sound, especially if there are plenty of reflections arriving from the side. When sound reaches the listener straight from the stage, the same signal is received at both ears because the head is symmetrical and the sound to both ears travels an identical path. When reflections come from the side, the signal at each ear is different, as sound to the furthest ear has to bend around the head. This means that the sound arrives later and is significantly altered. The brain senses that it is in a room because of the differences between the ear signals, and a feeling of being enveloped by the music occurs.

This need for lateral reflections influenced Marshall and Hyde to apply diffusers to the large overhead surfaces shown in Figure 3.25. The diffusers promote early arriving lateral reflections for spatial impression and clarity.

3.7 REDUCING THE EFFECTS OF EARLY ARRIVING REFLECTIONS IN LARGE SPACES

In Sections 3.2 and 3.3, it was discussed how early arriving reflections can cause problems in small spaces due to coloration. Problems also arise in large spaces, for example, in rooms with low ceilings. The overhead reflection can arrive soon after the direct sound for audience members at the rear of the room, and this can lead to coloration, as comb filtering may result. Figure 3.26 shows an application of diffusers in the Cinerama Theatre, Seattle,

Figure 3.26 Cinerama Theatre, Seattle, WA, with a diffusing ceiling (OptiCurve). (Courtesy of University of Salford.)

WA. Mainstream cinemas tend to be very dead spaces with the room effects added artificially through the surround sound reproduction system. The design brief for Cinerama, by Neil Grant of Harris Grant Associates, was to generate an acoustic with high envelopment and some reverberance, a first in modern commercial cinema design, by using diffusers on the ceiling and walls. While there is an extensive use of diffusers, measurements showed that the cinema satisfied THX design criteria. Diffusers were used to disperse reflections from the relatively low ceiling to minimize comb filtering. These were optimized curved diffusers (the design process is detailed in Chapter 11).

A secondary effect of using ceiling diffusers is to reduce the level of early arriving nonlateral reflections from the ceiling. This can increase the spatial impression, but whether this happens depends on the geometry of the room and the diffusers used.

3.8 STAGE ENCLOSURES

During musical performances, there is a need for surfaces or enclosures, conventionally called acoustical shells that surround the musicians. These shells reinforce and blend the sound that is projected toward the audience. It also heightens the ability of the musicians to hear themselves and others. Acoustical shells typically incorporate a rear wall, flared side walls, and an overhead canopy.

3.8.1 Overhead canopies

Overhead reflections from flat stage canopies can cause coloration if the canopy effectively covers the whole stage area. Overhead canopies might be used to hide the presence of a fly tower or may simply be an integral part of the stage canopy design. Figure 3.27[26] shows an example at Kresge Auditorium, Boston, MA, and Figure 3.28 shows another at First Baptist Church, Eugene, OR. The canopies provide reflections back to the stage, which are necessary for the musicians to hear themselves and others. Without early reflections from the stage shell, musicians find it difficult to blend their sound with that from other instruments. Canopies with little open area provide plenty of overhead reflections back to the musicians, but if the canopy elements are flat, there is a risk that the overhead reflections will be too strong and so cause coloration. The solution to this is to shape the canopy so that some breaking up of the reflected wavefronts occurs. The diffusers spatially and temporally disperse the reflections and so reduce coloration. The design of overhead stage canopies is discussed in Chapter 11.

Figure 3.27 Overhead stage canopy (Waveform) at Kresge Auditorium, Boston, MA. Acoustician: Rein Pirn, Acentech, Boston. (Courtesy of Daderot, commons.wikimedia.org/wiki/File:Kresge_Auditorium,_MIT_(interior_with_concert).JPG.)

Figure 3.28 Rear towers and overhead stage canopy (Overture) at First Baptist Church, Eugene, OR. Church designed by Steve Diamond. (Courtesy of Steve Diamond, AGI, Inc.)

Overhead stage canopies can also have a much greater open area. Figure 3.29 shows an example, where the canopy serves both the performers and the audience at Rivercenter for the Performing Arts, Columbus, GA. Again, the canopy is designed to promote better communication between musicians across the orchestral stage, leading to better ensemble among musicians and, consequently, better quality sound. This is achieved by ensuring an even distribution of reflected energy from each of the instruments on the stage to all musicians, with the reflected energy being delayed by about 20–30 ms.[27] The canopy design is more than just getting the right shape for the diffusers; the positions and angles of the elements must be optimized to get the best performance.

A canopy may also be used to reflect sound towards audience areas lacking sound energy. There is risk, however, that providing extra overhead reflections from the stage canopy could lead to coloration. Canopies with large open areas are usually designed so much of the sound will go through the stage canopy to the void above, to be returned to the audience or stage from the true ceiling of the auditorium, and so provide additional reverberation.

If the canopy elements are sparse and flat, then the pressure distribution will be uneven. For some receiver positions, there will be specular reflections from a canopy element, and so at mid–high frequency, a strong reflection results. For other receiver positions, the geometric reflection point misses the canopy elements, leading to a much quieter reflection. Figure 3.30[28] shows a comparative sound pressure level distribution from a canopy with open areas and plane surfaces and a canopy with open areas and diffusing surfaces. Consequently, canopies with spaces between the reflectors benefit from using diffusing elements, as they enable a

Figure 3.29 Rivercenter for the Performing Arts, Columbus, GA. Acoustician: Jaffe-Holden Acoustics. (Courtesy of Jaffe-Holden Acoustics.)

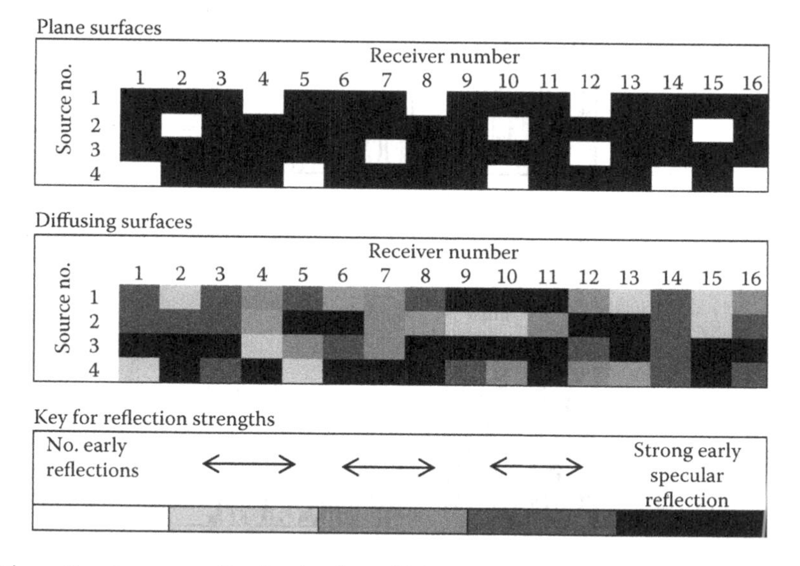

Figure 3.30 Schematic of pressure distribution from fairly open canopies with different surface treatments. (Adapted from Dalenbäck, B.-I., Kleiner, M., Svensson, P., *J. Audio Eng. Soc.*, 42, 793–807, 1994.)

more uniform coverage over the stage area by scattering sound to receivers, which would otherwise lack reflections due to the gaps between the canopy.

3.8.2 Rear and side of stage enclosures

In addition to the orientation of the shell surfaces, the nature of the surfaces is critical to good performance. A shell can contain reflecting and diffusing surfaces[29] and, less often, absorption.

Marshall et al.[30] suggested that early reflections among musicians greatly improve their sensation of playing as a group if the reflections

a. occur within a temporal window that is dependent on the nature of the musical programme material, typically between 17 and 35 ms after the direct sound;
b. include high-frequency content roughly between 500 and 2000 Hz, containing the attack transients that are cues for rhythm and expression; and
c. contain a balance of all the parts in the ensemble at all performance positions.

Condition (a) is easily met by spacing the shell an appropriate distance from the performers, while (b) and (c) depend more on the surface topology. Acoustical shells have used a wide variety of surfaces ranging from flat reflecting panels to various forms of surface irregularity, such as curved surfaces, poly-cylinders, fluted columns, and Schroeder diffusers. The current state-of-the-art is to utilize optimization to obtain the best shape and orientation, a technique discussed in Chapter 11. In Figure 3.31, the reflections from a flat and

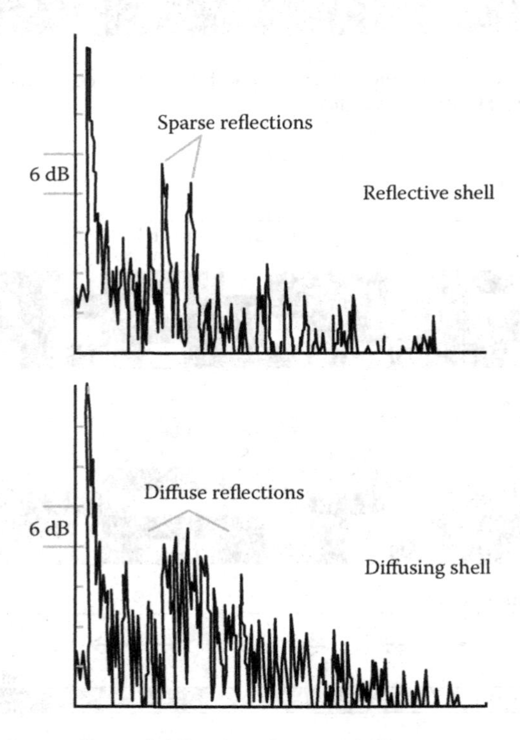

Figure 3.31 The reflections from a flat and diffusely reflecting shell.

diffuse shell are compared. The diffuse reflections are spread over time and are of reduced level, lowering the chances of coloration and harshness due to comb filtering. The diffusion also satisfied requirements (b) and (c).

D'Antonio[31] carried out a series of experiments using objective measures, as well as musicians' perceptual evaluations, to determine the appropriate combination and orientation of reflecting, diffusing, and absorbing surfaces. The study began by looking at small chamber groups. An example of one of the test arrangements with a string ensemble is shown in Figure 3.32. Five different microphone systems were used for each playing environment to obtain five simultaneously recorded signals:

1. A mannequin, with microphones at the entrance to the ear canal, was placed within the group to determine ensemble blend without self-masking.
2. Probe microphones (Figure 3.33) were inserted into the ear canal to determine ensemble blend with self-masking of the musician's instrument.
3. Headband microphones, located at the entrance to the ear canal, were also used to monitor ensemble blend with self-masking.
4. An omnidirectional microphone was placed within the group as a monophonic control.
5. Spaced omnidirectional microphones were placed at the front of the house.

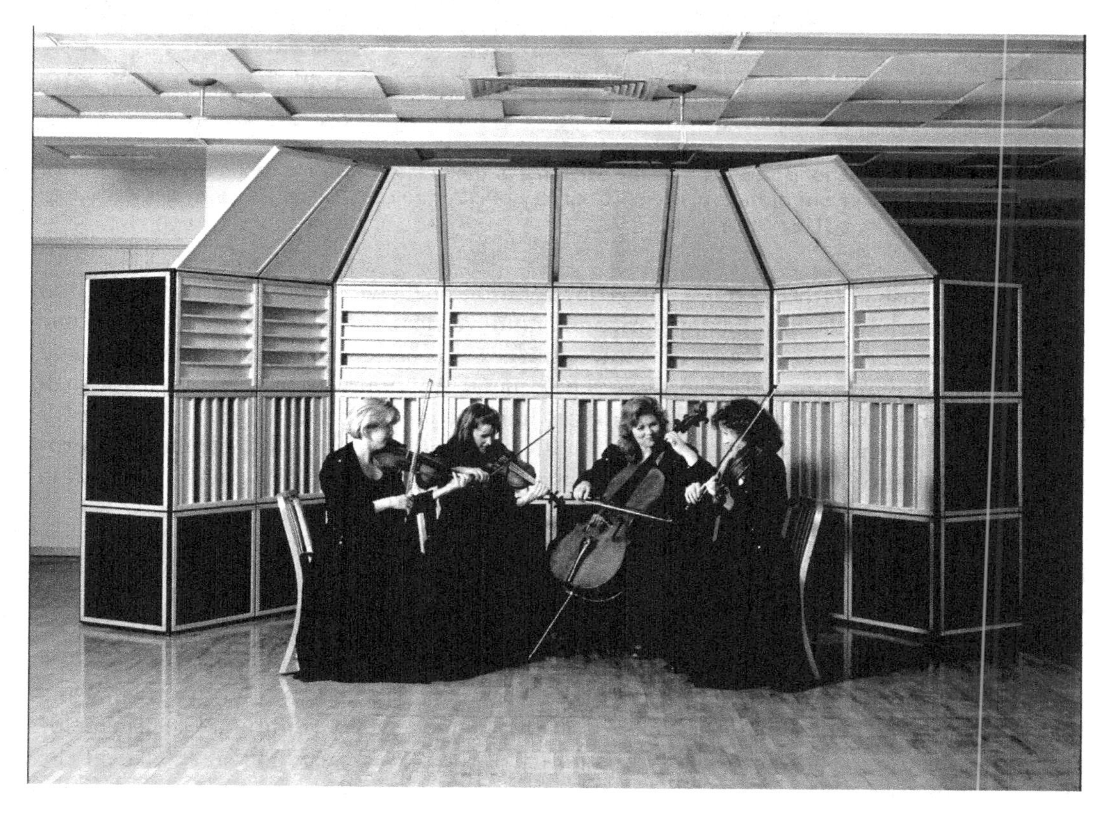

Figure 3.32 Cavani string quartet performing in front of VAMPS® shell at the Cleveland Institute of Music. (Courtesy of D'Antonio, P., Cox, T.J., *J. Audio Eng. Soc.*, 46, 955–976, 1998.)

Figure 3.33 Probe microphone inserted into the ear canal of the first violinist. Microphone is inserted to a point just in front of the ear drum. (Courtesy of D'Antonio, P., *Proc. 91st Convention Audio Eng. Soc.*, preprint 3118, 1991.)

The string ensemble preferred a mixed orientation shell with lower vertical wells and upper horizontal wells, producing horizontal and vertical diffusion, respectively. Some musicians preferred flat surfaces on the lower surfaces for better bass coupling. Mutual and self-hearing were unanimously improved. The general reaction was that the diffusive shell provides warmth and intimacy and minimizes harshness. The preferred shell distance for warmth and intimacy was approximately 0.9–1.8 m. The preferred shell distance for projected sound quality was 2.7–3.7 m. There was unanimous agreement that a height of 4.9 m was better than 2.4 m.

A brass quintet experienced harshness from a completely reflective shell and preferred lower diffusers with vertical wells and upper diffusers with horizontal wells. The preferred distance for mutual and self-hearing as well as projected sound quality was 2.7 m. A horn duo preferred a mixture of flat surfaces and diffusers at approximately 1.8 m.

Following the chamber group research, the requirements of a symphony orchestra were investigated. There is an inherent imbalance in an orchestra because the percussion and brass are naturally louder than the strings and woodwinds. Many traditional shells employ an average acoustic solution to satisfy a majority of the players, using fixed acoustical elements designed for existing musical formats and orchestral arrangements. Since each musical section has a different preference for its own local acoustical environment, and since musical format, orchestral arrangements, and conductor's preferences change, the benefit of a variable acoustical design was explored. The result was a modular framework, which would allow local acoustical environments in the rear of the orchestra, where the loudest instruments are located. A study was done with the Baltimore Symphony Orchestra at the Meyerhoff Symphony Hall, Baltimore, MD, using questionnaires and experimental measurements.

The various sections of the orchestra were asked to mark the quality of the acoustic on a 4-point scale before and after treatment. For instance, the oboe section was asked how the addition of the diffusing shell affected synchronicity, intonation, tone production, distant hearing, mutual hearing, and self-hearing, compared to the flat existing enclosure. All aspects were improved, with the average score increasing from 1.5 to 3.5. Over the whole orchestra, the average score increased from 1.9 to 3.3.

The shell was placed around the entire perimeter of the stage, similar to the shell shown in Figure 3.34. There was no change to the overhead circular disc canopy. The shell consisted of a lower open support that allowed sound to reflect from the hard existing wall behind the shell. The next 0.6 m high tier consisted of horizontally diffusing single plane Schroeder diffusers with the centre of this level at seated ear height. The next 0.6 m in height contained all vertical diffusion, resulting from the wells being oriented horizontally. A 0.6 m cantilever canopy was oriented on top of the vertically diffusing diffusers, at an angle of 45° with respect to the face of the diffusers. The purpose of the experiment was to blend the outer strings into the woodwinds, decrease the harshness of the brass, intensify the fullness and warmth of the strings, and control strong specular reflections to enhance the sense of ensemble and rhythmic performance of the musicians.[3] Through the questionnaires, musicians reported improved ensemble playing.

Figure 3.34 Diffusers around the stage of the Corning Glass Center, NY. (Copyright Paul Warchol Photography, http://www.warcholphotography.com.)

(a)

(b)

Figure 3.35 Two views of the pit at Nicholas Music Center, Rutgers University. (a) Side view, and (b) view from pit rail. Architects: Bhavnani & King Architects, NY. (Courtesy of RPG Diffusor Systems, Inc.)

3.8.3 Orchestra pits

Semienclosed stages are a challenging space for music performance. The ceiling is typically low, as is the volume. Sound levels can become excessive and ensemble hearing is often lacking. The situation is even worse within an orchestral pit in an opera or ballet house. These situations can be improved with the use of absorbers and diffusers with low–mid frequency absorption to control sound levels and diffusion to improve ensemble. Figure 3.35b shows the ceiling and walls of an orchestra pit treated with hemispherically scattering diffusers and single-plane diffusers, respectively. The pit rail, shown in Figure 3.35a, is also an important surface for communication between the stage and orchestra on the left. It is shown with upper diffusers having horizontal dividers to scatter sound in a vertical hemidisc and lower diffusers to provide horizontal scattering for the benefit of the conductor and orchestra. Diffusive concrete masonry units, such as those shown in Figure 7.14, are also becoming popular for use in orchestra pits.

3.8.4 Outdoor stage shells

For outdoor concerts, stage shells can provide beneficial reflections to increase the sound energy reaching the audience and to improve the ability for musicians to hear each other. Indeed, in many cases, the stage shell may provide the only reflecting surface apart from the floor to reinforce the sound. Figure 3.36 shows an example of a temporary, demountable shell designed for a chamber orchestra and 50–60 audience members.

Several authors have optimized the shape of an outdoor stage shell using ray tracing or image source models to examine the distribution of reflections. Palma et al.[32] derived the shape of a smooth concave outdoor shell using a genetic algorithm to try and create an even spatial distribution of rays across the audience. Boning and Bassuet[33] also counted reflection arrivals but found that getting the optimization to concentrate on the receivers that

Figure 3.36 The Tippet Rise Tiara music shell. Client: Tippet Rise Art Center, MT. Designers: Alban Bassuet and Willem Boning, Arup Acoustics. Construction: Gunnstock Timber Frames, Powell, WY.

received the fewest second- and third-order image source reflections was most successful in the design of their shell (shown in Figure 3.36). Di Rosario et al.[34] used a multiobjective optimization so they could optimize both the number of reflections and the total sound level amplification by their faceted, concave shell. Chapter 11 discusses the optimization of surfaces to create diffusers, a process that has many similarities to the processes used to design these stage shells.

3.9 BLURRING THE FOCUSING FROM CONCAVE SURFACES

Concave surfaces can cause focussing in a similar manner to concave mirrors. This can lead to uneven sound levels around a room, which is usually undesirable. Furthermore, as the concave surface concentrates energy in particular locations, there is a risk that this reflection will be significantly above the general reverberation of the space and so cause echo problems or coloration. A famous example of this is the Royal Albert Hall, where the large dome caused problems with long delayed and focussed reflections. Indeed, in 1871, the *Times* reported that the Prince of Wales struggled with his opening speech, 'the reading was somewhat marred by an echo which seemed to be suddenly awoke from the organ or picture gallery, and repeated the words with a mocking emphasis which at another time would have been amusing'.[35] A solution was finally found in 1968, when volume diffusers, the *mushrooms*, were used to stop the sound reaching the dome. They also provide earlier reflections to the audience (see Figure 3.37[36]).

Figure 3.37 Looking up at the volume diffusers used to prevent sound that can cause echoes reaching the dome in the Royal Albert Hall. (Courtesy of Egghead06, commons.wikimedia.org/wiki/File:AcousticDiscsRoyalAlbertHall.JPG.)

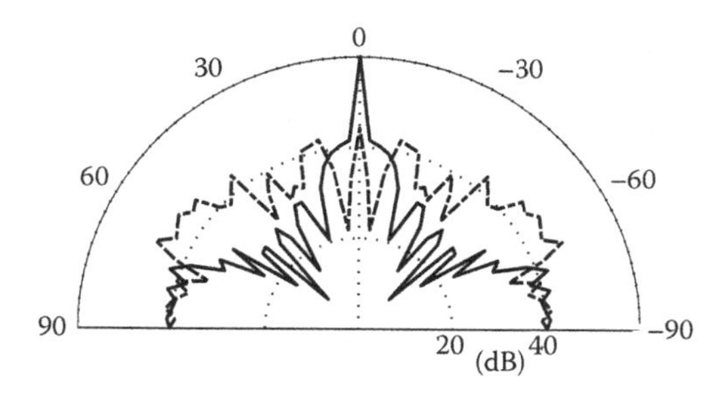

Figure 3.38 Reflection from a concave arc compared to an optimized curved diffuser at 3 kHz —— concave arc and - - - - optimized curved diffuser.

Assuming that a concave surface cannot be removed, the only treatments available are absorbers or diffusers. Both treatments will work; the choice depends on other acoustic considerations, such as reverberance. Figure 3.38 shows the polar response near a concave wall, which was to be part of a music rehearsal room at the Edwina Palmer Hall, Hitchin, England. The concentration of sound at the focus is clear. To overcome this problem, a diffuser was specified. An optimized diffuser was developed and applied as shown in Figure 3.39. Figure 3.38 shows that the dispersion of the focused energy is quite dramatic. It would have been possible to remove the echo with absorbers, but then, the musicians would not have received reflections from this wall. This would potentially have a detrimental effect on the ensemble. In this case, the wall reflection had to be preserved, but the focussing removed, and hence, diffusers were preferred over absorbers.

The acoustician, Raf Orlowski, commented,

> For architectural reasons, a concave form was developed for the hall, which obviously gave rise to concerns about focussing. Curve shape optimization was used to minimize focussing by the concave wall using a geometrical motif based on an amplitude modulated wave... Subjective listening tests to piano and clarinet music in the hall indicated a very uniform sound field with no evidence of focussing. Furthermore, both instruments produced an expansive sound with a very good balance between clarity and reverberance. Musicians found the hall easy to play in.[37]

See Chapter 11 for detailed discussions of focussing from concave surfaces and their treatment.

3.10 IN AUDIENCE AREAS

3.10.1 Coverage

Canopies of reflectors are commonly seen above audiences in concert halls. An example is shown in Figure 3.40. Similar to overhead stage canopies discussed previously, making these canopy elements diffusing can reduce coloration and make the distribution of sound more uniform across the audience.

The Philharmonie de Paris is a surround sound hall, with a large volume that has many reflecting surfaces well away from the outer boundary of the auditorium to provide strong early reflections, especially ones arriving at the audience from the side. Figure 3.41 is a view from a side balcony showing some of the overhead reflectors, balcony fronts, and walls near the audience that were shaped to provide early reflection coverage. The acoustic brief specified the need to achieve *early acoustic efficiency*.[38] To do this, the brief specified the surface area of reflectors that must be within 15 m of the stage and/or from a section of the audience

Figure 3.39 Optimized curved surface (Waveform) in the Edwina Palmer Hall. Acoustician: Raf Orlowski. (Courtesy of Arup Acoustics.)

and are orientated to provide reflections towards the audience ('1400 m^2; 500 m^2 of which are less than 15 metres from the stage').

Nowadays, digital architecture does not just use technologies to represent structures within a computer, it also creates algorithms to generate the shapes and forms of a building. Acoustic consultants are increasingly embracing the tools that have been developed to enable

Figure 3.40 Optimized curved audience canopy (Waveform Spline) in Northridge High School, Middlebury, IN. Architect: Moake Park Group, Inc. Acoustician: dBA Acoustics. (Courtesy of dBA Acoustics.)

digital architecture, for example, using the graphical algorithm editor Grasshopper[39] with the Rhinoceros 3D CAD programme.[40] For the Philharmonie de Paris, Scelo[41] used such tools to design the surfaces that provide first-order reflections to the musicians and the audience. The concept is to get a computer algorithm to find the right surface orientation and curvature, rather than this being a trial-and-error process laboriously done by the consultant. Like the optimization of outdoor stage shells described in Section 3.8.4, the initial design had to rely on the simplest ray tracing tools, before the designs being checked in a more sophisticated geometric room acoustic model. In the future, there is a need for the digital architecture tools to embrace more accurate acoustic models, preferably one that can properly model sound as a wave. Figure 3.29 is an example of a canopy optimized using a wave-based prediction model.

3.10.2 Diffuse fields

How does the amount of diffusers on the walls and ceiling relate to the general acoustic in a hall, including how diffuse a room is? If a room is roughly cubic, and the absorption evenly distributed, then a diffuse space can be created above the Schroeder frequency even with flat walls. If the absorption is not very evenly distributed, or the dimensions of the space are not very similar, then a nondiffuse space results and adding diffusers to the space will have a number of effects. In the case of very disproportioned rooms, such as underground stations, the reverberation time will decrease when diffusers are added because the mean free path is reduced, giving more absorption from the increased number of reflections per second (see Section 3.5).

Consider another case, a room with just absorption on the floor and flat hard surfaces for the walls and ceiling. A reverberation chamber with an absorbent sample on the floor or a

Figure 3.41 View from side balcony of Philharmonie de Paris. Lead acoustician: Harold Marshall.

large shoebox-shaped auditorium for classical music is like this. These are nondiffuse spaces and are likely to have a reverberation time different to that predicted by Eyring formulation (Equation 2.5)—the Eyring formulation would be correct for a diffuse space.[42] When the room is tall relative to its width and length, then some sound will be able to propagate in the horizontal plane without being absorbed by the highly absorbent floor, and consequently, a longer reverberation time can result. Adding surface diffusers to the walls that scatter sound vertically will cause sound paths to more often involve all the surfaces in a room, and consequently, the floor absorption will become more important and the reverberation time will decrease.

As diffusers are added and the reverberation time decreases, so will the sound level of the reverberant field. But it would be incorrect to assume that sound levels will ever become identical throughout a room. This is because even in ideal reverberant conditions, there is scatter in measured levels.[43] Furthermore, as illustrated by the Philharmonie de Paris, the early reflected sound can be changed by introducing reflectors and diffusers, which can have a significant effect on the sound level. On the other hand, as diffusers are added, the variation in the reverberation time across a space does decrease and decays become more linear[44–46] (see also the discussion of sound reproduction rooms in Section 3.2). Indeed, monitoring the difference between T_{15} and T_{30} predicted within a computer model, as well as the calculated statistical T_{60}, as acoustic treatment is altered, is one indication of how mixing an environment is, i.e., one where the reverberant field tends towards a stochastic process.

Sakuma[47] showed in a simulation study that a surface scattering coefficient of 0.4 is sufficient to create a diffuse space when the room is disproportionate or the absorption unevenly distributed. (The scattering coefficient measures the proportion of sound not specularly reflected—see Section 5.3.) A scale model study by Prodi and Visentin[48] on a proportionate room with unevenly distributed absorption found that the required scattering coefficient varied with frequency, being smaller in lower-frequency octave bands.

Architectural acoustics is used to having simple formulations that relate the amount of absorption in the room to the reverberation time. Is there an equivalent formulation for surface diffusers which relates objective parameters of a room to the number of diffusers applied? A number of authors have produced formulations that relate the reverberation time to the absorption and scattering coefficients.[49,50] But these are complex equations, and it is probably just as easy to gauge the change in reverberation caused by scattering using a geometric room acoustic model.

A few people have examined the effect of large-scale application of diffusers on the acoustics of concert halls and looked beyond reverberation. Fujii et al.[51] examined two similar halls, one with side wall columns, which promote scattering, and one without. Chiles[52] and Kim et al.[53] examined the effect of large-scale scattering in physical scale models; Chiles also examined computer models. (In some of the measurements in some of these studies, it appears that the diffusers may also have been partly absorbing the sound, and this makes interpreting some of the results more difficult.) For fan-shaped concert halls, adding diffusers on the side walls is beneficial because it increases early reflections from the side.

For shoebox-shaped auditoria, hemispherically scattering side wall and ceiling diffusers promote more early sound reflected back towards the stage than would be the case for flat walls. The same is true for single-plane diffusers that have corrugations that create scattering front-to-back in the hall. This gives an increase in clarity and sound level for the front seating and a decrease for the audience at the rear of the hall—generally, these would not be desirable effects. Cox[54] examined a wavy ceiling designed for a lecture theatre designed by Alvar Aalto using a Boundary Element Model and showed that the corrugations reduced the sound reflected from the ceiling towards the rear of the room and therefore was detrimental to speech intelligibility. For concert halls, it is possible to overcome these effects by using single-plane diffusers on the ceiling that just scatter laterally. For the lecture theatre, a flat ceiling would have been a better acoustic design.

Putting diffusers on the rear wall of a concert hall or around or close to the stage can help to promote a diffuse space with less effect on the early reflected sound distribution. However, this raises the interesting question as to whether a diffuse sound field is desirable within a concert hall. While a diffuse field makes the prediction of sound behaviour more straightforward, what is currently a subject of heated debate is whether this is desirable for the audience.

Furthermore, the situation is different for other auditoria layouts. For instance, in fan-shaped halls, the promotion of additional lateral reflections by introducing single-plane side wall diffusers would be desirable for promoting spaciousness. Incidentally, for rectangular halls, one study[53] showed that 1-IACC (internal aural cross-correlation[25]) is increased for front and side seats, as might be expected, whereas another study showed that IACC is not strongly affected.[51]

3.11 DIFFUSING AND ABSORBING LIGHTING

One way of making acoustic treatment more sustainable is to make devices that have dual purpose, such as a sound diffuser that also provides light diffusion. Dynamic lighting provides different illuminance and colour that can be adjusted to suit the task at hand[55] and can have a positive effect on both work productivity and school learning. Two scientific classroom learning studies were carried out at the University Hospital in Hamburg-Eppendorf Clinic for Child and Youth Psychology[56] and the University of Twente, the Netherlands.[57] The Hamburg study revealed an increase of 35% in reading speeds, an error reduction of 45%, and a decrease in hyperactivity of 76%. The Dutch study reported 18% higher concentration levels, higher long-term motivation levels, and a higher appreciation of the learning environment. Typically, dynamic lighting systems include automatically or manually adjusted setting levels for *energy* to start the day or after lunch (horizontal illuminance of 650 lux at desk level and a cold, blue-rich white light, 12,000 K), *focus* for testing (1000 lux, 6500 K), *calm* for relaxed ambience (300 lux, 2900 K), and *standard* (300 lux, 3000–4000 K).

D'Antonio created a low-voltage LED lighting fixture, which provides dynamic lighting, with a lens that is both translucent and sound diffusive (see Figure 3.42[58]). Speech intelligibility is a function of the signal-to-noise ratio (SNR). In a classroom, the signal consists of the direct sound and the fused early reflections, while the noise consists of external intrusion, HVAC, occupant noise, and reverberation. Since the consonants are an important component for speech intelligibility, and they occur in the 2000–6000 Hz bandwidth, it is not constructive to use absorbers on the entire ceiling. Acoustic ceiling tiles may excessively reduce the high-frequency consonant sounds and result in the masking of high-frequency consonants by low-frequency vowel sounds. Several researchers, including Bradley et al.[59] and Choi,[60] have found that simply adding large amounts of absorption to the room to achieve short reverberation times leads to reduced SNR values and degraded speech intelligibility, due to reduced sound levels at more distant seats. They have also found that perceptually important improvements can be made without the negative effects of excessive absorption and lower speech levels by appropriately including diffusers in the acoustical treatment of a classroom to add important early reflections. Choi has modelled several combinations of absorption and diffusion in various areas of a classroom, verifying the importance of early reflections. D'Antonio has proposed a classroom design to passively increase the signal and reduce the reverberation time to achieve an SNR in excess of 15 dB and provide sufficient reverberation control (see Figure 3.43).

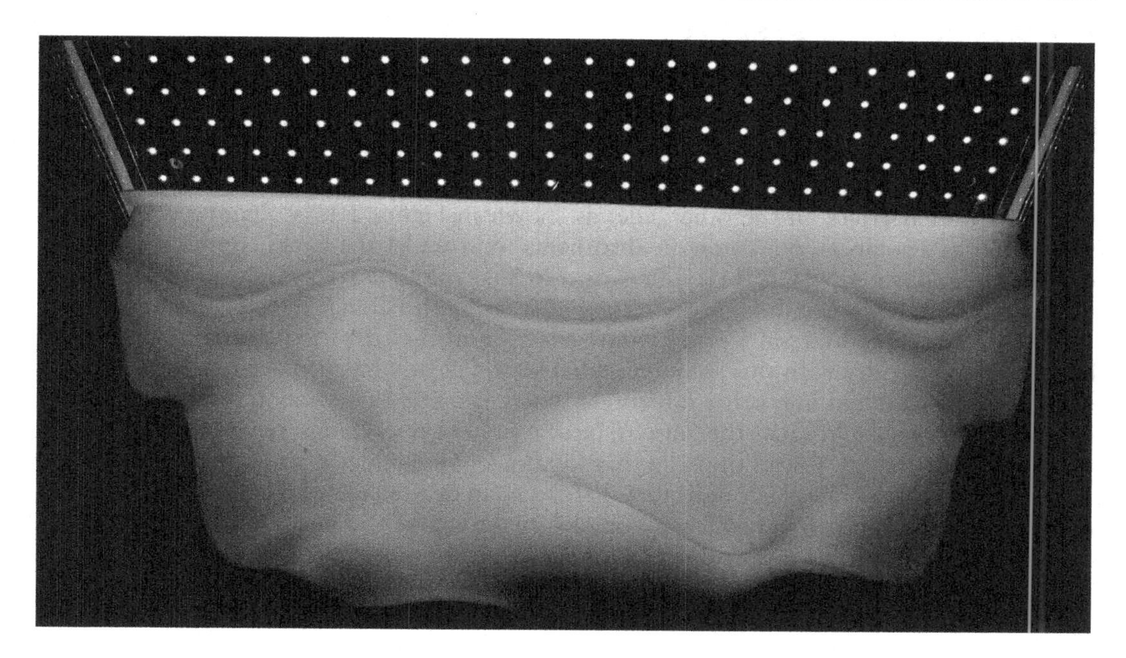

Figure 3.42 Combined dynamic lighting and sound diffuser (Lumaphon®).[58] Translucent diffuser moved to reveal LEDs that are normally hidden.

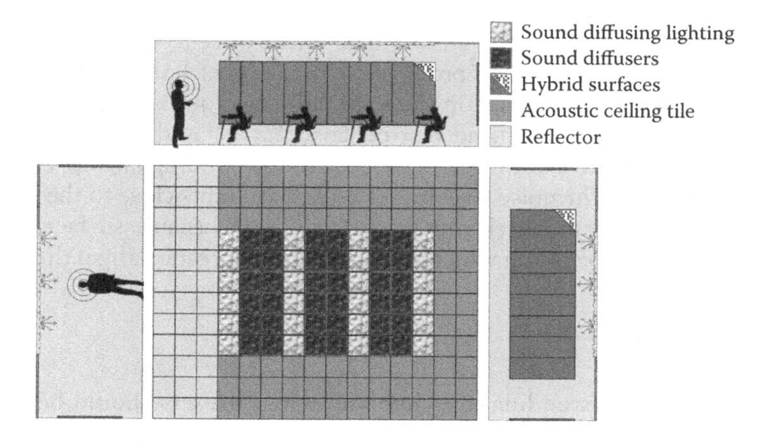

Figure 3.43 A classroom design exploiting sound diffusing lighting and other treatments.

The signal is increased passively through the temporal fusion of the direct and early ceiling reflections from the diffusers and diffuser lighting fixtures in the middle of the ceiling. The perimeter of the ceiling is treated with acoustical ceiling tile to control reverberation. The middle third of the side and rear walls are treated with hybrid surfaces, which absorb the mid–low frequencies and diffuse the high-frequency consonant information. If additional reverberation control is needed, the upper third of the side and rear walls can be made absorbing with fabric-wrapped panels. The front wall and ceiling above the teacher should be left reflective to project sound and redirect consonant information back into the classroom when the teacher faces away from the students.

3.12 BARRIERS AND STREETS

Barriers can be used to reduce noise propagation from roads and railway lines to neighbouring houses. One problem that roadside barriers suffer from is double reflections from high-sided vehicles, which enables some of the noise to bypass the barrier—this is shown in Figure 2.15a. Another problem is reflections from barriers on one side of the road, which then pass over barriers on the other side, as shown in Figure 2.15b. Chapter 2 discusses absorption as a solution, but common absorbents wear under the harsh conditions of high winds, salt, and water, which are common next to busy roads. Diffusers may offer a solution to this problem, as shown in Figure 2.15d; by dispersing the sound, the reflection problems are decreased. Dong and Sui[61] predicted a reduction of 3–5 dBA for a train and a QRD barrier. The difficulty with this solution is that sound energy has not been removed by the diffusers, just scattered into other directions. There is a risk that wind and other meteorological conditions could cause the noise to be scattered or refracted to noise-sensitive areas. Consequently, there is a need to make this solution robust under a wide range of weather conditions. Recent studies have indicated that trees can be used to reduce turbulence around roadside barriers, and so natural wind breaks might enable the performance of diffusing (and other) barriers to be more robust to changing meteorology.[62]

Various studies have been carried out into T-shaped barriers with Schroeder diffusers on the top that alter the sound diffraction and hence improve performance.[63] While the insertion loss can be increased by a few decibels, this is not a cost effective way of improving barrier performance, however.

Street canyons are roads with high-sided buildings on both sides, which form a semienclosed space where sound levels can build up and noise levels can become unacceptable. The surface structures of the buildings forming the street canyon influence the sound levels. For example, balcony fronts can have the unfortunate effect of reflecting sound back down to street level, thereby exacerbating noise problems. On the other hand, other types of surface roughness may have a role in breaking up the reflected sound on the building fronts, causing it to disperse and therefore be minimized at street level. Kang[64] and Onaga and Rindel[65] used geometric computer models to examine the influence of the building façades on noise levels in streets. Scattering increases the noise level by very small amounts close to the source, but further away, they reduce the sound level from traffic by 2–4 dB because surface roughness promotes more cross-street propagation which attenuates rapidly, especially if directed upwards!

3.13 CONCLUSIONS

In recent decades, an understanding of where and why diffusers should be used has been created. While this knowledge is still incomplete, in many common applications, it is now well established how diffusers should be used. This chapter has detailed some of these applications, as well as presented some of the key principles to be further developed later in this book. The remaining chapters concerning diffusers are as follows:

- Chapter 5 discusses the measurement of reflections from surfaces and the characterization in terms of scattering and diffusion coefficients.
- Chapter 9 discusses methods for predicting reflections, presenting both complex and simple methods in the frequency and time domains.
- Chapters 10, 11, and 12 deal with the design of key diffuser types: Schroeder, geometric, and hybrid diffusers.
- Chapter 13 discusses the role of scattering coefficients in geometric room acoustic models.

REFERENCES

1. P. D'Antonio and T. J. Cox, "Diffusor application in rooms", *Appl. Acoust.*, **60**(2), 113–42 (2000).
2. P. D'Antonio and T. J. Cox, "Two decades of sound diffusor design and development part 1: Applications and design", *J. Audio Eng. Soc.*, **46**(11), 955–76 (1998).
3. L. L. Beranek, *Concert and Opera Halls: How They Sound*, Springer-Verlag, New York, 69–74 (2004).
4. J. P. O'Keefe, T. J. Cox, N. Muncy, and S. Barbar, "Modern measurements, optimized diffusion, and electronic enhancement in a large fan-shaped auditorium", *J. Acoust. Soc. Am.*, **103**(5), 3032–3 (1998). Also *Proc. 16th ICA*.
5. B. M. Fazenda, J. A. S. Angus, and T. J. Cox, "Making Space, the SOS Guide to Control Room Design" (February 2015).
6. D. David and C. Davis, "The LEDE concept for the control of acoustic and psychoacoustic parameters in recording control rooms", *J. Audio Eng. Soc.*, **28**, 585–95 (1980).
7. M. R. Schroeder, "Binaural dissimilarity and optimum ceilings for concert halls: More lateral sound diffusion", *J. Acoust. Soc. Am.*, **65**, 958–63 (1979).
8. P. D'Antonio and J. H. Konnert, "The reflection phase grating diffusor: Design theory and application", *J. Audio Eng. Soc.*, **32**(4), 228–38 (1984).
9. P. D'Antonio and J. H. Konnert, "The RFZ/RPG approach to control room monitoring", *Proc. Audio Eng. Soc.*, preprint 2157 (I-6) (1984).
10. ITU-R BS.1116-1, "Methods for the subjective assessment of small impairments in audio systems including multichannel sound systems" (1997).
11. R. Walker, "A controlled-reflection listening room for multi-channel sound", *Proc. 104th Convention Audio Eng. Soc.*, preprint 4645 (1998).
12. Z. Davidovic and B. Petrovic, "Acoustical design of control room for stereo and multichannel production and reproduction—A novel approach", *Proc. 129th Convention Audio Eng. Soc.*, preprint 8295 (2010).
13. F. E. Toole, *Sound Reproduction*, Focal Press, Oxford, UK (2008).
14. F. E. Toole, "Loudspeakers and rooms for sound reproduction—A scientific review", *J. Audio Eng. Soc.*, **54**(6) 451–76 (2006).
15. P. D'Antonio, F. Becker, and C. Bilello, "Sound intensity and interaural cross-correlation measurements using time delay spectrometry", *J. Audio Eng. Soc.*, **37**(9), 659–73 (1989).
16. P. D'Antonio, "New types of acoustical materials simplify room designs", *Proc. 81st Convention Audio Eng. Soc*, preprint 2365 (1986).
17. J. A. S. Angus, A. C. Marvin, and J. Clegg, "The effect of acoustic diffusers on room mode density", *Proc. 96th Convention Audio Eng. Soc.*, preprint 3851 (1994).
18. X. T. Zhu, Z. M. Zhu, and J. C. Cheng, "Using optimized surface modifications to improve low frequency response in a room", *Appl. Acoust.*, **65**(9), 841–60 (2004).
19. S. Felix, M. Asch, M. Filoche, and B. Sapoval, "Localization and increased damping in irregular acoustic cavities", *J. Sound Vib.*, **299**(4–5), 965–76 (2007).
20. P. D'Antonio, *Two-dimensional Primitive Root Diffusor*, US patent 5, 401, 921 (1995).
21. P. D'Antonio, *Acoustical Diffusing and Absorbing Cinder Blocks*, US patent 5, 193, 318 (1993).
22. J. Kang, "Experimental approach to the effect of diffusers on the sound attenuation in long enclosures", *Bldg. Acoust.*, **2**, 391–402 (1995).
23. A. H. Marshall and J. R. Hyde, "Some practical considerations in the use of quadratic residue diffusing surfaces", *Proc. 10th ICA*, Sydney, paper E7.3 (1980).
24. A. H. Marshall, J. R. Hyde, and M. F. E. Barron, "The acoustical design of Wellington Town Hall: Design development, implementation and modelling results", *Proc. IoA (UK)*, Edinburgh (1982).
25. M. Barron, "The subjective effects of first reflections in concert halls—The need for lateral reflections", *J. Sound Vib.*, **15**, 475–94 (1971).
26. Daderot, "Kresge Auditorium, MIT (interior with concert)", license CC BY-SA 3.0, commons.wikimedia.org/wiki/File:Kresge_Auditorium,_MIT_(interior_with_concert).JPG.

27. M. Barron, *Auditorium Acoustics and Architectural Design*, 2nd edn, Taylor & Francis, Abingdon, UK (2009).
28. B.-I. Dalenbäck, M. Kleiner, and P. Svensson, "A macroscopic view of diffuse reflection", *J. Audio Eng. Soc.*, **42**, 793–807 (1994).
29. P. D'Antonio, *Variable Acoustics Modular Performance Shell*, US Patent 5, 168, 129 (1992).
30. A. H. Marshall, D. Gottlob, and H. Alrutz, "Acoustical conditions preferred for ensemble", *J. Acoust. Soc. Am.*, **64**(5), 1437–42 (1978).
31. P. D'Antonio, "Performance acoustics: The importance of diffusing surfaces and the variable acoustics modular performance shell", *Proc. 91st Convention Audio Eng. Soc.*, preprint 3118 (B-2) (1991).
32. M. Palma, M. Sarotto, T. Echenagucia, M. Sassone, and A. Astolfi. "Sound-strength driven parametric design of an acoustic shell in a free field environment", *Bldg. Acoust.*, **21**(1), 31–42 (2014).
33. W. Boning and A. A. Bassuet, "A room without walls: Optimizing an outdoor music shell to maintain views and maximize reflections", *Proc. IOA(UK) Auditorium Acoustics* (2015).
34. S. Di Rosario, S. Pone, B. Parenti, and E. Pignatelli, "ReS, Resonant String Shell, development and design of an acoustic shell for outdoor chamber music concerts", *Proc. IOA(UK) Auditorium Acoustics* (2015).
35. R. A. Metkemeijer, "The acoustics of the auditorium of the Royal Albert Hall before and after redevelopment", *Proc. IOA(UK)*, **19**(3), 57–66 (2002).
36. Egghead06, "Acoustic Discs Royal Albert Hall", commons.wikimedia.org/wiki/File:AcousticDiscsRoyalAlbertHall.JPG, accessed 29 December 2015.
37. R. J. Orlowski, "Shape and diffusion in the design of music spaces", *Proc. 17th ICA*, (2001).
38. E. Kahle, T. Wulfrank, Y. Jurkiewicz, and N. Faillet, "Philharmonie de Paris—The acoustic brief", *Proc. IOA(UK) Auditorium Acoustics* (2015).
39. Grasshopper: Algorithmic modelling for Rhino. http://www.grasshopper3d.com, accessed 27 December 2015.
40. Rhinoceros: Modelling Tools for Designers. http://www.rhino3d.com, accessed 27 December 2015.
41. T. Scelo, "Integration of acoustics in parametric architectural design", *Proc. ISRA*, **43** (2013).
42. H. Kuttruff, *Room Acoustics*, 6th edn, CRC Press, Oxon, UK (2009).
43. D. Lubman, "Precision of reverberant sound power measurements", *J. Acoust. Soc. Am.*, **56**, 523–33 (1974).
44. J. L. Davy, I. P. Dunn, and P. Dubout, "The variance of decay rates in reverberation rooms", *Acustica*, **43**, 12–25 (1979).
45. J. L. Davy, "The variance of impulse decays", *Acustica*, **44**, 51–6 (1980).
46. S. Chiles and M. Barron, "Sound level distribution and scatter in proportionate spaces", *J. Acoustic. Am. Soc.*, **116**(3), 1585–95 (2004).
47. T. Sakuma, "Approximate theory of reverberation in rectangular rooms with specular and diffuse reflections", *J. Audio Eng. Soc.*, **132**(4), 2325–36 (2012).
48. N. Prodi and C. Visentin, "An experimental evaluation of the impact of scattering on sound field diffusivity", *J. Acoust. Soc. Am.*, **133**(2), 810–20 (2013).
49. J. J. Embrechts, "A geometrical acoustics approach linking surface scattering and reverberation in room acoustics", *Acta Acust. Acust.*, **100**(5), 864–79 (2014).
50. E. Gerretsen, "Estimation methods for sound levels and reverberation time in a room with irregular shape or absorption distribution", *Acta Acust. Acust.*, **92**, 797–806 (2006).
51. K. Fujii, T. Hotehama, K. Kato, R. Shimokura, Y. Okamoto, Y. Suzumura, and Y. Ando, "Spatial distribution of acoustical parameters in concert halls: Comparison of different scattered reflection", *J. Temporal Des. Arch. Environ.*, **4**, 59–68 (2004).
52. S. Chiles, "Sound Behaviour in Proportionate Spaces in Auditoria", PhD thesis, University of Bath, UK (2004).
53. Y. H. Kim, J. H. Kim, and J. Y. Jeon, "Scale model investigations of diffuser application strategies for acoustical design of performance venues", *Acta Acust. Acust.*, **97**(5), 791–9 (2011).

54. T. J. Cox, "Comment on article Nico F. Declercq et al.: An acoustic diffraction study of a specifically designed auditorium having a corrugated ceiling: Alvar Aalto's lecture room", *Acta Acust. Acust.*, **97**(5), 909 (2011).
55. M. S. Mott, D. H. Robinson, A. Walden, J. Burnette, and A. S. Rutherford, "Illuminating the effects of dynamic lighting on student learning", *Sage Open*, 1–9, doi: 10.1177/2158244012445585 (2014).
56. C. Barkmann, N. Wessolowski, and M. Schulte-Markwort, "Applicability and efficacy of variable light in schools", *Physiol. Behav.*, **105**(3), 621–7 (2012).
57. P. J. C. Sleegers, N. M. Moolenaar, M. Galetzka, A. Pruyn, B. E. Sarroukh, and B. van der Zande, "Lighting affects students' concentration positively: Findings from three Dutch studies", *Light. Res. Technol.*, **45**(2), 159–75 (2013).
58. P. D'Antonio, *Combination Light Diffuser and Acoustical Treatment and Listening Room Including Such Fixtures*, US patent 8,967,823 (2015).
59. J. S. Bradley, H. Sato, and M. Picard, "On the importance of early reflections for speech in rooms", *J. Acoust. Soc. Am.*, **106**, 3233–44 (2003).
60. Y-J. Choi, "Effects of periodic type diffusers on classroom acoustics", *Appl. Acoust.*, **74**(5), 694–707 (2013).
61. F. Dong and F. Sui, "Acoustic performance of noise barriers with acoustic diffusers for high speed railway", *Proc. ISCV22* (2015).
62. T. Van Renterghem, D. Botteldooren, W. M. Cornelis, and D. Gabriels, "Reducing screen-induced refraction of noise barriers in wind by vegetative screens", *Acta Acust. Acust.*, **88**(2), 231–8 (2002).
63. M. R. Monazzam and Y. W. Lam, "Performance of profiled single noise barriers covered with quadratic residue diffusers", *Appl. Acoust.*, **66**(6), 709–30 (2005).
64. J. Kang, "Sound propagation in street canyons: Comparison between diffusely and geometrically reflecting boundaries", *J. Acoustic. Am. Soc.*, **107**(3), 1394–404 (2000).
65. H. Onaga and J. H. Rindel, "Acoustic characteristics of urban streets in relation to scattering caused by building facades", *Appl. Acoust.*, **68**, 310–25 (2007).

Chapter 4

Measurement of absorber properties

This chapter covers a variety of methods used to measure and characterize absorbers. Table 4.1 outlines the main techniques applied to existing materials and most used by acousticians. Indeed, for many, the only important measurement is that which gives the random incidence absorption coefficient in a reverberation chamber. While this may be the number that is needed for performance specifications in room design, other measurements are needed to understand, design, and model absorptive materials. For instance, the prediction of the random incidence absorption coefficient is problematic, and consequently, it is necessary to measure materials in a more controlled environment to allow direct comparison between theory and experiment.

The more controlled environment that is often used is the impedance tube, which allows normal incidence impedance and absorption to be determined. Less often, but nevertheless valuable, are the free field measurements on samples done in a hemianechoic chamber. The most common free field method uses a two-microphone approach, but this is often applicable only to isotropic, homogeneous samples. Consequently, attention turned to using more than two microphones; however, the measurements appear to be very sensitive to noise. These techniques can be adapted for *in situ* measurements, the next subject in this chapter.

Chapter 13 discusses how to convert between the absorption coefficients resulting from the different methods. It also examines how absorption coefficients can be applied to real room predictions as well as their use in geometric room acoustic models.

There is also a need to be able to characterize the propagation within the absorbent material, to enable theoretical modelling and development of new absorbents. For instance, the well-known Delaney and Bazley empirical model outlined in Chapter 6 requires the flow resistivity and porosity of the porous material to be known. For this reason, methods to measure all the key parameters that characterize the propagation within the absorbent are outlined.

4.1 IMPEDANCE TUBE MEASUREMENT FOR ABSORPTION COEFFICIENT AND SURFACE IMPEDANCE

The impedance tube, sometimes called a Kundt tube, enables both the normal incidence absorption coefficient and surface impedance to be measured. This is very useful as it enables measurement under well-defined and controlled conditions. Consequently, it is frequently used in validating prediction models for porous materials. This method has the advantage of needing only small samples (a few centimetres in diameter), and this makes it ideal for developers of materials, as the alternative is to construct large samples for reverberation chamber tests, which is more difficult and expensive. The final key advantage is that the impedance tube method can be carried out with relatively simple apparatus in a normal room and does not need a specialist test chamber. Problems with the method arise when the

Table 4.1 Principal measurement methods for the most common absorber properties

Apparatus	*Measures*	*Sample*	*Commonly used for*
Impedance tube	Normal incidence surface impedance, z, and absorption coefficient, α. Transmission coefficients for fabrics. Also can be used to extract porosity, tortuosity, and characteristic lengths via inverse methods.	Small	Absorber and prediction model development and evaluation
Two-microphone in hemi-anechoic	Surface impedance, z, and absorption coefficient, α for specific angles of incidence	Large	Research
Reverberation chamber	Random incidence absorption coefficient, α	Large	Performance specification
In situ	Surface impedance, z, for specific angles of incidence	–	Research
Flow resistivity rig	Flow resistivity, σ	Small	Absorber and prediction model development

absorption from the small sample is not representative of the behaviour of a large sample, as would happen with some resonant absorbers. For this reason, the method is most used with local reacting porous absorbers.

Figure 4.1 shows the set-up. A loudspeaker generates sound and a plane wave propagates down the tube before reflecting from the sample. The impedance of the sample alters how sound is reflected, and by measuring the resulting pressure, it is possible to calculate the normal incidence absorption coefficient and surface impedance of the sample. This is such a common technique that it has been enshrined in international standards.[1,2]

The necessity for plane wave propagation imposes limitations on the system:

1. The losses within the tube should be minimized so that the plane waves propagate without significant attenuation. Consequently, thick metal is a common construction material for a tube that works over the mid to high frequency ranges of most concern in building design and noise control. For impedance tubes that are to work at bass frequencies, more extreme construction is needed to prevent significant losses from the tube. For example, thicker walls are required or steel-lined concrete tubes can be used; these can work down to the limit of human hearing (20 Hz). Whatever the size, to minimize losses, the tube should be smooth on the inside and clean.

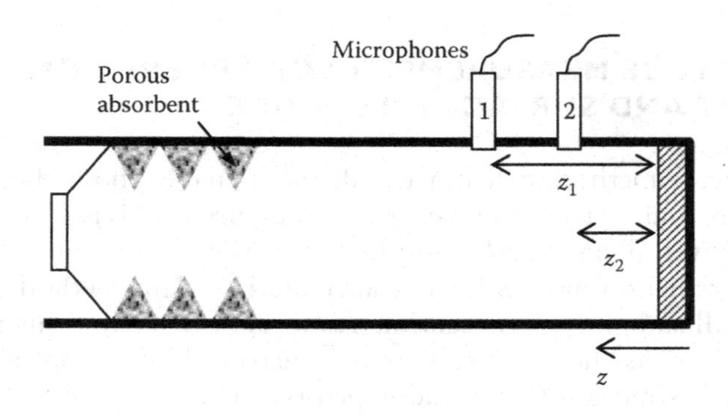

Figure 4.1 Set-up for impedance tube measurement using a two-microphone technique.

2. The tube should have constant cross-section over the measurement region where the sample and microphone positions are located. The actual shape is not that important; square and circular tubes are most popular. Although circular tubes are less prone to cross-mode problems within porous absorbents, square tubes are useful, as often, square samples are easier to make.
3. The loudspeaker should be a few tube diameters (or widths) from the first microphone position so that any cross-modes generated by the loudspeaker have decayed away. It is sometimes necessary to place absorbent at the loudspeaker end of the tube to reduce the effect of resonances within the impedance tube.
4. The microphone positions should not be too close to the sample so that any evanescent waves generated on reflection have had time to die away. For a homogeneous, isotropic sample, that means that the first measurement microphone should be at least half a tube diameter (or width) away. For samples that are structured and anisotropic, no microphones closer than two diameters away from the sample surface should be used.
5. The highest frequency, f_u, that can be measured in a tube is then determined by

$$f_u = \frac{c}{2d}, \tag{4.1}$$

where d is the tube diameter or maximum width and c is the speed of sound. This is a statement that there should not be any cross-modes in the tube; the first mode appears when half a wavelength fits across the tube. The limitation imposed by Equation 4.1 means that to cover a wide frequency range, several different impedance tubes of different diameter (or width) are required.

It is possible to measure at higher frequencies if multiple microphones are used across the width of the tube. The sound field within the tube can be considered to be a sum of the plane wave and higher modes, in a similar way to how a room sound field is decomposed into its modes. In a circular tube and an isotropic sample, one additional microphone enables the impedance of the sample for the plane wave and first cross-mode to be determined (circular symmetry can be exploited to reduce the number of additional microphones in this case). For a square tube, four microphones can be placed, as shown in Figure 4.2, and the measured signals summed.[3] In this case, the first and third cross-modes in the tube in each

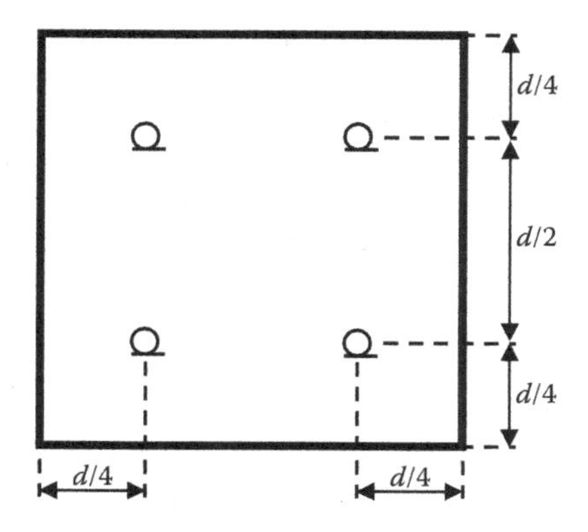

Figure 4.2 Set-up for impedance tube measurement at higher frequencies using four microphones.

direction cancel, while the microphones are at nodes of the second order mode, leaving the fourth-order mode to dominate. This quadruples the limit shown in Equation 4.1. Note that for frequencies where cross-modes dominate, the sample is not receiving plane waves, so the absorption coefficient measured is harder to relate to prediction models.

The experimental detail that is most critical is the requirement for the sample to be cut and mounted correctly. It is vital that the sample fits snugly into the tube. Any gaps around the edge must be filled and sealed; otherwise, there will be absorption by the edge of the sample and the measured absorption will be too high. Worse still, if the small gaps open up to an air cavity behind, a Helmholtz resonator could be formed, and the absorption overestimated by a large margin. The normal way is to use petroleum jelly (Vaseline), Plasticene, or mastic to fill the edges of the sample. It is also important not to wedge porous absorbers into the tube, as this changes the mechanics of the absorber frame and can lead to incorrect measurements due to the constrained vibration of the absorber frame.[4] (Rapid variations in the impedance data are seen around the resonant frequency of the frame.) Consequently, the sample should be 0.5–1 mm smaller than the diameter (or width) of the tube,[5] then the impedance and absorption coefficient measured are effectively the same as of those a sample with infinite lateral dimensions.

It is also important that the rear of the absorber is properly terminated. Air gaps between the absorber and the backing plate will lead to excess absorption being measured, unless of course it is planned to mount the absorber with an air gap, in which case the measurement would be correct.

The impedance tube is not often useable for extended reaction absorbers, unless the impedance tube happens to be the same size as the extended reaction device. For instance, the performance of a membrane absorber is usually dependent on the mounting of the membrane. It is not possible just to mount a smaller membrane absorber in an impedance tube, and expect the same absorption coefficient as the larger device.

4.1.1 Transfer function method

The pressure in the impedance tube is given by

$$p = A(e^{jkz} + Re^{-jkz}), \tag{4.2}$$

where R is the reflection coefficient; k is the wavenumber; the sample is assumed to be at $z = 0$; and A is a complex constant. The first term represents the incident wave, and the second, the reflected wave. Equation 4.2 has two unknowns that need to be found: the magnitude and the phase of the reflection coefficient. By measuring the pressure at two points in the tube, it is possible to set up and solve simultaneous equations for the reflection coefficient. This is the principle of the transfer function, often called the two-microphone method (although, as this is often used with one microphone which is moved, calling this a two-microphone method is nowadays rather misleading). From the reflection coefficient, it is possible to get the absorption coefficient (Equation 2.24) and normal incidence surface impedance (Equation 2.22).

This approach obtains the absorption coefficient and impedance for all frequencies (within limits) with only a couple of quick measurements. It is also a method where if something is done wrong, for example, the microphone positions are incorrect, then the formulations yield results that are clearly unphysical—it is easy to spot common measurement errors.

The transfer function between two microphone positions in the tube is measured (see Figure 4.1). Remembering that the transfer function is simply the ratio of pressures,

$H_{12} = p(z_2)/p(z_1)$, and applying Equation 4.2, the transfer function between microphone positions 1 and 2 is given by

$$H_{12} = \frac{e^{jkz_2} + Re^{-jkz_2}}{e^{jkz_1} + Re^{-jkz_1}}, \tag{4.3}$$

where z_1 and z_2 are the positions of the microphones shown in Figure 4.1. Rearrangement then directly leads to the complex pressure reflection coefficient:

$$R = \frac{H_{12}e^{jkz_1} - e^{jkz_2}}{e^{-jkz_2} - H_{12}e^{-jkz_1}}. \tag{4.4}$$

There are restrictions on the microphone spacing. If the microphones are too close together, the transfer function measured will be inaccurate because the change in pressure will be too small to be accurately measured. This leads to a lower frequency limit, f_l, for a given microphone spacing $|z_1 - z_2|$:

$$f_l > \frac{c}{20|z_1 - z_2|}. \tag{4.5}$$

Problems also arise if the microphone spacing becomes too wide. As the spacing approaches a wavelength, the simultaneous equations become impossible to resolve as the pressure measured at both microphones is identical. This leads to an upper frequency limit f_u due to microphone spacing given by

$$f_u < \frac{0.45c}{|z_1 - z_2|}. \tag{4.6}$$

Consequently, there are two upper frequency limits given by Equations 4.1 and 4.6, and the lowest figure should be taken.

The lower and upper frequency limits mean that to cover a reasonable number of octaves, it is often necessary to use more than two microphone positions in a tube; three positions are typically used. Three positions give three possible microphone spacings. By appropriately setting the frequency ranges for each of the spacings, it is possible to cover a wider frequency range.

There is a choice of methods for measuring the transfer function in Equation 4.3. A dual channel FFT analyzer can be used with a matched pair of microphones, using a white noise source. In that case, it is necessary to compensate for differences in the microphone responses by measuring once, then interchanging the microphones and measuring again. A more efficient method is to use a deterministic signal such as a maximum length sequence[6] or swept sine wave.[7] This means that one microphone can be used to measure the transfer function to each microphone position and the ratios of these transfer functions are used to obtain H_{12}. This negates the need for matched microphones. Using a deterministic signal rather than white noise also removes the need for time-consuming averaging.

It is important that any unused holes are blocked and that microphones are mounted flush to the tube sides. Knowing the microphone position accurately is vital.

Horoshenkov et al.[8] carried out a round robin test on impedance tube measurements involving seven laboratories. They used three samples: reconstituted porous rubber, reticulated foam, and fibreglass, and Figure 4.3 shows some of the results. The mean absorption coefficient is shown for each material, along with dotted lines indicating the 95% confidence limit in any one laboratory measurement.

The biggest errors are seen for the reconstituted porous rubber in the top graph. This is most likely due to differences in the various samples taken from the same block of material, so the large variations probably reflect true sample variation and not just experimental

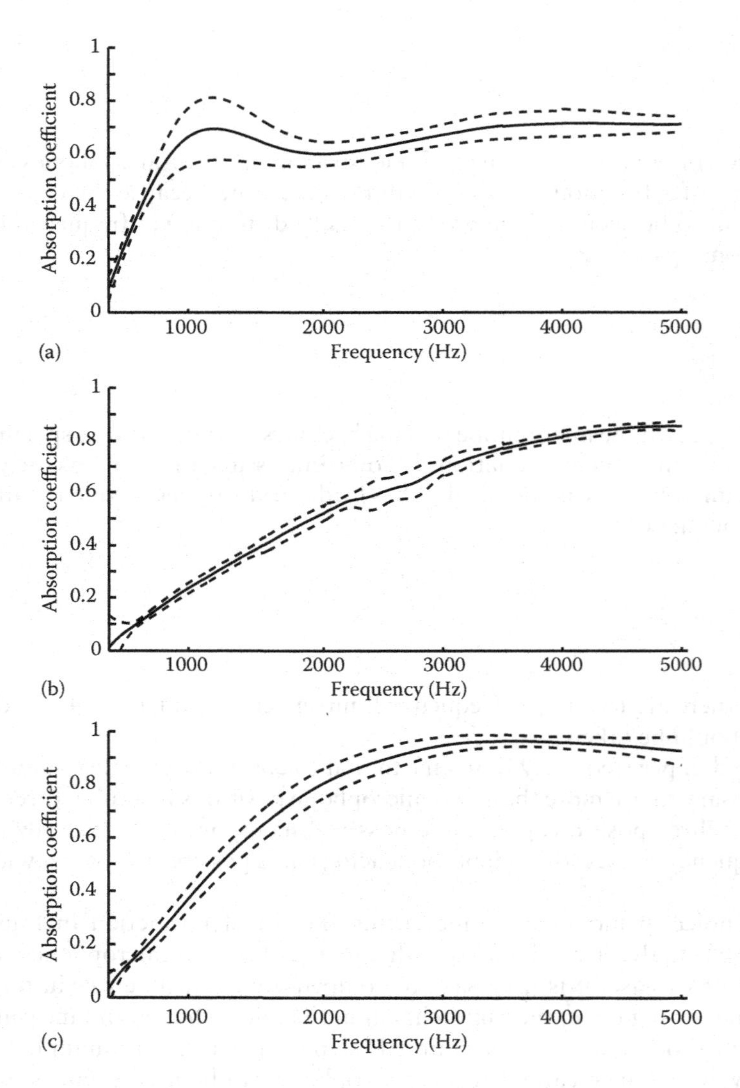

Figure 4.3 Comparison of impedance tube measurements of the absorption coefficient for three materials in up to seven laboratories. (a) Reconstituted porous rubber; (b) reticulated foam; and (c) fibreglass. The mean absorption coefficient is shown, along with dotted lines indicating the 95% confidence limit in any one laboratory measurement. <3500 Hz data from seven laboratories is used; 3500–4000 Hz from six laboratories; and >4000 Hz from four laboratories. (Adapted from Horoshenkov, K.V. et al., *J. Acoust. Soc. Am.*, 122, 345–353, 2007. Graphs kindly prepared by F.-X. Bécot.)

error. The fibreglass results shown in the bottom graph are similarly affected by sample variation. The foam samples were more consistent, but the effects of mounting were seen. Individual laboratory results showed patterns of minima and maxima indicative of structural resonances. This evidence is somewhat lost in producing the average plots in Figure 4.3, but the increase in errors around 2500 Hz are due to this effect.

4.1.2 Least mean square method

This is essentially an adaptation of the transfer function method. It is usual to use three microphone measurements to cover the frequency range that an impedance tube offers. There will be a region of frequency overlap where two sets of measurement results are applicable, and each will yield a slightly different answer. By applying a least mean square approach, it is possible to gain a formulation that produces one unambiguous result from the three (or more) microphones.[9] The complex reflection coefficient is given by

$$R = -\frac{\sum_{n=1}^{M-1}\sum_{m=n+1}^{M}\left[e^{-jk(L-z_n)}H_{nm} - e^{-jk(L-z_m)}\right]\left[e^{-jk(L+z_n)}H_{nm} - e^{-jk(L+z_m)}\right]^*}{\sum_{n=1}^{M-1}\sum_{m=n+1}^{M}\left|e^{-jk(L+z_n)}H_{nm} - e^{-jk(L+z_m)}\right|^2}, \tag{4.7}$$

where H_{nm} is the transfer function measured between microphones n and m, z_n and z_m are the distances of the nth and m^{th} microphones from the test specimen, and * indicates complex conjugate.

4.1.3 Transmission measurements

The impedance tube method can also allow the measurement of transmission through materials, and in 2009, this was incorporated into an American Society for Testing and Materials (ASTM) standard.[10] Figure 4.4 shows a typical arrangement. The right hand side might be an anechoic termination, generated by a thick porous layer of gradually increasing flow resistivity or an anechoic wedge. The quality of the anechoic termination is crucial for accurate measurement[11,12] because any reflected sound will pass back through the test material and be measured by microphones 1 and 2 as reflections from the front of the test sample. (This leads to a transmission coefficient spectrum that inaccurately oscillates with respect to

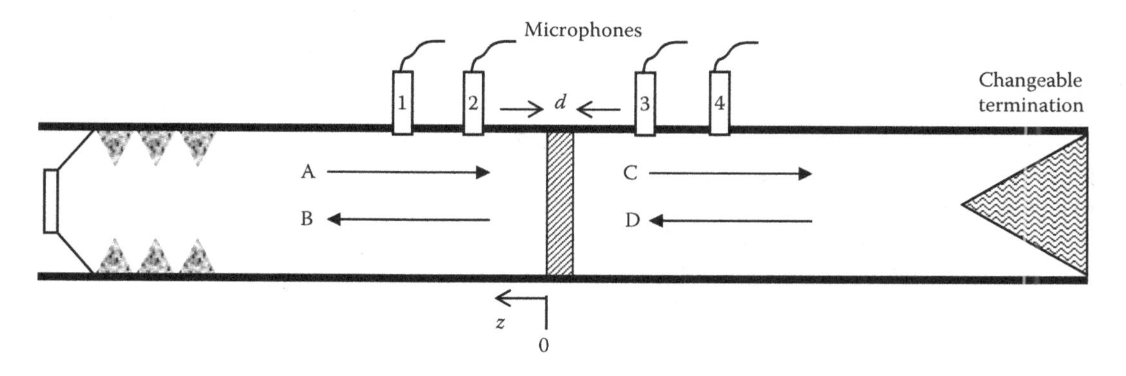

Figure 4.4 Set-up for transmission measurement in an impedance tube.

frequency due to interference.) Provided that the sample being tested has low flow resistivity and high porosity, a very good anechoic termination will yield accurate results; indeed, in theory, a fourth microphone is not even needed.

A more robust technique is the two-load method, which can cover a wider range of samples. Two test conditions are measured, one where the right hand termination is rigid and the other where it is open. The formulations for this method are derived below in Equations 4.8 through 4.17;[13] and alternative wave decomposition formulations are also available.[14]

The pressure amplitudes A, B, C, and D of the plane wave components in the tube are defined in Figure 4.4. To the left of the sample, the pressure and velocity are given by

$$p = Ae^{jkz} + Be^{-jkz} \tag{4.8}$$

and

$$u = \frac{1}{\rho c}(Ae^{jkz} - Be^{-jkz}), \tag{4.9}$$

and to the right of the sample:

$$p = Ce^{jkz} + De^{-jkz}, \quad u = \frac{1}{\rho c}(Ce^{jkz} - De^{-jkz}). \tag{4.10}$$

Using the pressures measured at the four microphones, $p(z_1)...p(z_4)$, it is possible to derive expressions for the pressure amplitudes:

$$\begin{aligned} A &= \frac{p(z_1)e^{-jkz_2} - p(z_2)e^{-jkz_1}}{2j\sin(k(z_1 - z_2))}, \quad B = \frac{p(z_2)e^{-jkz_1} - p(z_1)e^{-jkz_2}}{2j\sin(k(z_1 - z_2))}, \\ C &= \frac{p(z_3)e^{-jkz_4} - p(z_4)e^{-jkz_3}}{2j\sin(k(z_3 - z_4))}, \quad D = \frac{p(z_4)e^{-jkz_3} - p(z_3)e^{-jkz_4}}{2j\sin(k(z_3 - z_4))}. \end{aligned} \tag{4.11}$$

Once the amplitudes are calculated from the measured pressures, it is then possible to evaluate the pressures and velocities on the front and rear of the test sample, at $z = 0$ and $z = -d$, using Equations 4.8 through 4.10. The propagation through the test sample can be described by a transfer matrix, as described in Section 2.6.1:

$$\begin{bmatrix} p \\ u \end{bmatrix}_{z=0} = \begin{bmatrix} T_{11} & T_{12} \\ T_{21} & T_{22} \end{bmatrix} \begin{bmatrix} p \\ u \end{bmatrix}_{z=-d}, \tag{4.12}$$

where the pressures p and velocities u are known. To obtain the transmission coefficient, the four transfer matrix components, T_{11}, T_{12}, T_{21}, and T_{22}, which determine how the sample reacts to sound, are needed. By making two sets of measurements, with different termination conditions at the right hand end of the tube, enough independent equations are generated to enable the matrix components to be solved:

$$\begin{bmatrix} p^r & p^o \\ u^r & u^o \end{bmatrix}_{z=0} = \begin{bmatrix} T_{11} & T_{12} \\ T_{21} & T_{22} \end{bmatrix} \begin{bmatrix} p^r & p^o \\ u^r & u^o \end{bmatrix}_{z=-d}, \tag{4.13}$$

where a superscript r indicates the measurements with a rigid termination, and o, the measurements with the open end (roughly a pressure release case). It is vital that the termination impedances are very different across the frequency range of interest; otherwise, the additional equations derived will not be independent and the solution will become inaccurate. These sets of simultaneous equations can be solved to yield the transfer matrix elements.

These transfer matrix elements determine what happens to sound as it propagates through the test sample, but they involve velocities, and the transmission loss is defined purely in terms of a ratio of pressures. Consequently, some further work is needed to get the transmission loss from the matrix elements. The pressures at $z = 0$ and $z = -d$ from Equations 4.8 through 4.10 can be substituted into Equation 4.12. After some manipulation, it is possible to derive a new matrix that relates the pressure amplitudes in the system:

$$\begin{bmatrix} A \\ B \end{bmatrix} = \begin{bmatrix} \frac{1}{2}\left(T_{11} + \frac{T_{12}}{\rho c} + \rho c T_{21} + T_{22}\right)e^{-jkd} & \frac{1}{2}\left(T_{11} - \frac{T_{12}}{\rho c} + \rho c T_{21} - T_{22}\right)e^{jkd} \\ \frac{1}{2}\left(T_{11} + \frac{T_{12}}{\rho c} - \rho c T_{21} - T_{22}\right)e^{-jkd} & \frac{1}{2}\left(T_{11} - \frac{T_{12}}{\rho c} - \rho c T_{21} + T_{22}\right)e^{jkd} \end{bmatrix} \begin{bmatrix} C \\ D \end{bmatrix}. \tag{4.14}$$

This equation is true for any termination condition. From inspection of Figure 4.4, the transmission loss (TL) is given by $20\log_{10}(|A/C|)$, which is easiest to work out when there is a perfect anechoic termination, i.e., $D = 0$. Consequently, the transmission loss is given by

$$TL = 20\log_{10}\left(\frac{1}{2}\left|T_{11} + \frac{T_{12}}{\rho c} + \rho c T_{21} + T_{22}\right|\right). \tag{4.15}$$

These formulations are sufficient for carrying out the calculation, but with some algebraic manipulation, it is possible to derive a single equation for the transmission loss, which bypasses the need to explicitly calculate the transmission matrix elements. Equation 4.14 shows that it is possible to relate the pressure amplitudes by a simple matrix. Consider the amplitudes derived from the two different measurements in a matrix form:

$$\begin{bmatrix} A^r & A^o \\ B^r & B^o \end{bmatrix} = \begin{bmatrix} a_{11} & a_{12} \\ a_{21} & a_{22} \end{bmatrix} \begin{bmatrix} C^r & C^o \\ D^r & D^o \end{bmatrix}, \tag{4.16}$$

where the superscripts denote the different measurement conditions. The element a_{11} gives the transmission loss (again, by considering the case $D = 0$). So by manipulating the simultaneous equations represented in Equation 4.16 it is possible to show that

$$TL = 20\log_{10}(a_{11}) = 20\log_{10}\left(\frac{A^r D^o - A^o D^r}{C^r D^o - C^o D^r}\right). \tag{4.17}$$

So the transmission loss can be calculated directly from the pressure amplitudes under the different measurement conditions using this one formulation.

4.2 TWO-MICROPHONE FREE FIELD MEASUREMENT

The disadvantage of the impedance tube is that it does not readily allow oblique incidence measurement. In contrast, the two-microphone free field method allows this to be done. By its very nature, the test method needs a large sample, which can be difficult to produce. It also needs an anechoic or hemianechoic space for the measurement. This method, like the impedance tube, is of most use to porous absorber designers or modellers.

The method can be thought of as an extension of the impedance tube method in Section 4.1.1. The technique is most straightforward for homogeneous, isotropic materials. Consider a large sample of absorbent being irradiated by a loudspeaker a long way from the surface as shown in Figure 4.5. The measurement can be done in an anechoic or hemianechoic chamber; it can even be done in a large room provided that time windowing is used to remove unwanted reflections from other surfaces. It is assumed that plane waves are incident on the surface. Furthermore, for large isotropic, homogeneous samples, it can be assumed that the reflected sound is also a plane wave, in which case the equations set out in Sections 4.1.1 and 4.1.2 for the transfer function measurement in the impedance tube can be applied directly to the free field case.

Some practical details need careful consideration. Although, in theory, the sample should be infinite, in reality, the sample will be finite in extent and so edge diffraction becomes important. The diffraction from the edges at low frequencies causes the reflected wave to no longer be planar, and so the simple theories no longer apply. A rough lower frequency limit is when half a wavelength fits across the smallest sample dimension. Consequently, samples are typically several square metres in area. When large samples are not available, one solution is to bring the source closer to the surface, say 20 cm away, so the edge waves become less significant.[13] In which case, it is necessary to use spherical wave equations rather than the plane wave formulations given previously. The drawback of using spherical wave formulations is that the interpretation of the measured impedance becomes less straightforward.

Returning to the plane wave case, the first microphone is typically 5 mm from the surface, and the second is 15 mm from the surface; the lower and upper frequency limits discussed for the impedance tube related to microphone spacing are still relevant. The

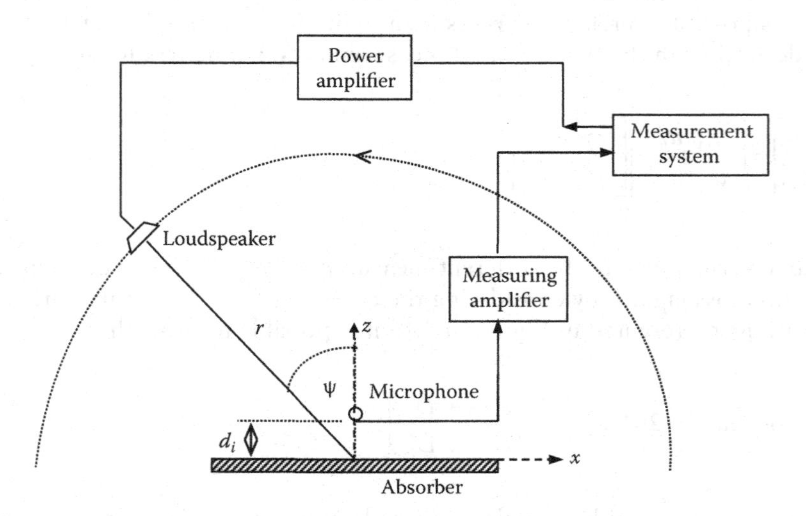

Figure 4.5 Experimental set-up for two- or multi-microphone free field measurement.

microphones must be small enough that they do not cause significant disturbance to the acoustic sound field. While it is possible to use two microphones, one microphone that is moved may be preferred as it provides less disturbance to the sound field and it removes the need for calibration. A deterministic test signal such as a maximum length sequence or swept sine must be used in this case.

The method can be extended to deal with oblique incidence, in which case the formulations should be rederived. For an incident angle of ψ, the transfer function between the two microphones positions is given by

$$H_{12} = \frac{e^{jkz_2\cos(\psi)} + Re^{-jkz_2\cos(\psi)}}{e^{jkz_1\cos(\psi)} + Re^{-jkz_1\cos(\psi)}}, \tag{4.18}$$

where it has been assumed that the two microphone positions have the same x and y displacement in the coordinate system defined in Figure 4.5. A rearrangement leads directly to the reflection coefficient

$$R = \frac{H_{12}e^{jkz_1\cos(\psi)} - e^{jkz_2\cos(\psi)}}{e^{-jkz_2\cos(\psi)} - H_{12}e^{-jkz_1\cos(\psi)}}. \tag{4.19}$$

The impedance can then be calculated using Equation 2.22, and from Equation 2.24, the absorption coefficient at that particular angle of incidence can be found.

Problems arise for large angles of incidence. As the angle of incidence increases, the effects of edge diffraction become more significant at higher frequencies. Consequently, very large angles of incidence are difficult to measure unless very large samples are available.

4.2.1 Multimicrophone techniques for nonisotropic, nonplanar surfaces

For nonisotropic or nonplanar surfaces, it is still possible to carry out a free field measurement using methods similar to those detailed in Section 4.2, although the system becomes more elaborate and sensitive to measurement error. The formulations for the two-microphone free field method have assumed that the dominant reflected wave is a plane wave, which is true for isotropic, homogeneous, infinitely large planar surfaces. As soon as the surface becomes rough, or there are impedance variations, then there is potential for nonplane wave propagation. For example, if a periodic impedance variation is considered, a set of grating or diffraction lobes in nonspecular directions are generated. (See, for example, Figure 10.3, where the Schroeder diffuser generates 11 lobes.) To measure the absorption in the periodic case, it is necessary to measure the magnitude and phase of each of these reflected waves. This requires a measurement using more than two microphone positions because there are additional unknowns—the magnitude and phase of each of the grating lobe waves—to be resolved. To measure the 11 propagating waves seen in Figure 10.3, 12 microphone positions would be needed. In reality, it is unlikely that this case could be measured, because multimicrophone systems become prone to measurement noise[15,16] as the number of microphones increases.

The next section details a multimicrophone method for surfaces with periodic impedance variation[15] to give an idea of how such a multimicrophone system might work. Although it is difficult to implement and get accurate measurements, it does enable incident angle dependent absorption to be measured.

4.2.2 Multimicrophone free field measurement for periodic surfaces

There will be multiple reflected waves, not just the plane waves considered in the two-microphone method. By using more than two microphone positions, it is possible to measure the amplitude and phase of these reflected waves. The set-up is shown in Figure 4.5. The pressure at the mth microphone, p_m, measured at coordinates (x_m, z_m), is given by

$$p_m(x_m, z_m) = P_i e^{j(-x_m k_x + z_m k_z)} + \sum_{n=-\infty}^{\infty} A_n e^{j(-x_m \beta_n - z_m \gamma_n)}. \tag{4.20}$$

The first term on the right hand side is an incident plane wave p_i and the second term is the scattered or reflected pressure. In this equation,

$$\begin{aligned} k_x &= k\sin(\psi), \\ k_z &= k\cos(\psi), \\ \beta_n &= k_x + n\frac{2\pi}{W}, \\ \gamma_n &= -jk\sqrt{\left(\sin(\psi) + n\frac{\lambda}{W}\right)^2 - 1}. \end{aligned} \tag{4.21}$$

A_n is a complex coefficient describing the amplitude of each of the reflected waves, P is a constant, ψ is the incident angle, k is the wavenumber, λ is the wavelength, and W is the width of one period. The surface is periodic and (assumed) infinite so that a Fourier representation of the reflected sound field is used. Chapter 8 gives more details of the theory.

The scattered pressure is an infinite sum of waves, with complex amplitudes A_n. Not all of these waves will propagate into the far field. The waves that are confined to the near field, the evanescent waves, need not be modelled, which means that the sum over n is finite. The upper and lower limits for the sum in Equation 4.20 are determined by

$$\left[\sin(\psi) + n\frac{\lambda}{W}\right]^2 \leq 1. \tag{4.22}$$

Let the lower limit be denoted n_1 and the upper limit n_2; the number of coefficients to be determined must be small for this measurement to work.

The absorption coefficient is found by taking one minus the ratio of the reflected to incident energy, which gives:

$$\alpha(\psi) = 1 - \left|{}^{A_0}\!/\!{}_{P_i}\right|^2 - \frac{1}{\cos(\psi)} \sum_{n=n_1,\neq 0}^{n_2} \left|{}^{A_n}\!/\!{}_{P_i}\right|^2 \sqrt{1 - (\sin(\psi) + n\lambda / W)^2} \,. \tag{4.23}$$

When only the $n = 0$ term exists, a two-microphone approach can be used as the only radiating wave is the plane wave term. When more than one term is present, $|n_1| \vee |n_2| > 0$, then more microphones are needed. In this case, $N = (|n_1| + |n_2| + 2)$ receiver points need to be measured since the transfer functions between two measured positions will be used.

In the data processing, it is convenient to use the transfer functions between adjacent measurement positions $H_{m,m+1}$:

$$H_{m,m+1} = \frac{p_m}{p_{m+1}} = \frac{P_i e^{j(-x_m k_x + z_m k_z)} + \sum_{n=n_1}^{n_2} A_n e^{j(-x_m \beta_n - z_m \gamma_n)}}{P_i e^{j(-x_{m+1} k_x + z_{m+1} k_z)} + \sum_{n=n_1}^{n_2} A_n e^{j(-x_{m+1} \beta_n - z_{m+1} \gamma_n)}}. \tag{4.24}$$

This means that $N - 1$ simultaneous equations can be obtained in terms of A_n/P_i:

$$\begin{aligned} &\sum_{n=n_1}^{n_2} \left(\frac{A_n}{P_i}\right)\left(e^{j(-x_m \beta_n - z_m \gamma_n)} - H_{m,m+1} e^{j(-x_{m+1} \beta_n - z_{m+1} \gamma_n)}\right) \\ &= H_{m,m+1} e^{j(-x_{m+1} k_x + z_{m+1} k_z)} - e^{j(-x_m k_x + z_m k_z)}. \end{aligned} \tag{4.25}$$

Once the simultaneous equations have been formed, these can be solved to give A_n/P_i, which can be substituted into Equation 4.23, and gives the absorption coefficient after a little manipulation.

The choice of measurement positions is critical. If the microphone is only allowed to traverse the z direction, then critical frequencies occur where there are insufficient unique simultaneous equations to resolve the coefficients. These critical frequencies manifest themselves as frequencies for which nonsensical absorption coefficients are obtained. These critical frequencies can be avoided by changing the microphone position in z and x. The typical spacings used give microphones 5–10 cm apart.

The multimicrophone method is very sensitive to evanescent (nonpropagating) waves. The microphone must be far enough away from the sample to prevent the measurement of evanescent waves, as these have been neglected in the previous theories, but if the microphone is too far from the surface, diffraction from the sample edges will cause the measured pressures to be inaccurate. The multimicrophone method is much more noise sensitive than the two-microphone method. Very accurate microphone positioning is needed. Others looking at multiple microphone techniques have found similar noise sensitivity.[16]

Figure 8.11 shows an example measurement result. It is compared to two prediction models. Good accuracy is achieved with the multimicrophone measurement system in this case. At low frequencies, only two microphone positions are used as there is only one plane wave reflection. At mid–high frequencies, three microphone positions are needed as an additional reflected wave is present.

4.3 REVERBERATION CHAMBER METHOD

In most applications, the sound will be incident on an absorber from a multitude of incident angles at once. It is not efficient to laboriously measure absorption coefficients for all angles of incidence in the free field and reconstruct these into a random incidence absorption coefficient (although this can be done, as is discussed in Chapter 13). Consequently, a quicker method is needed, and this is afforded by the reverberation chamber method.[17] The random incidence absorption coefficient is the parameter used most in the design of spaces to specify the absorption performance of materials. It is well known and defined; however, it is notoriously difficult to predict. So while the random incidence absorption coefficient is needed to enable room design, it is not very useful for those interested in validating prediction models.

The reverberation chamber test requires large sample sizes and a specialist room and so is expensive to undertake. It also gives only absorption coefficients; the impedance cannot be directly measured. Consequently, developers of absorptive materials will often use the impedance tube to build up an understanding of material properties on small samples before undertaking reverberation tests.

The reverberation time of a room is dependent on the total absorption in the room (see Equation 2.1). Consequently, by measuring the reverberation time of a room before and after a sample of absorbent is introduced, it is possible to calculate the random incidence absorption coefficient. It is necessary to have defined acoustic conditions for the test, and the normal technique is to try and generate a diffuse field. A *diffuse field* can be roughly defined as requiring the reflected sound energy to have minimal variation throughout the room and the energy to be propagating evenly in all directions. To achieve this, reverberation chambers often use diffusers in the volume of the room, and the chamber walls are often skewed (splayed). Furthermore, the room should be of a certain minimum size, and room dimensions should be irrationally related to reduce the influence of room modes.

Despite these measures, a diffuse field is not completely achieved, and consequently, the reverberation time is position dependent. For this reason, it is normal to use multiple source and receiver positions and to average the results to reduce the effect of nondiffuseness. The source is normally placed in the corner of a room, pointing into the corner, because it maximally excites the modes of the room and reduces the amount of direct radiation from the loudspeaker to the test sample. Receivers should be at least 1 m from the room boundaries, room diffusers, and the sample and should be chosen to obtain a diverse sampling of the room volume. Even with all these precautions, the measured absorption coefficients are often more inaccurate at low than high frequencies due to modal effects.

The reverberation time before the sample is introduced is given by

$$T_0 = \frac{55.3V}{c\alpha_0 S + 4Vm_1}, \tag{4.26}$$

where V is the room volume, c is the speed of sound, α_0 is the average absorption coefficient of the empty room, and S is the surface area of the room; m_1 allows for air absorption in the room, and typical values are shown in Table 4.2.

Table 4.2 Air absorption constant m_1 at 20°C and normal atmospheric pressure in 10^{-3} m^{-1}

Relative humidity (%)	Frequency (Hz)							
	63	*125*	*250*	*500*	*1000*	*2000*	*4000*	*8000*
20	0.06	0.16	0.32	0.6	1.5	4.96	17.2	50
30	0.044	0.14	0.33	0.58	1.15	3.25	11.26	38.76
40	0.035	0.12	0.32	0.6	1.07	2.58	8.39	29.94
50	0.028	0.1	0.3	0.63	1.07	2.28	6.83	24.24
60	0.024	0.088	0.28	0.64	1.11	2.14	5.9	20.48
70	0.021	0.077	0.26	0.64	1.15	2.08	5.32	17.88

Complete formulations for air attenuation can be found in ISO 9613-2.[18] For the highest frequencies, more exact values are available that allow for variation of the air absorption across an octave band for different reverberation times.[19] In most cases, however, the values in Table 4.2 will be sufficient.

The reverberation time after the sample is introduced is given by

$$T_1 = \frac{55.3V}{c(\alpha_0[S - S_s] + \alpha_s S_s) + 4Vm_1}, \tag{4.27}$$

where S is the surface area and α_s is the absorption coefficient of the sample. By rearranging Equations 4.26 and 4.27, it is possible to obtain the absorption coefficient of the sample. If ISO 354:2003 is followed, the factor $[S - S_s]$ is approximated to S, which simplifies the end formulation.

Standards are based on Sabine's formulations, as used above. It is well known that this equation becomes inaccurate for highly absorbent samples and overestimates the absorption coefficient, in which cases other formulations like the ones derived by Eyring can be used (given in Equation 2.5). The bias error is often relatively small, with one study showing that it is of the order of about 3% for a highly absorbing mineral wool sample.[20] Furthermore, while it can be argued that using other reverberation time formulations produces more correct answers, the databases of absorption coefficients available to designers have been derived from Sabine's formulation. Consequently, while it is known that Sabine's equation produces a systematic (or bias) error, it continues to be industry practice to use it. Appendix A gives a table of typical measured absorption coefficients for common materials.

The reverberation times can be measured by interrupted noise, maximum length sequences, or swept sine waves. Maximum length sequences allow rapid measurement, but problems in getting sufficiently long decays can arise due to non-linearities in loudspeakers.[21]

To get an accurate measurement, it is necessary to have a big difference between T_0 and T_1. This necessitates a large sample area of 10–12 m^2. Even with such a large sample, the accuracy is compromised due to edge effects. Figure 4.6[22] shows the average absorption coefficient for two types of porous absorbent measured in 13 laboratories. Even averaged across all laboratories, at some frequencies, the absorption coefficient exceeds unity. Sound is diffracted by the edges of the sample, which usually leads to excess absorption around 200–500 Hz. It is normal practice to cover the edges of the sample and to use rectangular samples to reduce edge effects. Nevertheless, even with the edges covered, absorption coefficients greater than 1 can be measured. Section 13.1 discusses this further, including how these edge effects might be compensated for in real room predictions.

Another issue is caused by the diffusers that are hung in the middle of reverberation chambers to increase diffusivity. They can cause problems by reducing the mean free path in the room—the average time between reflections—and hence the reverberation time. An example of the reverberation chamber with free hanging and boundary diffusers was shown in Figure 3.24.

Given the standard deviations of the reverberation times T_0 and T_1, it is possible to calculate the random experimental error. The standard deviations are calculated from the set of reverberation times for all source and receiver combinations. If the standard deviation of the reverberation time measure T_0 is σ_0, then the 95% confidence limit is given by

$$\delta_0 = \frac{2\sigma_0}{\sqrt{n}}, \tag{4.28}$$

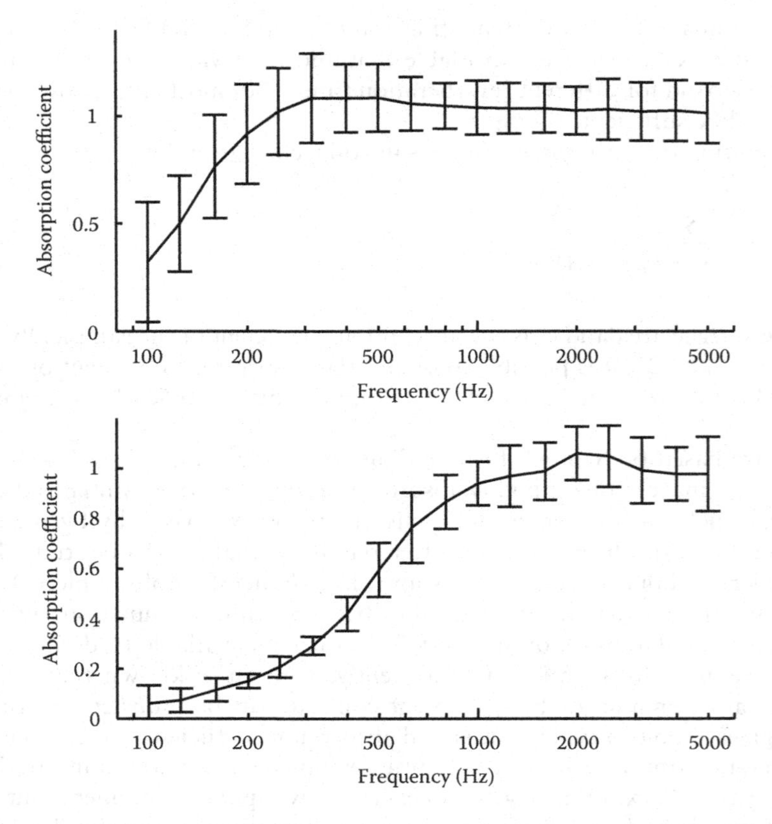

Figure 4.6 Comparison of absorption coefficients for two different samples measured in 13 different laboratories. The mean absorption coefficient across all laboratories is shown for each sample, along with error bars indicating the 95% confidence limit in any one laboratory measurement. The top graph is for a 100 mm mineral wool absorber in a wooden casing covered with nonwoven fleece and the bottom is a 25-mm-thick foam absorber. (Adapted from Jeong, C.H., Chang, J.H., *Acta Acust. Acust.*, 101, 99–112, 2015.)

where n is the number of source and receiver pairs. (It is assumed that n is sufficiently large that two standard uncertainties is equivalent to the 95% confidence limit.) A similar relationship exists for T_1. The accuracy (95% confidence limit) of the empty room average absorption coefficient is given by

$$\delta_{\alpha_0} = \left| \frac{55.3V}{cT_0^2 S_0} \right| \delta_0, \tag{4.29}$$

where the effect of inaccurate estimation of air absorption has been assumed to be smaller than the effect of reverberation time variation between measurement positions. The accuracy of sample absorption is given by

$$\delta_{\alpha_s} = \sqrt{\left| \frac{55.3V}{cT_1^2 S_s} \right|^2 \delta_1^2 + \left(\frac{c[S_0 - S_s](\delta_{\alpha_0})^2}{S_s} \right)^2}. \tag{4.30}$$

While good repeatability within a laboratory can be achieved, there are reproducibility problems between laboratories. The measured absorption coefficients can vary greatly from laboratory to laboratory (see Figure 4.6). The error bars indicate the 95% confidence limits in one of the laboratory measurements. This shows that when the sample was sent to a laboratory, 95% of the time, the absorption coefficient would be within ±0.1 of the true mean value at 1 kHz. A statistical calculation shows that if a sample is sent to two laboratories, then it would be expected that 73% of the time, the two absorption coefficients would differ by 0.1 or less, but 3% of the time, the difference would be 0.2 or more.

The reasons for inconsistencies between the laboratories are well documented. The main reason is the lack of a diffuse field in the reverberation chamber, especially when a highly absorbing sample is being measured. For example, with the sample on the floor, horizontally propagating sound will decay relatively slowly because it does not reflect from the floor where the sample lies. While diffusers can help to direct sound up and down to increase diffuseness, it seems that current standards are not sufficiently rigorous.

Currently, an ISO working group is examining possible solutions to the reproducibility problem. Correcting for edge effects by measuring samples of different sizes would make the tests much more expensive, so it is hard to see that approach being adopted. Calculating the change in the mean free path caused by volume diffusers using a geometric room acoustic model to produce an effective room volume and surface area is more promising, although a proper method has yet to be derived. Alternatively, diffusers can be placed on the boundaries rather than hanging in the middle of the room. In this way, the volume and surface area of the room can be determined accurately.

Some favour the use of a standard reference sample that would have to be measured within some tolerance for a laboratory to qualify. Others have suggested using the reference sample for calibration. To make it robust, this could be constructed from a microperforated absorber with an airspace behind, partitioned with honeycomb to make it more local reacting. A small portion of the sample could be measured in an impedance tube, and assuming local reaction, applying Paris's formula, Equation 13.1, would give the *true* random incidence absorption coefficient. By comparing the measured value in the reverberation chamber with this true value, a frequency-dependent correction factor can then be worked out to be applied to measurements in that laboratory. There are problems with this approach, such as the fact that the edge effect depends not just on frequency but also on the absorption coefficient. So the correction is strictly only right for samples with an absorption coefficient spectrum similar to the calibration sample.

There are other approaches as well. A better estimation of the random incidence absorption coefficient could be obtained by using a geometric room acoustic model to properly model the sound distribution in the reverberation chamber.[23] Another is to fit a model for the porous absorbent to the reverberation chamber result and hence allow the true random incidence absorption coefficient to be calculated.[24] Chapter 13 takes the discussion of random incidence absorption coefficients further, examining how the values measured in laboratories are used in predictions, including their use in geometric room acoustic models. Section 13.1 also quantifies the effects of edge absorption and sample size. Seating is a common absorbent with very significant edge absorption, and the measurement of this particular surface is dealt with in the next section.

It is also possible to measure discrete objects in the reverberation chamber, for example, people.[17] They are arranged randomly around the room, and a total absorption per object calculated from the reverberation times. Sakagami et al.[25] detail a method for measuring absorbents hung in the reverberation chamber, as might happen in the case where a porous absorbent is hung in a factory to reduce reverberant noise levels.

4.3.1 Measurement of seating absorption

The reverberation time in a concert hall is dominated by the absorption of the seating and audience, and it is essential that these can be measured or predicted accurately in the early stages of design. The wrong estimation of seating absorption has been blamed for acoustic problems in many halls, and consequently, a measurement procedure is given here. Davies et al.[26] compared seating absorption coefficients in a reverberation chamber and in concert halls for 10 different cases and showed that the Kath and Kuhl method[27] is best for estimating seating absorption in a real hall. Beranek[28] showed that it is best to calculate seating absorption coefficients based on absorption per unit floor area rather than by absorption per seat.

The aim of measuring the absorption coefficients of a small sample of seats in a reverberation chamber is to predict the total absorption that a larger area of the same seats will exhibit when installed in an auditorium. There are problems, however, because a small sample of chairs in the reverberation chamber (say 24) is unrepresentative of the larger block of seating in the auditorium because edge effects are overemphasized in the reverberation chamber.

The Kath and Kuhl method involves placing the seating in the corner of the reverberation chamber with their intended row spacing. The edges are obscured with barriers for some of the measurements. The barriers need to be massive and stiff to reduce low-frequency absorption. The barriers should be at least as high as the seating, and higher if any audience is present for an occupied measurement. Excessive extra height (say, more than 100 mm above the top of the seating for the unoccupied case) should be avoided. The set-up is schematically shown in Figure 4.7. The array is mirrored in the adjacent walls of the chamber, thus effectively increasing its size, but it is not effectively infinite as Kath and Kuhl suggested. Diffraction effects will still be present and so the measured absorption coefficient may still vary with sample size.

The concept is to separately measure three absorption coefficients by carrying out tests with and without barriers. The measurements are as follows:

- For an *infinite* array with no edges, yields an absorption coefficient α_∞, with side and front barriers in place;
- For the front edges, α_f, by measuring with the side barrier only in place and combining the result for α_∞; and
- For the side edges, α_s, by measuring with the front barrier only in place, and combining with the result for α_∞.

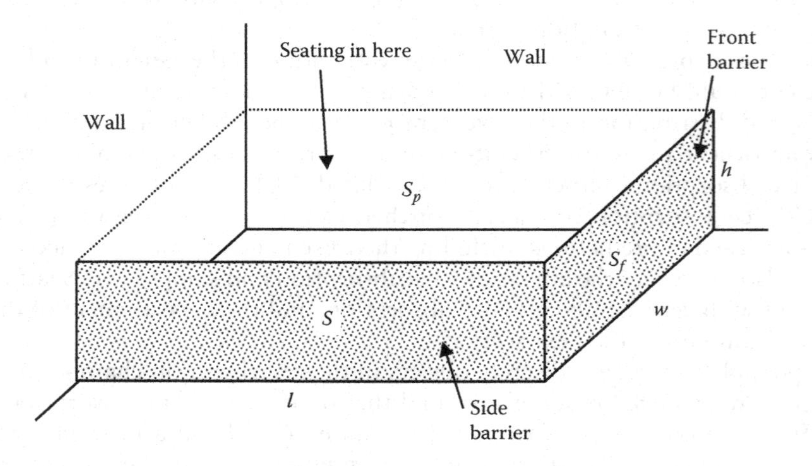

Figure 4.7 Set-up for the Kath and Kuhl method for measuring seating absorption.

Then in the hall, if the areas of the front edges, S_f, side edges, S_s, and plan area, S_p, are known, the absorption coefficient of the audience block is given by the following:

$$\alpha = \alpha_\infty + \alpha_f \frac{S_f}{S_p} + \alpha_s \frac{S_s}{S_p}. \tag{4.31}$$

In the reverberation chamber, the absorption coefficients, α_∞, α_f, and α_s, are determined by the following formulations. First, with both the front and side barriers in place, the infinite array absorption coefficient is obtained:

$$\alpha_\infty = \frac{A_1}{(l+\lambda/8)(w+\lambda/8)}, \tag{4.32}$$

where A_1 is the total absorption of the sample with both barriers in place, and the $\lambda/8$ terms correct for pressure doubling at the chamber walls, where λ is the wavelength of the centre frequency in the octave band.

With an additional measurement of the total absorption A_2 with the front barrier missing, the absorption of the front edge is determined:

$$\alpha_f = \frac{A_2 - A_1}{h(w+\lambda/8)}. \tag{4.33}$$

Finally, with an additional measurement of the total absorption with the side barrier missing A_3, the absorption of the side edge is determined:

$$\alpha_s = \frac{A_3 - A_1}{h(l+\lambda/8)}. \tag{4.34}$$

The corner placing of the seats is advantageous because it increases the effective size of the array. However, there is a disadvantage: the pressure in a reverberant field is increased at the boundaries,[29] so the absorption coefficients measured will be higher than those found when the sample is in the centre of the chamber. To compensate for this, Kath and Kuhl[30] proposed that the absorber areas used in the calculations be increased by strips of width $\lambda/8$. This extra absorbing area accounts for the increase in measured total absorption due to the increase of up to 3 dB in sound pressure level close to the wall. In a corner, there is an increase of up to 6 dB, and a correction of $(\lambda/8)^2$ is needed. This is the reason for the extra terms in the denominators of the above formulations.

Figure 4.8 shows a comparison of the measured absorption coefficient in the auditorium and a prediction from the reverberation chamber results. Good prediction accuracy is achieved for this case. Discrepancies found by Davies et al.[26] for other halls were likely to be due to the nondiffuseness of the auditoria and indefinable changes in the hall between without and with seating measurements.

Bradley[31] also suggested a seating absorption measurement method that attempts to take account of the variation of seating absorption with sample size. This involves making measurements on five or six differently sized arrays of seats and then extrapolating to the expected absorption for large seating blocks in auditoria. It assumes a linear relationship between the absorption coefficient and the perimeter to area ratio, hence is known as the

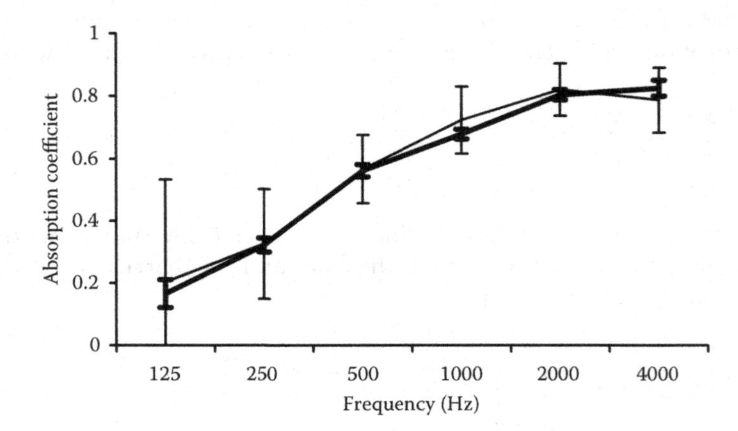

Figure 4.8 Random incidence absorption coefficient measured in an auditorium compared to a prediction based on a Kath and Kuhl reverberation chamber measurement. ▬▬ prediction for full auditorium seating based on reverberation chamber measurement and —— full auditorium measurement. (Data from Davies, W.J., Orlowski, R.J., Lam, Y.W., *J. Acoust. Soc. Am.*, 96, 879–888, 1994.)

P/A method. Although this is accurate, it requires more tests than does the method using barriers.

4.4 *IN SITU* MEASUREMENT OF ABSORPTIVE PROPERTIES

There is great interest in being able to measure the absorption coefficient and surface impedance of products *in situ*. To take one example, in geometric room acoustic models, the absorption coefficients of surfaces are required, but how can these be determined if the room is already built? It is for this sort of problem that *in situ* techniques have been developed. For those interested in the context of *in situ* measurement, the papers by Nocke and Mellert[32] and Brandão et al.[33] give a comprehensive historical review of the literature.

One possible technique is to use a portable impedance tube, as documented in an international standard for measuring road surfaces.[34] Another approach is to adopt the two-microphone free field method outlined in Section 4.2. If the surface to be tested is large, homogeneous, isotropic, and planar, and the unwanted reflections from other surfaces can be removed by time gating (windowing), then this process has been shown to work and give accurate results.[35] This process will fail if the unwanted reflections cannot be removed, or nonplane wave reflections are significant, which might arise if the surface is too small or if the surface to be tested is inhomogeneous.

Other techniques try to separate the incident and reflected sound from a surface by arrival times. This is schematically shown in Figure 4.9. An impulse response is measured and the reflected and incident sounds are isolated by applying rectangular windows. From these, the reflection coefficient is calculated. Achieving this, however, requires considerable distance between the microphone and the surface, which means that problems often arise because of edge reflections from the test sample and unwanted reflections from other surfaces. Very large surfaces are needed; otherwise, there is poor accuracy at low frequencies. Consequently, while this technique is potentially accurate, its range of applicability is limited.

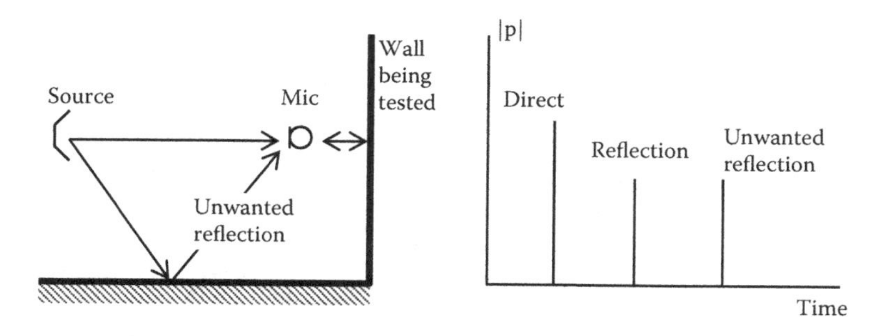

Figure 4.9 Measurement of *in situ* absorption properties using time gating.

A more promising technique was developed by Mommertz,[36] as this exploits a subtraction technique that enables the microphone to be placed close to the test surface. The technique appears in standards for measuring noise barriers[37] and road surfaces.[38] This allows measurement from 250 Hz to 8 kHz for normal incidence on plane surfaces greater than 4 m^2. The low frequency accuracy is compromised for oblique incidence or smaller samples. The test arrangement is shown in Figure 4.9. The sound source is connected to the microphone by a rod to ensure that the distance between the source and microphone remains constant. Precise positioning is vital if the measurement method is to be accurate. The impulse response between the source and microphone is measured with the microphone close to the test surface and separately in the free field. These two impulses can be subtracted, which leaves the reflected sound and unwanted interfering reflections; these parasitic reflections can be removed by time windowing. To allow the subtraction, a deterministic test signal such as a maximum length sequence or swept sine must be used.

Placing the microphone very close to the surface ensures that the interfering reflections are maximally spaced from the wanted reflections, consequently allowing more accurate measurements. By deconvolving the loudspeaker's free field impulse response from the *in situ* measured impulse response, the length of the direct and reflected sound in the impulse responses can be shortened. This can help make the gating process more accurate. Mommertz advocates doing this deconvolution by preemphasizing the test signal—an approximate inverse of the loudspeaker impulse response is used to prefilter the maximum length sequence before it is sent to the loudspeaker. Alternatively, this deconvolution could be done as part of the postprocessing before the windows are applied. If overlap still exists between the wanted and parasitic reflections, then a window with a smooth transition should be used, like a half-Hanning.

If the microphone is very close to the surface, a simple ratio of the reflected and incident spectra can be taken to give the complex pressure reflection coefficient and, from there, absorption coefficients and surface impedance. If the microphone is not close to the surface, a correction for spherical spreading and propagation phase must be made to the incident and reflected spectra before the ratio is taken.

Figure 4.10 shows a typical result. For normal incidence, the reflection coefficients of the sample obtained using the *in situ* method match those obtained using an impedance tube method. For oblique incidence and at low frequencies (<800 Hz), the method fails with the reflection coefficient exceeding one. This occurs because there is an implicit assumption of plane waves in the methodology. At low frequencies, the edges of the test sample create other

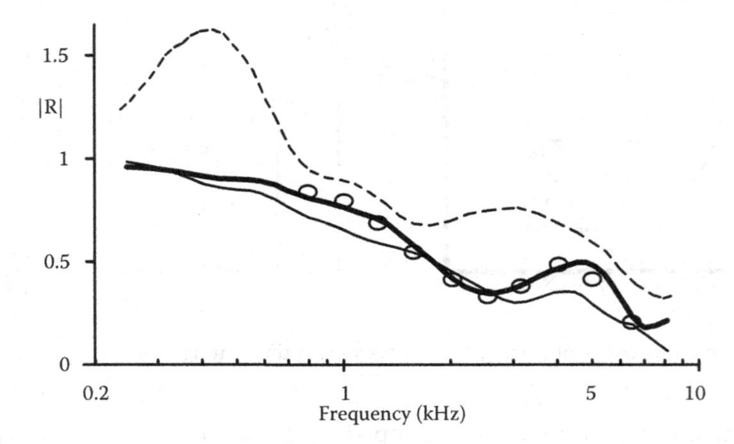

Figure 4.10 Measurements of reflection coefficient magnitude. *In situ* method: ▬▬ $\psi = 0°$; —— $\psi = 45°$; ---- $\psi = 81°$; and ○ impedance tube method. (Data from Mommertz, E., *Appl. Acoust.*, 46, 251–264, 1995.)

types of reflected waves, which then render the technique inaccurate. One solution to this is to consider spherical wave reflection.[39]

Problems arise with this *in situ* method if the acoustic medium changes between the free field and sample measurement. For instance, Mommertz gives an example of a temperature change of 1°C leading to an error of 0.03 in absorption coefficient. One way to reduce this error is to realign oversampled versions of the impulse responses before subtraction.[40]

The final *in situ* method detailed uses an alternative approach. No attempt is made to separate the incident and reflected sound; this removes some of the geometric restrictions on the measurement system. Making it work accurately, however, requires a good theoretical model of the sound field close to the test surface. The idea of the method is as follows: given measurements of the sound field in the vicinity of the test surface and a theoretical model for the sound propagation, it is possible to apply a numerical optimization or root finding scheme to derive the unknown properties of the test sample. The measurements of the sound field could be pressures from microphones and/or particle velocities from PU probes.[33]

Figure 4.11 shows a typical microphone set-up used by Nocke.[41] The transfer function between the source and receiver is measured, and from this, the angle dependent impedance and absorption coefficient is derived by numeric inversion.

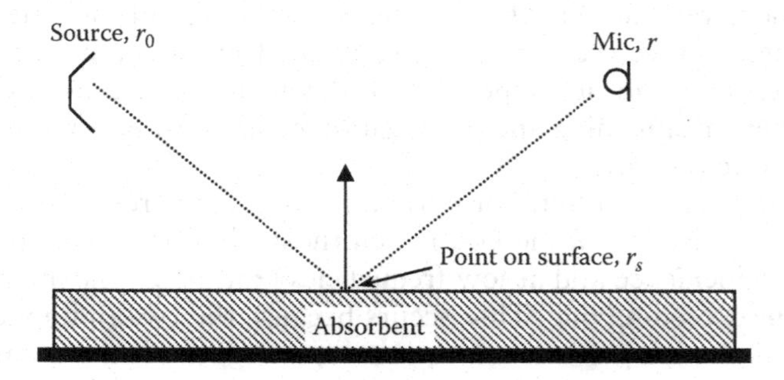

Figure 4.11 Typical set-up for *in situ* measurement. (After Nocke, C., *Appl. Acoust.*, 59, 253–264, 2000.)

To simplify the description, consider a plane homogeneous sample. The pressure above the absorber is given by

$$p(\mathbf{r}) = p_i(\mathbf{r}) - F(k, \mathbf{r}, \mathbf{r}_0, R(\mathbf{r}_s), p_i(\mathbf{r}_s)), \tag{4.35}$$

where $p_i(\mathbf{r})$ is the pressure direct from the source to the receiver, $\mathbf{r}$ is the receiver position, $\mathbf{r}_0$ is the source position, $\mathbf{r}_s$ is the position of a point on the surface, $F()$ is a function that gives the pressure at the microphone due to reflections from the surface, k is the wavenumber, and R is the reflection coefficient.

There are various prediction theories that can be used to model the function $F()$, which gives the reflected energy. Researchers have tried various spherical wave reflection theories (see Chapter 2 for one such formulation) or alternatively use a slightly simpler formulation as derived by Di and Gilbert.[42] It would also be possible (and easier) to use simple plane wave formulations provided that the surface is large so that edge diffraction is not significant or homogenous and the source is not too close to the surface so the incident wave can be considered to be plane.

Given that the incident and reflected pressures have been measured and that the receiver and source positions given in Equation 4.35 are known; the only unknown in the equation is the complex reflection coefficient R, as a function of the vector on the surface $\mathbf{r}_s$. This reflection coefficient can therefore be found by using an iterative procedure. For simplicity, assume that R is the same for the whole surface. Given the measured pressure is $p_m(\mathbf{r})$ and the predicted pressure $p(\mathbf{r})$, then a numerical optimizer can be tasked with the procedure of minimizing the mean square error between the measured and predicted pressures, $|p_m(\mathbf{r}) - p(\mathbf{r})|^2$ by finding the value of R, which gives minimal error. There are a variety of numerical optimization methods that can be used for this process, such as genetic algorithms.

Taherzadeh and Attenborough[43] found numerical optimization formulated this way was rather slow when calculating the impedance of the ground from excess attenuation measurements because spherical wave reflection coefficients must be used for the typical measurement geometries used. They advocated the use of a root-finding algorithm, which is more efficient because it exploits the Newton Raphson algorithm, using calculated derivatives. This efficient formulation greatly reduces computation time. Whichever numerical process is used, there is a risk of finding the wrong solution because the measured pressures do not always map to a single, unique reflection coefficient. A simple solution to this is to make measurements with a couple of different geometries to ensure that the correct answer is found.

In theory, this method works for samples where R varies across the surface. As a number of different surface reflection coefficients have to be derived, more microphone positions are needed. In this case, the optimization problem becomes slower to solve and the risk of getting incorrect solutions from the optimizer increases. The experience with multiple microphone techniques is that the more receiver positions are, the more problems with noise sensitivity and evanescent waves. Consequently, it might be anticipated that resolving a large number of different surface reflection coefficients might prove to be problematical.

Nocke[41] restricts himself to deriving an average absorption coefficient for inhomogeneous surfaces. Figure 4.12 shows a typical result showing the *in situ* method compared to impedance tube measurements. By using the spherical wave formulation, accurate results are achieved down to 80 Hz, but this requires a very large sample of 16 m^2 to prevent edge effects being significant. The upper frequency limit measured was 4 kHz, presumably limited by the accuracy of the microphone positioning. Kruse[44] examined the method for typical ground surfaces and found the results to be inaccurate below about 400 Hz and for hard ground. More measurements utilizing different geometries are needed to overcome these problems.

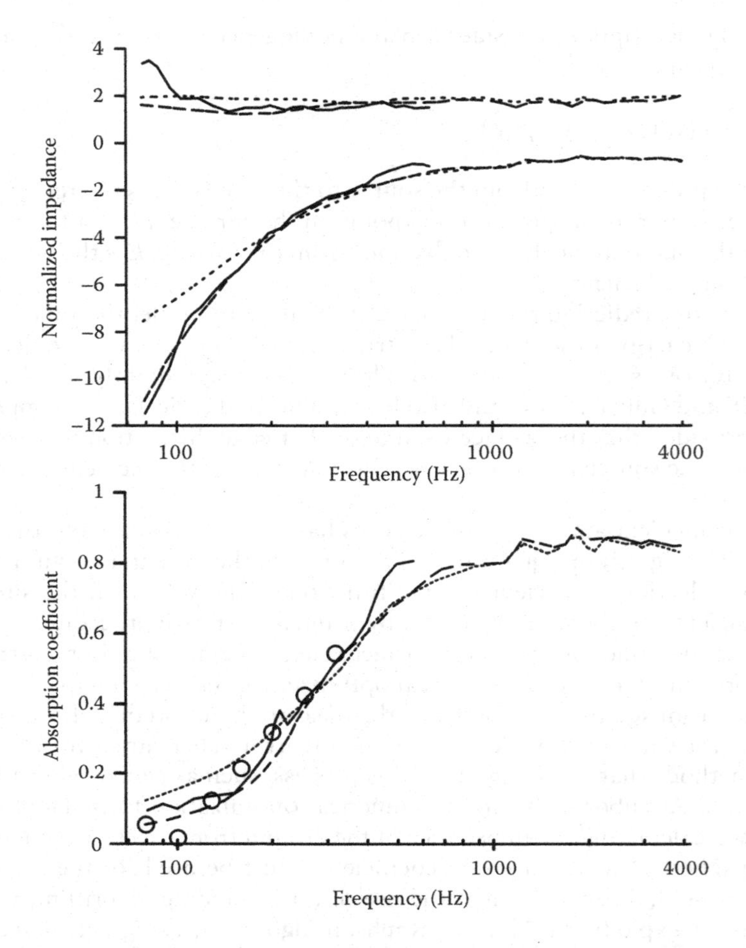

Figure 4.12 Comparison of *in situ* and impedance tube measurements for a fibrous absorber: —— impedance tube A; ○ impedance tube B (on absorption coefficient graph only); _ _ _ *in situ* ψ = 10°; and ..... *in situ* ψ = 12°. (Data from Nocke, C., *Appl. Acoust.*, 59, 253–264, 2000.)

4.5 MEASUREMENT OF INTERNAL PROPERTIES OF POROUS ABSORBENTS

The remaining sections of this chapter are devoted to the measurement of properties within absorbents, characterizing the sound propagation within the porous materials either in terms of propagation constants and characteristic impedances or finding key parameters, such as flow resistivity, tortuosity, porosity, and characteristic lengths. These are key measurements for those involved in the development or modelling of porous materials.

Many different techniques will be outlined, but a good approach that reduces the need for building many test rigs and yet acceptably accurate results is to use a direct measurement to gain the flow resistivity and then use an inverse method to obtain the other parameters.

Direct measurements using specialist apparatus is often seen as best because parameters are then obtained without using acoustic measurements and so values are independent and robust. Another approach is to look at the low or high frequency asymptotic behaviour of parameters readily obtained from the impedance tube. For example, this analytical approach allows the flow resistivity to be estimated from the low-frequency effective density. A final approach, an inverse method, is to take surface impedance measurements and then get parameter values by fitting prediction models for porous absorbent behaviour to the measured data. This is done by adjusting the parameters within the prediction models using numerical optimisation.

4.5.1 Measurement of flow resistivity

The flow resistivity of a porous absorber is one of the most important defining characteristics. Once the flow resistivity is known, simple empirical models can be used to find the characteristic impedance and wavenumber and, from there, the surface impedance and absorption coefficient. The importance of this parameter is discussed in more detail in Chapter 6. For now, three non-acoustic measurement techniques will be considered.

The methods follow directly from the definition of flow resistivity. Consider a slice of the porous material of thickness d subject to a mean steady flow velocity U. The d.c. pressure drop across the sample, ΔP, is measured, and from these quantities, the flow resistivity σ is given by

$$\sigma = \frac{\Delta P}{Ud}. \tag{4.36}$$

The measurement of flow resistivity has been enshrined into International Standards.[45] In the direct flow method, a steady air supply pushes air through the porous material. Sensors are used to measure the air flow and pressure drop relative to atmospheric pressure, and hence, the flow resistivity is obtained. This is shown in Figure 4.13.

It is important that flow rates are kept small because, otherwise, the relationship between the pressure drop and velocity becomes non-linear. Flow rates between 5×10^{-4} and 5×10^{-2} ms^{-1} are recommended by Bies and Hansen.[46] Ingard[47] produced results that show that the flow rate should not be greater than 0.1 ms^{-1} to get results consistent with the velocity amplitude of typical sound waves in absorbents (0.01 ms^{-1}). (However, note that at this flow

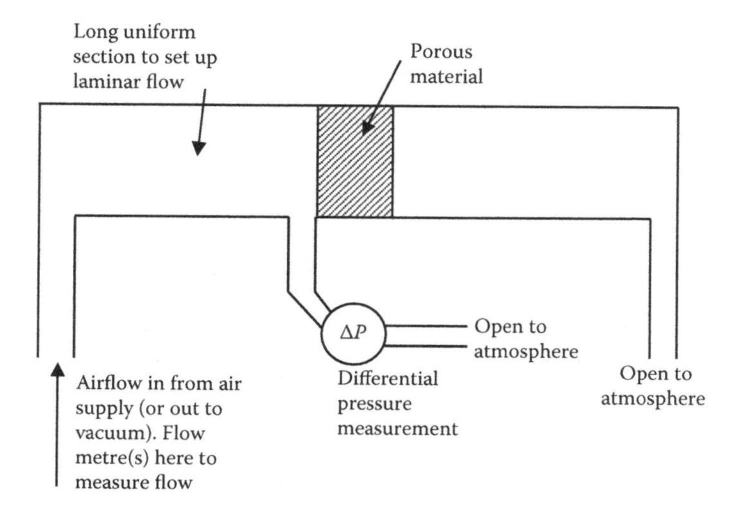

Figure 4.13 Set-up for direct airflow method for measuring flow resistivity.

rate, the flow resistivity varies with flow rate following either a linear[48] or quadratic relationship.[49]) The flow resistivity is calculated from

$$\sigma = \frac{\rho_0 \Delta P A}{md}, \tag{4.37}$$

where ρ_0 is the density of air, A is the cross-sectional area of the specimen, and m the air mass flow rate (kg^{-1}).

In the alternative airflow method, a piston is used to generate a low-frequency, alternating airflow through the test specimen; the piston should move at a frequency of $f = 2$ Hz. The set-up is shown in Figure 4.14. The root mean square (rms) airflow velocity is then

$$u_{rms} = \frac{\pi f h A_p}{\sqrt{2}A}, \tag{4.38}$$

where h is the peak–peak displacement of the piston, A_p is the cross-sectional area of the piston, and A is the cross-sectional area of the porous material. The standard recommends 0.5 mms^{-1} $< u_{rms} <$ 4 mms^{-1}. A condenser microphone is used to measure the rms pressure relative to atmospheric pressure.

The problem with these experimental techniques is that there is a great variation between results from different laboratories. Garai and Pompoli[50] found that repeatability within a laboratory is good, with an error of about 2.5% with repeat measurements on one sample and about 5% with five samples cut from the same material. Reproducibility between laboratories, on the other hand, is a problem with errors around 15%.

There are alternative methods for obtaining the flow resistance. Ingard[47] devised a measurement system that does not require blowers and flow instrumentation and so greatly simplifies the apparatus required. The set-up is shown in Figure 4.15. A piston within a tube falls under gravity and pushes air through the porous absorbent. When the piston has reached terminal velocity, the pressure drop ΔP across the sample is given by

$$\Delta P = \frac{Mg}{A_p}, \tag{4.39}$$

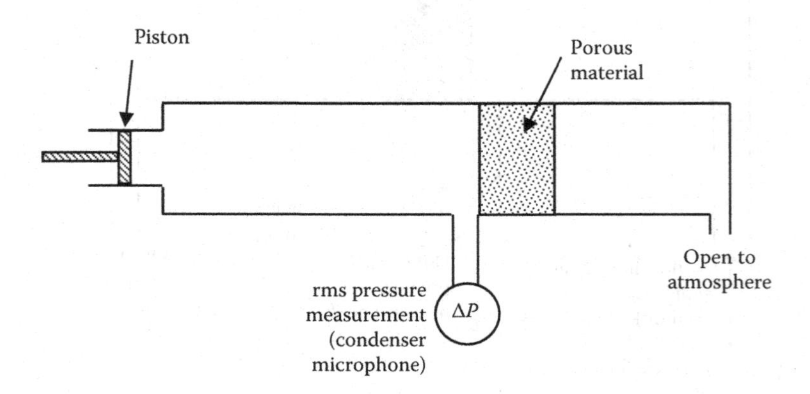

Figure 4.14 Set-up for alternating airflow method for measuring flow resistivity.

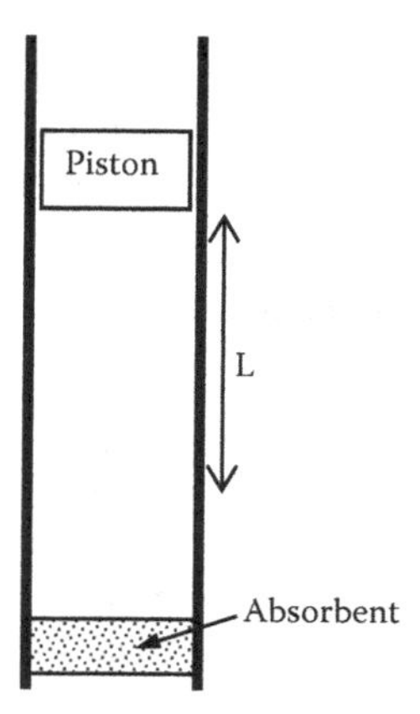

Figure 4.15 Set-up for Ingard's method for measuring flow resistivity.

where M is the mass of the piston, g is the acceleration due to gravity, and A_p is the cross-sectional area of the piston or tube. The flow velocity U is given by

$$U = \frac{A_p}{S_s} v, \tag{4.40}$$

where S_s is the cross-sectional area of the sample and v is the terminal velocity of the piston found by timing how long it takes the piston to travel a set length. Equations 4.39 and 4.40 can be combined with Equation 4.36 to give the flow resistivity. It is first necessary, however, to carry out two calibration measurements because there will be frictional forces between the piston and the tube walls and there will be leakage between the piston and the tube wall. In total, three timed falls of the piston are made, as summarized in Table 4.3.

The time taken for the piston to fall in the tube with no sample over a set length is measured in two cases: first with the tube open, t_0, and second with one end of the tube closed (it does not matter which), t_1. Then a calibration factor is found:

$$C = \frac{1 - t_0/t}{1 - t/t_1}, \tag{4.41}$$

where t is the time it takes the piston to drop over the same measurement distance L with the sample present. Then the flow resistance is then given by

$$\sigma = \frac{CMg\cos(\psi) S_s t}{LA_p^2}, \tag{4.42}$$

where ψ is the angle of the tube to the vertical.

Table 4.3 Measurements needed for flow resistivity measurement using falling piston method

Time measured	*Condition*
t	Sample in tube, both ends of tube open
t_0	No sample in tube, both ends of tube open
t_1	No sample in tube, one end of tube closed (either end)

Ingard[47] reports using a tube with an inner diameter of 0.08 m, a length of 1.2 m, and L = 0.6 m. The piston was 264 g and 10 cm long. The gap between the cylinder and the tube wall was about 0.2 mm. The system does not work for materials with very small or large flow resistivities due to problems with accurate timing and calibration.

4.5.2 Measurement of flow impedance

Strictly speaking, porous materials are not just resistive but contain reactance as well.[47] The reactance comes from additional mass due to viscous boundary layer effects and constrained flow. Consequently, while Section 4.5.1 has given methods for flow resistance, there could be interest in measuring the flow impedance to get the reactance. As the resistive term dominates, this type of measurement is not that commonly undertaken. While these methods give alternative methods for measuring the flow resistivity, they are not the most independent and robust because they use sound waves.

This flow impedance is measured within an impedance tube.[51] Figure 4.16 shows an arrangement that can be used; the tube is about 5 cm in diameter. A pure tone plane wave is produced by the sound source. The frequency is adjusted so that the distance w is an odd integer multiple of a quarter of a wavelength:

$$w = n\lambda/4, \tag{4.43}$$

where n is the number of quarter wavelengths in the length w, and n must be odd. There are an infinite set of frequencies that satisfy this condition, but a low frequency (say less than 100 Hz) is needed to get accurate results, as the sample should be thin compared to wavelength. This forces the pressure and surface impedance at the rear of the sample to be 0. The flow impedance z_f is then a ratio of the pressures measured at microphones 1 and 2:

$$z_f = j\rho_0 c_0 (-1)^{(n-1)/2} \frac{p_1}{p_2}, \tag{4.44}$$

where p_1 and p_2 are the complex pressures for microphones 1 and 2. This formulation can be derived as follows. As the impedance at the rear of the sample is zero, the flow impedance equals the difference in the pressures on either side of the sample divided by the particle

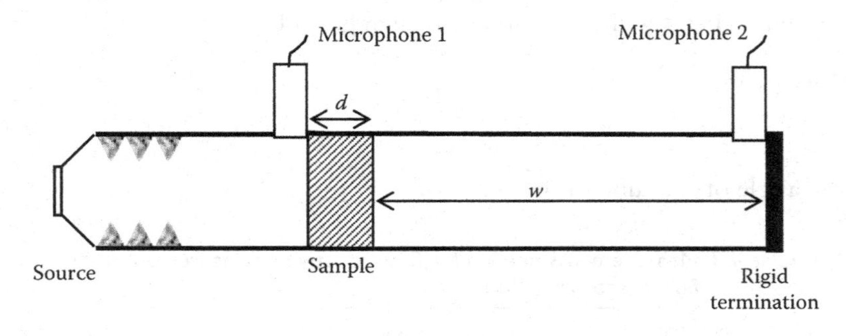

Figure 4.16 Apparatus for determining flow impedance using an impedance tube. (After Ingard, U.K., Dear, T.A., *J. Sound Vib.*, 103, 567–572, 1985.)

velocity (which should be the same on both sides of the sample). This impedance can be derived, using the transfer matrix method outlined in Section 2.6.1. The calculation yields a complex impedance, of which the real part is the flow resistance and the imaginary part the flow reactance.

Others have also produced variants on this type of measurement.[52] Section 4.1.3 detailed a method for measuring the transmission through a sample using an impedance tube, four microphones, and two different tube terminations (see Figure 4.4). Once the transmission matrix coefficients in Equation 4.12 are known, it is possible to calculate the pressure difference across the sample by considering the case $p(z = -d) = 0$ at low frequency. Provided the sample is thin compared to wavelength, the flow impedance is given by T_{12}.

For most common porous absorbers, it is also possible to derive the flow resistivity, σ, from the asymptotic behaviour of the effective density, ρ_e, at low frequency using the following formulation[53]:

$$\sigma = -\lim_{\omega \to 0}(Im\{\rho_e\}\omega), \tag{4.45}$$

where ω is the angular frequency. Using impedance tube measurements,[54] the characteristic impedance and wavenumber of the material are measured; these can be used to calculate the effective density using Equations 6.24 and 6.25. Equation 4.45 then gives an estimation of the flow resistivity with an accuracy of typically about 15%.[53]

4.5.3 Measurement of wavenumber and characteristic impedance

Given a theoretical model for the propagation of sound in a porous absorbent and some measurements on a simple experimental set-up, it is possible to derive the wavenumber and the characteristic impedance of porous absorbers.[55–57] For example, a numerical fit can be carried out between the experimental data and the theoretical model to find values for unknown parameters in the theoretical model. Such a technique is similar to that used for *in situ* measurements discussed at the end of Section 4.4 and is also used to gain other porous parameters, such as tortuosity as discussed in Section 4.5.5. There are various arrangements that can be used for this measurement. The most convenient methods are probably those that use the impedance tube.

Smith and Parrott[55] review two possible methods, of which the two thicknesses method is most convenient and so will be described here. The surface impedance is measured for two different thicknesses of the absorbent with a rigid backing. For a thickness of d_1, the surface impedance is z_1:

$$z_1 = -jz_c \cot(kd_1), \tag{4.46}$$

where z_c is the characteristic impedance of the sample. A similar relationship gives the surface impedance $z_2 = -jz_c \cot(kd_2)$ for a depth d_2. These relationships were derived in Section 2.6.1. For simplicity, assume that $d_2 = 2d_1$; typically, d_1 would be a couple of centimetres. The equations for z_1 and z_2 can then be rearranged using trigonometric identities and solved to give the characteristic impedance z_c and wavenumber k:

$$z_c = \sqrt{z_1(2z_2 - z_1)}, \tag{4.47}$$

and

$$k = \frac{-j}{2d_2} \ln \left(\frac{1 + \sqrt{\frac{2z_2 - z_1}{z_1}}}{1 - \sqrt{\frac{2z_2 - z_1}{z_1}}} \right). \tag{4.48}$$

It is necessary to choose depths so that $z_1 \neq z_2$; otherwise, the results become unreliable. This happens when $kd_1 = kd_2 \pm n\pi$, where n is an integer. Unfortunately, there is no way to test for this before the measurements, as the wavenumber is unknown. Consequently, this must be checked for in analysing the results, and if necessary, another set of thicknesses measured.

Another approach is to exploit the transmission measurement approach described in Section 4.1.3. The transmission measurement enables the transfer matrix that describes how sound propagates through a sample to be determined. Combining Equation 4.18 with the general formulation for sound propagation through an impedance layer (Equation 2.29) yields the following:

$$\begin{bmatrix} T_{11} & T_{12} \\ T_{21} & T_{22} \end{bmatrix} = \begin{Bmatrix} \cos(kd) & j\rho c \sin(kd) \\ j\sin(kd)/\rho c & \cos(kd) \end{Bmatrix}, \tag{4.49}$$

where k is the wavenumber, ρc is the characteristic impedance, and d is the depth of the material. The material properties can be derived from Equation 4.52:

$$k = \frac{1}{d}\cos^{-1}(T_{11}),$$

$$\text{or} \tag{4.50}$$

$$k = \frac{1}{d}\sin^{-1}\left(\sqrt{-T_{12}T_{21}}\right),$$

and

$$z = \sqrt{\frac{T_{12}}{T_{21}}}. \tag{4.51}$$

Results from Song and Bolton[11] indicate that a 5-cm-thick sample is reasonable for common porous materials.

Doutres et al.[58] used a three-microphone technique to gain the wavenumber and characteristic impedance. The set-up is shown in Figure 4.17, where the third microphone is flush mounted to the rigid termination behind the porous absorber. The rigid termination forces the particle velocity at the rear to be 0, enabling Equation 2.29 to be rearranged to give the

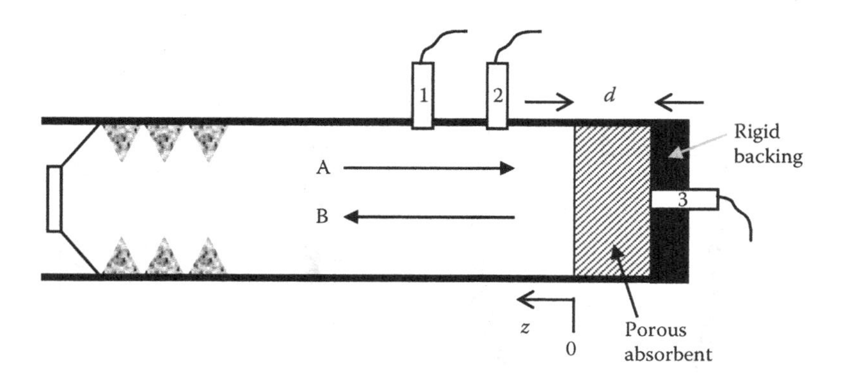

Figure 4.17 Set-up for determining wavenumber and characteristic impedance using three-microphone technique.

wavenumber, k, of the porous absorbent in terms of the ratio of the pressures between the front and back of the absorbent:

$$k = \frac{1}{-d}\cos^{-1}\left\{\frac{p(z=0)}{p(z=-d)}\right\}. \tag{4.52}$$

By considering the pressure in the tube, Equation 4.2, the ratio of the pressures can be deduced from the reflection coefficient calculated via Equation 4.4 and a transfer function between a front and rear microphone, $H_{32} = p(z_2)/p(z = -d)$:

$$\frac{p(z=0)}{p(z=-d)} = \frac{1+R}{e^{jkz_2} + \mathrm{Re}^{-jkz_2}} H_{32}. \tag{4.53}$$

The characteristic impedance of the porous absorbent, z, can then be calculated from the measured surface impedance, z_s, via a rearranged version of Equation 2.32:

$$z = jz_s \tan(kd). \tag{4.54}$$

4.5.4 Measurement of porosity

The porosity of interest here is a ratio of the pore volume involved in sound propagation to the total volume; this is the *open porosity*. While it is possible to measure the density of the sample and compare this to the density of the matrix material, this will not necessarily yield the correct porosity for sound waves because it will include the porosity of closed pores, which are relatively inaccessible to sound. For specialist absorbers, such as mineral wool, the porosity is close to 1, and so the value is often assumed rather than measured. Table 6.5 gives a table of typical porosities and Section 6.4.2 discusses the significance of this parameter in more detail.

One possible approach is to take tomographic pictures of the material structure using a scanning electron microscope or x-rays and then calculate the porosity via an image analysis.[59] Three other methods, however, are more common: one uses the isothermal compression of air and the other two use mass measurements of the sample when subjected to different pressures.

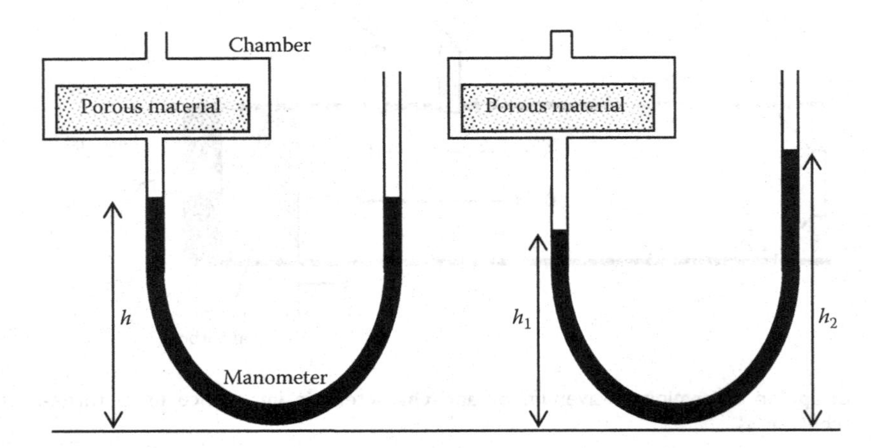

Figure 4.18 Apparatus for measuring porosity. (After Beranek, L.L., *J. Acoust. Soc. Am.*, 13, 248–260, 1942.)

Figure 4.18 shows a set-up used by Beranek[60] that exploits the isothermal compression of the air volume within and external to a porous absorber. This summary is partly taken from Cremer and Müller.[61] More recent improvements in the technique are outlined later.

There is a chamber of known volume connected to a U-shaped manometer. The material to be tested is placed in the chamber. The valve at the top of the chamber can be open or closed. With the valve open, the liquid in both legs will have the same height, h. This height is measured. The valve is then closed and the pressure in the vessel is increased by raising the right leg of the manometer. The surface of the liquids is now at h_1 and h_2 in the two legs and these heights are measured. The difference in the liquid levels in the two legs $(h_2 - h_1)$ is the increase in pressure in the sample in metres of water, ΔP. This needs to be converted to SI units:

$$\Delta P = \rho_w(h_2 - h_1)g, \tag{4.55}$$

where ρ_w is the density of the liquid in the manometer and g is the acceleration due to gravity. The height difference multiplied by the cross-sectional area of the manometer tube, S_s, is the reduction in the volume, ΔV, in the chamber:

$$\Delta V = S_s(h - h_1). \tag{4.56}$$

Assuming that this is an isothermal system, the product of the pressure and volume is constant ($PV = nRT$). This gas law is needed in a differential form for the derivation:

$$\Delta PV + \Delta VP = 0. \tag{4.57}$$

By considering the volumes of air being compressed in the chamber, both within and external to the test sample, and remembering that the porosity ϕ gives the ratio of the pore to total volume of the sample, Equation 4.57 can be expanded to

$$\Delta P(V - V_a + \varepsilon V_a) + \Delta VP_0 = 0, \tag{4.58}$$

where P_0 is atmospheric pressure, V_a is the volume of the material being tested in the chamber, and V is the volume of the chamber. Rearranging then gives the porosity:

$$\phi = \frac{P_0}{V_a}\frac{\Delta V}{\Delta P} + 1 - \frac{V}{V_a}. \tag{4.59}$$

Leclaire et al.[62] improved this method by using a reference chamber to reduce the influence of temperature and atmospheric pressure changes. Calibration needs to be done only once, and so this also reduces the number of operations in the measurement process. However, the method is still relatively slow, taking 20 minutes for calibration and 15 minutes per sample for the measurement. Porosities are accurate to about 5%.

Champoux et al.[63] produced a more elaborate apparatus, which exploited the same physical principle, but without the liquid in the manometer. A micrometre drive produces precise small changes in volume, and the pressure differences are measured, using a differential pressure transducer. As the formulations assume isothermal conditions, care must be taken to ensure that temperatures are stable. The paper discusses the various precautions needed to ensure that the system is isothermal and isolated from atmospheric pressure fluctuations, such as the use of a heat sink, thermal insulation material, and a large air reservoir. Porosities measured over a wide range of materials were accurate to better than 1%.

Panneton and Gros[64] produced a method that employs a completely different approach and uses equipment that is readily available in many laboratories. Two masses are required: the mass when the sample is in air, M_a, and the sample mass when in a vacuum, M_v. The porosity is given by

$$\phi = 1 - \frac{M_v - M_a}{\rho_0 V}, \tag{4.60}$$

where ρ_0 is the density of air and V is the sample volume. Precise measurement scales are needed because the difference in the two masses on the numerator will be a few tenths of a gram typically; however, this approach can yield porosities accurately provided that the sample volume is large enough so the mass difference is measured accurately. Salissou and Panneton[65] produced a related method that measures the mass of the sample in a vacuum and when under high-pressure conditions.

There are a variety of approaches for getting the porosity using sound waves and porous absorber models (particularly the semi-phenomenological model outlined in Section 6.5.3). For instance, the porosity can be estimated from the ratio of the atmospheric pressure to the dynamic bulk modulus at low frequency. Other approaches are given in the next section, which looks at tortuosity measurement, because the porosity is found as a by-product of the process. On the whole, these are not the best approaches because they rely on prediction models. They are, however, very useful in laboratories that lack the equipment for the measurement approaches outlined previously.

4.5.5 Measurement of tortuosity

Tortuosity, α_∞, can be defined in terms of how the effective density, ρ_e, of a porous material is changed when it is impregnated by an ideal nonviscous fluid:

$$\rho_e = \alpha_\infty \rho_0, \tag{4.61}$$

where ρ_0 is the equilibrium density. This naturally leads to a measurement method using fluid saturation. For a nonconducting porous absorber, which does not get damaged by being soaked, the tortuosity can be measured by saturating the material with a conducting fluid and measuring the electrical resistivity.[66,67] The electrical resistance of the conducting fluid is measured alone, r_f, and then the electrical resistance of the porous material is impregnated with the conducting fluid, r_a. Then the tortuosity can be found from

$$\alpha_\infty = \frac{\phi r_a}{r_f}. \tag{4.62}$$

This is the best approach for getting the tortuosity because it does not involve sound and so is independent of any acoustic prediction models.

Using ultrasonic sound waves is another possibility, and as the test frequencies are beyond the audible range, this approach has a certain amount of independence from any prediction model using the measured values. Starting from a semi-phenomenological model shown in Section 6.5.3, and considering sound propagation at high frequency, it is possible to show that the classical tortuosity, α_∞, is given by[68]

$$\alpha_\infty = \lim_{\omega\to\infty}\left(\frac{c_0}{c(\omega)}\right)^2, \tag{4.63}$$

where c_0 is the speed of sound in air and c is the wave speed in the porous material. By measuring the time of flight of short ultrasonic impulses, say at 50–100 kHz, the tortuosity can be derived. The increase in time of flight is measured when the porous material is introduced to estimate the ratio in the previous formulation. The sample must be thin (typically a few millimetres thick) because there is an assumption of negligible absorption and dispersion, and the sample must have relatively small grains, pores, or fibres (<1 mm). It can be difficult to make such thin samples that are representative of large amounts of the material. A powerful source is required because sound waves are easily attenuated at these frequencies. Allard et al.[68] found errors of 1%–8% for the ultrasonic method in comparison to the electrical resistivity measurement in plastic foams.

The porosity can also be derived from these ultrasonic measurements, but the errors are rather large. More accurate estimations of the porosity can be achieved—while simultaneously obtaining the tortuosity—by measuring the reflection of ultrasonic waves from materials at two incident angles.[69]

For porous materials with larger grains, pores, or fibres (>1 mm), the wavelength of the ultrasonic waves becomes comparable to these structural elements, and consequently, the assumptions behind the previous method breaks down. Umnova et al.[70] used lower frequencies (~3–20 kHz) produced using laser-generated sparks and larger samples, which were several centimetres thick. Both reflected and transmitted waves were measured and used to derive the tortuosity and porosity. Tortuosity is estimated from Equation 4.63 and porosity from the high frequency limit of the reflection coefficient using

$$\phi = \sqrt{\alpha_\infty}\left.\frac{1-|R(\omega)|}{1+|R(\omega)|}\right|_{\omega\to\infty}. \tag{4.64}$$

Tortuosity errors were small for three of the samples tested (<4%), and for porosity, the errors ranged from 1% to 20%.

Tortuosity can also be deduced from the real part of the characteristic impedance for mixes where the grain size and pores are large.[71] Above some critical frequency, the real part of the characteristic impedance $Re(z_c)$ approaches asymptotically to a limit value[72]:

$$\alpha_\infty = Re\left(\frac{z_c}{\rho_0 c_0}\right)^2 \phi^2 \bigg|_{\omega\to\infty}. \tag{4.65}$$

Bonfiglio and Pompoli[53] tested 10 types of porous absorbents and examined the errors from these type of analytical formulations. They found errors of 15% for porosity and 41% for tortuosity. The next section on characteristic length measurement includes a method using two gases that also yields tortuosity.

4.5.6 Measurement of characteristic lengths

There are a few methods available for gaining the viscous (Λ) and thermal (Λ') characteristic lengths. However, getting accurate results is problematic. Leclaire et al.[73] used ultrasound frequencies, two gases (air and helium), and a semi-phenomenological model similar to that given in Section 6.5.3 to measure the characteristic lengths. Using the formulations in that section, it is possible to show that the complex wavenumber in the high frequency limit is given by[74]

$$k = \omega\sqrt{\alpha_\infty \frac{\rho_0}{\gamma P_0}}\left\{1 + (1-j)\frac{\delta_v}{2}\left(\frac{1}{\Lambda} + \frac{\gamma - 1}{N_p \Lambda'}\right)\right\}, \tag{4.66}$$

where ω is the ultrasonic angular frequency, α_∞ is the tortuosity, p_0 is the density of air, γ is the ratio of the specific heat capacities, P_0 is atmospheric pressure, N_p is the Prandtl number, and δ_v is the size of the viscous boundary layer.

To get the two characteristic lengths, it is necessary to generate two different equations for the wavenumber. This can be done by measuring the system with two gases, say air and helium or, for better accuracy, air and argon.[75] Argon gives a better signal-to-noise ratio than helium does for two reasons. First, there is better coupling between the ultrasonic transducers and the gas as the characteristic impedance of argon is closer to that of air, and second, the viscous boundary layer for argon is closer to that of air. Measuring with two gases also enables the tortuosity to be found. Time of flight and attenuation of ultrasonic pulses are measured, and from this, the characteristic lengths are determined. Defining the accuracy is difficult because no easy reference values are available to compare values to, but errors of the order of 10%–20% are typical.

Panneton and Olny[54] also started with the semi-phenomenological model in Section 6.5.3 but worked within the audible frequency range. (They also examined the Wilson relaxation model of Section 6.5.4.) By rearranging the semi-phenomenological formulation for the real and imaginary parts of the effective density and assuming the porosity and flow resistivity are known from other measurements, it is possible to derive expressions for tortuosity and viscous characteristic length. The effective density is measured in the impedance tube. Because the technique works within the audible frequency range and relies on the semi-phenomenological model, the method lacks independence from prediction models. However, the formulations are such that some checks on applicability of the model are possible, which enables the reliability of the results to be tested. Simulations show that the tortuosity is accurate to within a few percentage, but the viscous characteristic length has a bias error of up to 20%. Working with equations for the bulk modulus enables the thermal characteristic length to be determined in a similar manner.[76] Bonfiglio and Pompoli[53] provide a review of the various formulations for these estimations and found very large errors for the estimation of characteristic lengths based on asymptotic formulations (>500% for the viscous characteristic length and 83% for the thermal characteristic length).

4.5.7 Inverse methods for multiple material parameters

A variety of methods for gaining different parameters have been suggested, but no single technique discussed previously yields all of them accurately. By measuring both sound reflected from and transmitted through porous material at ultrasonic frequencies, it is possible to gain all the required parameters from a single measurement.[77] However, to get both characteristic lengths, it is necessary to assume a given ratio between the two, say $\Lambda'/\Lambda = 2$.

A series of measurements of transmitted and reflected pressure are made, and then a numerical fit is carried out, using a model that predicts the sound propagation in the porous material. The prediction model could be the semi-phenomenological model outlined in Section 6.5.3 in the high-frequency limit or a relaxation model. The fit is carried out directly in the time domain. The values for the internal acoustic properties of the material—porosity, tortuosity, and characteristic lengths—are changed until the least mean square error between the measured and predicted reflected and transmitted pressures is minimized across the measurement bandwidth. This fitting process is essentially a constrained numerical optimization, for which a number of different algorithms can be used.

The quality of the results is dependent on how well the prediction model can predict the material behaviour. For instance, if the pores are too big for the prediction model used in the measurement bandwidth, then the numerical optimization will yield incorrect answers.

A similar procedure can be carried out in the audible frequency range. Again, a series of reflection and transmission measurements are undertaken, but this time in an impedance tube. Then a numerical fit is carried out by changing the values for the internal acoustic properties of the material until the mean squared error between the measured and predicted surface impedance (and maybe transmission coefficients) is minimized across the measurement bandwidth using a constrained numerical optimization. At the end of the fitting process, parameters such as flow resistivity will have been deduced.[78]

Again, this technique relies on the prediction model being able to represent the material behaviour correctly. For instance, if a frame resonance is prominent in the measurements, this is going to make the results incorrect. There is also a risk that the numerical optimization will not find the correct parameter values, either because a local minimum is found or because there is not a unique set of solutions for a particular measured impedance.

Bonfiglio and Pompoli[53] evaluated the accuracy of this method on 10 types of porous absorbers. This approach was most successful in predicting porosity and tortuosity with average errors of 1%–2% and 13%–19%, respectively. It is less successful for flow resistivity (33%–43% error) and the characteristic lengths (31%–52% error). Although error values can be large, all the parameters except flow resistivity were more accurately estimated using optimisation approaches rather than calculations based on asymptotic formulations.

4.6 SUMMARY

This chapter has reviewed the methods used to measure and characterize absorbing materials. Chapter 6 gives details about how these measured parameters are exploited in porous absorber models. The next chapter details the measurement of reflectors and diffusers.

REFERENCES

1. ISO 10534-2:1998, "Acoustics—Determination of sound absorption coefficient and impedance in impedance tubes. Part 1: Method using standing wave ratio".
2. ISO 10534-2:1998, "Acoustics—Determination of sound absorption coefficient and impedance in impedance tubes. Part 2: Transfer-function method".
3. W. Schneider, M. Leistner, F. Zickmantel, and R. Tippkemper, "Large scale impedance tubes", *Proc. Joint Congress CFA/DAGA*, 1, 469–70 (2004).
4. B. H. Song and J. S. Bolton, "Investigation of the vibrational modes of edge-constrained fibrous samples placed in a standing wave tube", *J. Acoust. Soc. Am.*, **113**, 1833–49 (2003).
5. N. Kino and T. Ueno, "Investigation of sample size effects in impedance tube measurements", *Appl. Acoust.*, **68**, 1485–93 (2007).
6. D. D. Rife and J. Vanderkooy, "Transfer function measurement with maximum-length sequences", *J. Audio Eng. Soc.*, **37**(6), 419–43 (1989).
7. A. Farina, "Simultaneous measurement of impulse response and distortion with a swept-sine technique", *Proc. 108th Convention Audio Eng. Soc.*, 18–22 (2000).
8. K. V. Horoshenkov et al., "Reproducibility experiments on measuring acoustical properties of rigid-frame porous media (round-robin tests)", *J. Acoust. Soc. Am.*, **122**(1), 345–53 (2007).
9. Y. Cho, "Least Squares Estimation of Acoustic Reflection Coefficient", PhD thesis, University of Southampton, UK (2005).
10. ASTM E2611-09, "Standard test method for measurement of normal incidence sound transmission of acoustical materials based on the transfer matrix method".
11. B. H. Song and J. S. Bolton, "A transfer-matrix approach for estimating the characteristic impedance and wave numbers of limp and rigid porous materials", *J. Acoust. Soc. Am.*, **107**(3), 1131–52 (2000).
12. G. Pispola, K. V. Horoshenkov, and F. Asdrubali, "Transmission loss measurement of consolidated granular media", *J. Acoust. Soc. Am.*, **117**(5), 2716–19 (2005).
13. J. F. Allard and N. Atalla, *Propagation of Sound in Porous Media: Modelling Sound Absorbing Materials*, John Wiley & Sons, Chichester, UK (2009).
14. Y. Salissou and R. Panneton, "A general wave decomposition formula for the measurement of normal incidence sound transmission loss in impedance tube", *J. Acoust. Soc. Am.*, **125**(4), 2083–90 (2009).
15. T. Wu, Y. W. Lam, and T. J. Cox, "Measurement of non-uniform impedance surface by the two microphone method", *Proc. 17th ICA*, Rome (2001).
16. W. Lauriksa, G. Jansensa, J. F. Allard, L. De Geeterea, and G. Vermeira, "Evaluation of free field techniques for the measurement of the surface impedance of sound absorbing materials", *Proc. 17th ICA*, Rome (2001).
17. BS EN ISO 354:2003, "Acoustics—Measurement of sound absorption in a reverberation room".
18. ISO 9613-2:1996, "Acoustics—Attenuation of sound during propagation outdoors. Part 2: General method of calculation".
19. R. H. C. Wenmaekers, C. C. J. M. Hak, and M. C. J. Hornikx, "The effective air absorption coefficient for predicting reverberation time in full octave bands", *J. Acoust. Soc. Am.*, **136**(6), 3063–71 (2014).
20. M. L. S. Vercammen, "Improving the accuracy of sound absorption measurement according to ISO 354", *Proc. ISRA* (2010).
21. J. S. Bradley, "Optimizing the decay range in room acoustics measurements using maximum length-sequence techniques", *J. Audio Eng. Soc.*, **44**(4), 266–73 (1996).
22. C. H. Jeong and J. H. Chang, "Reproducibility of the random incidence absorption coefficient converted from the Sabine absorption coefficient", *Acta Acust. Acust.*, **101**(1), 99–112 (2015).
23. C. H. Jeong, "Non-uniform sound intensity distributions when measuring absorption coefficients in reverberation chambers using a phased beam tracing", *J. Acoust. Soc. Am.*, **127**(6), 3560–8 (2010).
24. C. H. Jeong, "Converting Sabine absorption coefficients to random incidence absorption coefficients", *J. Acoust. Soc. Am.*, **133**(6), 3951–62 (2013).

25. K. Sakagami, T. Uyama, M. Morimoto, and M. Kiyama, "Prediction of the reverberation absorption coefficient of finite-size membrane absorbers", *Appl. Acoust.*, **66**, 653–68 (2005).
26. W. J. Davies, R. J. Orlowski, and Y. W. Lam, "Measuring auditorium seat absorption", *J. Acoust. Soc. Am.*, **96**, 879–88 (1994).
27. U. Kath and W. Kuhl, "Messungen zur schallabsorption von personen auf ungepolsterten stuhlen", *Acustica*, **14**, 49–55 (1964).
28. L. L. Beranek, "Audience and chair absorption in large halls. II", *J. Acoust. Soc. Am.*, **45**, 13–19 (1969).
29. R. V. Waterhouse, "Interference patterns in reverberant sound fields", *J. Acoust. Soc. Am.*, **27**, 247–58 (1955).
30. U. Kath and W. Kuhl, "Einfluss von streuflache und hallraumdimensionen auf den gemessenen schallabsorptionsgrad", *Acustica*, **11**, 50–64 (1961).
31. J. S. Bradley, "Predicting theater chair absorption from reverberation chamber measurements", *J. Acoust. Soc. Am.*, **91**, 1514–24 (1992).
32. C. Nocke and V. Mellert, "Brief review on *in situ* measurement techniques of impedance or absorption", *Proc. Forum Acusticum, Sevilla* (2002).
33. E. Brandão, A. Lenzi, and S. Paul, "A review of the *in situ* impedance and sound absorption measurement techniques", *Acta Acust. Acust.*, **101**(3), 443–63 (2015).
34. BS ISO 13472-2:2010, "Acoustics—Measurement of sound absorption properties of road surfaces *in situ*. Part 2: Spot method for reflective surfaces".
35. G. Dutilleux, T. E. Vigran, and U. R. Kristiansen, "An *in situ* transfer function technique for the assessment of the acoustic absorption of materials in buildings", *Appl. Acoust.*, **62**(5), 555–72 (2001).
36. E. Mommertz, "Angle-dependent *in situ* measurements of reflection coefficients using a subtraction technique", *Appl. Acoust.*, **46**(3), 251–64 (1995).
37. BS CEN/TS 1793-5:2003, "Road traffic noise reducing devices—Test method for determining the acoustic performance. Part 5: Intrinsic characteristics—*In situ* values of sound reflection and airborne sound insulation".
38. BS ISO 13472-1:2002, "Acoustics—Measurement of sound absorption properties of road surfaces *in situ*. Extended surface method".
39. C. Nocke, V. Mellert, T. Waters-Fuller, K. Attenborough, and K. M. Li, "Impedance deduction from broad-band point-source measurements at grazing incidence", *Acustica*, **83**(6), 1085–90 (1997).
40. P. Robinson and N. Xiang, "On the subtraction method for *in-situ* reflection and diffusion coefficient measurements", *J. Acoust. Soc. Am.*, **127**(3), EL99-104 (2010).
41. C. Nocke, "*In situ* acoustic impedance measurement using a free-field transfer function method", *Appl. Acoust.*, **59**(3), 253–64 (2000).
42. X. Di and K. E. Gilbert, "An exact Laplace transform formulation for a point source above a ground surface", *J. Acoust. Soc. Am.*, **93**(2), 714–20 (1993).
43. S. Taherzadeh and K. Attenborough, "Deduction of ground impedance from measurements of excess attenuation spectra", *J. Acoust. Soc. Am.*, **105**(3), 2039–42 (1999).
44. R. Kruse, "Application of the two-microphone method for *in situ* ground impedance measurements", *Acta Acust. Acust.*, **93**, 837–42 (2007).
45. BS EN 29053:1993, ISO 9053:1991, "Acoustics—Materials for acoustical applications. Determination of airflow resistance".
46. D. A. Bies and C. H. Hansen, *Engineering Noise Control: Theory and Practice*, 4th edn, CRC Press, Abingdon, UK (2009).
47. U. K. Ingard, *Notes on Sound Absorption Technology*, Noise Control Foundation , Poughkeepsie, NY (1994).
48. O. Umnova, K. Attenborough, E. Standley, and A. Cummings, "Behavior of rigid-porous layers at high levels of continuous acoustic excitation: Theory and experiment", *J. Acoust. Soc. Am.*, **114**(3), 1346–56 (2003).
49. Y. Auregan and M. Pachebat, "Measurement of the nonlinear behaviour of acoustical rigid porous materials", *Phys. Fluids*, **11**, 1342–5 (1999).

50. M. Garai and F. Pompoli, "A European inter-laboratory test of airflow resistivity measurements", *Acta Acust. Acust.*, **89**(3), 471–8 (2003).
51. U. K. Ingard and T. A. Dear, "Measurement of acoustic flow resistance", *J. Sound Vib.*, **103**, 567–72 (1985).
52. J. D. McIntosh, M. T. Zuroski, and R. F. Lambert, "Standing wave apparatus for measuring fundamental properties of acoustic materials in air", *J. Acoust. Soc. Am.*, **88**, 1929–38 (1990).
53. P. Bonfiglio and F. Pompoli, "Inversion problems for determining physical parameters of porous materials: Overview and comparison between different methods", *Acta Acust. Acust.*, **99**(3), 341–51 (2013).
54. R. Panneton and X. Olny, "Acoustical determination of the parameters governing viscous dissipation in porous media", *J. Acoust. Soc. Am.*, **119**(4), 2027–40 (2006).
55. C. D. Smith and T. L. Parrott, "Comparison of three methods for measuring acoustic properties of bulk materials", *J. Acoust. Soc. Am.*, **74**, 1577–82 (1983).
56. C. K. Amédin, Y. Champoux, and A. Berry, "Acoustical characterisation of absorbing porous materials through transmission measurements in the free field", *J. Acoust. Soc. Am.*, **102**(2), 1982–94 (1997).
57. J. Tran-vana, X. Olny, F. C. Sgard, and Y. Gervais, "Global inverse methods for determining the acoustical parameters of porous materials", *Proc. 17th ICA*, Rome (2001).
58. O. Doutres, Y. Salissou, N. Atalla, and R. Panneton, "Evaluation of the acoustic and non-acoustic properties of sound absorbing materials using a three-microphone impedance tube", *Appl. Acoust.*, **71**(6), 506–9 (2010).
59. W. Maysenhölder, M. Heggli, X. Zhou, T. Zhang, E. Frei, and M. Schneebeli, "Microstructure and sound absorption of snow", *Cold Reg. Sci. Technol.*, **83**, 3–12 (2012).
60. L. L. Beranek, "Acoustic impedance of porous materials", *J. Acoust. Soc. Am.*, **13**, 248–60 (1942).
61. L. Cremer and H. A. Müller, *Principles and Applications of Room Acoustics*, Applied Science Publishers (translated by T. J. Schultz), Barking, UK (1978).
62. P. Leclaire, O. Umnova, K. V. Horoshenkov, and L. Maillet, "Porosity measurement by comparison of air volumes", *Rev. Sci. Instrum.*, **74**(3), 1366–70 (2003).
63. Y. Champoux, M. R. Stinson, and G. A. Daigle, "Air-based system for the measurement of porosity", *J. Acoust. Soc. Am.*, **89**, 910–6 (1991).
64. R. Panneton and E. Gros, "A missing mass method to measure the open porosity of porous solids", *Acta Acust. Acust.*, **91**, 342–8 (2005).
65. Y. Salissou and R. Panneton, "Pressure/mass method to measure open porosity of porous solids", *Can. Acoust.*, **36**(3), 142–3 (2008).
66. R. J. S. Brown, "Connection between formation factor for electrical resistivity and fluid-solid coupling factor in Biot's equations for acoustic waves in fluid-filled porous media", *Geophysics*, **45**, 1269–75 (1980).
67. E. Sarradj, T. Lerch, and J. Hübelt, "Input parameters for the prediction of acoustical properties of open porous asphalt", *Acta Acust. Acust.*, **92**, 85–96 (2006).
68. J. F. Allard, B. Castagnede, M. Henry, and W. Lauriks, "Evaluation of tortuosity in acoustic porous materials saturated by air", *Rev. Sci. Instrum.*, **65**, 754–5 (1994).
69. Z. E. A. Fellah, S. Berger, W. Lauriks, C. Depollier, C. Arisetgui, and J. Y. Chapelon, "Measuring the porosity and the tortuosity of porous materials via reflected waves at oblique incidence", *J. Acoust. Soc. Am.*, **113**(5), 2424–33 (1983).
70. U. Umnova, K. Attenborough, H.-C. Shin, and A. Cummings, "Deduction of tortuosity and porosity from acoustic reflection and transmission measurements on thick samples of rigid-porous materials", *Appl. Acoust.*, **66**, 607–24 (2005).
71. N. N. Voronina and K. V. Horoshenkov, "A new empirical model for the acoustic properties of loose granular media", *Appl. Acoust.*, **64**, 415–32 (2003).
72. K. Attenborough, "Acoustical characteristics of rigid fibrous absorbents and granular materials", *J. Acoust. Soc. Am.*, **73**(3), 785–99 (1983).
73. P. Leclaire, L. Kelders, W. Lauriks, M. Melon, N. Brown, and B. Castagnède, "Determination of the viscous and thermal characteristic lengths of plastic foams by ultrasonic measurements in helium and air", *J. Appl. Phys.*, **80**(4), 2009–12 (1996).

74. D. Lafarge, J. F. Allard, B. Brouard, C. Verhaegen, and W. Lauriks, "Characteristic dimensions and prediction at high-frequencies of the surface impedance of porous layers", *J. Acoust. Soc. Am.*, **93**(5), 2474–8 (1993).
75. N. Kino, "Ultrasonic measurements of the two characteristic lengths in fibrous materials", *Appl. Acoust.*, **68**, 1427–38 (2007).
76. X. Olny and R. Panneton, "Acoustical determination of the parameters governing thermal dissipation in porous media", *J. Acoust. Soc. Am.*, **123**(2), 814–24 (2008).
77. Z. E. A. Fellah, F. G. Mitri, M. Fellah, E. Ogam, and C. Depollier, "Ultrasonic characterization of porous absorbing materials: Inverse problem", *J. Sound Vib.*, **302**, 746–59 (2007).
78. Z. E. A. Fellah, M. Fellah, W. Lauriks, and C. Depolier, "Direct and inverse scattering of transient acoustic waves by a slab of rigid porous materials", *J. Acoust. Soc. Am.*, **113**, 61–72 (2003).

Chapter 5

Measurement of reflections

Scattering and diffusion

The last two decades have seen a concerted effort to develop and standardize methods for measuring and characterizing the scattering from surfaces.[1] Without measurements of surface reflections, it is impossible to confidently design and apply diffusers. Without proper characterization, geometric room acoustics models produce inaccurate results, and it is impossible to write performance specifications for diffusers.

This chapter starts by contrasting the scattering and diffusion coefficients that have been enshrined in international standards. It then details measurement methods for both coefficients. This chapter concludes by describing some other techniques for characterizing surfaces that might be useful supplements to the standard methods.

5.1 DIFFUSION COEFFICIENTS VS SCATTERING COEFFICIENTS

The reflection from any given surface varies with frequency, as well as source and receiver location. To allow efficient design and also prediction in a geometric room acoustic model (GRAM), it is necessary to condense this information into a single number that sums up how a surface spatially distributes reflected sound in any one-third octave band. There are two coefficients enshrined in International Standards*:

- The scattering coefficient, s, is a ratio of energy scattered in a nonspecular manner to the total reflected energy. This was developed to be used in geometrical room acoustic modelling programs and is standardized in part 1 of ISO 17497.[2]
- The diffusion coefficient, d, is a measure of the uniformity of the reflected sound and appears in part 2 of ISO 17497.[3] This is used in diffuser design and allows acousticians to compare the effectiveness of surfaces and give performance specifications.

The difference between the definitions may appear subtle, but it is significant. Table 5.1 contrasts the two coefficients, giving an overview of how they are used and the restrictions that apply to the measurement standards.

There are two definitions because of the diverging needs of room acoustic modelling programs and diffuser designers. When sound reflects from a surface in most GRAMs, some of the energy is treated as a specular reflection, with the rest being put into an algorithm to model the diffuse reflections. Consequently, models just need a coefficient that quantifies how much of the reflection is not in the specular direction. The diffuse reflection algorithms

* The terms *scattering* and *diffusion* are used in different ways and are interchanged in different subject fields. In this book, the nomenclature that is used within the measurement standards has been adopted.

Table 5.1 Overview of standardized diffusion and scattering coefficients

	Scattering, ISO 17497-1	*Diffusion, ISO 17497-2*
Use	GRAM	Diffusion evaluation and performance specifications
Type	Random incidence in reverberation chamber	Free field method in anechoic chamber or reflection free zone
Predict coefficients?	Possible but slow	Yes and quicker
Sample requirements		
Absorption	Low, $\alpha < 0.5$	Not limited
Scale	Full or model scale, best done at model scale	Full or model scale, best done at model scale
Sample depth	Shallow	Not limited

are mostly relatively simple; see Chapter 13 for more details. In the reverberant tail of a room impulse response, there are so many reflections present, that the inaccuracies in the way the diffuse reflections from a particular surface are redistributed will nearly always be masked. Problems arise, however, when considering the first-order reflection from a surface, because incorrect modelling changes the early reflections within an impulse response, and these are perceptually much more important.

Many applications of diffusers in rooms are concerned with the first reflection from a surface; examples include overhead reflections from stage canopies or diffusers in small critical listening environments creating a reflection free zone. To evaluate the quality of a diffuser in such an application, it is necessary to know more than just how much energy is moved away from the specular reflection direction.

One example of this would be an anisotropic surface such as a column or balcony front that scatters predominately in a single plane. This is illustrated by Figure 5.1, where the scattering coefficients from ISO 17497-1 for a single plane and a hemispherical diffuser are compared. The single-plane device produces a high value for the scattering coefficient, even though it is plane and extruded in one direction. To use a simplistic analysis, even if the scattering coefficient in the plane of maximum dispersion is 1, the scattering coefficient in the extruded direction must be close to 0. Therefore, it might be expected that the hemispherical coefficient would be somewhere around a half, and yet a value of 1 is obtained.

A measurement of diffuser quality using a diffusion coefficient should determine how well a surface reflects sound in all directions, rather than just examining the ability of a device to move energy away from the specular reflection direction. It is important that a diffusion coefficient differentiates between redirection and dispersion. Figure 5.2 shows the reflection from a flat surface and a set of battens that redirects most of the reflection in another direction (in this case it happens to be back towards the source). The normalized diffusion coefficient measures the battens to have little dispersion capabilities, as would be anticipated. The scattering coefficient, however, sees the redirection of energy from the specular reflection angles as scattering and therefore gives a high coefficient for the battens, even though this is only achieving redirection, not dispersion.

Diffusers are often applied to treat first-order reflections, for example, to prevent echoes from the rear wall of concert halls. If all the diffuser achieves is redirection, there is a risk that the echo problem will simply move to another place in the hall. If the diffuser achieves spatial dispersion, however, this has the potential to reduce the echo without creating new

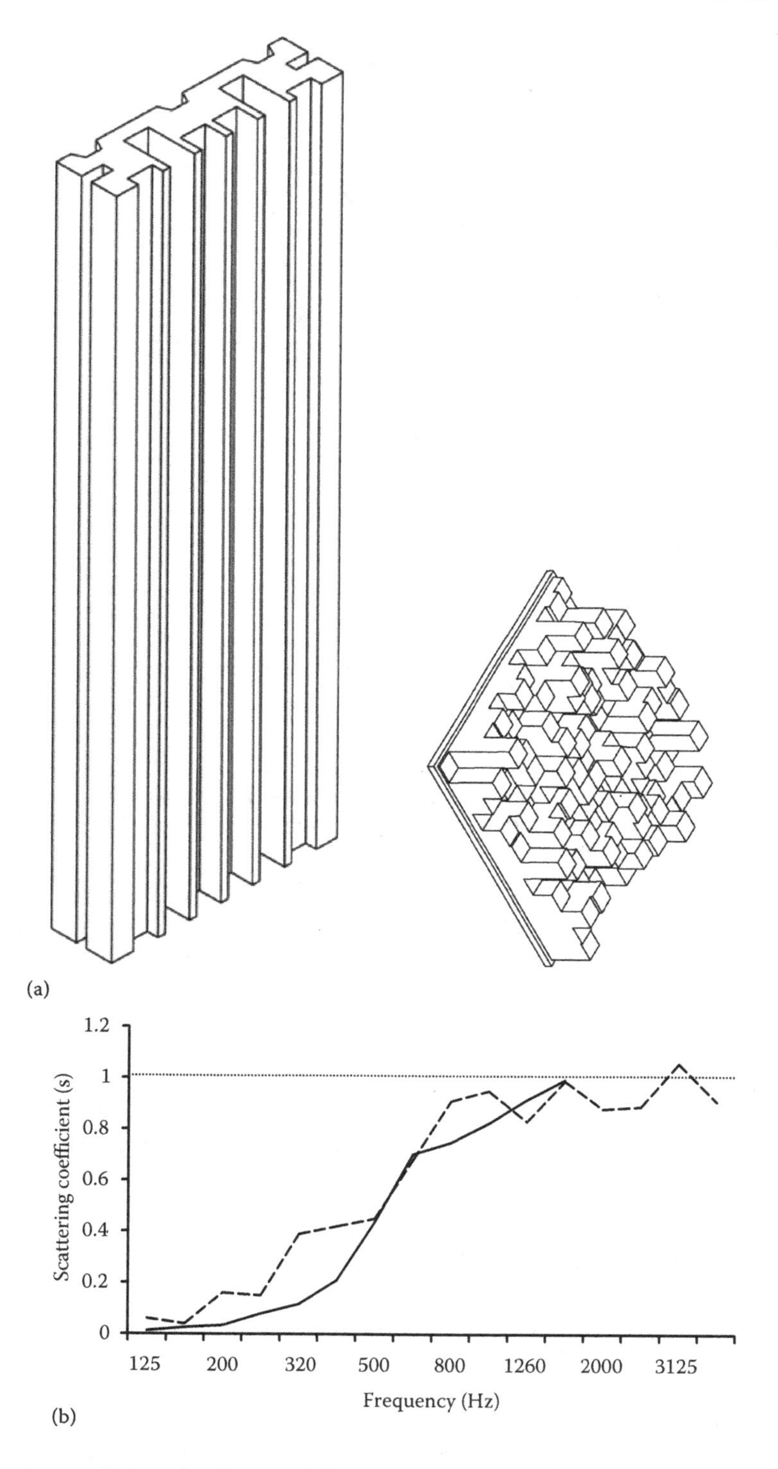

Figure 5.1 Scattering coefficients for the two different diffusers shown. (a) The left diffuser is a single-plane device (FlutterFree®); the right diffuser is a hemispherical device (Skyline®). (b) Multiple periods of each were used: - - - - single plane and —— hemispherical.

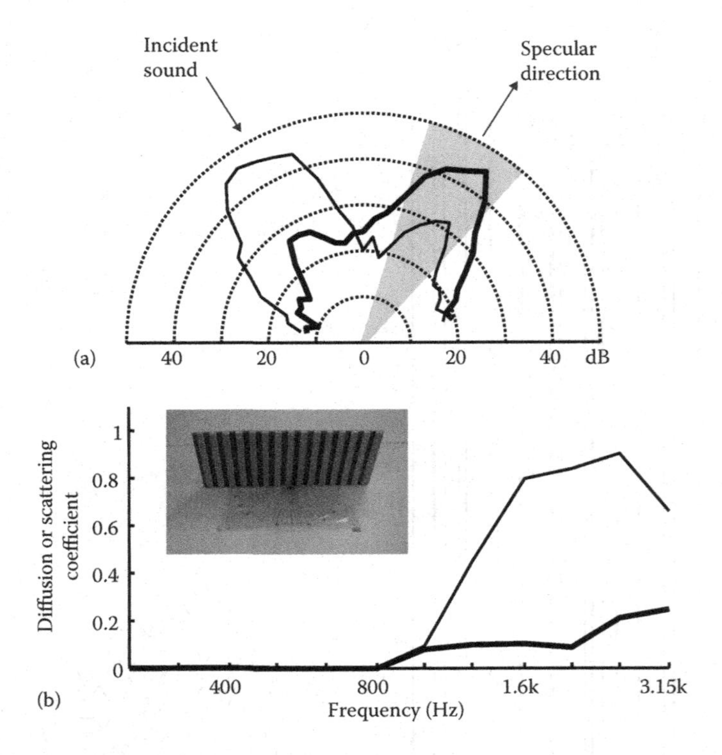

Figure 5.2 (a) The polar response for ▬▬ a plane surface and —— a set of 15 battens at 2500 Hz. (b) The diffusion and scattering coefficient frequency responses for the battens: —— correlation scattering coefficient and ▬▬ normalized diffusion coefficient. The surface made from battens is shown as an insert photo.

difficulties elsewhere. For a diffusion coefficient to be useful to designers, it must rank diffusers correctly according to quality. The scattering coefficient does not always succeed in doing this and hence why a separate coefficient was developed for quality.

In summary, the scattering coefficient gives a rough estimate of the scattering process sufficient for higher-order reflections in GRAMs. The scattering coefficient should not be used to evaluate the worth of surfaces when designing or specifying diffusers. The scattering coefficient is only concerned with how much energy is moved from the specular direction; it does not measure the quality of dispersion. For this reason, diffusing surfaces need to be evaluated using the diffusion coefficient when the quality of spatial dispersion is important. The diffusion coefficient should not, however, be blindly used in GRAMs as its definition is incompatible with current algorithms.

5.2 THE DIFFUSION COEFFICIENT

When Schroeder introduced his revolutionary diffuser designs, which are described in Chapter 10, he also introduced a possible measure for complete diffuse reflection. This was different from Lambert's law, the common redistribution law used in GRAMs. Schroeder defined optimum diffuse reflection as being when the grating lobes produced by a periodic phase grating, all have the same energy. Since the 1970s, many other types of diffusers have been produced, and the idea of measuring the similarity of the lobe energy is no longer a useful criterion, not the least because non-periodic surfaces do not have major grating lobes. Consequently, new definitions to measure the diffuseness of reflections have been developed.

The current state-of-the-art in diffuser design is numerical optimization, as described in Chapters 10 and 11. Using diffusion coefficients in a numerical optimization has enabled designs to move away from the rigid geometric constructs imposed by phase grating diffusers. This has enabled designs in which both acoustic and visual requirements can be considered and their conflicting requirements resolved. It is now possible to make diffusers that blend in with architectural forms rather than appearing as add-ons, and this is important for acoustic treatments to be acceptable to architects. Enabling diffuser optimization was one of the main drivers behind the development of a diffusion coefficient.

When a designer requires absorbing surfaces in a space, a performance specification in terms of the absorption coefficient is used to ensure quality and compliance with design requirements. Standardization of the diffusion coefficient allows similar specifications for diffuse reflection. Without standardization, the industry is vulnerable to published performance data that have no basis in fact and diffusers that do not perform as intended.

The evaluation criteria developed can also monitor the reflections of accidental diffusers. Surface diffusion is often applied in a haphazard fashion because there is a poor understanding of when and where to apply diffusers. For instance, discussions with consultants produce contradictory examples where it is claimed that too much or too little surface diffusion has resulted in acoustic aberrations. A priori to developing a better understanding of where diffusers are needed is an index to measure their quality. This measure is provided by the diffusion coefficient.

5.2.1 Measuring polar responses

The measurement of free field polar responses of the sound scattered from a surface is the first step to getting the diffusion coefficient. Polar responses are also valuable in their own right, not the least because they provide evidence that proper measurements have been carried out. The polar response of a loudspeaker can be determined by measuring the sound energy distribution on a semicircle or hemisphere surrounding the source. With care, this concept can be translated from loudspeaker measurements and used for the backscattering from architectural surfaces. While polar responses tell designers much about how a surface reflects sound, they contain a considerable amount of data and a different polar response is required for each frequency band and angle of incidence. This is one of the reasons why a diffusion coefficient has been developed, because it condenses the polar response data.

In order to characterize a diffuser's performance, it is necessary to measure or predict how the surface reflects sound. (Chapter 9 details prediction methods.) This is most often done by looking at how the scattered energy is spatially distributed. This spatial distribution for a given angle of incidence is conventionally described by polar responses in one-third octaves. An ideal diffuser produces a polar response that is invariant to the angle of incidence, the angle of observation, and the frequency (within its operational bandwidth). Figure 5.3 shows a measured polar response for normal incidence, at 2 kHz, from an array of 2D number theoretic diffusers, one of which is also shown.

A source is used to irradiate the test surface while measurement microphones at radial positions in front of the surface record the pressure impulse response. The microphone positions usually map out a semicircle. Various methods can be used to get the impulse responses. The most common uses a maximum length sequence (MLS) signal. Other possibilities include swept sine waves and impulses. As time variance and non-linearity are usually not an issue, MLS signals are the most efficient to use. Once the impulse responses have been measured, time gating is used to separate the reflections from the direct sound.

Polar response measurements can be made in a single plane using a 2D goniometer on a semicircle,[4] as shown in Figure 5.4, or over a number of different planes using a 3D

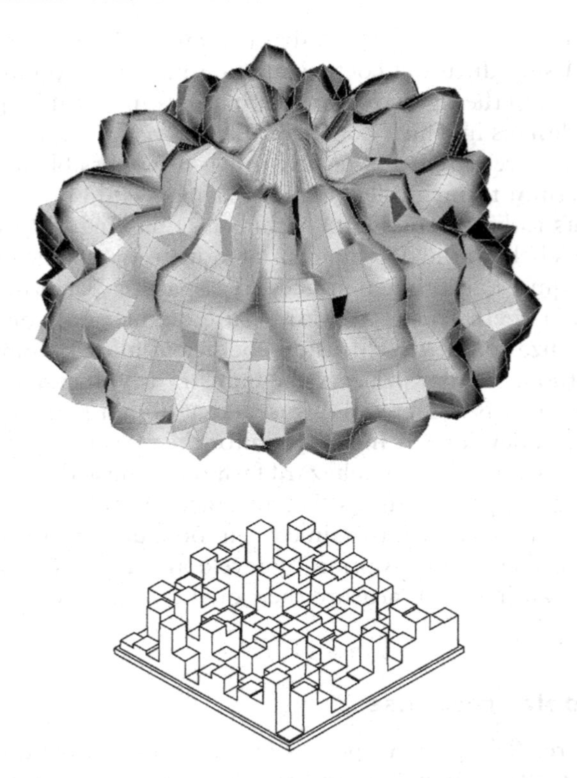

Figure 5.3 Three-dimensional polar balloon measured from a Skyline diffuser, which is shown below the polar response.

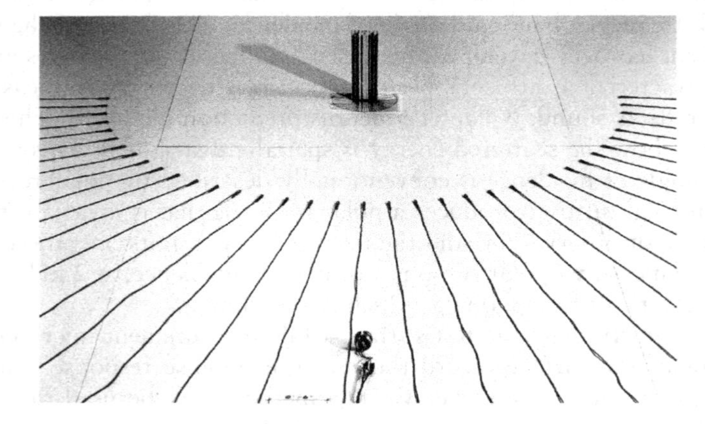

Figure 5.4 A system for measuring the scattering from a surface in a plane using a boundary plane measurement. The diffuser (a Schroeder diffuser) is shown top-middle. There are 37 microphones arranged on an arc. The source loudspeaker is in the bottom middle of the picture. (Reproduced from AES-4id-2001, *J. Audio Eng. Soc.*, 49, 149–165, 2001, with permission from the Audio Engineering Society.)

goniometer, as shown in Figure 5.5.[5] The choice of single-plane or multiplane measurement depends on the type of diffuser and the fact that a hemispherical measurement system is difficult and expensive to construct. Figure 5.6a illustrates that an extruded diffuser (known as a single-plane or 1D diffuser) produces scattering in one plane, and consequently, a single-plane evaluation is appropriate.[3,6] If the surface produces scattering in multiple planes, as shown in Figure 5.6b, then a hemispherical evaluation is done by carrying out a couple of single-plane measurements in orthogonal directions. Figures 5.4 and 5.5 show 1:5 scale

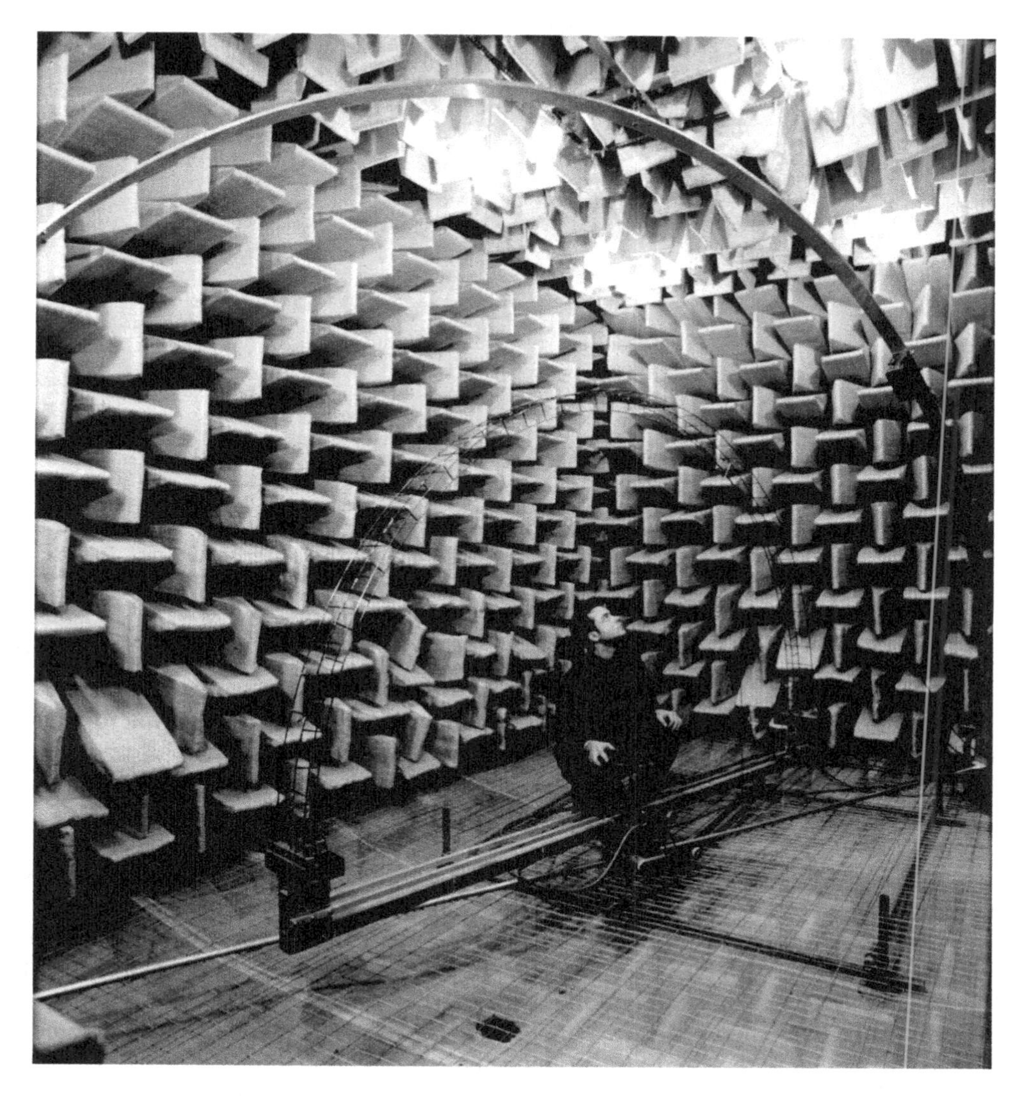

Figure 5.5 A system for measuring the scattering from a surface over a hemisphere. The diffuser being tested is the small pyramid in the centre. The source arc is most obvious; the receiver arc is acoustically transparent and is thus more difficult to see. (Reproduced from AES-4id-2001, *J. Audio Eng. Soc.*, 49, 149–165, 2001, with permission from the Audio Engineering Society.)

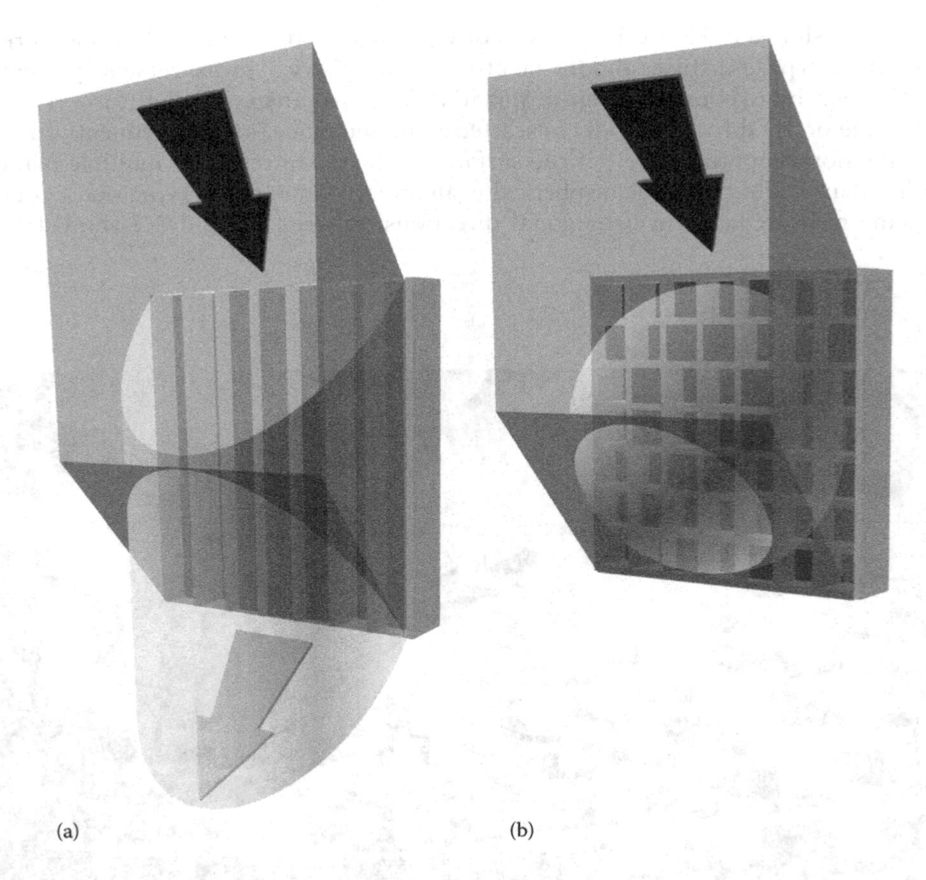

Figure 5.6 Reflection from a single-plane (a) and hemispherical diffuser (b). (After D'Antonio., P., Cox, T.J., *Appl. Acoust.*, 60, 113–142, 2000.)

measurement systems. Scale model measurements are more practical because otherwise the source and receiver radii become too large (see Figure 5.8 and its description).

In Figure 5.4, a fixed microphone array is used. In Figure 5.5, a single microphone is moved on a lightweight scaffolding across the surface of a hemisphere. It is also possible to use a boom arm to rotate a microphone on a single arc, as shown in Figure 5.7.

The single-plane measurement can be made in an anechoic chamber,[7] but it is also possible to use a boundary layer technique[4]; this latter technique is shown in Figure 5.4. This can be done in a large room provided that the ceiling and walls are sufficiently far away from the test set-up to allow parasitic room reflections to be removed by time gating. The sample is shown in the top middle, along with 37 pressure zone microphones arranged on a semicircle 1 m from the sample at 5° intervals. The source is at normal incidence, with the loudspeaker located at the bottom middle 2 m from the sample.

The measurement geometry is shown in plan view in Figure 5.8, and from this geometry, it is possible to calculate the size of the non-anechoic room needed for the measurement. In this case, the loudspeaker (L), microphone (M), and diffuser (D) are placed on a flat, hard surface on the floor. The microphone radius is denoted by R and the loudspeaker radius is $2R$. Figure 5.8 illustrates the ellipsoidal area (dashed) necessary to make a reflection free zone measurement. Consider the respective sound paths of the direct sound (LM = R), the

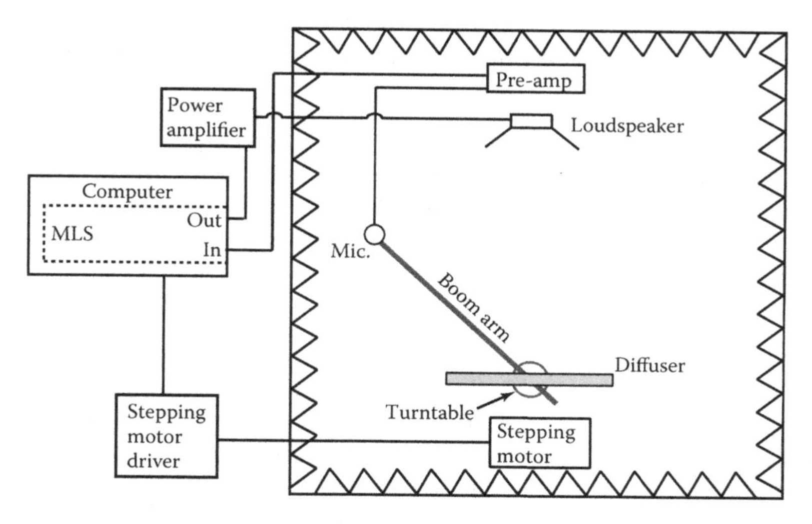

Figure 5.7 A schematic of a measurement system.

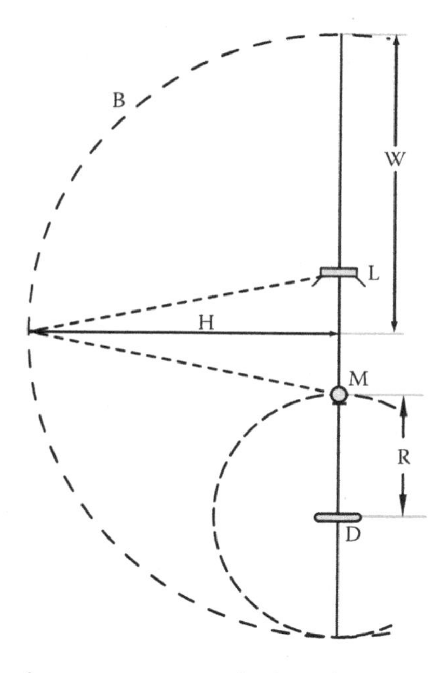

Figure 5.8 Plan view of reflection free zone geometry for boundary measurement technique. Loudspeaker (L), microphone (M), diffuser (D), microphone radius (R), room boundary (B), and ellipse axes (H) and (W).

scattered sound (LDM = $3R$), and the limiting reflection from the second-order reflection from the speaker (LDLM = $5R$). Therefore, the reflection free zone, which measures only scattering from the sample, is $4R$. It is possible to determine the ellipsoidal area from the measurement geometry. If the limiting path is $5R$, then the total path travelled from the loudspeaker to the room boundary, B, and back to the microphone is also equal to $5R$ = LBM (LB + BM). The minor axis of the ellipsoid, H, equals $2.45R$ and the major axis, W, equals $2.5R$. If $R = 5$ m, then this requires a room 12.2 m high by 24.4 m wide by 25 m long. Since this is an unreasonably large room, measurements have been done at 1:5 scale in which $R = 1$ m.

In the boundary layer measurement, the floor acts as a mirror image, and what is measured is effectively the diffuser paired with a mirror image of itself, as illustrated in Figure 5.9, where a section is shown. In Figure 5.9a and b left column, a diffuser is shown in a real test condition and its image source equivalent. The diffuser is extruded in the vertical direction, and so it can be seen that what is effectively measured is a sample twice as high, with the source and receiver at the midpoint. In Figure 5.9a and b right column, a nonextruded shape (arc) is shown in the real test configuration and its image source equivalent. For this sample, the actual sample and its mirror image is being measured.

As the source and receivers are not located exactly on the boundary, there is an upper frequency limit for this measurement. The longest and shortest propagation paths must not differ by more than half a wavelength. In terms of the geometry shown in Figure 5.9, the path difference $|r_1 - r_2 - r_3| \ll \lambda/2$, where r_1, r_2, and r_3 are defined in the figure.

A potential problem with this measurement process is that the microphone(s) will get in the way of the sound propagating from the source to the panel. For this reason, the fixed microphone array or the boom arm used to move a single microphone must be small enough not to cause significant reflections or disturbance to the sound field. Where possible, supports should be located not in a direct line between source and diffuser and be covered in absorbent material or be acoustically transparent. Fixed microphone arrays have been constructed using small pressure zone microphones. This potential interference by the

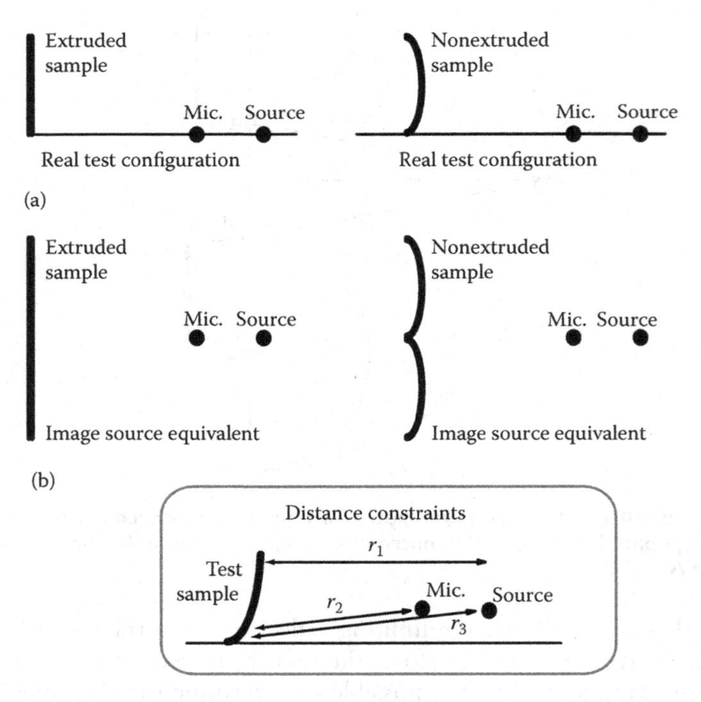

Figure 5.9 (a) Boundary plane measurement and (b) the equivalent image source configuration. Left column: extruded sample; right column: nonextruded surface. The insert shows distances used for calculating the maximum frequency.

microphones on the sound field makes diffuser measurements more awkward than measuring the polar responses from loudspeakers.

The single-plane measurement is quick and easy to carry out. In contrast, measurements with a hemispherical goniometer require considerably more complex engineering to achieve the acoustically transparent microphone positioning in an anechoic chamber. Hemispherical measurements are also more time-consuming due to the increase in the number of microphone positions required. A spatial resolution of 5° in azimuth and elevation requires 1296 measurement positions for every angle of incidence. (5° was chosen because tests have shown that this is a sufficient resolution to gain the polar response accurately without overburdening measurements.) Consequently, for hemispherical evaluation, it is much easier to use the approach suggested by Farina,[8] which is to measure two semicircular arcs rather than a whole hemisphere. Incidentally, if symmetry in the sample exists, and the source lies in the plane of symmetry, it is possible to reduce the number of measurements.

Figure 5.10 illustrates the sequence of events to determine the scattered impulse response at a particular observation angle, for a given angle of incidence. As an illustration, consider

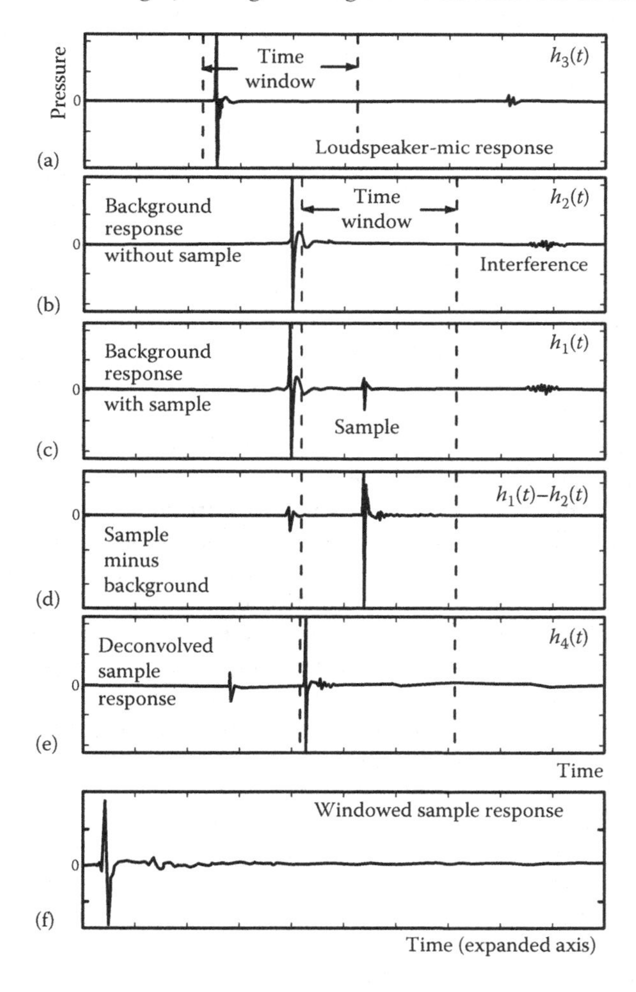

Figure 5.10 (a)–(f) Data reduction process to extract the scattered impulse response from a test sample at a given observation angle.

the measurement in a single plane illustrated in Figure 5.4. To obtain the impulse response of a sample under test, it is necessary to deconvolve the loudspeaker/microphone response at each scattering angle, $h_3(t)$. It is also necessary to minimize any room interference and reflections from microphone supports and wires within the time window of interest. To obtain the loudspeaker mic response (Figure 5.10a) at each scattering angle, the loudspeaker is placed at the sample position and rotated so its on-axis response is coincident with the on-axis response of each microphone for each angle.

The loudspeaker is then placed in its normal source position, without any sample present, and the background response without sample, $h_2(t)$, at each angle is measured using simultaneous acquisition. A vertical dotted line representing the time window of 10 ms used to isolate the reflections is also shown. The sample under test is then placed in position and the scattered sound is measured, obtaining the background response with sample, $h_1(t)$.

Since the microphones are stationary and the measurement process is rapid, the respective background response can be subtracted from each microphone position, prior to deconvolution. This is illustrated as *sample minus background* in Figure 5.10d. The direct sound is significantly decreased and is negligible in the time window encompassing the scattered sound. The room interference is also decreased. The loudspeaker/microphone response can now be deconvolved as illustrated in deconvolved sample response, $h_4(t)$, which is calculated using

$$h_4(t) = IFT\left\{\frac{FT[h_1(t) - h_2(t)]}{FT[h_3(t)]}\right\}, \tag{5.1}$$

where FT and IFT are the forward and inverse Fourier transforms, respectively. The data within the time window are gated to isolate the windowed sample response.

This process assumes that the system remains time invariant for the with and without sample measurements. As the speed of sound changes with temperature, this can create problems between the two sets of measurements, especially for full-scale measurements. One way of mitigating this difficulty is to use modern instrumentation that is capable of making multimicrophone measurements simultaneously and therefore decrease the time between the with and without measurements.

If the speed of sound does change, it is possible to shift one of the measurements in time to align the direct sound peaks and so achieve better subtraction. To achieve this accurately, it is first necessary to oversample the impulse response to allow more precise alignment in time. Robinson and Xiang[9] used a cross-correlation calculation and trial and error to work out the best oversampling rate and shift to minimize the residual of the direct sound after subtraction. Hughes[10] proposed using the microphones close to the source to gauge the correction needed for the most critical receivers at grazing angles, where subtraction is most problematic because the direct and reflected sound arrive almost simultaneously.

The data are further postprocessed to provide frequency responses, polar responses, and finally diffusion coefficients, as shown in Figure 5.11.

At the top, Figure 5.11a shows the 2D boundary measurement geometry with the exciting loudspeaker at an angle of incidence of −60° with respect to the normal. Also shown are the 37 receiving microphones. A flat non-absorbing sample is being measured. Below that, Figure 5.11b shows the impulse response for the microphone at 0°, with the scattered data outlined in a box, corresponding to the time window in Figure 5.10. The scattered data are windowed for all of the angles of observation and concatenated in Figure 5.11c. A Fourier transform is then applied to each of the impulse responses to get the frequency responses, shown in Figure 5.11d. Five of the 37 frequency responses are only shown for clarity. The frequency response energy is summed over one-third octave bands and three of the polar responses are shown in

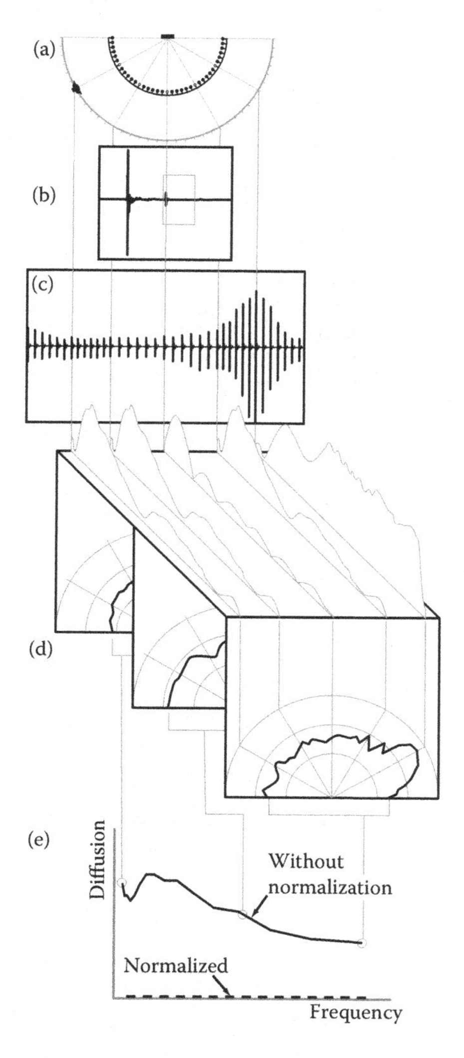

Figure 5.11 Data processing from a flat reflector at −60° incidence. (a) Goniometer; (b) impulse response for mic at 0°; (c) windowed and concatenated scattered data; (d) five of the scattered frequency responses and three of the 1/3rd octave polar responses; and (e) unnormalized and normalized diffusion coefficient.

Figure 5.11d. The visible polar response at high frequency is narrow and directed in the specular direction of +60°, as would be expected. The polar responses can then be further processed to give a diffusion coefficient, which is plotted versus frequency in Figure 5.11e. Without normalization, the diffusion coefficient decreases as the frequency increases. This happens because the width of the panel becomes increasingly large compared to the wavelength and the reflection becomes more specular. To remove these finite-panel effects, a normalized diffusion coefficient is used, and because a flat reflector is being measured, this is 0 for all frequencies.

In Figure 5.12, the same procedure is shown for a diffusing sample.[5] The polar responses are more semicircular and the diffusion coefficient without normalization is closer to 0.6, the maximum value for complete diffuse reflection on real surfaces. By normalizing the diffusion coefficient with the flat reflector value, the frequency at which surface scattering becomes important is clearer and the results are easier to interpret. Normalization also

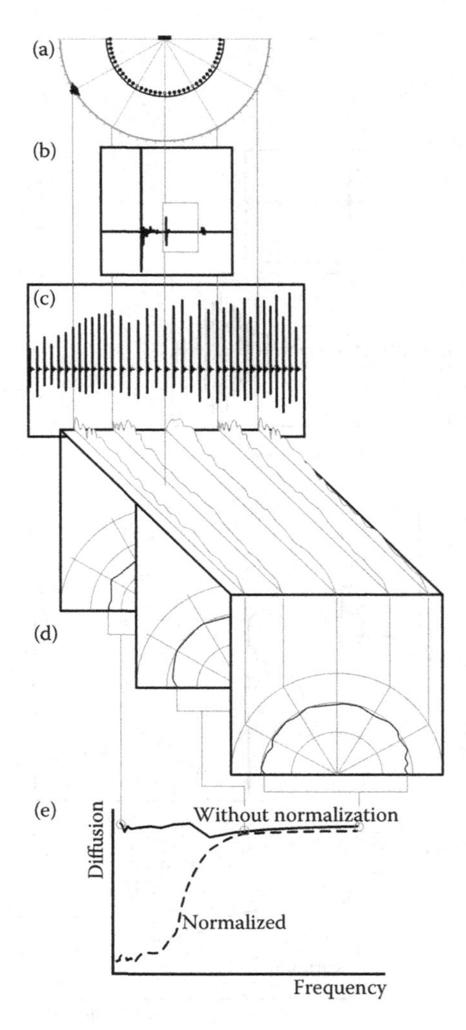

Figure 5.12 (a)–(e) Summary of data processing technique from a diffuser at −60° incidence. (Reproduced from AES-4id-2001, *J. Audio Eng. Soc.*, 49, 149–165, 2001, with permission from the Audio Engineering Society.)

removes edge diffraction effects, so that the normalized diffusion coefficient represents the uniformity of reflection from the surface topology only.

In Figure 5.13, a presentation format for a sample of three semicylinders is shown. The top row shows a photo of the three semicylinders on the left, the diffusion coefficient of the test sample compared to a reference flat reflector in the middle, and the normalized diffusion coefficient on the right. The remaining images are the one-third octave polar responses of the test sample compared with a reference flat reflector. Round-robin testing on a reference sample has been shown to be effective in comparing the results from different laboratories. Three semicylinders, which can easily be fabricated from PVC pipe as a test sample, can be used to calibrate different goniometers.

Figure 5.14 compares the measured and predicted polar responses from a surface measured both on a single plane and a hemisphere.[11] The agreement between theory and measurement is good; this is a measurement process that yields accurate results.

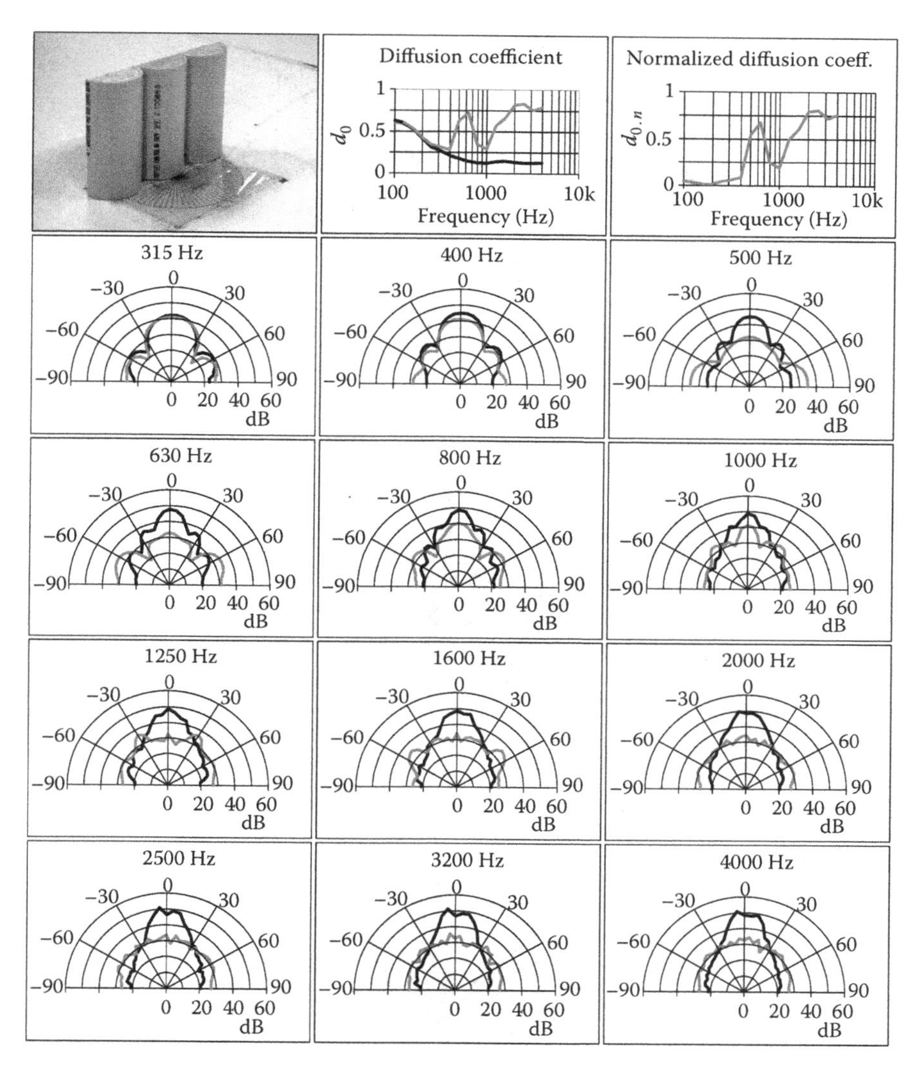

Figure 5.13 Presentation format for diffusing surfaces at normal incidence: —— diffuser and —— reference reflector.

5.2.1.1 Near and far fields

All free field measurements suffer from the problem that the relative levels within the polar response are dependent on the distances from the source and receiver to the surface, unless the source and receivers are in the far field. (The *far field* being where the scattered pressure falls by 6 dB per distance doubling for 3D geometries and 3 dB per distance doubling in 2D geometries.) Unfortunately, in most room applications, it is usual for sources and receivers to be in the near rather than the far field, unless the test surface is small. Figure 5.15 shows the scattering from a plane surface for a variety of receiver distances.[11] As the receiver approaches the surface, the scattered pressure is more evenly distributed over the polar response. A plane surface appears to be a very good diffuser when measurements are made close to the surface. In fact, close enough to the surface, the reflection is provided by an approximate image source that radiates the same energy to all receivers except for minor

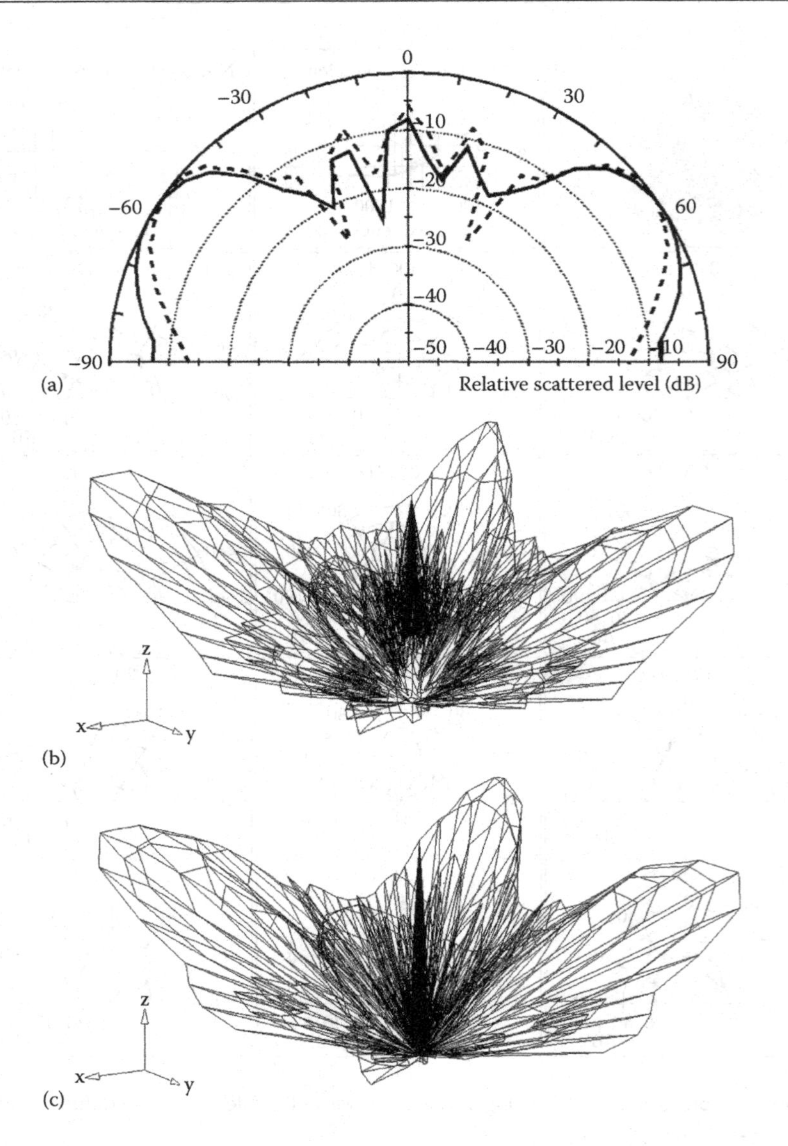

Figure 5.14 Comparison of measured and predicted polar responses for a square-based pyramid and normal incidence. (a) —— measurement (2D) using boundary plane and - - - - single-plane BEM prediction (1 kHz). (b) 3D measurements (2 kHz). (c) 3D BEM prediction (2 kHz). (After Hargreaves, T.J., Cox, T.J., Lam, Y.W., D'Antonio, P., *J. Acoust. Soc. Am.*, 108, 1710–1720, 2000.)

deviations due to spherical spreading and path length differences. This seemingly contradicts conventional wisdom in room design that a plane surface is a poor diffuser.

To understand this contradiction, it is necessary to review why plane surfaces can cause colouration and echoes. Problems occur with directional sources such as trumpets, for example. In this case, the reflected energy will be concentrated over a narrow solid angle, leading to a risk of detrimental effects such as echoes, coloration, or image shift for receivers within this solid angle. The results shown in Figure 5.15 were produced using an omnidirectional source. Furthermore, the plane surface does not spread out the reflected impulse response in time, and as discussed in Chapter 11, this is another important reason why it is not a diffuser.

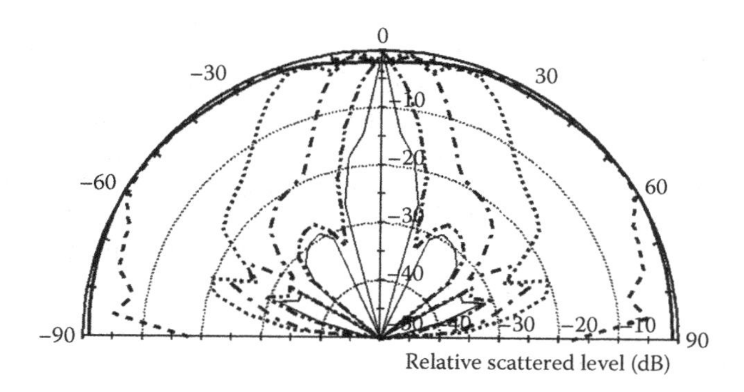

Figure 5.15 Effect of receiver arc radius on the polar response of a 1 m square plane panel. Single-plane BEM prediction, 5 kHz, normal incidence, source distance = 100 m. Receiver distances: ▬▬ 0.1, – – – 0.5, ········· 1, -·-·-· 2, —··— 5, —— 100 m. (After Hargreaves, T.J., Cox, T.J., Lam, Y.W., D'Antonio, P., *J. Acoust. Soc. Am.*, 108, 1710–1720, 2000.)

The solution usually adopted is for scattering measurements based on polar distributions to be taken in the far field, even when this is further than any real listeners would ever be. Then some receivers will be outside the specular zone (which is defined in Figure 5.16)[11], and it is possible to measure the energy dispersed from receivers in the specular zone to those outside the zone. In the spatial domain and in the far field, the effect of a diffuser should be to move energy from the specular zone to other positions. So, unless receivers are placed both outside and within the specular zone, measuring energy levels alone will not detect the effects of diffuse reflections.

There are standard formulations for approximately calculating the required distance for measurements to be in the far field.[12] These criteria apply to both sources and receivers, but for now, only receivers are considered to simplify explanations—the source will be assumed to be at infinity. There are two criteria to satisfy: the receiver radius should be large compared to wavelength and the differences between path lengths from points on the surface to

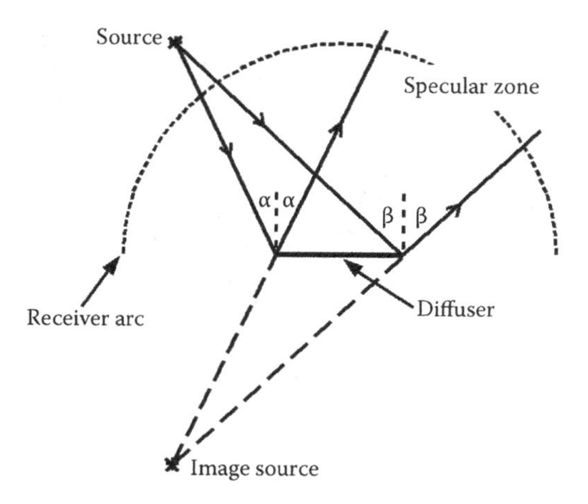

Figure 5.16 Definition of the specular zone—the region over which a geometric reflection occurs. (Although the specular zone is strictly a high-frequency construction, practice has shown it to be a useful concept for the geometries and frequencies typically used in diffuser design.) (After Hargreaves, T.J., Cox, T.J., Lam, Y.W., D'Antonio, P., *J. Acoust. Soc. Am.*, 108, 1710–1720, 2000.)

the receiver should be small compared to wavelength. With the geometries and frequencies used for measuring diffuser scattering, it is the latter criterion that is most exacting.

Unfortunately, the common far field formulations are not applicable to the case of oblique receivers, where significant destructive interference occurs. Problems arise because the amount of destructive interference is very sensitive to the relative magnitudes of the waves coming from the secondary sources on the scattering surface (assuming the scattering is modelled following Huygen's principle). Consequently, the receiver distance required to achieve the true far field for oblique receivers is often so large that measurements cannot be accommodated in normal test facilities. Figure 5.17 shows the scattering from a surface as a function of distance; it takes a receiver distance of hundreds of metres to reach a completely stable far field polar response.[11]

Fortunately, a pragmatic approach may be taken, as knowing the minima in a polar response to very exact detail is not necessary. This is particularly true if a diffusion coefficient is going to be evaluated, as this involves reducing the many scattered pressure values in a polar response to a single figure of merit. Consequently, errors from the slight misrepresentation of notches in the polar response will have negligible effect on the diffusion coefficient value. The situation is also less critical when one-third octave bandwidths are used, as is normal practice. So, the true far field does not have to be obtained. It is sufficient to ensure that the majority of receivers are outside the specular zone so that the diffuser's ability to move energy out of the specular zone can be measured. Then a reasonable approximation to the far field polar response can be obtained.

ISO 17497-2 recommends that 80% of receivers are outside the specular zone, ideally in revisions of the standard, this figure, should be referenced to Fresnel zones.[13] In Figure 5.18, the diffusion coefficients for two surfaces, as a function of receiver distance, are shown.[11] The point where 80% of the receivers are outside the specular zone is shown. The plane panel example is one of the worst-case scenarios and the error introduced into the diffusion coefficient is only 0.1. Furthermore, this is a single frequency prediction. Once summing across one-third octave bands is used, this error approximately halves. Consequently, a reasonable approximation to the true far field polar responses and diffusion coefficients can be obtained.

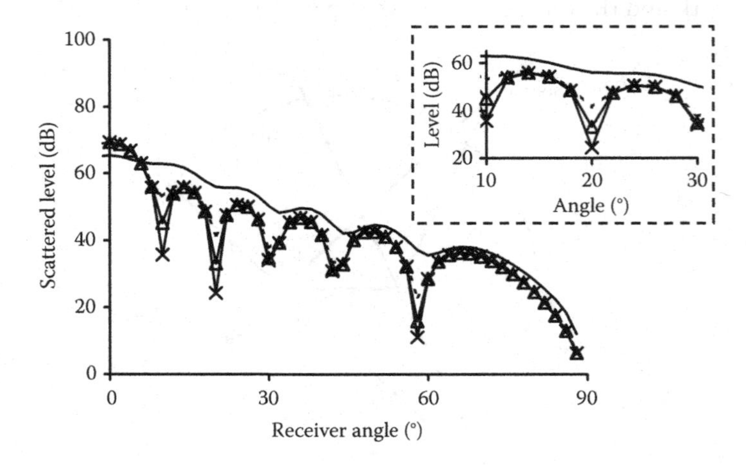

Figure 5.17 Variation of the scattered polar response with receiver distance to illustrate the extent of near field. Receiver angle on a linear scale for clarity; insert graph is an enlargement of a section of the main graph. 1 m plane surface at 1 kHz using 2D BEM predictions. A distance correction of $1/\sqrt{r}$ has been used to correct for cylindrical wave spreading. Receiver distances: —— 2.94, ---- 12, ▲ 32, and ✱ 100 m. (After Hargreaves, T.J., Cox, T.J., Lam, Y.W., D'Antonio, P., *J. Acoust. Soc. Am.*, 108, 1710–1720, 2000.)

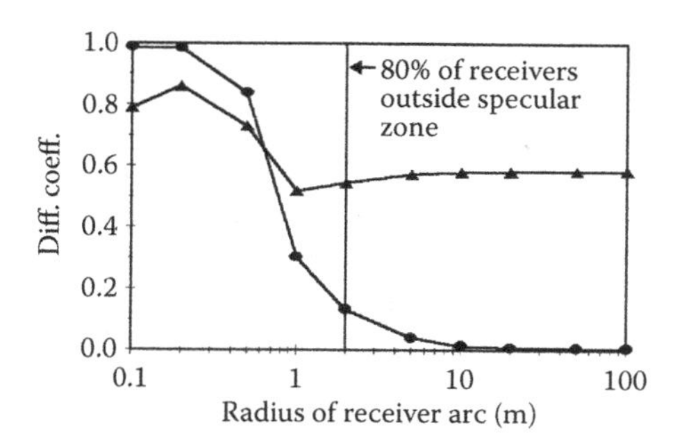

Figure 5.18 Effect of receiver arc radius on the diffusion coefficient. Single-plane BEM predictions, normal incidence, source distance = 100 m. -●- 1 m wide plane panel, 5 kHz; -▲- 1 m wide random binary panel, 400 Hz. (After Hargreaves, T.J., Cox, T.J., Lam, Y.W., D'Antonio, P., *J. Acoust. Soc. Am.*, 108, 1710–1720, 2000.)

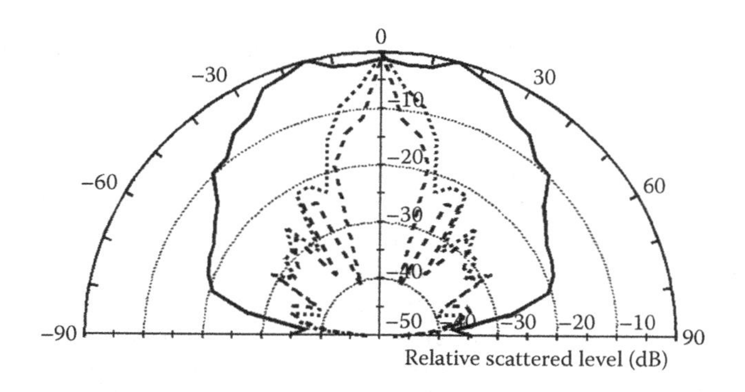

Figure 5.19 Effect of receiver arc radius on the polar response of a concave arc. Single-plane BEM predictions, 2 kHz, normal incidence, source distance = 10 m. —— near field, – – – focal distance, and - - - - far field. (After Hargreaves, T.J., Cox, T.J., Lam, Y.W., D'Antonio, P., *J. Acoust. Soc. Am.*, 108, 1710–1720, 2000.)

For some surfaces, it is not sufficient to just measure in the far field. For concave reflectors and others that might have significant focussing closer to the surface, it is necessary to monitor in the near, as well as the far field. This is illustrated in Figure 5.19, where the scattering from a concave surface is shown as a function of distance.[11] It can be seen that receivers very close to the surface detect a good diffuser, but a little further out, the reflected sound is highly focussed. In the far field, some diffusion is created. In summary, a pragmatic approach requires receivers to be both inside and outside the specular zone and measurements at application realistic distances are also needed to check for focussing with concave surfaces.

5.2.1.2 Sample considerations

It is important to test a sample that is representative of the entire structure to be applied in real applications. For instance, one period of a Schroeder diffuser should not be tested if the intention is to apply the surface periodically because the scattering from a periodic and single diffuser will be very different. Where the whole sample cannot be tested because of geometric constraints on source and receiver distances, the following techniques are suggested in ISO 17497-2[3]: for a periodic sample, at least four complete repeat sequences should be included so

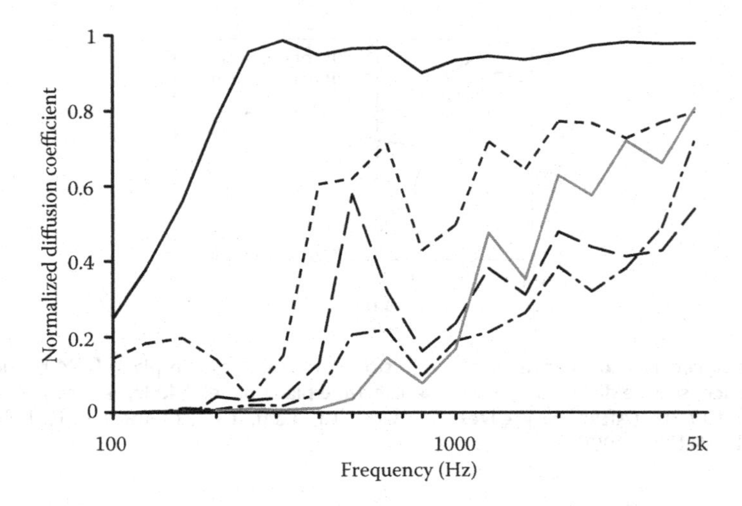

Figure 5.20 The normalized diffusion coefficient for various sets of semicylinders. The number of cylinders in each set is —— 1, - - - - 2, — — — 4, -·-·-·- 6, and —— 12.

the effects of lobing from repetition is measured (although the width of the diffraction lobes depends somewhat on the number of repeat units in the sample[14]). Figure 5.20 shows the normalized diffusion coefficient from five different sets of semicylinders. The plots show that one cylinder is not representative of the scattering from an array of semicylinders.

For random surfaces, representative samples of the surface roughness should be tested, large enough so that surface effects, rather than edge diffraction, are more prominent in the polar responses. Sometimes, it can be useful to change the source and receiver radii as this allows wider test samples to be measured while still maintaining 80% of receivers outside the specular zone.

Some have said that the diffusion coefficient method will not work for large surfaces with small surface roughness and that it is intended to be used for single diffusers only. This is not true, and the evaluation method can conceptually be used on any sized surface. The problem is that when the surface becomes large, the measurement becomes impractical because it is impossible to get far enough away from the surface. In this case, the evaluation can still be done, but only with the use of prediction or scale models.

When scale models are used, it is important that a representative test sample is constructed. The absorption properties should be the same for the full scale surface at full scale frequencies and the test surface at the equivalent model scale frequency. When considering absorption from samples, losses due to viscous boundary layer effects should be included. This consideration can limit the usable model scales because viscous boundary layer effects do not scale in the same way as physical dimensions.

3D printers mean that non-absorbing scale samples can easily be fabricated. A photo of samples made using a rapid prototyping system is shown in Figure 5.21. The software renders a 3D CAD file into thin layers. The printer lays down a thin layer of powder and a print head with resin rasterizes this layer and successive layers, depositing resin where required.

5.2.2 Calculating the diffusion coefficient

The diffusion coefficient is a frequency-dependent figure of merit derived from the polar distribution. This is evaluated in one-third octave bandwidths, which has the advantage

Figure 5.21 Samples generated using a 3D printer.

of smoothing out some of the local variations in the polar responses, so the diffusion coefficient is based more on the overall envelope. There have been various statistical operations suggested to calculate a diffusion coefficient: standard deviation,[15–17] directivity,[18,19] specular zone levels[20,21] and spherical harmonics,[22] percentiles, and autocorrelation.[11] In any such data reduction, there is a risk of losing essential detail. It has been shown that the autocorrelation coefficient offers significant advantages over other statistical techniques.

The autocorrelation function is commonly used to measure the similarity between a signal and a delayed version of itself, looking for self-similarity in time. It is also possible to use the autocorrelation to measure the scattered energy's spatial similarity, with receiver angle. A surface that reflects sound uniformly to all receivers will produce high values in the spatial autocorrelation function. Conversely, surfaces that concentrate reflected energy in one direction will give low values. To form the diffusion coefficient, the circular autocorrelation function is first calculated and then an average is taken. This process can be simplified to a single equation. For a fixed source position, the diffusion coefficient, d_ψ, can be calculated using

$$d_\psi = \frac{\left(\sum_{i=1}^{n} 10^{L_i/10}\right)^2 - \sum_{i=1}^{n}\left(10^{L_i/10}\right)^2}{(n-1)\sum_{i=1}^{n}\left(10^{L_i/10}\right)^2}, \tag{5.2}$$

where L_i is a set of sound pressure levels in decibels in a polar response, n is the number of receivers, and ψ is the angle of incidence. This equation is valid only when each receiver position samples the same measurement area. This is automatically achieved for single-plane measurements on a semicircle with an even angular spacing between receivers. (The fact that receiver positions at ±90° actually sample half the area of the other receivers makes an insignificant difference to the diffusion coefficient.)

Figure 5.22a shows the diffusion coefficient for a few commercial products and a reference flat surface. At low frequency, edge diffraction causes the diffusion coefficient to increase with decreasing frequency because the sample acts as a point source reflecting the same energy in all directions. While there is a clear physical explanation for this effect, it does lead to confusion and so a normalized diffusion coefficient was introduced. The result of normalization is shown in Figure 5.22b. This gives the more intuitive response, with surfaces producing little diffusion at low frequency. It also more clearly illustrates the frequency where diffuse reflection begins. The normalized diffusion coefficient, $d_{\psi,n}$, is calculated using the following formulation:

$$d_{\psi,n} = \frac{d_\psi - d_{\psi,r}}{1 - d_{\psi,r}}, \tag{5.3}$$

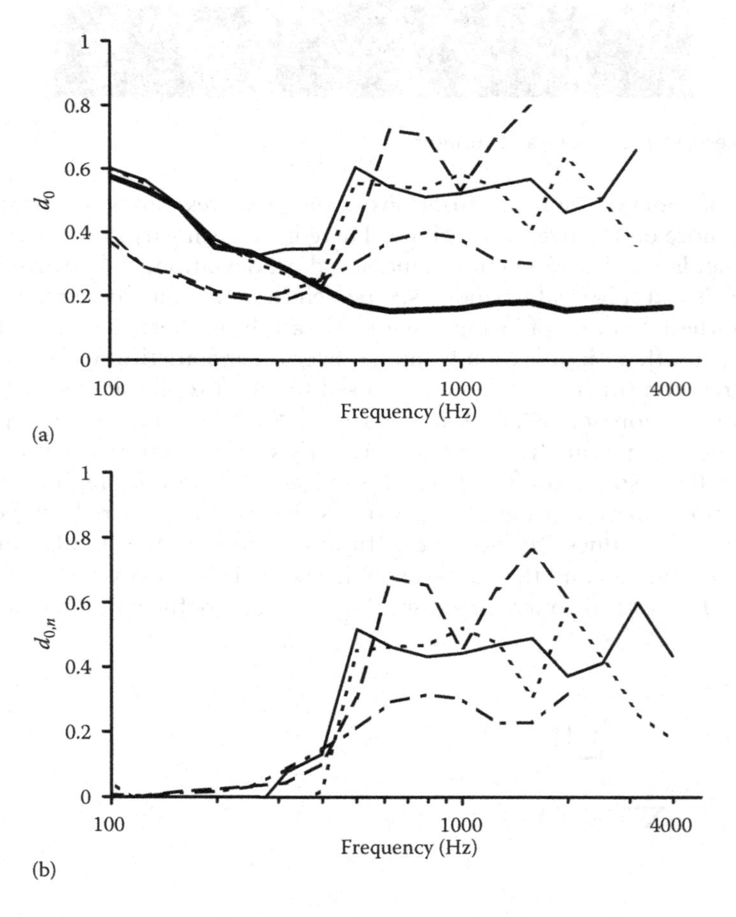

Figure 5.22 Diffusion coefficient for four commercial products: (a) not normalized and (b) normalized. The thick line in the top plot is the diffusion coefficient for a flat reflector.

where d_ψ and $d_{\psi,r}$ are the diffusion coefficients calculated using Equation 5.2 for the test sample and a reference flat surface of the same overall size as the test sample. At low frequency, sometimes, the normalized diffusion coefficient dips below 0 due to experimental uncertainties. In these cases, the negative values should be set to 0.

5.2.3 Obtaining polar responses

Section 5.2.1 already discussed how to measure the polar responses and Chapter 9 outlines possible prediction models. ISO 17497-2[3] recommends a receiver every 5°, with the source at 10 m and receiver arc radii 5 m (equivalent full scale). It recommends that to obtain a random incidence diffusion coefficient, source positions should be measured and averaged with a maximum angular separation of 10°. In reality, measuring 0°, ±30°, and ±60° is usually sufficient.

In many applications, the source position is well known. In performance spaces, for example, this is the location of the stage. In that case, it makes more sense to evaluate the diffusion coefficient for this specific angle of incidence, as the first-order effects of a diffuser are of primary importance.

ISO 17497-2 stipulates that different radii polar responses might be used to check for focussing effects. If measurements are made at different radial distances from the surface, it is best to calculate the diffusion coefficient for each radius and then average those results.

5.2.4 Discussion

The diffusion coefficient ranks diffusers correctly and has a clear physical basis in the autocorrelation function. However, it is not known exactly how the diffusion coefficient values relate to subjective response, which would be useful in evaluating the merits of diffusers.

The experimentally measured and theoretically predicted diffusion coefficient values tend to be small. This can be seen by glancing at the coefficient values given in Appendix B. Values for the diffusion coefficient can, in theory, spread over the entire range from 0 to 1. A value close to 0 has been measured for a concave surface designed to focus sound on a single receiver. A value of 1 can be measured for a small single semicylinder. But a single semicylinder on its own is not much use because it cannot cover a wide surface area. As soon as more complex surfaces are introduced, such as a set of semicylinders, the diffusion coefficient is reduced because of inevitable lobing caused by constructive and destructive interference (see Figure 5.20). This is unavoidable in extended structures, and so the diffusion coefficient is rarely close to 1 for usable and realistic surfaces. This is why it is important to measure application realistic samples, as the reflection from a single object is not representative of the response from a periodic or modulated array.

5.2.5 Diffusion coefficient table

Appendix B gives calculated values for normalized diffusion coefficients, for various surfaces, following the procedure in ISO 17497-2. The predictions were carried out using a 2D boundary element model as described in Chapter 9. All surfaces were modelled as thin panel extrusions and the rear of the surfaces were not enclosed. Therefore, these represent the diffusion coefficient for single-plane devices such as semicircular arcs. The source and receiver were 100 and 50 m, respectively, from the surface. Each one-third octave band polar response was found by averaging seven single frequency responses. The random incidence diffusion coefficient values were found by an arithmetic average of the diffusion coefficient values for 10 different angles of incidence (Paris's formulation was not applied). The table in Appendix B reports three incident angles: normal, 57°, and random incidence.

The first part of the table shows the effect of changing the number of diffusers in an array, illustrating that the diffusion coefficients for single and multiple devices are very different. The rest of the table keeps all the test sample widths the same, at about 3.6 m, to allow comparisons. When considering other diffusers in the table, it is important to compare like with like. For example, ensure that the diffusers being compared have the same maximum depth.

The second section shows the effect of diffuser depth on the diffusion coefficient. To do this, a set of semiellipses are used. As might be expected, as the semiellipses get deeper, they start diffusing at a lower frequency—although period width becomes a limiting factor for deeper devices, as discussed in Section 10.3. The third section shows the diffusion coefficients for different triangles; the scattering from triangles is discussed in detail in Chapter 11. The fourth section gives examples of what happens when ellipses are mounted on a flat baffle with spaces between—as might be expected, the exposed flat sections reduce the dispersion generated at high frequencies.

The fifth part shows the performance that can be obtained from optimized curved surfaces, the design of which is discussed in Chapter 11. The table demonstrates that the optimization design process is very effective. The sixth part gives results for flat and planar hybrid surfaces, where absorptive patches are used to generate dispersion at high frequency; these devices are discussed in detail in Chapter 12. The seventh and last part gives data for a variety of reflection phase gratings: simple Schroeder diffusers, fractal designs, and optimized surfaces. These designs are discussed in Chapter 10. The Schroeder diffusers were modelled by meshing the entire surface so the predictions are valid above (and below) the limit where plane wave propagation becomes less dominant in the wells.

5.3 THE SCATTERING COEFFICIENT

GRAMs must incorporate scattering algorithms to achieve reasonable prediction accuracy. For example, without surface scattering, GRAMs tend to over predict reverberation time.[23–25] This is especially true in spaces where absorption is unevenly distributed, as what happens in many concert halls, or where rooms are highly disproportionate, as what happens in many factories. Moreover, for acoustic parameters that are highly dependent on early reflection prediction accuracy, such as early lateral energy fraction and clarity, there can be great sensitivity to the modelling technique used for diffuse reflections and to the value of scattering coefficients assumed.[26] In the first round-robin study of room acoustic models,[25] three models were found to perform significantly better than others. These three models produced results approximately within one subjective difference limen, while the less successful models produced predictions inaccurate by many difference limen. What differentiated the three best models from the others was the inclusion of a method to model surface scattering.

How scattering is incorporated into GRAMs is detailed in Chapter 13. This process is inevitably approximate, because the models cannot explicitly represent the true wave nature of sound.[27] Room models use a scattering coefficient to determine the proportion of the energy that is reflected in a specular manner and the proportion that is scattered. (A note of caution, some computer models confusingly call these *diffusion coefficients*.) A key driving force behind the standardization of ISO 17497-1 was to produce a method to enable scattering coefficients to be determined in a rigorous manner.

The scattered energy in a room model is usually distributed according to Lambert's cosine law.[28] Lambert's law is used because it fits with the philosophy of the room models that are based on high-frequency techniques. The law is correct for high-frequency, point incoherent scattering. Diffusion coefficients used by diffuser designers are based on uniform energy distribution. Uniform energy distribution is a possible design goal because the reflected sound from surfaces displays strong interference effects at the most important acoustic frequencies. Indeed, this coherence is explicitly exploited in many diffuser designs—just try and explain how a Schroeder diffuser works without referring to interference! Consequently, Lambert's law is inapplicable for evaluating diffusers. Conversely, using diffusion coefficients measured according to ISO 17497-2 in GRAM risks incorrect results, unless the model has been explicitly designed to use this coefficient.

5.3.1 Principle

The scattering coefficient evaluation requires the separation of the reflected sound into specular and scattered components. The specular component is the proportion of energy that is reflected in the same way as would happen for a large plane surface. The scattered components give the energy reflected in a nonspecular manner. This is illustrated in Figure 5.23.[29] The coefficient has a clear physical meaning and the definition is very useful for GRAMs that have separate algorithms dealing with specular and scattered components. With this definition, it is then possible to define a scattering coefficient, s, as the proportion of energy not reflected in a specular manner.

This definition takes no account of how the scattered energy is distributed but assumes that in most rooms, there is a large amount of mixing of different reflections. Therefore, any inaccuracies that arise from this simplification will average out. This is probably a reasonable assumption for the reverberant field, where there are many reflections. It could well be troublesome for the early sound field, however, where the impulse response can be dominated by a few isolated reflections and the correct modelling of these is essential for gaining accurate predictions or auralizations. Section 5.1 has already illustrated how scattering coefficients can give misleading results for the first-order reflections from redirecting and anisotropic surfaces.

The scattering coefficient depends on frequency and the source location. Similar to the random incidence absorption coefficient obtained in reverberation chambers, an angular average of the scattering coefficient—the random incidence scattering coefficient—can be defined.

As a general assumption, the surface under test is assumed to be wide and shallow. The method will not work for isolated items and deep surfaces, as it is trying to measure the scattering from surface roughness and not the edges. It also has problems when the surface absorption is medium to high, as the coefficient estimation becomes inaccurate. For example, if the absorption coefficient exceeds 0.5, then it is impossible to gain accurate measurements below 800 Hz for highly scattering surfaces.[30] This is because the natural variation in reverberation times in the reverberation chamber makes the measurement uncertainty too large.

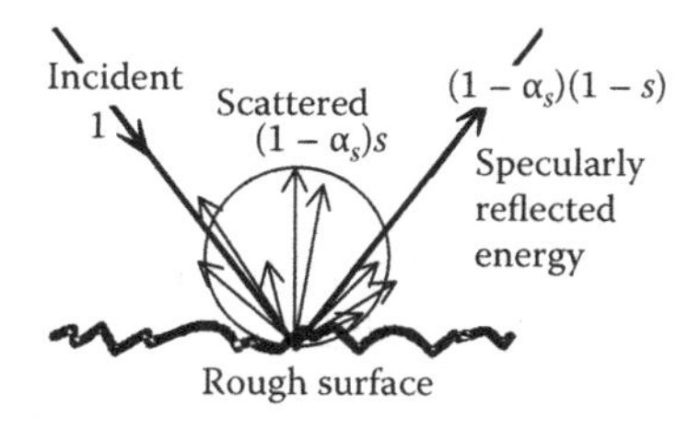

Figure 5.23 Definitions used for scattering coefficient. (After Vorländer, M., Mommertz, E., *Appl. Acoust.*, 60, 187–200, 2000.)

5.3.2 Rationale and procedure

The energies of reflections (normalized with respect to a reflection from a non-absorbing flat surface) are expressed as follow[31]:

$$\begin{aligned} E_{spec} &= (1-\alpha_s)(1-s) \equiv (1-\alpha_{spec}), \\ E_{total} &= 1-\alpha_s, \end{aligned} \tag{5.4}$$

where E_{spec} is the specular reflected energy, E_{total} is the total reflected energy, s is the scattering coefficient, α_s is the absorption coefficient, and α_{spec} is the apparent specular absorption coefficient.

The apparent specular absorption coefficient warrants further description. It is the energy dispersed from specular reflection directions. This energy is not dissipated to heat, instead it is reflected into nonspecular directions. Rearranging the previous formulations gives the following equation for the scattering coefficient:

$$s = \frac{\alpha_{spec} - \alpha_s}{1-\alpha_s} = 1 - \frac{E_{spec}}{E_{total}}. \tag{5.5}$$

The measurement of this quantity is easiest to explain in the free field, although it is in the reverberation chamber where this process is most useful and powerful. The set-up is shown in Figure 5.24.[29] The specular absorption coefficient is found by rotating the test sample, while phase locked averaging the reflected pulses. Figure 5.25 shows three bandpass filtered pulses for different orientations of a corrugated surface.[29] The initial parts of the reflections are highly correlated. These are the specular components of the reflection and remain unaltered as the sample is rotated. In contrast, the later parts of the three reflected pulses are not in phase and depend strongly on the specific orientation; this is the scattered component. By averaging the reflected pressure while rotating the sample, the scattered components are averaged to 0, and only the specular energy remains.

Transferring this procedure to the reverberation chamber, the measurement technique is as follows. A circular test sample is placed on a turntable and rotated. While the turntable is turned, the room impulse response is repeatedly measured. The later parts of the impulse response, which are due to the scattering from the surface, will cancel out, and the averaged impulse response contains only the specular reflection component. This impulse response

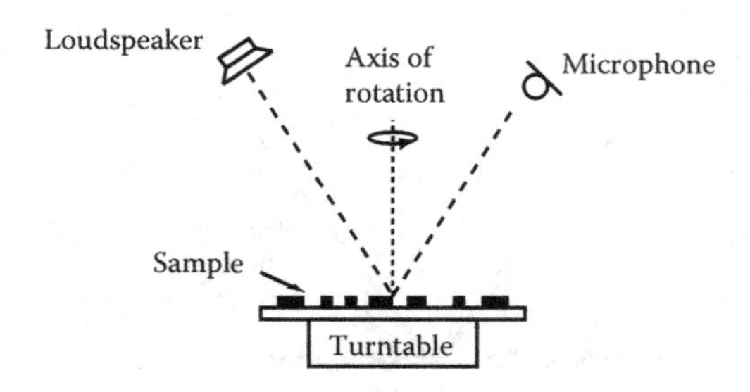

Figure 5.24 Set-up used for measuring the scattering coefficient. (After Vorländer, M., Mommertz, E., *Appl. Acoust.*, 60, 187–200, 2000.)

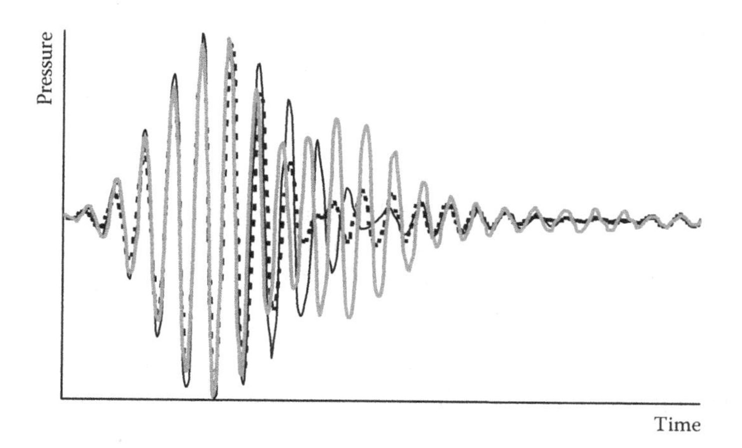

Figure 5.25 Band-limited reflected pulses for different sample orientations. (After Vorländer, M., Mommertz, E., *Appl. Acoust.*, 60, 187–200, 2000.)

is then backwards integrated to give the reverberation time due to the specular reflection component. The reverberation time with the sample stationary (not rotating) can also be obtained, and this decay is due to all reflections—specular plus diffuse. By manipulating these reverberation times, it is possible to derive the specular and total reflected energy and, from Equation 5.5, the scattering coefficient.

In reality, four reverberation times are needed. It is difficult to get a perfectly flat and circular turntable, especially for full-scale measurements.[32–34] Consequently, the imperfections in the turntable must be compensated for by additional measurements. The four reverberation times that must be measured are shown in Table 5.2.

Once these reverberation times are measured, the following formulations are used to get the scattering coefficient. The random incidence absorption coefficient, α_s, of the sample is calculated using

$$\alpha_s = 55.3\frac{V}{S}\left(\frac{1}{c_2T_2} - \frac{1}{c_1T_1}\right) - \frac{4V}{S}(m_2 - m_1), \tag{5.6}$$

where V is the volume of the reverberation room, S is the area of the test sample, c_1 is the speed of sound in air during the measurement of T_1, c_2 is the speed of sound in air during the measurement of T_2, m_1 is the energy attenuation coefficient of air in m^{-1} during the measurement of T_1 (see Sections 2.2.1 and 4.3), and m_2 is the energy attenuation coefficient of air during the measurement of T_2.

Table 5.2 The measurement conditions for the four different reverberation times

Reverberation time	*Test sample*	*Turntable*
T_1	Not present	Not rotating
T_2	Present	Not rotating
T_3	Not present	Rotating
T_4	Present	Rotating

The specular absorption coefficient, α_{spec}, is calculated using the following formulation:

$$\alpha_{spec} = 55.3\frac{V}{S}\left(\frac{1}{c_4 T_4} - \frac{1}{c_3 T_3}\right) - \frac{4V}{S}(m_4 - m_3), \tag{5.7}$$

where c_3 is the speed of sound in air during the measurement of T_3, c_4 is the speed of sound in air during the measurement of T_4, m_3 is the energy attenuation coefficient of air during the measurement of T_3, and m_4 is the energy attenuation coefficient of air during the measurement of T_4.

Finally, the random-incidence scattering coefficient, s, is calculated using Equation 5.5. These reverberation times are measured using the standard procedures in ISO 354,[35] so multiple source receiver pairs are needed to average out spatial variation within the reverberation chamber. The measurement must use a deterministic source signal to allow the phase locked pressure averaging. Originally, MLSs were favoured, but these are sensitive to time variance. This is not a problem at the model scale, but at full scale, the turntable must move slowly and time invariance cannot be guaranteed. The signal periodicity must be longer than the reverberation time in the room, which means that full-scale measurements can take an hour, as typically 72 measurements in a single rotation are needed. For this reason, swept sine waves are better because they are less sensitive to time variance errors.[32,34] Measurements can be speeded up by simultaneous acquisition on all microphones and radiating the MLS signal out of all required sources at the same time during the measurements with rotation.

At full scale, not only is the measurement slow but also there are also considerable logistical problems in fabricating a 3.6-m-diameter flat turntable. For instance, the doors of the reverberation chamber are probably too small for the turntable to go through in one piece and so it needs to be dismantled. Yet a completely flat and strong turntable when assembled must be made. Furthermore, a powerful, slow, and quiet motor is required.[32,34] For these reasons, model scale measurements are to be preferred for speed and efficiency.

5.3.3 Sample considerations

The sample must be shallow and not be too absorbing. Since the measurement method is intended to measure surface roughness, the results are only reliable when the structural depth of the sample is small compared to the size of the specimen. An empirically derived limit is $h \leq d/16$, where d is the diameter of the turntable and h the structural depth. Figure 5.26 shows 1:10 samples of four commercial products. The measured scattering coefficients are very sensitive to edge conditions.[34,36] Even with shallow surfaces, the edge corrugations can cause scattering coefficients to be larger than 1 at high frequencies. For round samples, there needs to be a smooth and rigid border covering the edge corrugations.

Most diffusers are rectangular, and if these are mounted on top of a round base plate, then scattering from the rectangular edges results in additional scattering and misleading results. The solution is to recess square samples within a circular base plate, as shown in Figure 5.27. Figure 5.28 shows measurements for the different mounting conditions, indicating that the recessed sample mounting gives the best results.

As Figure 5.1 illustrated, the scattering coefficient overestimates the scattering for single-plane devices such as extruded surfaces. This happens because the topology changes dramatically when the surface is rotated and, consequently, the surface is seen as being very good at scattering. A more strict measure would be two coefficients in two orthogonal directions, as is done for the diffusion coefficient. But then, most current GRAMs can deal

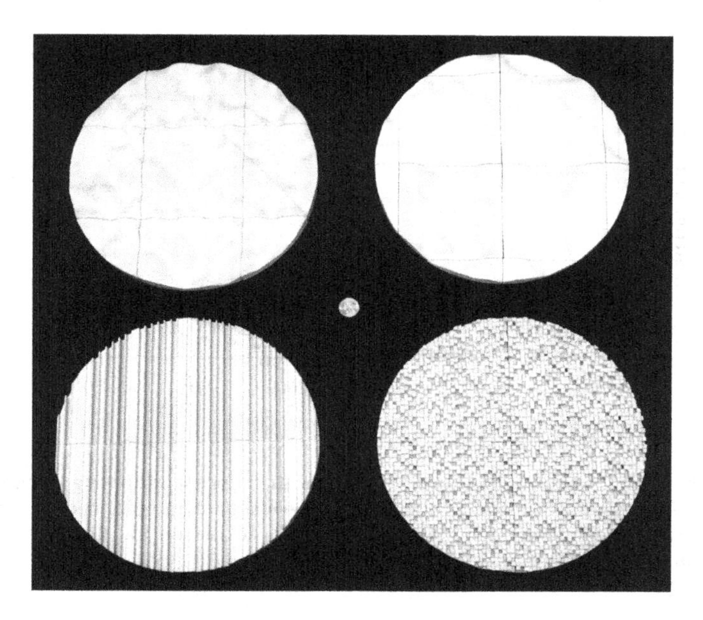

Figure 5.26 Four 1:10 scale models of commercial diffusers used to measure scattering coefficients. A small coin is placed in the centre of the photo to indicate scale.

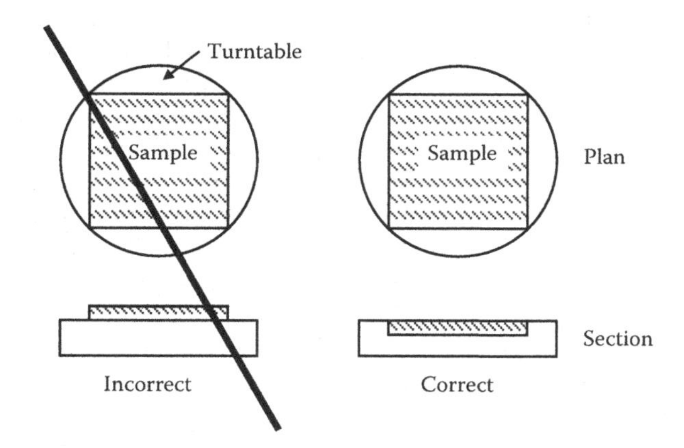

Figure 5.27 Mounting condition for noncircular samples.

with only single hemispherical scattering coefficient, so this more strict evaluation is incompatible with those models.

5.3.4 *In situ* measurement

Arrays of microphones and loudspeakers can be used to measure the scattering coefficient *in situ*.[37] Instead of rotating the surface, the source and receivers are simultaneously rotated. This then yields a scattering coefficient for a particular angle of incidence. The microphones and loudspeakers need to be sufficiently far from the surface to allow the reflected sound to be isolated by time windowing. Both arrays also need to have a carefully engineered directivity to minimize the effect of other room reflections.

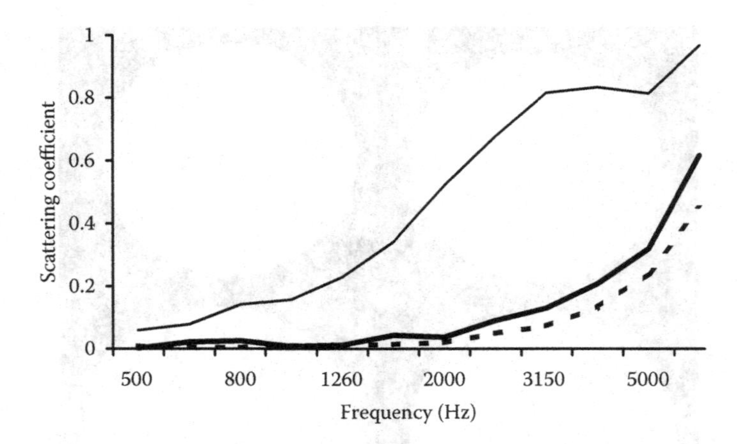

Figure 5.28 Scattering from a sinusoidal-shaped sample with different mounting conditions and sample shapes: —— square sample, proud edges; ▬▬ square sample, recessed edges; and - - - circular sample. (Modified from Gomes, M.H.A., Vorländer, M., Gerges, S.N.Y., *Proc. Forum Acusticum Sevilla*, 2002.)

5.3.5 Predicting the scattering coefficient

Predicting the random incidence scattering coefficient is slow because of the need to evaluate the sound for a large number of sample orientations. Furthermore, it would be anticipated that the reverberation chamber would introduce inaccuracies due to nondiffuseness and other effects commonly seen in random incidence absorption measurements. Nevertheless, it is possible to carry out predictions for the free field scattering coefficient as a function of incidence angle and use Paris's formula (Equation 13.1) to get approximate random incidence values.

The free field scattering coefficient follows a similar principle to the random incidence coefficient. The measurement is done in an anechoic chamber and the receiver is placed in the specular reflection direction, as was shown in Figure 5.24. The surface is again rotated and the reflected pulses phase locked averaged. The energy remaining after the averaging is the specular energy. From this, the scattering coefficient is obtained.

This measurement process can be mimicked in a boundary element method (BEM). Hargreaves[38] did this for a sample of rectangular battens and obtained accurate predictions, but this was done only for a few single frequencies.

If a simpler numerical model is used, the prediction time can be greatly decreased. Indeed, for Schroeder diffusers, it is possible to draw up a very simple formulation for the scattering coefficient. The far field scattering from a Schroeder diffuser can be predicted using a simple Fourier model as described in Chapters 9 and 10. The Fourier model is not exact, but it does give reasonably accurate predictions of the scattering from the surfaces except at low frequencies and large angles of incidence or reflection. Under this approximate model, the pressure scattered from the surface, p_s, is given by

$$p_s(\psi,\theta) \approx A\sum_{np=1}^{N_p}\sum_{n=1}^{N} e^{-2jkd_n} e^{jknw\left[\sin(\theta)+\sin(\psi)\right]}, \tag{5.8}$$

where ψ is the angle of incidence, θ is the angle of reflection, N_p is the number of periods, N is the number of wells in a period, w is the well width, k is the wavenumber, d_n is the depth of the nth well, and A is a constant.

This approximate theory enables a simple formulation for the scattering coefficient to be derived. In addition, the polar response can also be calculated from Equation 5.8 and so the diffusion coefficient can be found. Hence, a comparison of the diffusion and scattering coefficients can be made.[30] The scattering coefficient is not the ISO coefficient, however. Instead, it is a free field version of the Mommertz and Vorländer coefficient[29,31] derived following a similar philosophy.

The free field scattering coefficient is evaluated by finding the invariant energy, E_{spec}, in the specular direction ($\psi = -\theta$), when the surface is moved. Equation 5.8 is a single-plane formulation, so it is natural to translate the surface. The surface has been assumed infinitely large so that edge effects are not significant. In this case, the averaging is done by translation over a complete period. In this ideal case, the scattering does not change with translation because the receiver is in the specular reflection direction and all the terms, which vary when the surfaces moves, cancel out. The specular or invariant energy can be shown to be

$$E_{spec} \approx \left| A' \sum_{np=1}^{N_p} \sum_{n=1}^{N} e^{-2jkd_n} \right|^2, \tag{5.9}$$

where A' is a constant.

The invariant energy is placed in a ratio with the energy from a flat plane surface for normalization purposes.[31] This then gives a specular reflection coefficient, R_{spec}:

$$R_{spec} = \frac{\left| A' \sum_{np=1}^{N_p} \sum_{n=1}^{N} e^{-2jkd_n} \right|^2}{\left| A' N_p N \right|^2}. \tag{5.10}$$

This then represents the proportion of energy that is reflected in a specular manner by the surface, and so the scattering coefficient can be readily evaluated:

$$s = 1 - \left| \frac{1}{N} \sum_{n=1}^{N} e^{-2jkd_n} \right|^2. \tag{5.11}$$

The scattering coefficient is independent of angle of incidence, and so in this special case, the random incidence coefficient is numerically identical to the free field case. This is similar to the prediction model that Embrechts et al.[39] produced for Gaussian rough surfaces.

Equation 5.11 shows that to get the greatest scattering, the reflection coefficients $\exp(-2kjd_n)$ must have a phase evenly spaced from 0 to 2π (in an Argand diagram, the reflection coefficients are evenly spaced around the circumference of a circle of radius one). This is achieved for the modified versions of the primitive root diffuser (PRD)[40,41] at integer multiples of a design frequency, as discussed in Chapter 10. This is illustrated in Figure 5.29 (the line labelled *correlation scattering coefficient* can be ignored for now).[30] At multiples of the design frequency, the scattering coefficient, using Equation 5.11, is 1, except at the flat plate or critical frequency of $(N - 1)f_0 = 3$ kHz (see Section 10.3 for more on critical frequencies). This complete scattering at the design frequency and multiples thereof simply means that no energy is reflected in the specular direction. It does not necessarily say how

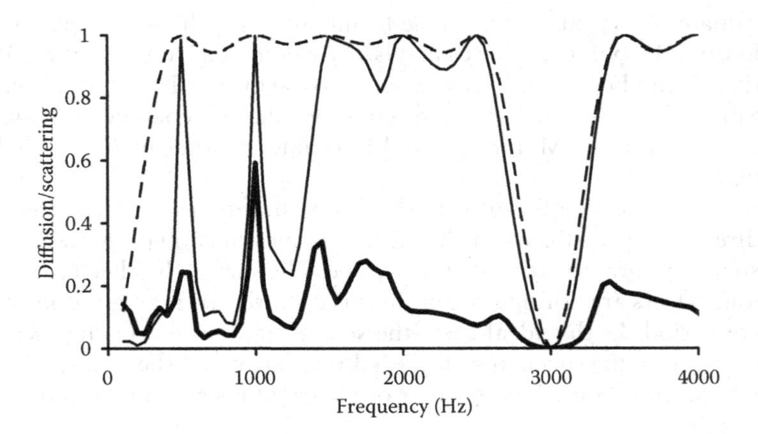

Figure 5.29 Various scattering and diffusion coefficients for a PRD with a design frequency of 500 Hz: - - - scattering coefficient using Equation 5.11, ▬▬ normalized diffusion coefficient, and —— correlation scattering coefficient. (Modified from Cox, T.J., D'Antonio, P., *Proc. 17th ICA*, Italy, 2001.)

good the dispersion produced is. This is why diffusion coefficients are numerically less than the free field scattering coefficient.

The free field scattering coefficient is given by the sum of the reflection coefficients squared—there is no dependence on the order of the wells in the diffuser. Although the shape of the polar response changes when the order of the wells is changed, the energy in the specular reflection direction does not. Consequently, while the diffusion coefficient will vary if the order of the wells is changed, the scattering coefficient remains the same. This is another illustration of why the diffusion coefficient is a stricter test of reflection quality.

5.3.6 The correlation scattering coefficient (from polar responses)

Mommertz presented a method for evaluating a scattering coefficient from polar responses. This correlates the scattered pressure polar responses from the test surface and a reference flat surface[42] to give a scattering coefficient. This will be called the correlation scattering coefficient δ_c. The coefficient is given by

$$\delta_c = 1 - \frac{\left| \sum_{i=1}^{n} p_1(\theta_i) p_0^*(\theta_i) \right|^2}{\sum_{i=1}^{n} \left| p_1(\theta_i) \right|^2 \sum_{i=1}^{n} \left| p_0(\theta_i) \right|^2}, \tag{5.12}$$

where p_1 is the pressure scattered from the test surface, p_0 is the pressure scattered from the flat surface, * denotes complex conjugate, θ_i is the receiver angle of the ith measurement position, and n is the number of measurements in the polar response. (An alternative description of this coefficient was given by Embrechts et al.,[39] who described it in terms of an least mean squares [LMS] problem.)

This is not the same as the ISO coefficient or the free field scattering coefficient. This is illustrated in Figure 5.29, where the scattering coefficient from Equation 5.11 is compared to the correlation scattering coefficient for a PRD. This difference arises because the coefficient definition is different. The free field Mommertz and Vorländer method measures the amount of energy moved from the specular direction when the surface is moved. The correlation scattering coefficient measures the dissimilarity between the test and flat surface scattering over a polar response. In the case of randomly rough surfaces, the two coefficients will be similar, but for diffusers with distinct polar responses, this is not the case.

Although the correlation scattering coefficient is not identical to the ISO scattering coefficient, it does illustrate and help contrast the performance of diffusion and scattering coefficients.[30] A useful property of the correlation scattering coefficient is that it is readily predicted. Consequently, it is possible to compare prediction and measurement in a 2D polar response for a single cylinder and a set of cylinders. Predictions were carried out using a BEM (see Chapter 9) and measurements in a 2D goniometer. Figure 5.30 compares the predicted and measured correlation scattering coefficients and a good match is achieved. This provides evidence that the coefficient can be predicted and that the measurement system used is robust. Problems might occur, however, in measurements where exact microphone position replication is not achieved, for example, if a moving boom arm is used or where time variance cannot be maintained between the reference surface and diffuser measurements.

Kosaka and Sakuma[43] predict the correlation scattering coefficient using a 3D BEM and explored some of the practical requirements for accurate results. The receiver and sources should be at least a diameter away from the test sample to reduce errors due to near field effects. They found that a 5° spatial resolution was required and that two single frequency evaluations per one-third octave band were useful for a rough evaluation of scattering coefficients. At least 10 periods need to be tested for a periodic sample.

Embrechts et al.[44] examined a sine-shaped surface and found significant differences between prediction and measurements. At high frequencies, they mainly attribute the differences to a number of experimental issues leading to an overestimation of the scattering. Low-frequency errors were thought to arise from the surface, producing little scattering, as the measurements became inaccurate and uncertainties inherent in decay slope evaluation mean a measured value of exactly 0 for the scattering coefficient can never be obtained.

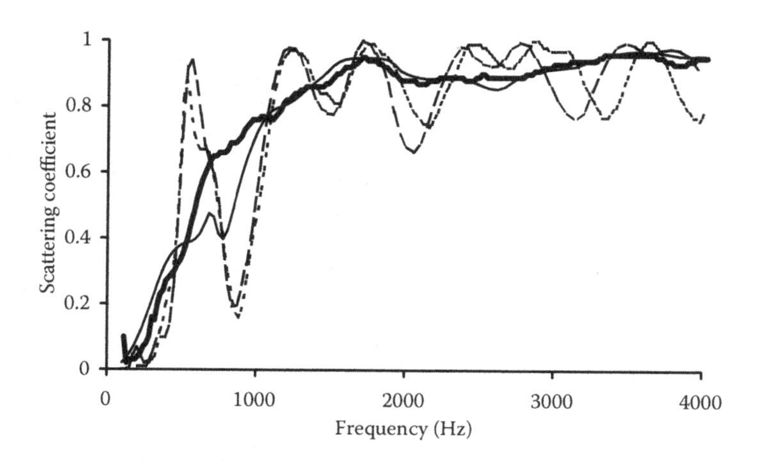

Figure 5.30 Comparison of predicted and measured correlation scattering coefficients: —— prediction, one cylinder; ▬▬ measurement, one cylinder; – – – prediction, four cylinders; and - - - - measurement, four cylinders.

A sample of a single-plane quadratic residue diffuser (QRD) has also been tested at a 1:5 scale. One period of the sample is shown on the left in Figure 5.1a. About four periods were used in the final sample. The sample was measured in the 2D goniometer first with the QRD wells perpendicular to the measurement arc and then with the wells parallel to the measurement arc. For each sample orientation, three incidence angles (0°, 30°, and 60°) were measured. These measurement results were then averaged to give an approximate random incidence coefficient. A similar sample was measured using the ISO 17497-1 method in a model reverberation chamber. The results are compared in Figure 5.31. There is reasonable correspondence between the random incidence and correlation scattering coefficients, although in two frequency bands, the results are significantly different. For example, for the 3.2 kHz octave band, the random incidence measurement exceeds 1, something that cannot happen with the correlation scattering coefficient. However, considering that one measurement is done in a diffuse field and the other in the free field, the match is actually quite good, better than many have obtained when diffuse and free field absorption coefficients are compared.

Also shown in Figure 5.31 is the scattering coefficient predicted using the simplest Fourier model, Equation 5.11. The prediction accuracy is surprisingly good considering that the Fourier theory makes many assumptions that are not entirely correct for this type of surface.

Figure 5.31 also shows the normalized diffusion coefficient for the diffuser. In this case, the scattering and diffusion coefficients agree as to the frequency at which significant dispersion begins (≈500 Hz). The diffusion coefficient is numerically much less than the scattering coefficients, as discussed earlier.

The polar response measurement system can also be used to illustrate some other key differences between scattering and diffusion coefficients. For example, in Section 5.1, the case of redirection was discussed. Another example is shown in Figure 5.32, which illustrates the effects of a concave focussing surface. The surface is designed to focus the sound on one microphone in the receiver arc. The diffusion coefficient interprets the concave surface as being worse at diffusing sound than the plane surface. The normalized diffusion coefficient is less than 0 for virtually every frequency. The correlation scattering coefficient, however, interprets the focussed polar response as being different from the plane surface

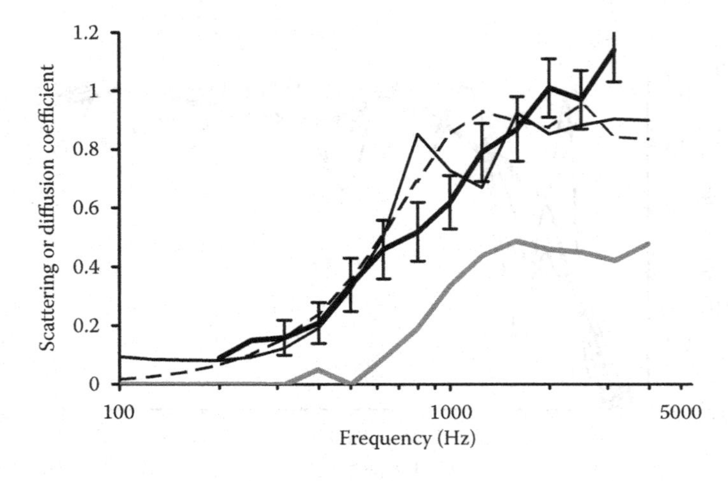

Figure 5.31 Various diffusion and scattering coefficients for a Schroeder diffuser. Diffuser was shown on the left in Figure 5.1a. —— correlation scattering coefficient, ▬▬ random incidence ISO scattering coefficient (with error bars), ▬▬ normalized diffusion coefficient, and - - - - scattering coefficient predicted using Equation 5.11.

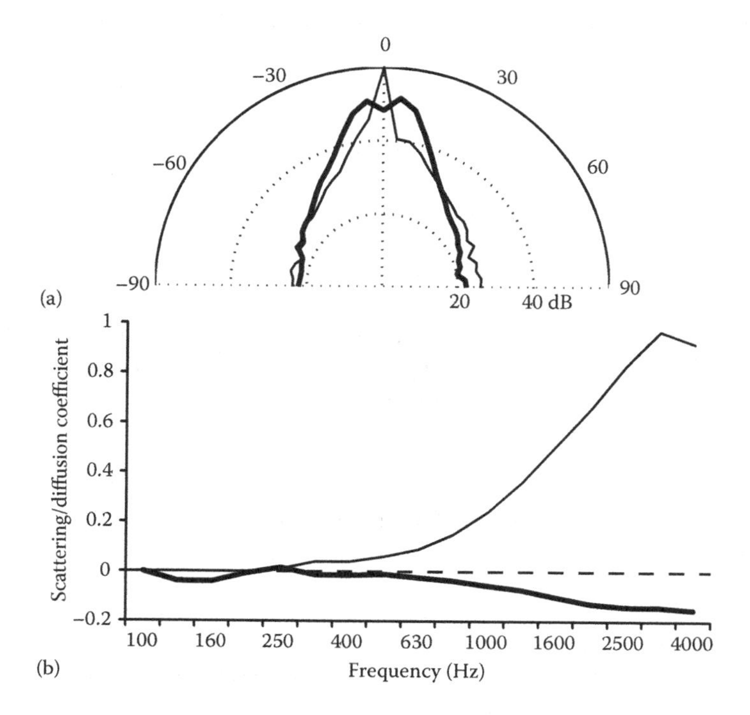

Figure 5.32 (a) Polar responses for ▬▬ flat surface and —— concave sample. (b) Diffusion and scattering coefficients for concave sample: ▬▬ correlation scattering coefficient and —— normalized diffusion coefficient.

and interprets this as increased scattering. This illustrates that scattering coefficients should not be used to interpret single surface items but should be used only for large surfaces with roughness. Furthermore, it shows that when evaluating diffusers, it is necessary not only to test the far field but also to test at receiver positions where aberrations such as focussing may occur.

5.3.7 Scattering coefficient table

Appendix C includes three tables of correlation scattering coefficients. Tables C.1 and C.2 were created for single-plane diffusers (3 m × 3 m) predicted using a 3D BEM. These show how the normal and random incidence correlation scattering coefficients vary between different surface topologies (sinusoidal, triangular, and rectangular) for a variety of different heights. For rectangular battens, the effect of spacing the battens out is also illustrated for random incidence. Below 250 Hz, the scattering coefficients should be taken to be 0.

Table C.3 considers single-plane diffusers but predicted using a 2D BEM. Some details of the geometry were given in Section 5.2.5, along with the rationale behind the choice of surfaces used in the prediction. The table includes all the surfaces included in the diffusion coefficient table, except hybrid surfaces. The formulation for the correlation scattering coefficient needs to be revised for surfaces that partially absorb, because the current formulation interprets any absorption as being scattering.

The random incidence coefficients tend to have raised values at low frequencies, especially for deep surfaces; values of up to 0.2 at 100 Hz are calculated for many surfaces. At oblique incidence, edge scattering becomes important and the edges of the test samples are very different from the reference flat surface. Better results might be obtained if a reference surface with the same overall thickness as the test sample was used. Furthermore, because the rear of the test surfaces was not enclosed in a box, the scattering from the rear of the surface may also be having some effect at low frequency. A pragmatic solution would be to set all scattering coefficients to 0 for frequencies where a quarter wavelength is larger than the surface depth.

The scattering coefficient does not discriminate between different diffusers in a consistent manner—see the values for the optimized curved surfaces compared to the semiellipses, for example. The coefficient also interprets redirection as scattering. For instance, a 45° triangle returns a strong reflection back to a normal source, as discussed in Chapter 11, yet the scattering coefficient interprets this as dispersion. Nevertheless, with care, this published table of scattering coefficients can be used by geometric room acoustic modellers, as discussed in Chapter 13.

Appendix D also gives a list of random-incidence scattering coefficients that have been measured according to ISO 17497-1.

5.4 OTHER METHODS FOR CHARACTERIZING DIFFUSE REFLECTIONS

There have been other methods developed to characterize the scattering from surfaces. The first method is a pragmatic approach, which has similarities to *in situ* absorption methods—maybe it will be developed into an *in situ* method for diffusion coefficients? The second method tries to characterize the effect of diffuse reflections by investigating the change in the diffuseness of a space. Finally, evaluating the comb filtering in the total sound field and temporal diffusion are discussed.

5.4.1 Measuring scattering coefficients by solving the inverse problem

The original development of this concept was by Farina.[45] The sound field in the vicinity of a diffuser is measured using a deterministic signal to gain the impulse response. It is necessary to make the measurements over many different spatial positions; these could be on an arc, as was done for the diffusion coefficient measurement, or if more convenient, these could be on a straight line parallel to the diffuser surface. The reflected impulse response is isolated by time windowing, as was done in Section 5.2.1, and then the frequency response found by using a Fourier transform. The frequency response is then normalized to a measurement of the incident sound field without the diffuser present to make the measurement independent of the source frequency response and sound power level.

The process is to predict the measured scattering from the surface using a GRAM and compare the predicted and measured polar response. The scattering and absorption coefficients within the GRAM are varied until the error between the predicted and measured polar responses are minimized. This is a trial and error process, which can be solved using an exhaustive search. The number of combinations to be tried is rather small if we assume the absorption and scattering coefficients only need to be varied in increments of say 0.01, and consequently, it is possible to simply do a complete check of every possible combination to find the correct coefficients.

This is a pragmatic approach. This process will probably give the coefficient most appropriate within the GRAM for randomly rough surfaces. Problems may arise, however, because the scattering coefficient will be dependent on the GRAM used, so this does not give a robust parameter for all models.

Problems arise when the scattered polar response does not match any of the possible polar responses generated by the GRAM. Although Farina's paper shows good matches being achieved, many other diffusers have responses that do not match GRAMs well. For instance, polar responses with a small number of distinct lobes (see Figure 11.14 or 11.15 for examples) are unlike anything from a GRAM. In this case, the best matched scattering coefficient might be nonsensical. This will happen with periodic devices at low to mid frequencies, as well as triangles and pyramids. This could be cured, however, by using more accurate dispersion patterns in the GRAM.

A further problem may occur from an ambiguity between the effects of scattering and absorption. There may be several good fits to the polar response, resulting from different combinations of absorption and scattering coefficients.

5.4.2 Temporal evaluation

The diffusion coefficient only monitors how the energy reflected from a surface is distributed spatially. The phase and the temporal response have been neglected. With a single cylinder or hemisphere, it is possible to produce a good spatial distribution, but without temporal dispersion (spreading of the impulse response). With most diffusers that have a complex profile, however, good spatial dispersion goes hand-in-hand with good temporal dispersion. Consequently, while current diffuser evaluation concentrates on energy dispersion, in the future, it might also become necessary to look at the phase in the polar response or the reflected impulse response. Pertinent comments about this point can be found in Chapter 11, where the scattering from cylinders is discussed. The scattering coefficients implicitly deal with time dispersion because of their definition.

One approach for evaluating the temporal dispersion is to look at decay characteristics. Redondo et al.[46] carried out a backwards integration[47] of the impulse responses of the sound scattered off various surfaces to examine the decay time. Figure 5.33 (left column) shows the impulse responses for a plane surface and various semicylinders. Figure 5.34 shows the decay curve for the four surfaces. Both the plane surface and the single semicylinder produce little temporal dispersion. This is reflected in the graph, as the sound energy decays rapidly. The four semicylinders have two strong reflection arrivals (because of symmetry, there are two, rather than four, arrivals; see Figure 5.33), and these arrivals appear as sudden drops in the decay curve. The surface that creates most temporal dispersion, the random array of semicylinders, has the longest decay time, as might be expected. The perceptual significance of the decay times, however, has not been quantified.

In reality, the listener hears a fusion of the direct sound from the source and the reflection from the surface. This has led some to suggest that the total field (direct plus scattered) is an appropriate way of measuring the effectiveness of diffusers to reduce coloration.

The autocorrelation function offers a way of checking for the randomness of signals, including impulse responses.[48,49] The weighted autocorrelation, s'_{xx}, is given by

$$s'_{xx}(\tau) = b(\tau) s_{xx}(\tau) , \tag{5.13}$$

where s_{xx} is the normal autocorrelation function, τ is the delay, and b is a weighting function, which is shown in Figure 5.35.[48] This weighting function makes allowance for the time dependence of a reflection's audibility. If a strong single reflection occurs at time τ_0, then it will cause audible coloration if

$$s'_{xx}(\tau_0) > 0.063 s'_{xx}(0) . \tag{5.14}$$

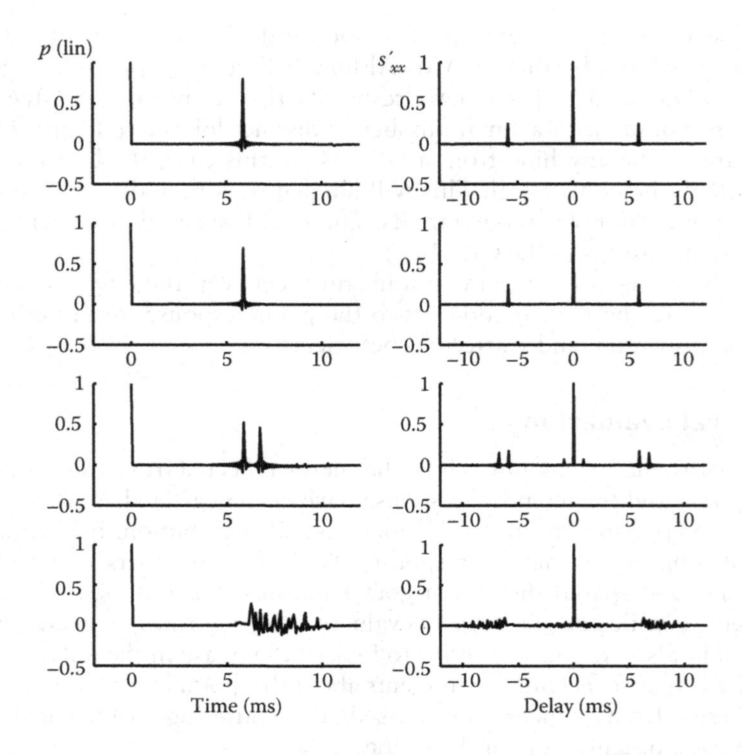

Figure 5.33 Left column: impulse responses. Right column: weighted autocorrelation functions for the scattering from various surfaces. From top to bottom: plane surface, one semicylinder, four semicylinders, and random semicylinders.

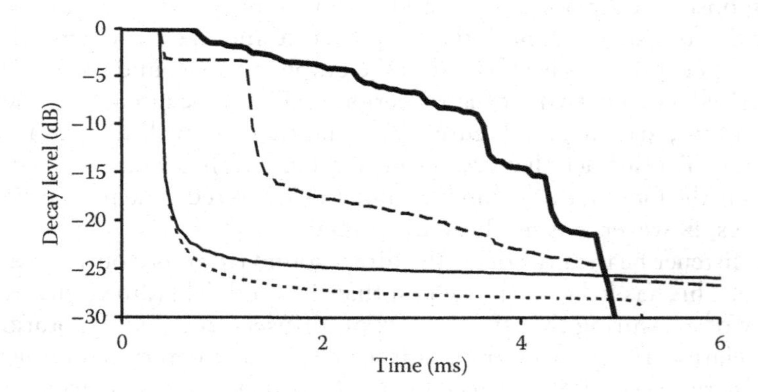

Figure 5.34 The decay of sound reflected from four surfaces: ---- flat surface, —— one semicylinder, ——— four semicylinders, and ▬▬ random array of semicylinders. The corresponding impulse responses can be seen in Figure 5.33.

The results from examining the various impulse responses for a plane surface and semicylinders are shown in Figure 5.33 (right column). The maximum value in the sidebands of the autocorrelation function does decrease as more semicylinders are introduced and the sound becomes less specular. However, all of the plots exceed the threshold indicated by Equation 5.14. So coloration probably occurs in all cases. Furthermore, Equation 5.14 is based on the threshold of audibility, so it cannot be used to analyze whether there is an

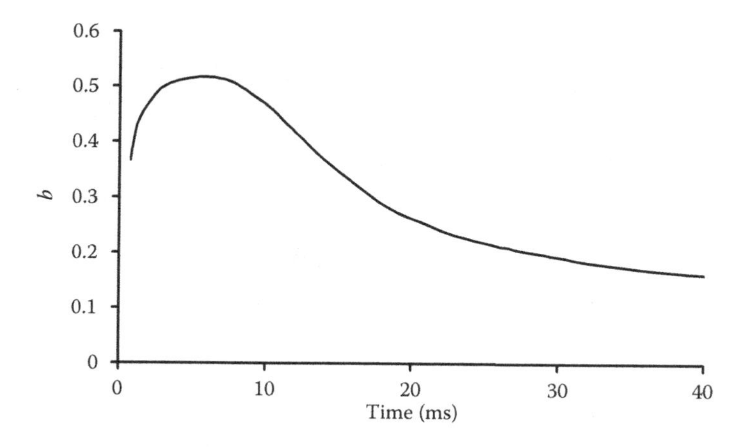

Figure 5.35 Weighting curve for examining coloration. (After Bilsen, F.A., *Acustica*, 19, 27–32, 1968.)

audible difference between the four cases because they are all above threshold. Maybe the maximum value in the sidebands of the autocorrelation can be used as a measure of the amount of coloration, but that is untested. Robinson et al.[50] examined how temporal diffusion changes echo perception with a small number of subjects, finding a change in the echo threshold for speech but not for the music motif tested. The results indicated that it was the total energy in the reflection that is most important and not the peak pressure.

Using an understanding of how the ear processes the total sound field, a more complete but complex method can be used. In the following discussions, the delay times given are only rough guides, as transition between different perceptual attributes is quite gradual.

When there is a short delay between the incident and reflected sound, the dominating feature is the variation in the frequency response—comb filtering—examples of which can be found in Chapter 11 and heard in most bathrooms. This happens for delays less than 25 ms[28,51] or a path length difference of 8.5 m.

For delays between 25 and 50 ms, equivalent to path length differences of 8.5 and 17 m, the principle audible effect caused by the interference between direct sound and reflections is temporal fluctuations. The temporal fluctuations are relatively slow, typically a few hundred Hz or less, and so to see the effect, the low-frequency envelope needs to be examined. To illustrate this, impulse responses from various diffusers were convolved with some music and the low-frequency envelopes were calculated. In Figure 5.36, the dashed lines show the low-frequency envelope for a snippet of saxophone music. This consisted of three notes, with heavy vibrato on the last note. The low-frequency envelope was calculated via the Hilbert transform[28] and then a fourth-order low-frequency Butterworth filter with a −3 dB point of 100 Hz was applied.

When an additional reflection from the plane surface is added (solid line, top plot), the corruption of the envelope by the temporal fluctuations is apparent. When the impulse response involving either diffuser is considered instead (solid lines, middle and bottom plots), something closer to the envelope of the original music is recovered. The PRD does better than a QRD in this case because it suppresses the reflection more effectively in the specular reflection direction. More work is needed to turn the temporal modulation effects into a perceptually validated metric.

For delays beyond 50 ms, then strong reflections have the potential to be heard as separate sounds. Then criteria for the audibility of echoes could be considered[28] or, where there are large numbers of diffusers, some measure of the evenness of the reverberation.[52] Such

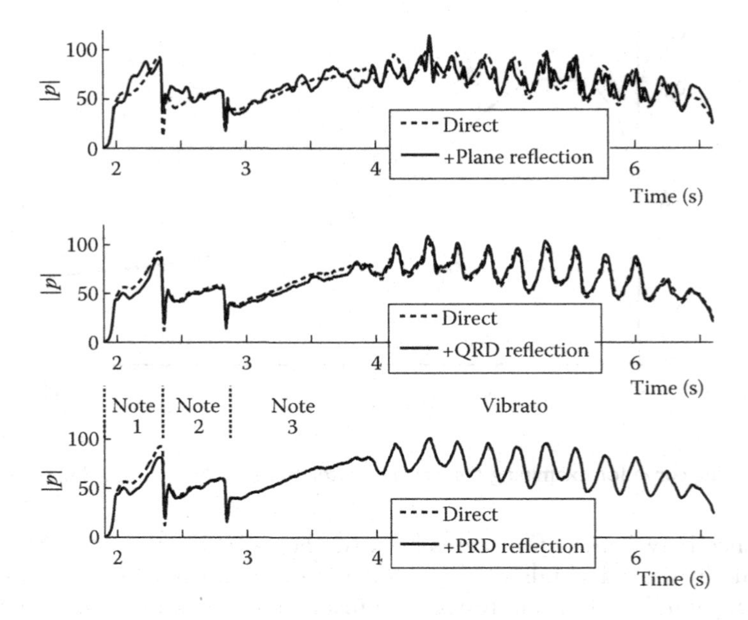

Figure 5.36 Low-frequency envelope for a short snippet of saxophone playing. Direct sound means the original music source is just heard (dashed lines). The other three solid lines are the direct sound plus one reflection from a surface as indicated on the graphs. Specular reflection direction.

criteria might be applied to monitor the effectiveness of diffusers *in situ* when used to control echoes, but the delay times are so long that they are not of interest for measuring diffusers within the laboratory.

As the above has shown, as the source and receiver positions change, the auditory effects alter because of the changing path lengths. The problem with total sound field evaluation is that it is so dependent on the geometry. The depth and frequencies of the minima and maxima in the comb filtering are strongly dependent on the delay time and relative level between the direct and reflected sound and, hence, on the source and receiver distances.

The total sound field might be analyzed in the critical bandwidths of the ears to examine whether the minima and maxima in the spectra are audible.[53] To be able to hear the effects of comb filtering, the analysis bandwidth of the ear, the width of the critical bands, must be of a similar size to the frequency spacing of the minima and maxima. Provided delay times are <25 ms, so that frequency response variation dominates what is heard by the listener, then there are a number of options for quantifying the coloration. The most promising are those that attempt to mimic the processing of the ear.

The sound is initially passed through a bank of auditory filters that mimic the action of the cochlear, which breaks sound into critical bands. This can be done according to the Bark scale[54] or similar. Figure 5.37 shows such an analysis for the plane surface and the random array of cylinders where equivalent rectangular bandwidths have been used.[55] Based on this representation, there are two evaluation methods: one based on the spectrum and the other on the autocorrelation.[56] Consider the former, which uses the ratio of the minimum and maximum energy in the spectrum. This ratio is examined to see if it exceeds some threshold value; this essentially quantifies the unevenness of the spectrum. Figure 5.37 shows an interesting result, however. Above 500 Hz, where the cylinders are diffusing, the spectrum for the plane surface has less variation than that for the diffuser, which contradicts expectation. When processing musical signals, the brain looks for regular harmonic structures to

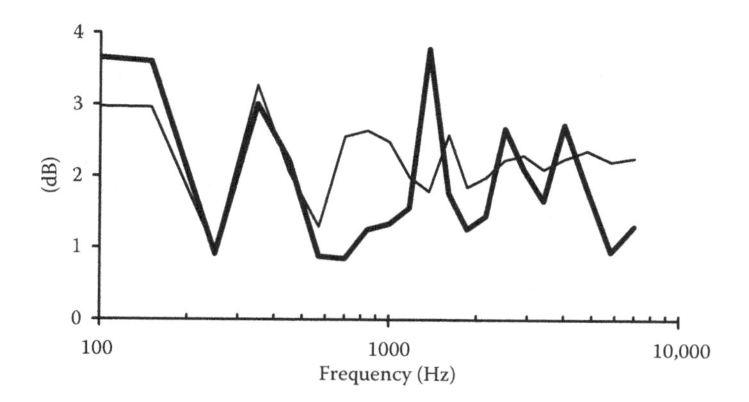

Figure 5.37 Level in the critical bands for total sound field: —— plane surface and ▬▬ random semicylinders.

determine pitch, and these periodic structures are also important for timbre. Consequently, further work is needed to understand how the brain processes the regular periodic patterns that arise from comb filtering in comparison to the more random total field generated by a diffusing structure.

5.5 SUMMARY

This chapter has mapped out some of the measurement techniques used for reflections from diffusers and other surfaces. Having two coefficients gives potential for much confusion. It is important that practitioners and researchers appreciate the difference because using the wrong coefficient could lead to poor predictions or bad designs.

REFERENCES

1. T. J. Cox, B.-I. L. Dalenbäck, P. D'Antonio, J. J. Embrechts, J. Y. Jeon, E. Mommertz, and M. Vorländer, "A tutorial on scattering and diffusion coefficients for room acoustic surfaces", *Acta Acust. Acust.*, **92**, 1–15 (2006).
2. ISO 17497-1:2004, "Acoustics—Sound-scattering properties of surfaces—Part 1: Measurement of the random-incidence scattering coefficient in a reverberation room".
3. ISO 17497-2:2012, "Acoustics—Sound-scattering properties of surfaces—Part 2: Measurement of the directional diffusion coefficient in a free field".
4. P. D'Antonio, J. H. Konnert, and P. Kovitz, "The disc project: Experimental measurement of the directional scattering properties of architectural acoustic surfaces", *Wallace Clement Sabine Centennial Symposium*, New York, 141–4 (1994).
5. AES-4id-2001: "AES information document for room acoustics and sound reinforcement systems—Characterisation and measurement of surface scattering uniformity", *J. Audio Eng. Soc.*, **49**, 149–65 (2001).
6. P. D'Antonio and T. J. Cox, "Diffusor application in rooms", *Appl. Acoust.*, **60**(2), 113–42 (2000).
7. T. J. Cox, "Objective and Subjective Evaluation of Reflection and Diffusing Surfaces In Auditoria", PhD thesis, University of Salford (1992).
8. A. Farina, personal communication.

9. P. Robinson and N. Xiang. "On the subtraction method for *in-situ* reflection and diffusion coefficient measurements". *J. Acoust. Soc. Am.*, **127**, EL99 (2010).
10. R. J. Hughes, "Volume Diffusers for Architectural Acoustics", PhD thesis, University of Salford (2011).
11. T. J. Hargreaves, T. J. Cox, Y. W. Lam, and P. D'Antonio, "Surface diffusion coefficients for room acoustics: Free field measures", *J. Acoust. Soc. Am.*, **108**(4), 1710–20 (2000).
12. L. E. Kinsler, A. R. Frey, A. B. Coppens, and J. V. Sanders, *Fundamentals of Acoustics*, 4th edn, John Wiley & Sons, New York (2000).
13. J. H. Rindel, "Attenuation of sound reflections due to diffraction", *Proc. Nordic Acoustical Meeting*, NAM86 (1986).
14. P. D'Antonio and J. H. Konnert, "The reflection phase grating diffusor: Design theory and application", *J. Audio Eng. Soc.*, **32**(4), 228–38 (1984).
15. T. J. Cox, "Optimization of profiled diffusers", *J. Acoust. Soc. Am.*, **97**(5), 2928–41 (1995).
16. T. J. Cox, "Designing curved diffusers for performance spaces", *J. Audio. Eng. Soc.*, **44**, 354–64 (1996).
17. P. D'Antonio, "The disc project: Directional scattering coefficient determination and auralization of virtual environments", *Proc. Noise-Con*, 93, 259–64 (1993).
18. J. Angus, A. C. Marvin, J., Clegg and J. F. Dawson, "A practical metric for evaluating sound diffusers", *Proc. 98th Convention Audio Eng. Soc.*, preprint 3955 (D5) (1995).
19. D. Takahashi, "Development of optimum acoustic diffusers", *J. Acoust. Soc. Jpn. (E)*, **16**(2), 51–8 (1995).
20. T. J. Cox, "Diffusion parameters for baffled diffusors", *Proc. 99th Convention Audio Eng. Soc.*, paper 4115 (1995).
21. Y. W. Lam, "A boundary integral formulation for the prediction of acoustic scattering from periodic structures", *J. Acoust. Soc. Am.*, **105**(2), 762–9 (1999).
22. J. A. S. Angus, "Diffuser assessment using surface spherical harmonics", *J. Acoust. Soc. Am.*, **104**(3), 1857–8 (1998).
23. Y. W. Lam, "On the parameters controlling diffusion calculation in a hybrid computer model for room acoustic prediction", *Proc. IoA(UK)*, **16**, 537–44 (1994).
24. M. Hodgson, "Evidence of diffuse surface reflections in rooms", *J. Acoust. Soc. Am.*, **89**, 765–71 (1991).
25. M. Vorländer, "International round robin on room acoustical computer simulations", *Proc. 15th ICA*, **II**, 689–92 (1995).
26. Y. W. Lam, "Diffuse reflection modeling methods", *J. Acoustic. Soc. Am.*, **100**(4), 2181–92 (1996).
27. B. Dalenbäck, M. Kleiner, and P. Svensson, "A macroscopic view of diffuse reflection", *J. Audio. Eng. Soc.*, **42**, 793–807 (1994).
28. H. Kuttruff, *Room Acoustics*, 6th edn, CRC Press, Oxon, UK (2009).
29. M. Vorländer and E. Mommertz, "Definition and measurement of random-incidence scattering coefficients", *Appl. Acoust.*, **60**(2), 187–200 (2000).
30. T. J. Cox and P. D'Antonio, "Contrasting surface diffusion and scattering coefficients", *Proc. 17th ICA*, 6B.09.01, Italy (2001).
31. E. Mommertz and M. Vorländer, "Measurement of scattering coefficients of surfaces in the reverberation chamber and in the free field", *Proc. 15th ICA*, **II**, 577–80 (1995).
32. J. J. Embrechts, "Practical aspects of the ISO procedure for measuring the scattering coefficient in a real-scale experiment", *Proc. Forum Acusticum Sevilla*, RBA-06-001-IP (2002).
33. L. De Geetere and G. Vermeir, "Investigations on real-scale experiments for the measurement of the ISO scattering coefficient in the reverberation room", *Proc. Forum Acusticum Sevilla*, RBA-06-004 (2002).
34. M. Vorländer, J. J. Embrechts, L. De Geetere, G. Vermeir, and M. Gomes, "Case studies in measurement of random incidence scattering coefficients", *Acta Acust. Acust.*, **90**, 858–67 (2004).
35. BS EN ISO 354:2003, "Acoustics—Measurement of sound absorption in a reverberation room".
36. M. H. A. Gomes, M. Vorländer, and S. N. Y. Gerges, "Aspects of the sample geometry in the measurement of the random-incidence scattering coefficient", *Proc. Forum Acusticum Sevilla*, RBA-06-002-IP (2002).

37. J. Ducourneau, A. Faiz, and J. Chatillon, "New device for measuring mapping of sound scattering coefficients of vertical uneven surfaces in a reverberant workplace", *Appl. Acoust.*, **90**, 21–30 (2015).
38. T. J. Hargreaves, "Acoustic Diffusion and Scattering Coefficients for Room Surfaces", PhD thesis, University of Salford (2000).
39. J. J. Embrechts, D. Archambeau, and G. B. Stan, "Determination of the scattering coefficient of random rough diffusing surfaces for room acoustics applications", *Acta Acust. Acust.*, **87**, 482–94 (2001).
40. E. Feldman, "A reflection grating that nullifies the specular reflection: A cone of silence", *J. Acoust. Soc. Am.*, **98**(1), 623–34 (1995).
41. T. J. Cox and P. D'Antonio, "Acoustic phase gratings for reduced specular reflection", *Appl. Acoust.*, **60**(2), 167–86 (2000).
42. E. Mommertz, "Determination of scattering coefficients from reflection directivity of architectural surfaces", *Appl. Acoust.*, **60**(2), 201–4 (2000).
43. Y. Kosaka and T. Sakuma, "Numerical examination on scattering coefficients of architectural surfaces using the boundary element method", *Acoust. Sci. Tech.*, **26**(2), 136–44 (2005).
44. J. J. Embrechts, L. De Geetere, G. Vermeir, M. Vorländer, and T. Sakuma, "Calculation of the random-incidence scattering coefficients of a sine-shaped surface", *Acta Acust. Acust.*, **92**(4), 593–603 (2006).
45. A. Farina, "A new method for measuring the scattering coefficient and the diffusion coefficient of panels", *Acustica*, **86**(6), 928–42 (2000).
46. J. Redondo, R. Pico, B. Roig, and M. R. Avis, "Time domain simulation of sound diffusers using finite-difference schemes", *Acta Acust. Acust.*, **93**(4), 611–22 (2007).
47. M. R. Schroeder, "New method of measuring reverberation time", *J. Acoust. Soc. Am.*, 37(6), 409–12 (1965).
48. F. A. Bilsen, "Thresholds of perception of repetition pitch. Conclusions concerning coloration in room acoustics and correlation in the hearing organ", *Acustica*, **19**, 27–32 (1968).
49. S. Sato and J. Y. Jeon, "Evaluation of the scattered sound field by using the autocorrelation function of impulse responses", *Proc. IoA(UK)*, **28**(2), 210–7 (2006).
50. P. W. Robinson, A. Walther, C. Faller, and J. Braasch, "Echo thresholds for reflections from acoustically diffusive architectural surfaces", *J. Acoust. Soc. Am.*, **134**(4), 2755–64 (2012).
51. P. Rubak, "Coloration in room impulse responses", *Proc. Joint Baltic-Nordic Acoustics Meeting, Åland* (2004).
52. K. Srodecki, "Evaluation of the reverberation decay quality in rooms using the autocorrelation function and the cepstrum analysis", *Acustica*, **80**, 216–25 (1994).
53. F. A. Everest and K. C. Pohlmann, *The Master Handbook of Acoustics*, 5th edn, TAB Electronics Technical Library, New York (2009).
54. E. Zwicker, "Subdivision of the audible frequency range into critical bands", *J. Acoust. Soc. Am.*, **33**(2), 248 (1961).
55. B. C. J. Moore and B. R. Glasberg, "Suggested formulae for calculating auditory-filter bandwidths and excitation patterns", *J. Acoust. Soc. Am.*, **74**, 750–3 (1983).
56. A. N. Salomons, "Coloration and Binaural Decoloration of Sound Due to Reflections", PhD thesis, University of Delft (1995).

Chapter 6

Porous sound absorption

Typical porous absorbers are carpets, acoustic tiles, acoustic (open cell) foams, curtains, cushions, cotton, and mineral wools such as fibreglass. These are materials where absorption is mainly caused by the sound propagation that occurs in the network of interconnected pores; this creates viscous and thermal effects that cause acoustic energy to be dissipated. As discussed in Chapter 2, porous materials are widely used to treat acoustic problems, such as in cavity walls to reduce noise transmission and in cacophonous rooms to reduce reverberation. This chapter will detail the physical processes creating the absorption and models for predicting performance.

The first section gives a qualitative description of the use of porous absorbers; this will be followed by example materials in Section 6.2. Porous absorbers range from standard materials, such as mineral wool, to more specialized treatments, such as absorbent plaster. Sections 6.4 and onwards then outline the methods needed to predict the absorption coefficient. These sections start by outlining how the sound propagation within a porous absorbent might be modelled in terms of characteristic parameters of the material. There are empirical and semi-phenomenological approaches, and both are detailed. This chapter then proceeds to show how these acoustic parameters are combined with mounting conditions, to enable the absorption coefficient and surface impedance to be predicted, which is ultimately what is required in design.

6.1 ABSORPTION MECHANISMS AND CHARACTERISTICS

For the most common porous absorbents, losses due to vibrations of the material are usually less important than the dissipation as sound moves through the air in the pores. When sound propagates in small spaces, such as the interconnected pores of a porous absorber, energy is lost. This is primarily due to viscous boundary layer effects. Air is a viscous fluid, and consequently, sound energy is dissipated via friction with the pore walls. There is also a loss in momentum due to changes in flow as the sound moves through the irregular pores. The boundary layer in air at audible frequencies is submillimetre in size, and consequently, viscous losses occur in a small air layer adjacent to the pore walls. As well as viscous effects, there will be losses due to thermal conduction from the air to the absorber material. At high frequency the effects are usually adiabatic (no heat enters or leaves the system), and at low frequency, isothermal (temperature remains constant).

For the absorption to be effective, there must be interconnected air paths through the material; an open pore structure is needed. The difference in construction between an open and closed pore system is shown schematically in Figure 6.1.[1]

Figures 6.2 and 6.3 show the absorption coefficients for two porous absorbers, illustrating the effect of material thickness. The porous absorber is mounted on a rigid backing. These curves follow a characteristic shape with most absorption at high frequency and only small losses for the lowest frequencies. These curves can shift in frequency and move up and

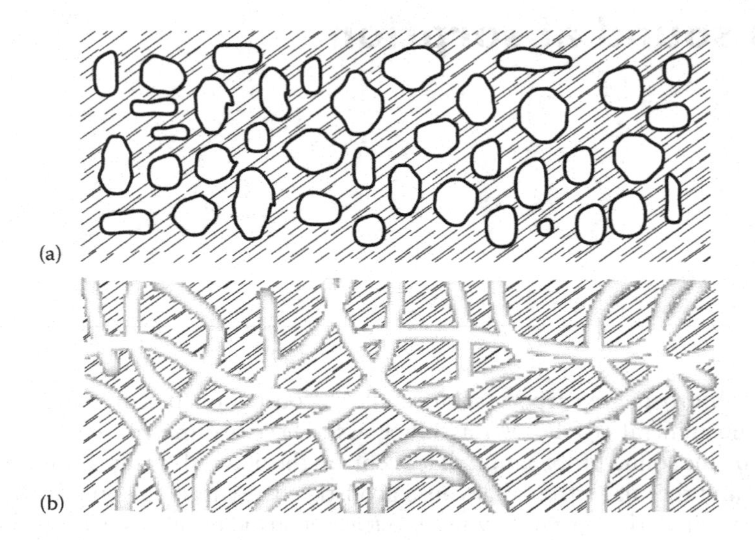

Figure 6.1 Illustration of the difference between closed (a) and open (b) pore structures. (Adapted from Cremer, L., Müller, H.A., *Principles and Applications of Room Acoustics Vol. 2*, Applied Science Publishers, 1978.)

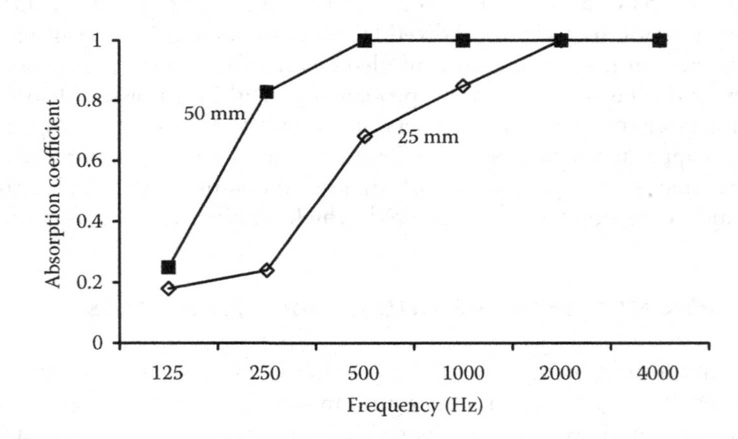

Figure 6.2 Random incidence absorption coefficient for mineral wool of two different thicknesses on a rigid backing.

down in absorption, depending on the characteristics of the particular material and how it is mounted.

For a rigid-backed absorbent, as the thickness of the porous material increases, more absorption at low frequency results (up until a critical depth, beyond which more thickness provides no additional absorption[2]). For the porous absorber to create significant absorption, it needs to be placed where the particle velocity is high. The particle velocity close to a solid room boundary is usually 0, and so the parts of the absorbent close to the boundary generate little absorption. It is the material furthest from the backing surface that are usually most effective, and this is why thick layers of porous material are needed to absorb low frequencies. A rough rule of thumb sometimes quoted is that the material needs to be at least a tenth of a wavelength thick to cause significant absorption[3] and a quarter of a wavelength to have the potential to absorb all incident sound. Consequently, substantial absorption

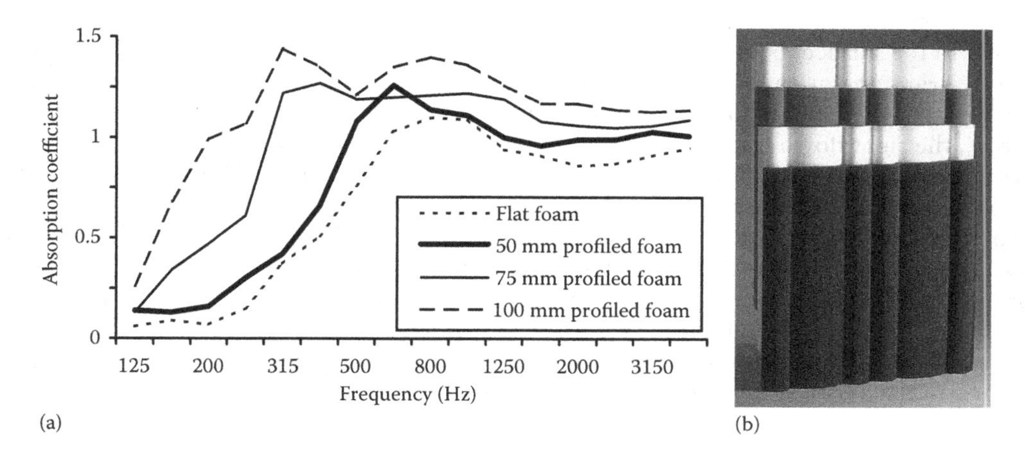

Figure 6.3 (a) Random incidence absorption coefficients for the profiled acoustic foam on a rigid backing for different depths shown in (b) (ProFoam).

cannot be achieved by simply applying a thin layer of paint. As the material very close to the boundary is absorbing relatively little, it is possible to simply space porous material away from a wall with an air gap behind, to improve performance. Figure 6.3 shows a simple way of achieving this by shaping acoustic foam, forming variable depth air cavities,[4] along with the absorption coefficients measured in a reverberation chamber.

The amount of energy absorbed by a porous material varies with the angle of incidence, as illustrated in Figure 6.4. For a mineral wool with high absorption, as illustrated by the

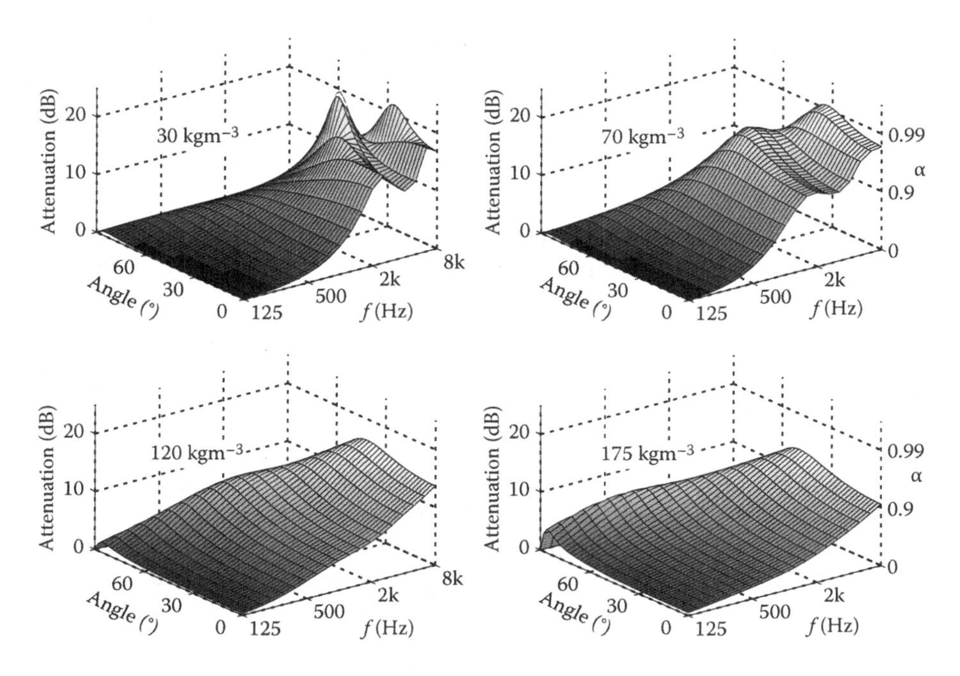

Figure 6.4 Attenuation of specular reflection from mineral wool as a function angle of incidence and frequency. Each plot represents a different mineral wool density as marked on the charts. On two of the charts, the equivalent absorption coefficient is also marked; 38 mm of mineral wool on a rigid backing, predicted using an empirical model.

top right plot, the absorption first increases as the angle of incidence moves away from the normal to the surface, before tailing off as the source moves around to grazing. The performance varies most with angle of incidence for the least dense mineral wools, however. The graphs also show how important it is to get the right density of mineral wool, or to be more correct, the right flow resistivity. With too high a flow resistivity, the impedance mismatch between the air and the absorbent causes the sound to reflect from the front face and the absorption reduces because not enough of the acoustic wave enters the material.

For low frequencies, where the wavelength is large, one has to go a considerable distance from the wall to reach a point where the particle velocity is significant. This makes porous absorbers inefficient and not particularly useful at low frequency. At bass frequencies, resonant structures will produce greater absorption from a given depth, as discussed in Chapter 7. Getting broadband passive absorption, across the frequencies of most interest in architectural acoustics, usually requires a combination of resonant and porous absorption.

6.2 SOME MATERIAL TYPES

There are many different porous materials that can make acoustic absorbers, and they fall into three rough categories as illustrated in Figure 6.5.[5] Fibrous materials, an example of which is shown in Figure 6.5c, form a complex interconnected network of air passages between the fibres. Synthetic or natural materials can be used, with mineral wools such as glass fibre being a common type. Foams absorb via small interconnecting pockets of air; they have a cellular structure, as shown in Figure 6.5a. Granular absorbers form voids between mesoscopic elements, as the image of the grains of sand illustrate (Figure 6.5b). Porous asphalt, ground, and other aggregates are common examples of granular porous absorbers. Higher absorption can be gained from granular material if they have microscopic pores within the grains,[6] as shown in Figure 6.5d from a scanning electron microscope (SEM). The additional absorption is created by having double porosity (activated carbon is one material with this property, see Section 6.2.8).

6.2.1 Mineral wool

Mineral wool is made from materials such as sand, basaltic rock, and recycled glass. The raw materials are melted at high temperature and then spun or pulled into woolly filaments. The filaments are bonded together to give the product its physical shape, with roughly 1%–5% of the final product weight being binder.[7] Glass fibre is made up of the same raw ingredients as normal glass such as sand, limestone, and soda ash, and the manufacturing process is similar to rock or basalt wool, although rock and basalt wool products tend to be heavier. The acoustic absorption is determined by the fibre composition, fibre orientation, fibre dimensions, product density, and the quantity and nature of the binder used. The mineral wool can be in the form of semirigid boards or loose blanket. Compressed mineral fibre board is the basis of the ubiquitous absorbing ceiling tiles that get mounted in T-bar grids.

Manmade mineral wools are cheap to manufacture and can be partially recycled. However, the manufacturing process uses considerable energy, which greatly increases the environmental impact of the material. The acoustic performance of mineral wool varies with density, and for low densities, absorption is reduced. Density alone is not sufficient to predict acoustic performance; however, because the fibre diameter is also a crucial factor, as the formulations later in this chapter show. Mineral wool is often laid down in layers and so is anisotropic (the acoustic properties vary with the angle of the sound propagation—see Section 6.8). For this reason, the acoustic properties vary depending on whether sound is

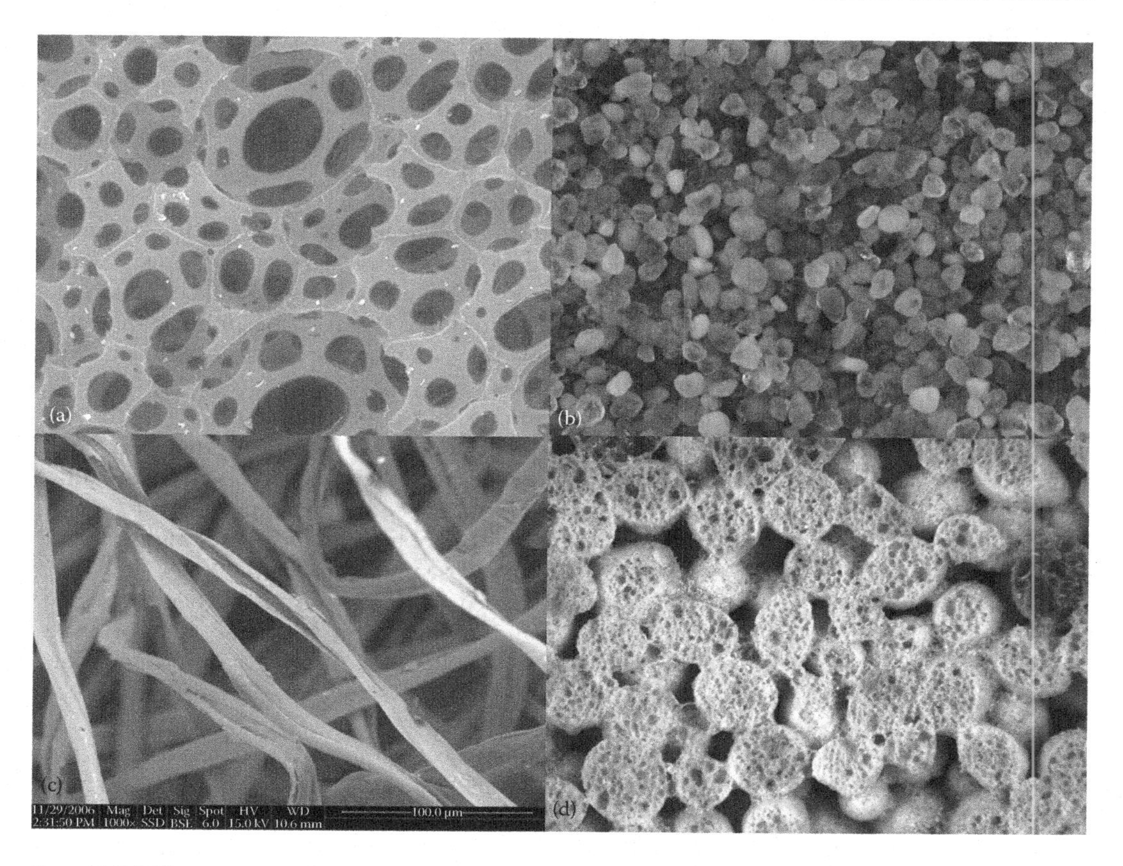

Figure 6.5 Different types of porous absorbers. (a) SEM image of white foam; (b) granular material (sand) under a microscope; (c) SEM image of cotton fibres; and (d) SEM image of a double porosity material. (SEM images courtesy of Jesus Alba, Del Rey Romina, and Vicente Jorge Sanchis, Universitat Politècnica de Valencia and Henkel KGaA. Sand image courtesy of Siim Sepp, commons.wikimedia.org/wiki/File:Sand_from_Gobi_Desert.jpg Licensed under CC BY-SA 3.0.)

incident parallel or perpendicular to the fibres, although for simplicity, this is often ignored in prediction models.

There have been concerns about the long-term health effects of manmade vitreous fibres (MMVFs), which have helped develop a market for non-MMVF and fibreless absorbers. MMVFs are known to be irritants, causing skin, eye, and upper respiratory tract irritation; the irritation is usually caused by mechanical action rather than an allergic reaction. Of particular concern are fibres with diameters of <3 μm, which can reach the pulmonary regions of the lung and form the greatest cancer hazard. In the past, the lack of scientific data led authorities to give cautionary classifications to MMVF. However, more recent studies have failed to find evidence that MMVFs are a significant risk to health. In 2002, The International Agency for Research on Cancer, which is part of the World Health Organization, reclassified mineral wool as category 3: "not classifiable as to their carcinogenicity to humans", meaning there is no strong evidence that mineral wool is carcinogenic to humans.[8,9] Subsequent studies (e.g., Reference 10) have failed to find evidence that MMVFs increase lung and head and neck cancers among workers exposed to rock and glass wool. Confounding exposure to asbestos hampers such studies.

6.2.2 Foam

Foams are cellular structures that can be open or closed cell. Open cell foams, ones with high reticulation, have pores that are interconnected and significant absorption can result. Closed cell structures, on the other hand, do not readily permit the passage of sound into the air pockets, and so the absorption is much lower. It is possible, however, to perforate closed foam structures at the end of manufacture and so provide moderate absorption by interconnecting the pores.[11] Consequently, it is important to check that open cell foam is used where sound absorption is needed. As the reticulation rate decreases, as the pores become less interconnected, material properties such as flow resistivity, tortuosity, and characteristic length ratio increase.[12] (These parameters partly determine the acoustic performance and will be described in Sections 6.4 and 6.5.2.) While this can increase low-frequency absorption, performance is poorer at mid–high frequencies.

Polyurethane (PU) foam is probably most common in architectural acoustics, although some companies produce melamine foams and researchers have explored other polymers. The fire rating of acoustic foams needs to be checked, especially when they are used in buildings. Metal and ceramic foams have fire resistance but are usually too expensive for general architectural use. They can find application where there are high temperatures and/or harsh environmental conditions such as acoustic liners in aero-engine mufflers.

Metallic foams with dead-end pores have attracted interest from researchers trying to model and enhance porous materials. These materials can show increased absorption at frequencies determined by the average length of the dead-ends.[13] A study[14] on porous absorbents with periodic dead-end pores has indicated that these structures can offer enhanced absorption at lower frequencies.

For completely open foams with high reticulation, decreasing the cell size increases the flow resistivity (and reduces the characteristic lengths).[12] This means that for thin samples, smaller cells give an appropriate flow resistivity for high absorption. For thicker layers, larger cell structures are needed. The size of the cells can be controlled during manufacture, for example, by changing the salt size and salt-to-polymer ratio.[15]

6.2.3 Sustainable materials

Section 1.2 outlined many of the general issues involved in choosing sustainable treatments.

Sheep wool has been suggested as a possible replacement for mineral wool and has the advantage of having a much lower impact on global warming than manmade mineral fibres.[16] However, in forms that are easy to manufacture, it is inherently a low-density (10–100 kgm^{-3}), low-flow-resistivity material (500–15,000 rayls m^{-1}), and so it needs to be relatively thick (5–10 cm) to achieve high absorption.[17] It is useful in sound insulation applications, filling cavity walls, where infill material needs to have only moderate absorption to be effective.

Crops such as cotton and hemp[18] can be made into absorbers and have the advantage of removing carbon dioxide from the atmosphere while they are growing. But this then requires land that could otherwise be devoted to growing food. When considering such natural fibres, the environmental impact during the growing of the crop needs consideration. For example, cotton requires far more pesticides, herbicides, land, and water than hemp does.[19] Any use of toxic chemicals to make natural products fire retardant or resistant to attacks from fungi and parasites needs to be included in an environmental audit. Also, the source of energy during processing needs consideration. Some *green* products use less sustainable manmade materials such as PU and polyester fibres as binders.

Oldham et al.[20] investigated cotton, flax, ramie, wool, jute, hemp, and sisal as possible sources of plant fibres to use in porous absorbers. Of the samples they tested, the ones that had smaller diameter fibres and were capable of being compacted to form a denser material performed best. The samples of cotton, flax, and ramie they measured had absorption coefficients comparable to that of mineral wool.

Oldham et al. also showed that thick layers of unshredded straw or reeds with roughly 5 mm diameter stems could be made into absorbers. Imagine the reeds laid horizontally on the floor of a reverberation chamber. In this configuration, slit-like voids are created because the stems have different diameters. Alternatively, the reeds can be set up vertically with the cut ends exposed. In this case, absorption is caused by quarter wave resonance within the narrow voids between the stems. Figure 6.6 compares the random incidence absorption coefficient for these two cases.

There is great interest in trying to make absorbers from recycled materials, whether that be cloth, metal, foams, wood, plastics, or rubber. To take one example, researchers have been investigating recycling tyres as absorbers.[21] The rubber can be painted and will survive the harsh environment around roads better than standard fibrous materials can. The tyre rubber is broken up into small granules, and these are then bonded together with a binder. The key is to use enough binder to hold the granules together without blocking the air pores, which are crucial to absorption, and to get the right grain size and shape. To take another example, Swift and Horoshenkov[22] showed that loose granulated mixes of waste foam with particle sizes <5 mm can be pressed into consolidated, elastic, porous media with a high proportion of open and interconnected pores and good absorption properties. Pfretzschner[21] tested rubber granular diameters ranging from 1.4 to 7 mm. Example absorption coefficients are shown in Figure 6.7, mostly measured in an impedance tube. They found that, for a given sample thickness, the absorption coefficient increases when the diameter of the grains decreases.

The behaviour of granular materials is different from that of fibrous ones such as mineral wool because grains pack together differently, usually resulting in a lower porosity for granular materials in comparison to fibrous ones. The absorption varies more with frequency for granular materials than for fibrous absorbents, as shown in Figure 6.7 in comparison to Figures 6.2 and 6.3. There is a critical thickness for the granular material, beyond which additional depth does not increase absorption. The broadband absorption for thick samples of the granular material shown in Figure 6.7 is limited to around 0.8, whereas with fibrous absorbent, the value can rise to unity.

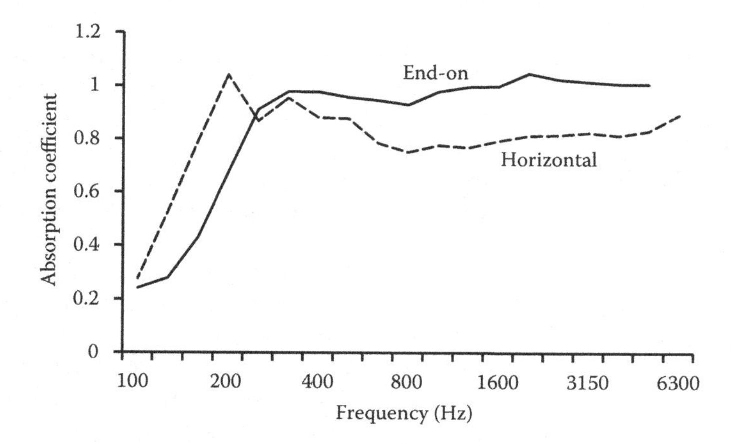

Figure 6.6 Random incidence absorption coefficient for reed samples where the stems are parallel or perpendicular to the floor. Samples are 14–15 cm thick. (After Oldham, D.J., Egan, C.A., Cookson, R.D., *Appl. Acoust.*, 72, 350–363, 2011.)

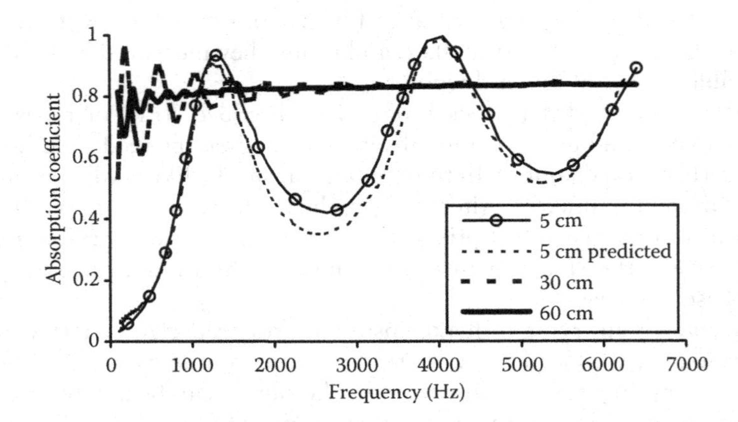

Figure 6.7 Absorption from granulated rubber 1 to 3 mm in size for different thicknesses. Measured in the impedance tube unless otherwise indicated in legend. (Modified from Pfretzschner, *J., Proc. Forum Acusticum*, 2002.)

Venegas[23] derived an analytical expression for limiting the high-frequency value of the absorption coefficient, α_{max}, of

$$\alpha_{max} = \frac{4\xi}{(\xi+1)^2}, \tag{6.1}$$

where

$$\xi = \frac{\phi}{\sqrt{\alpha_\infty}}, \tag{6.2}$$

where ϕ is the open porosity (the fractional air volume available to sound waves; see Section 6.4.2) and α_∞ is the tortuosity (the complexity of air paths through the absorbent; see Section 6.5.2.2).

Consequently, fibrous materials often give better acoustic absorption, but the recycled granular materials might be favoured for other reasons, such as sustainability. The performance of the recycled granular absorbent can be improved by forming wedges rather than flat boards or by having double porosity (see Section 6.2.8). Also included in Figure 6.7 is a prediction of the absorption coefficient using a porous absorber model, showing that good accuracy from such theories can be obtained; a similar theory is outlined later in this chapter.

In specifying or developing an environmentally friendly or sustainable absorbent, many issues have to be considered. It is preferable that the whole life cycle of the product is considered. Products made from recyclable materials are welcomed, but issues such as energy consumption during manufacture and the environmental impact of binders need to be considered. Another issue is whether the product is recyclable at the end of its life. An example of a product that achieves this, formed from either bonded stone aggregates (Quietstone) or a lighter weight recycled glass (Quietstone Lite), is shown in Figure 6.8. Both are impact resistant, highly absorptive, weather resistant, noncombustible, and durable. Due to its high density, the heavier material can be used in noise barriers. Quietstone Lite can be

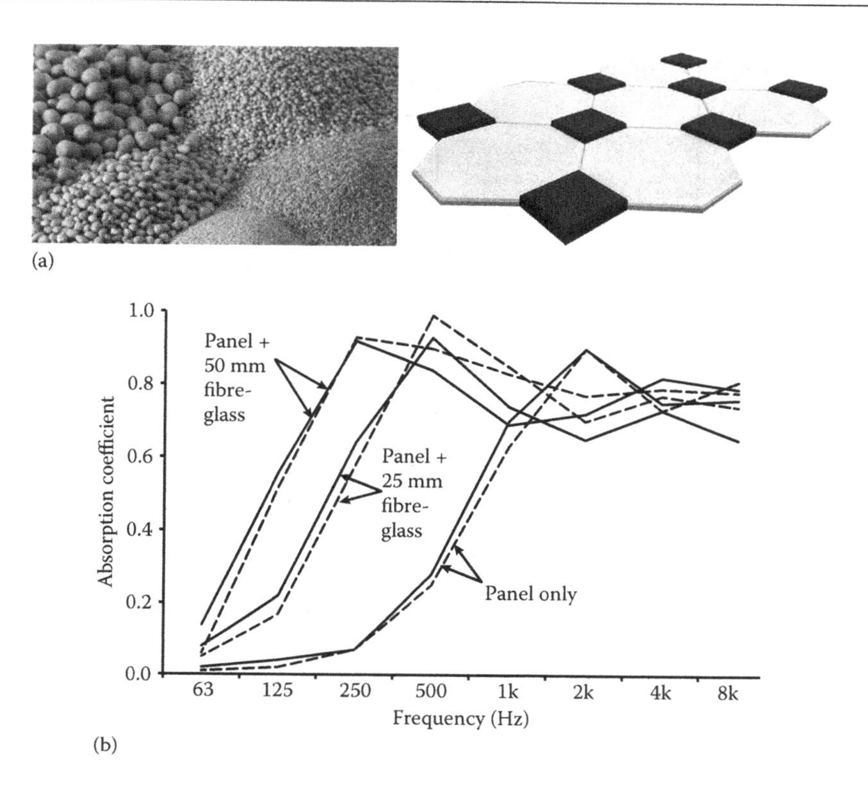

Figure 6.8 (a) 100% sintered recycled glass and an example of geometric shapes cut from an absorber made from the glass. (b) Random incidence absorption coefficients for Quietstone (solid lines) and Quietstone Lite (dashed lines) for three different mounting conditions as indicated.

lightly painted, cut into shapes, and used as the absorbent core for absorbent plaster systems described in Section 6.2.6. The absorption coefficients for both forms are shown in Figure 6.8.

6.2.4 Curtains (drapes)

Curtains or drapes are essentially porous absorbers. Most of the time, the deeper the folds are, the greater the absorption,[24] as this means that there is more resistive material and the absorbent is further from the rigid backing where the particle velocity is greater. This is illustrated in Figure 6.9, where the same curtain is hung with different fullness. It is also possible to increase the absorption by hanging the curtain away from the rigid surface and so placing the resistive material where the particle velocity is higher, hence producing more absorption. Increasing the density of the material generally increases the absorption produced,[25] as shown in Figure 6.10. Appendix A gives further data showing that the absorption coefficient varies greatly depending on the type of curtain and mounting.

Thin, transparent curtains are often used in modern interiors, but they normally achieve minimal acoustic absorption. New fire-retardant, translucent, and lightweight curtain fabrics are available with woven microslits, however. They provide absorption coefficients of 0.5–0.6 from 500 Hz upwards when installed in a ripple fold in front of a wall or windows.

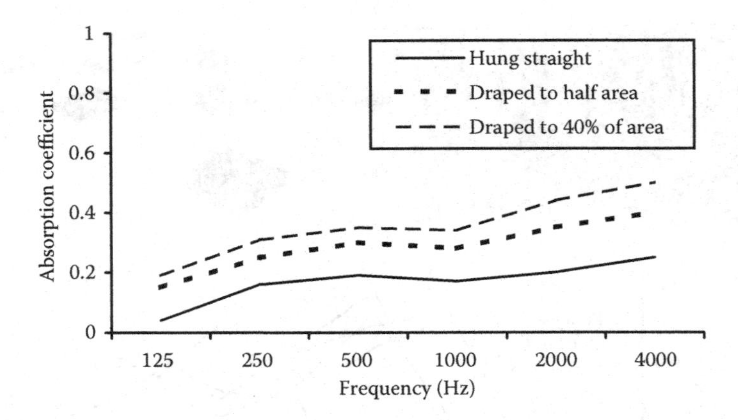

Figure 6.9 Curtain (drape) random incidence absorption coefficients with different fullness of draping. (Data from Harris, C.M. (ed), *Handbook of Noise Control*, 2nd edn, McGraw-Hill, 1991.)

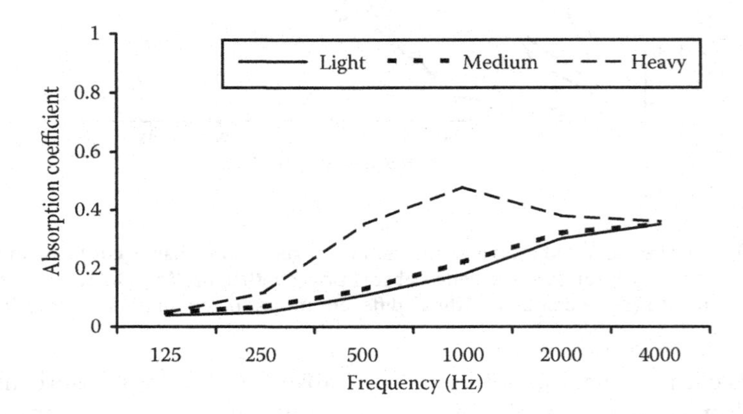

Figure 6.10 Curtain (drape) random incidence absorption coefficients for different material weights. (Data from L.L. Beranek, *Acoustics*, McGraw-Hill, 1954.)

6.2.5 Carpets

If present, carpet usually contributes a large proportion of the high-frequency absorption present in many rooms. The amount of absorption depends on the type of carpet and underlay.[24,26] A carpet is a porous material and so has little absorption at low frequencies but causes significant attenuation at higher frequencies. If the underlay is open cell, then its presence increases the thickness of the porous material and so boosts the absorption. Some underlay types are open cell, such as old-fashioned felt hair and foam rubber. Sponge rubber, however, can be open or closed cell. The absorption is also dependent on the type of carpet, for instance, the way the pile is constructed. Appendix A gives a large number of absorption coefficients for different carpet types taken from the literature. Everest[26] notes that the absorption coefficients reported for carpet vary quite considerably between different publications. This is illustrated in Figure 6.11, where minimum, mean, and maximum absorption coefficients in the literature are shown. This emphasizes the need for measurement of the carpet to be used, rather than assuming that an average value from the literature will be accurate.

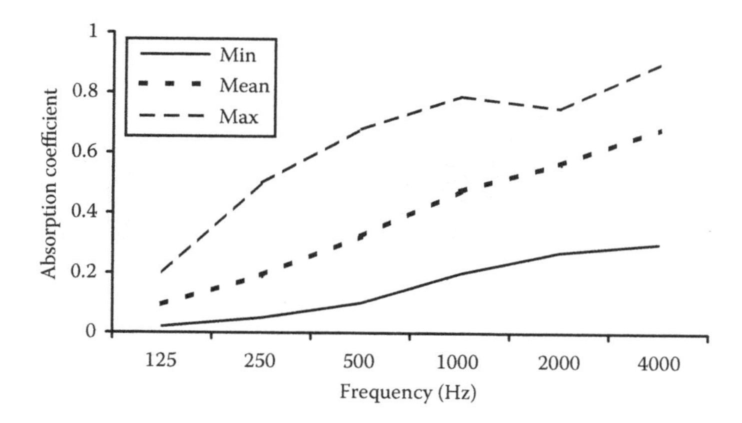

Figure 6.11 The minimum, mean, and maximum values for carpet absorption from the literature; random incidence absorption coefficients.

6.2.6 Acoustic plaster

An absorptive, smooth, seamless, and durable plastered finish is useful in satisfying the aesthetic requirements of architects and interior designers. There are several approaches to achieving a seamless appearance that looks like plaster. Current products use an absorptive substrate, whether that is mineral wool, fibreglass, or a noncombustible recycled glass panel, and acoustically transparent surface layer that looks like plaster. This surface layer is typically composed of some aggregate, such as marble particles, inorganic granulate, and cotton, with a binder that allows the pores between the aggregate to remain open when the binder dries. The installation of seamless areas is possible up to a maximum of 200 m^2. The maximum length or width should not exceed about 15 m. Expansion joints are needed for larger areas and intersecting boundary surfaces, to avoid cracking.

All of these systems are applied on-site. To ensure that a seamless surface is achieved with a perfect optical finish, it is necessary to start with a ceiling that is perfectly level and clean. Once the absorptive substrate panels are glued or mechanically attached to the surface and levelled, the seams between the panels must be filled and sanded smooth. Next, depending on the desired smoothness of the finish, one or more surface layers of the plaster-like coating are applied. When the porous substrate is mineral wool and a smooth surface is desired, typically a base coat is hand applied or sprayed on, hand trowelled and sanded, followed by a finish coat, which is applied and meticulously hand trowelled. If the final finish can be rougher, then a fine layer of the final coat may be sprayed on. When the absorptive substrate is recycled glass board, the seam fill process is followed by several thin base coats sprayed on, undercoat, and finish coat with the desired degree of smoothness. Some sanding may be required to remove rough areas prior to the final coat. As should be apparent from this description, the main disadvantage of acoustic plasters is that they are slow to apply. Another issue is that any damage from impact usually requires extensive areas to be refinished. The final acoustical performance is also dependent on the skill of the applicator to a certain extent.

A graphic illustrating an absorbing substrate with a succession of acoustically transparent layers with the granule size decreasing with each layer is shown in Figure 6.12. The top layer granules are so small that they provide the appearance of a smooth, seamless conventional gypsum-plastered surface. In fact, the top layer does seal the surface a little and acts as a thin membrane. As might be expected, this produces additional low-frequency absorption, and a little loss of absorption at high frequencies, in comparison to the substrate alone.

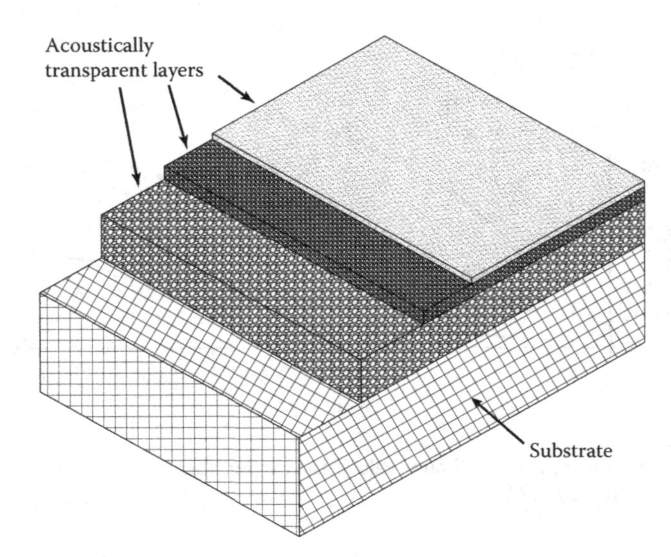

Figure 6.12 A schematic illustrating an absorbing substrate with a succession of acoustically transparent layers with the granule size decreasing with each layer.

6.2.7 Aerogels

Aerogels are highly porous solids formed from a gel in which the liquid is replaced with a gas. They have very high porosities, being 95% air, and so can be used as porous absorbents. Unfortunately, the lightest aerogels are fragile, and the denser ones are less suitable as acoustic absorbents because they have too high flow resistivity due to the small pores within the material. However, it is possible to use them in a granular form,[27] and provided that layers are built up to provide impedance matching to an incident wave, good absorption can be obtained.

6.2.8 Activated carbon

Activated carbon is made by first carbonizing some organic matter, such as coconut husks in an inert atmosphere, and then oxidizing this material by exposure to carbon dioxide or steam. Activated carbon has a large surface area and is widely used in chemical filtration and purification.

At low frequencies, a layer of activated carbon provides much stronger sound attenuation than other porous absorbers do. For instance, Figure 6.13 shows the absorption coefficient for a layer of activated carbon compared to a sample of sand. The excess absorption is due to reductions in surface reactance rather than changes in surface resistance.

The exact mechanism by which activated carbon produces strong low-frequency absorption is still being researched, however. At mid–high frequencies, much of the behaviour can be explained by a double porosity model.[6] Figure 6.14 shows SEM images at two magnifications. The voids between the grains of activated carbon form mesoscopic pores (millimetre in size). But the grains themselves contain much smaller microscopic pores (micrometre in size).

Pressure diffusion within the microscopic pores alters the macroscopic bulk modulus of the material. (The bulk modulus of a porous material measures how the air resists changes

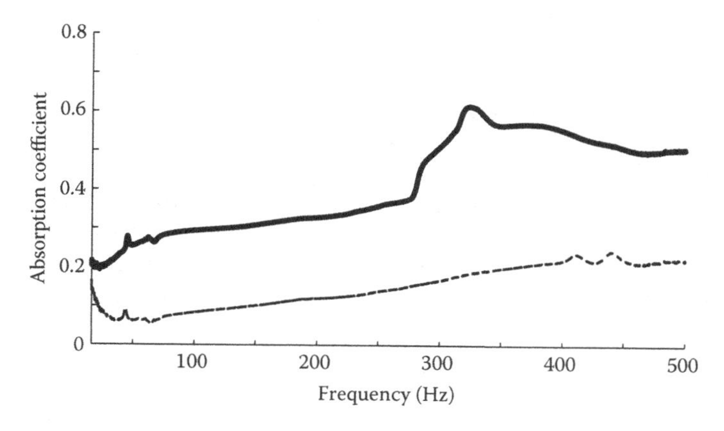

Figure 6.13 The low frequency absorption coefficient of: ——— activated carbon and -------- sand. Normal incidence measured in an impedance tube.

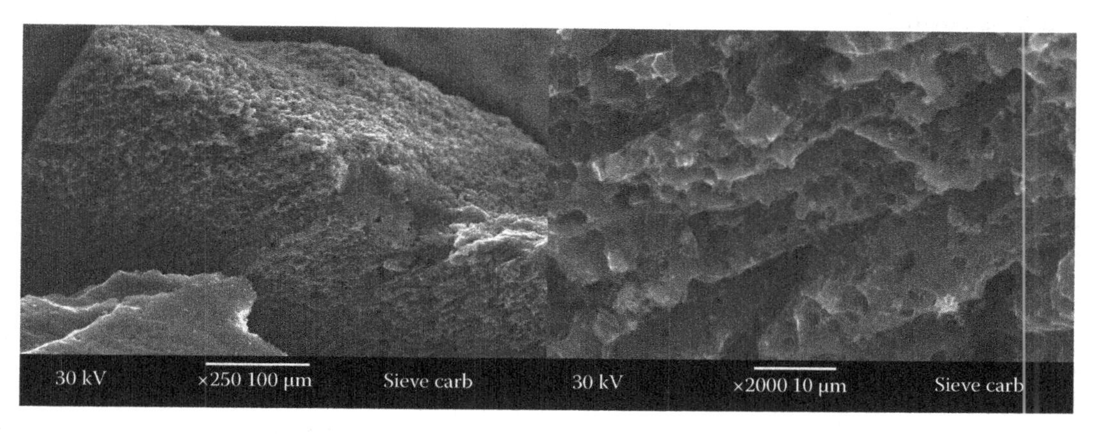

Figure 6.14 SEM images of activated carbon. The left image contains mostly one grain, whereas the right image is at higher magnification and shows the microscopic pores within a grain.

in volume under compression and hence is important in determining how sound waves propagate.) Within a certain frequency range, the average pressure in the mesoscopic pores and that inside the microscopic pores present a phase mismatch, and this leads to dissipation of sound energy. The pressure inside the grains is governed by a diffusion equation and depends on the parameters of the inner-grain structure (e.g., permeability, bulk modulus, or wave number) and on the mesoscopic geometry. This phenomenon seems to be exclusive to double porosity materials with high permeability contrast, i.e., materials for which the ratio between the characteristic size of the microscopic pores and that of mesoscopic pores is in the order of a 10^{-3}.[28]

But double porosity alone cannot explain the low-frequency behaviour. This is more likely to be caused by the adsorbent (note the *d*) properties of activated carbon. Gas adsorption is a process whereby gas molecules adhere to a surface due to the Van der Waals potential. Adsorbed molecules form a film on the surface of the carbon with a thickness of one or several molecular layers. If the gas pressure increases, as happens during sound wave compressions, then the number of adsorbed molecules increases. Desorption is the opposite process, whereby the molecules leave the surface and return to the surrounding atmosphere, and this happens during sound wave rarefactions. Adsorption and desorption requires energy, which comes from the sound wave, and hence generates some additional acoustic attenuation. Venegas and Umnova[29] have developed a triple porosity model including sorption effects that explains the behaviour of activated carbon quite well.

Bechwati et al.[30] measured the effective porosity of a sample of activated carbon below 50 Hz and found that at low frequency, it exceeded 1, attributing this apparently unphysical value to sorption of gases such as water vapour. Because the density of the gas is being changed by the adsorption and desorption cycle, if activated carbon is placed in a box, the compliance of the enclosure changes (see Section 7.5.6 for an application of this to resonant absorbers).

To make this sorption process effective, it appears necessary to have a structure with at least two scales of porosity—large pores are needed to allow sound to enter the material, and much smaller pores are needed to facilitate the adsorption and desorption. This is achieved by activated carbon.

6.2.9 Ground

When sound propagates outdoors, a decrease in sound pressure level at low frequencies is often measured. This dip is due to the interference between the direct sound straight from source to receiver and the reflection from the ground. This is known as the ground effect. The frequency of the dip depends on the positions of the source and receivers relative to themselves and the ground, as well as the acoustic properties of the ground itself. The calculation of the frequency and depth of the ground effect is enshrined in many guides and standards for calculating environmental noise.[31]

The surface impedance of earth is significantly altered by the soil/earth composition, roughness, degree of compaction, and moisture content. For example, the mid-frequency absorption of soil can vary between 0.2 and 0.9 depending on the moisture content.[32] Cultivated farmland offers significant decreases in A-weighted sound pressure levels for broadband sources in comparison to grassland, as subsoiling, discing, and ploughing change the flow resistivity of the surface and the variation of the soil properties with depth. In addition, ploughing can result in periodic surface roughness that leads to a diffraction effect, which can further attenuate sound. Attenborough et al.[33] showed that ploughed ground can produce up to 10 dB greater attenuation (A-weighted) for a broadband sound at a 50 m range than would be expected from *ISO-soft* ground. Earth is normally modelled as a porous absorber (see Section 6.6.3). Section 8.5 examines the related subject of absorption by vegetation.

6.3 COVERS

Often, porous absorbers are covered in cloth or plastic. For example, a thin impervious membrane might be used to wrap a fibrous absorbent when it is used as a duct lining, to prevent fibres being lost at the ventilation system. Another example is a thin layer of light reflecting material, which, when placed in front of mineral wool, makes a sound absorbing mirror.[34] Such impervious membranes will reduce the high-frequency absorption. At low frequencies, the membrane's acoustic mass is small and the sound passes through largely unaltered—although a small increase in absorption may occur. At high frequencies, the membrane's acoustic mass is high and it will prevent some or all of the sound wave entering the porous material. This membrane effect is why porous absorbers should not be painted except with a nonbridging paint. Most paints will block the pores, prevent sound waves freely entering the structure, and so altering the absorption.

Within rooms, porous absorbers are often finished by cloth wrapping to protect the absorbent material and make it look better. Nowadays, it is possible to print images on the cloth to create a striking effect, as Figure 2.5 showed. The cloth potentially has little effect on the absorption, provided it freely allows sound to enter the porous material. If glue is used to fix the cloth to the front face of the absorber, however, care must be taken to ensure that the high-frequency absorption is not unintentionally reduced because the glue prevents sound entering the porous material.

There are occasions where acoustically semi-transparent covers can favourably improve the absorption.[35] Traditionally, acousticians might seek a textile that is transparent to incident sound so that it does not modify the absorption of the porous material core being covered. The need for transparency excluded many fabrics with acrylics or other backings, as well as densely woven fabrics. However, a better target in many cases is semi-transparency, as this can equalize the absorption coefficient of the core porous material by attenuating high frequencies and boosting mid and low frequencies.

This was extensively investigated in an impedance tube study. The set-ups used are shown in Figure 6.15. By measuring the absorption coefficient in front of an anechoic wedge, the transparency of the fabric can be gauged. The anechoic wedge alone has an average absorption coefficient of 0.98 from 125 to 4000 Hz. Consequently, a completely transparent fabric will have an absorption coefficient of 1. The transparency is taken as an average of the absorption coefficients across the four octave bands from 250 to 2000 Hz. (An alternative technique would be to follow the method described in Section 4.1.3 and measure the transmission coefficient.) Following the measurement of transparency, the absorption coefficient of a 50 mm fibreglass panel with and without fabric applied is obtained with a rigid termination, as shown in Figure 6.15b.

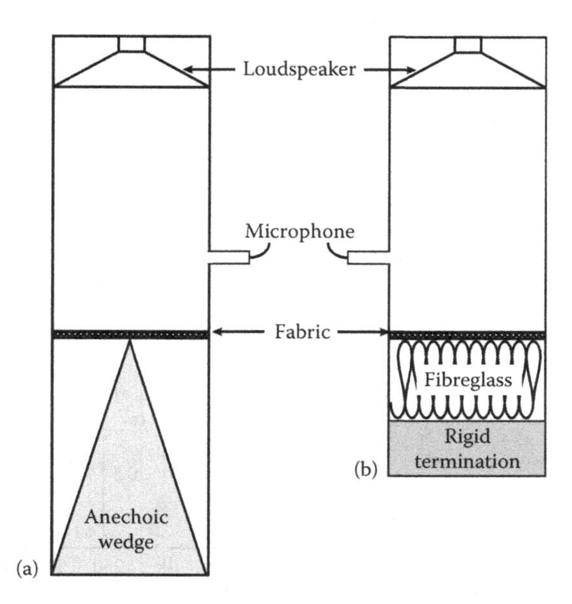

Figure 6.15 (a) Impedance tube with fabric under test using an anechoic termination. (b) Measurement of the absorption coefficient when the fabric is applied in front of a 50 mm fibreglass panel with a rigid termination.

In Figure 6.16, the results for a range of fabrics, with transparencies ranging from 95% to 40%, are shown. The fabric transparencies are shown on each plot, which includes a comparison between the absorption coefficient of the 50 mm fibreglass with and without the fabric. The gray area in each plot illustrates the range of absorption coefficients measured for several fabrics with a given transparency. For example, the 85% transparency graph includes the results from 18 fabrics. As the transparency of the fabric decreases, the high-frequency absorption reduces and the mid- to low-frequency absorption increases due to the diaphragmatic resonance of the fabric.

If the desire is to achieve a lot of high-frequency absorption with a low-frequency roll-off, then a highly transparent fabric is a good choice. If a more uniform absorption coefficient is desired, however, then a backed fabric or dense weave with a 75% transparency would be more appropriate. Or, if even more low-frequency absorption is desired, then the lowest transparency fabrics could be used.

Porous materials are often mounted behind perforated panels to protect the absorbent from damage. If the perforated sheet does not have a very open structure, the mass effect of the air in the holes will increase the absorption at low frequency, while at high frequency, the absorption will decrease absorption (see Figure 6.17). A commonly quoted guideline is that a greater than 20% open area means that the perforated sheet has little effect.[31] The transfer matrix techniques outlined in Chapter 7 can be used to predict the effects of perforated sheets on absorption. For membrane-wrapped porous material behind a perforated sheet, it is important that the membrane and perforated sheets are not in contact, otherwise the absorption is decreased,[31] except in the case of plate resonators (Section 7.2.7).

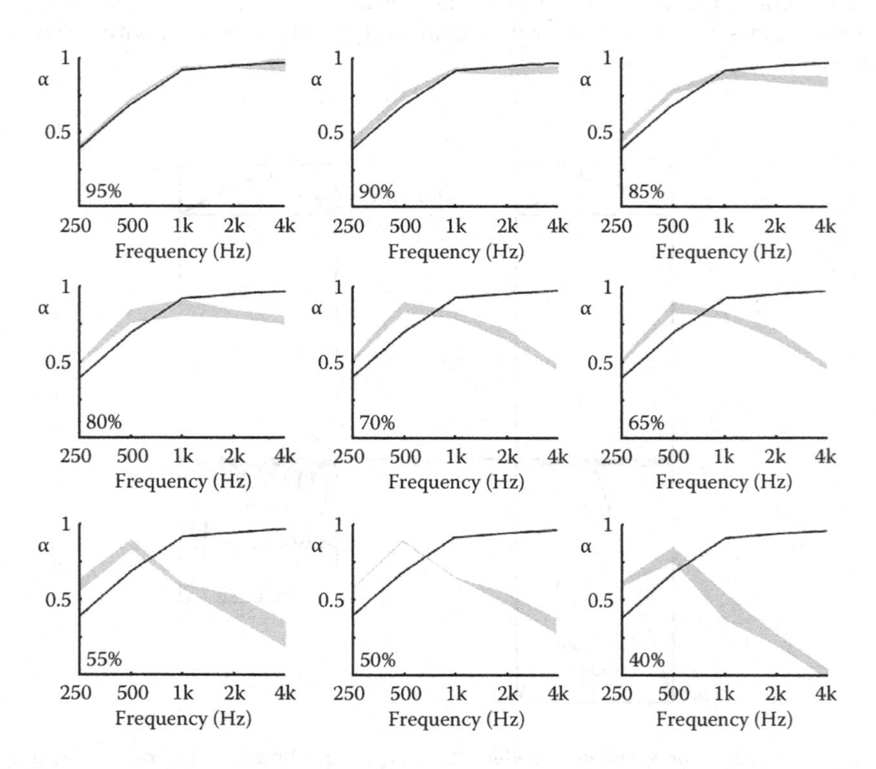

Figure 6.16 Comparison of the normal incidence absorption coefficient of a 50 mm fibreglass panel with and without applied fabrics with various transparencies as indicated in the bottom left of each plot.

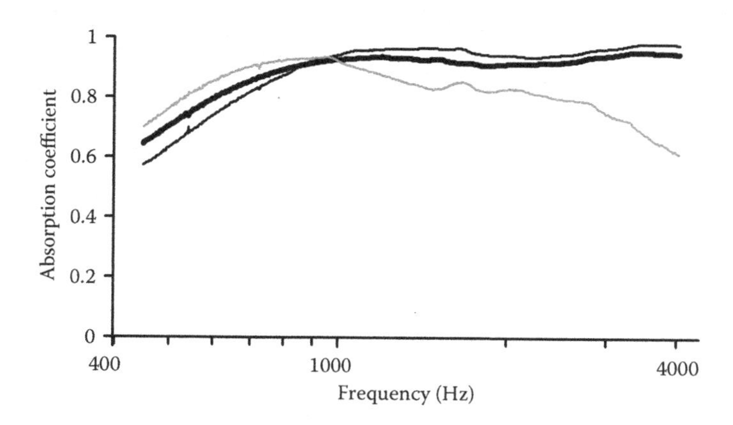

Figure 6.17 Measured normal incidence absorption coefficient for fibreglass with different finishes: ——— uncovered, ▬▬▬ cloth covering, and ——— perforated wood covering.

6.4 BASIC MATERIAL PROPERTIES

The rest of this chapter is devoted to the modelling of porous absorbents. Given a particular material, how can the surface impedance and absorption coefficient be estimated? The mathematical models also give insight into how absorption is produced and enable better products to be designed. To get the surface impedance and absorption coefficient, the characteristic impedance and wavenumber of the porous absorbent need to be known (or equivalently, other pairs of material properties such as the effective density and bulk modulus, or the viscous and thermal permeabilities, or the dynamic tortuosity and compressibility). To get these, some material properties are needed that determine how sound propagates through the absorbent. The most basic parameters used to characterise a porous absorbent are shown in Table 6.1. The first two parameters in Table 6.1 are usually the most important, flow resistivity and porosity, and so are considered in the next sections. The other material properties are used in more complex models as detailed in Sections 6.5.3 and 6.5.4.

6.4.1 Flow resistivity

Flow resistivity is a measure of how easily air can enter a porous absorber and the resistance that air flow meets through a structure. (Sometimes, this is called *airflow resistivity*, but in the context of airborne acoustics, the prefix *air* is redundant). It therefore gives some sense of how much sound energy is lost due to boundary layer effects within the material.

Table 6.1 Porous absorber material properties

Property	*Evaluates*
Flow resistivity, σ	Ease with which air enters and propagates through porous absorbent.
Open porosity, ϕ	Fractional amount of air volume within the absorbent readily available to sound waves.
Tortuosity, α_∞	Complexity of air paths through the absorbent.
Viscous and thermal characteristic lengths, Λ and Λ'	Allows modelling of the different thermal and viscous effects in different pore shapes.
Static viscous and thermal permeabilities, q_0 and q_0'	Need for most complex semi-phenomenological models.

A slice of the porous material of thickness d is subject to a mean steady flow velocity U. It is assumed that the flow velocity is small. The pressure drop ΔP is measured. From these quantities, the flow resistivity σ is defined as follows:

$$\sigma = \frac{\Delta P}{Ud}. \tag{6.3}$$

The flow resistance σ_s is defined as follows:

$$\sigma_s = \frac{\Delta P}{U} = \sigma d. \tag{6.4}$$

The flow resistivity is effectively the resistance per unit material thickness. If the flow velocity is not small, then non-linear factors must be considered. For instance, the flow resistivity formulations quoted in Equations 6.5 and 6.6 will break down for high-pressure sound waves, say when the sound pressure level exceeds 130 dB.[36]

It is important to check the units used with flow resistance and resistivity as two systems have been in use. The unit of flow resistance in MKS units is $\mathrm{Nm^{-3}s}$, often referred to as 1 rayl, and should nowadays be used. Older texts may use CGS rayls. To convert a CGS rayl to a MKS rayl, multiply by 10.

Flow resistivity is one of the most important parameters determining the absorption properties of conventional porous materials—if not the most important. It is the parameter that varies most between common porous materials and is the most important to determine. Section 4.5.1 details how flow resistivity is measured. Table 6.2 gives some example values (for ground, see Table 6.7).

There are several empirical and semiempirical formulations in the literature that can be used to estimate flow resistivity. For fibreglass, the following empirical relationship derived by Bies and Hansen[31,37] can be used:

$$\sigma = 7.95\times 10^{-10}\left(\frac{\rho_m^{1.53}}{a^2}\right), \tag{6.5}$$

where a is the fibre radius and ρ_m is the bulk density of the absorbent. For mineral and glass fibre, see Table 6.3 for typical diameters. Bies and Hansen showed that fibrous materials have an approximately linear relationship between flow resistivity and density, but this is not necessarily true for foam. For fibrous materials, Bies and Hansen measured samples with resistivity values ranging from 2000 to 200,000 rayls $\mathrm{m^{-1}}$, and for one type of foam with flow resistivity between 2000 and 40,000 rayls $\mathrm{m^{-1}}$.

A number of other empirical relationships have been reported that deal with different porous absorbents, such as granular materials, and larger fibre sizes. These are summarized in Table 6.4. These formulations often require the radius of the fibres, and Table 6.3 gives some indicative values for different materials. Flow resistivity for ground can be found in Table 6.7.

Formulations to predict the effect of extreme heat on the properties of fibrous porous absorbers can be found in Reference 38. As the temperature increases, so does the flow resistivity. A simple formulation to allow for this is the following:

$$\sigma(T) = \sigma(20)\left(\frac{273+T}{293}\right)^{1.65}, \tag{6.6}$$

Table 6.2 Examples of flow resistivities reported in various acoustic studies

Material	*Flow resistivity (rayl m⁻¹)*
Fibrous materials	
Coir[39]	1400–1600
Felt[40,41]	26,000–73,300
Hemp fibres[41]	6200
Mineral wool	1000–150,000
Polyester fibres[42,43]	880–44,400
Wood fibres[44]	11,000–180,000
Cellular materials	
Melamine foam[41,45]	10,500–17,500
Metal foam[46]	50,000
Plastic foam[40]	3600
Polyimide foam[47]	1000
Polylactide and polyethylene glycol foam[48]	6700–13,000
PU foam[41,47]	4500–12,900
PU foam, fully reticulated[12,49]	380–3200
PU foam, partially reticulated[12]	3000–42,000
Granular materials	
Consolidated foam granulates[50]	152,000
Coustone (Quietstone)[50]	31,500
Dry sand[51]	85,000–314,000
Glass beads, 0.68 and 1.68 mm in diameter[50]	13,000 and 43,200
Gravel[52]	10,000
Lead shot[36]	1373
Open cell synthetic rubber[41]	123,500
Perlite[53]	4300–32,800
Porous pavement[54]	2000–15,000
Vermiculite[53]	7600–135,000
Other	
Asphalt[52]	30,000,000
Nitrile foam granulate[53]	2800
Porous aluminium[55]	205
Porous ceramic[56]	44,500
Wood shavings[44]	2500–54,000

Table 6.3 Typical fibre diameters reported in various acoustic studies

Material	*Diameter (μm)*
Bagasse[19]	20
Bamboo[19]	14
Ceramic[19]	2–6
Coir[39]	156–370
Cotton[19,20]	8–33
Flax[19,20]	19–22
Glass and mineral wools[19,44,57]	3–22
Graphite[19]	5–10
Hemp[19,20,57]	22–94
Jute[19,20]	20–81
Kenaf[19,58]	21–78
Kevlar[19]	12
Polyester[19,42,58]	3–48
Poly(lactic acid)[57]	62
Polypropylene[19,57]	5–25, 63
Ramie[20]	37
Rice paddy[59]	8–20
Sisal[20]	213
Sugar cane[60]	11–23
Wood[19,44]	16–38
Wool[20]	14, 37–63

where T is the temperature in °C and $\sigma(20)$ is the flow resistivity at 20°C. When using this with later formulations for impedance, it is necessary to also adjust the speed of sound and gas density for temperature:

$$c(T) = c(20)\sqrt{\frac{273+T}{293}}, \tag{6.7}$$

and

$$\rho(T) = \rho(20)\frac{293}{273+T}. \tag{6.8}$$

For any thin materials that might be used to cover porous absorbents, it is better to define the properties in terms of the flow resistance. Wire, glass fibre, and more normal cloths can be produced with a wide variety of resistance values (see Section 6.3).

6.4.2 Open porosity

Porosity gives the fractional amount of air volume within the absorbent. It is a ratio of the total pore volume to the total volume of the absorbent. For sound, what is of interest is the open porosity, the air volume that is readily available to sound waves. (As only open porosity is discussed, the word *open* will be dropped for conciseness.) Good absorbers tend to have high porosity; for example, most mineral wools have a porosity of about 0.98. But in designing an

Table 6.4 Empirical relationships for flow resistivity (see also Equation 6.5)

Material	*Flow resistivity (rayls m⁻¹)*
Parallel to the fibres, all fibres having the same radii[7]	$\sigma = \dfrac{3.94\eta(1-\phi)^{1.413}[1+27(1-\phi)^3]}{a^2\phi}$
Perpendicular to the fibres, all fibres having the same radii[7]	$\sigma = \dfrac{10.56\eta(1-\phi)^{1.531}}{a^2\phi^3}$ $6 \le a \le 10\ \mu m$ $\dfrac{6.8\eta(1-\phi)^{1.296}}{a^2\phi^3}$ $20 \le a \le 30\ \mu m$
Random fibre orientation, all fibres having the same radii[7]	$\sigma = \dfrac{4\eta}{a^2}\left[\dfrac{0.55(1-\phi)^{4/3}}{\phi} + \dfrac{\sqrt{2}(1-\phi)^2}{\phi^3}\right]$
Random fibre radius distribution with a mean radius of a and random fibre orientation	$\sigma = \dfrac{3.2\eta(1-\phi)^{1.42}}{a^2}$ fibreglass $\sigma = \dfrac{4.4\eta(1-\phi)^{1.59}}{a^2}$ mineral fibre
Polyester fibrous materials[42] $18 \le 2a \le 48\ \mu m$ $12 \le \rho_m \le 60\ kgm^{-3}$ $900 \le \sigma \le 8500$ rayls m⁻¹	$\sigma = \dfrac{25.989\times10^{-9}\rho_m^{1.404}}{(2a)^2}$
Polyester fibre[61] $6 \le 2a \le 39\ \mu m$ $28 \le \rho_m \le 101\ kgm^{-3}$ $4000 \le \sigma \le 70{,}000$ rayls m⁻¹	$\sigma = \dfrac{15\times10^{-9}\rho_m^{1.53}}{(2a)^2}$
Sheep wool[17] $22 \le 2a \le 35\ \mu m$ $13 \le \rho_m \le 90\ kgm^{-3}$	$\sigma = \dfrac{490\times10^{-6}\rho_m^{1.61}}{2a}$
Wood materials with short fibres[44] $2a \approx 30\ \mu m$	$\sigma = 20.8\,\rho_m^{1.57}$
Loose granular material[53]	$\sigma = \dfrac{400(1-\phi^2)(1+\phi^5)\mu}{\phi D^2}$
Consolidated granular material[62]	$\log_{10}(\sigma) = -1.83\log_{10}(D) - 0.96$
Perpendicular to the fibres, all fibres having the same radii[63]	$\sigma_\perp = \dfrac{1}{a^2}\dfrac{16\eta(1-\phi)}{(-2\log(1-\phi)-2\phi-\phi^2)}$
Parallel to the fibres, all fibres having the same radii[63]	$\sigma_\parallel = \dfrac{\sigma_\perp}{2}$
Granular material with particle radius r[64]	$\sigma = \dfrac{3\eta}{r^2}\dfrac{\beta^3(3+2\beta^5)}{(2+3\beta^5)-\beta(3+2\beta^5)}$; $\beta = (1-\phi)^{1/3}$

Note: η is the viscosity of air (1.84×10^{-5} Nsm⁻²) and ϕ is the porosity; $\phi = 1 - \rho_m/\rho_f$, where ρ_f is the density of the fibres or the grain material and ρ_m is the bulk density of the material. D is the characteristic particle dimension: $D^2 = V_g/0.5233$, where V_g is the number of the particles in a unit volume.

Table 6.5 Typical porosity values for some materials

Material	*Typical porosities*
Fibrous materials	
Felts[1,41]	0.83–0.97
Hemp[41]	0.99
Mineral wool[1,7,41]	0.92–0.99
Polyester[43]	0.96
Wood fibre board[7]	0.65–0.80
Wood wool board[7]	0.50–0.65
Cellular materials	
Open cell acoustic foams (e.g., PU)[7,12,41,65]	0.93–0.995
Open cell ceramic foams[19]	0.80–0.90
Open cell metal foams[19,46]	0.75–0.95
Open cell synthetic rubber[41]	0.83
Polylactide foams[48]	0.82–0.88
Partially reticulated foams[12]	0.97–0.98
Reticulated vitreous carbon[66]	0.91–0.97
Granular materials	
Consolidated foam granulates[50]	0.61
Coustone[50]	0.4
Expanded perlite	0.98
Glass beads (1.6 and 0.7 mm)[50]	0.34, 0.38
Gravel and stone chip fill[7]	0.25–0.45
Lead shot, random loose packing[55]	0.36–0.45
Nitrile foam granulate[53]	0.91
Open porous asphalt[54,67]	0.13–0.29
Perlite (granular)[53]	0.60–0.78
Porous render[1]	0.60–0.65
Pumice concrete[7]	0.25–0.50
Pumice fill[7]	0.65–0.85
Rubber crumb[53]	0.44–0.54
Sand[51]	0.39–0.44
Sintered metal[7]	0.10–0.25
Vermiculite (granular)[53]	≈0.65–0.68
Other	
Aerogel[19]	>0.75
Asphalt[52]	0.1
Brick[1]	0.25–0.30
Ceramic filters[1]	0.33–0.42
Firebrick[1]	0.15–0.35
Marble[1]	≈0.005
Porous ceramic[56]	0.43
Sandstone[1]	0.02–0.06

absorber, it is possible to trade off porosity against flow resistivity (and to a lesser degree the structural factors outlined later), but note, the parameters are not independent. When determining the porosity, closed pores should not be included in the total pore volume, as these are relatively inaccessible to sound waves (closed pores are most commonly found in foams, even ones designed to be open celled). Although the porosity is a key parameter, for many common porous absorbents, the value does not vary greatly and is close to unity. Table 6.5 gives some typical values. Porosity values for the ground can be found in Table 6.7.

6.5 MODELLING PROPAGATION WITHIN POROUS MATERIALS

Two approaches are found to be most useful for modelling sound propagation through porous materials. The first is an empirical approach as exemplified by the work by Delany and Bazley. They measured a large number of porous materials and used curve fitting to arrive at relationships describing how the characteristic impedance and propagation wavenumber vary with flow resistivity. When applied to an existing material type, this empirical technique is the simplest to use and can be very effective.

A second approach is to formulate the problem using a semi-phenomenological approach. For instance, the propagation within the pores can be modelled semianalytically by working on a microscopic scale (or analytically for idealised geometries). This approach results in more complex theoretical models than the empirical approach. It is this type of modelling that holds the best chance of enabling the development of new porous absorbers to be undertaken without resorting to a completely experimental approach. Section 6.5.2 details some of the additional material properties that are needed in this approach, and Section 6.5.3 gives details of some of the models for rigid-framed materials.

Those looking for a simple solution are advised to use the empirical approach given in Section 6.5.1 before skipping to Section 6.6 to see how these can be applied to predict the absorption coefficient.

6.5.1 Macroscopic empirical models such as Delany and Bazley

When predicting the absorption of porous absorbents, it is necessary to know the characteristics of the material in terms of the characteristic impedance and (complex) wavenumber. Section 6.6 details how these material properties can then be used to get the surface impedance and absorption coefficient of a porous absorbent. MATLAB® scripts can be downloaded from the URL in Reference 68 for three of the empirical models outlined.

The empirical models take a macroscopic view and the details of the propagation through every pore are not considered. The impedance and wavenumber are found empirically. For fibrous absorbent materials, Delany and Bazley[69] undertook a large number of impedance tube measurements and derived empirical relationships relating the impedance and wavenumber to the flow resistivity. These relationships are widely used as they give reasonable estimations across a wide frequency range.

The characteristic impedance, z_c, is given by

$$z_c = \rho_0 c_0 (1 + 0.0571X^{-0.754} - j0.087X^{-0.732}), \tag{6.9}$$

and the wavenumber, k, by

$$k = \frac{\omega}{c_0}(1 + 0.0978X^{-0.700} - j0.189X^{-0.595}), \tag{6.10}$$

where ρ_0 and c_0 are the density and speed of sound in air, respectively, and ω is angular frequency. X is given by

$$X = \frac{\rho_0 f}{\sigma}, \tag{6.11}$$

where f is the frequency and σ is the flow resistivity of the fibrous material.

A good empirical match was achieved, but there are restrictions on the applicability of these formulations:

- The porosity, ϕ, should be close to 1, which most purpose-built fibrous absorbers achieve.
- $0.01 < X < 1.0$, which means the formulations works only over a defined frequency range.
- The limits of the flow resistivity in the measurements were $1000 \leq \sigma \leq 50{,}000$ rayl m^{-1}.

Figure 6.18 shows a large number of measurements undertaken by Mechel and Grundmann (reported in English in Reference 7), which give a visual indication of the accuracy of the empirical fit. These graphs show the normalized propagation constant (jk/k_0) and normalized characteristic impedance ($z_n = z_c/\rho_0 c_0$) for the measurements, alongside a thin dashed line giving the Delany and Bazley empirical values. The results shown are for glass fibre; similar accuracy is obtained for basalt and rock wool.

It is known that the Delany and Bazley model gives erroneous low frequency behaviour,[70] as this is outside its range of applicability. But often, the absorption is low at these

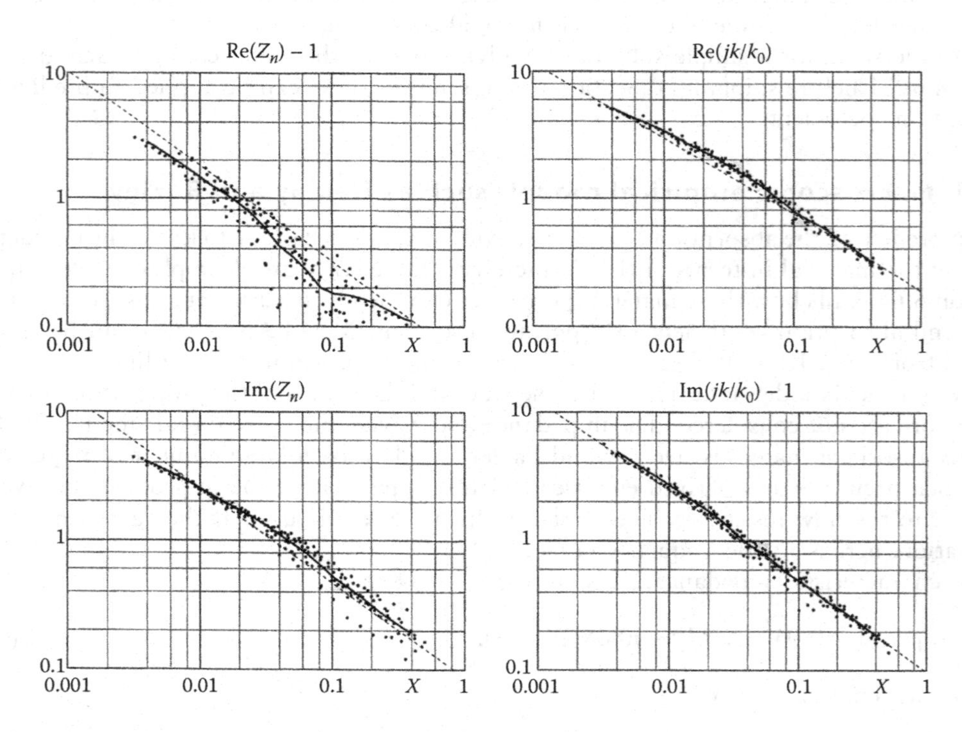

Figure 6.18 Normalized propagation constant and characteristic impedance for glass fibre. Points show measured data; the dashed line, the Delany and Bazley approximation; and the solid line, a running mean. (After Mechel, F.P., *Formulas of Acoustics*, Springer, 2002.)

frequencies, and so the inaccurate predictions are often not a problem. This is not always the case, however, and so various authors have produced improved relationships. For instance, Mechel and Grundmann produced a more complex set of empirical relationships.[7] They derive separate formulations for mineral fibre and glass fibre, and this enables improvement over the Delany and Bazley model. For many cases, however, the difference in the predicted quantities is small. Figure 6.18 shows the average for the Mechel and Grundmann measurements, which is presumably close to their empirical predictions, compared to the measurements and the Delany and Bazley predictions.

The Mechel and Grundmann empirical model can be most simply given as follows:

$$\left.\begin{matrix} jk/k_0 \\ \hline zn \end{matrix}\right\} = X^{-1}\beta_{-1} + X^{-1/2}\beta_{-1/2} + \beta_0 + X^{1/2}\beta_{1/2} + X_{\beta_1} + X^{3/2}\beta_{3/2}\,. \tag{6.12}$$

The coefficients are given in Table 6.6. The range of applicability is $0.003 < X < 0.4$. Other authors have also produced improved empirical models, especially to expand the applicability to other materials such as PU[49,71,72] and reticulated carbon foams[66] and other fibrous materials with larger fibre diameters[42] and granular material.[53]

One popular revision of the Delany and Bazley model was derived by Miki.[73] However, at low frequency, this can also produce effective densities that are physically incorrect.[74] The Wilson relaxation model outlined in Section 6.5.4, using the simplifications suggested in the paragraph below Equations 6.35 and 6.36, provide similar results to other empirical models and has the advantage that at low frequency the acoustic parameters are limited to correct physical values.

The ground is often modelled as a porous absorber. The Delany and Bazley formulations have been used quite extensively to obtain the characteristic impedance and wavenumber, even though these equations were developed for use with glass fibre rather than loose granular materials. For this, flow resistivity values for typical ground surfaces are needed, and Table 6.7 gives the appropriate values. While the Delany and Bazley model

Table 6.6 Coefficients for Mechel and Grundmann empirical model of porous absorbers

Coefficients	$-jk/k_0$	z_n
Mineral fibre (basalt or rock wool)		
β_{-1}	−0.003 557 57 − *j* 0.000 016 489 7	0.0026786 + *j* 0.003 857 61
$\beta_{-1/2}$	0.421 329 + *j* 0.342 011	0.135 298 − *j* 0.394 160
β_0	−0.507 733 + *j* 0.086 655	0.946 702 + *j* 1.476 53
$\beta_{1/2}$	−0.142 339 + *j* 1.259 86	−1.452 02 − *j* 4.562 33
β_1	1.290 48 − *j* 0.082 0811	4.031 71 + *j* 7.560 31
$\beta_{3/2}$	−0.771 857 − *j* 0.668 050	−2.869 93 − *j* 4.904 37
Glass fibre		
β_{-1}	−0.004 518 36 + *j* 0.000 541 333	−0.001 713 87 + *j* 0.001 194 89
$\beta_{-1/2}$	0.421 987 + *j* 0.376 270	0.283 876 − *j* 0.292 168
β_0	−0.383809 − *j* 0.353 780	−0.463 860 + *j* 0.188 081
$\beta_{1/2}$	−0.610 867 + *j* 2.599 22	3.127 36 + *j* 0.941 600
β_1	1.133 41 − *j* 1.74819	−2.109 2 0 − *j* 1.323 98
$\beta_{3/2}$	0	0

Source: Mechel, F.P., *Formulas of Acoustics*, Springer, Section G, 2002.

Table 6.7 Effective flow resistivity values for ground surfaces and other parameters

Surface	*Effective flow resistivity (rayl m⁻¹)*	*Water content (%)*	*Porosity*	*Porosity decay ζ (m⁻¹)*
Snow				
Dry snow, newly fallen 0.1 m over about 0.4 m older snow	1×10^4–3×10^4	–	–	–
Snow	5×10^3–16×10^3		0.5–0.9	
Sugar snow	2.5×10^4–5×10^4	–	–	–
Snow (new)	1×10^3–1×10^4	–	0.64–0.88	0–3
Snow (old crusted)	8×10^3–3×10^4	–	0.5	0–100
Snow (Great Himalaya snowpack)	1×10^3–2.3×10^4		0.52–0.89	
Forests and crops				
In forest, pine, or hemlock	2×10^4–8×10^4	–	–	–
Forest floor	7×10^3–20×10^4	–	0.4–0.8	0–40
Mineral layer beneath mixed deciduous forest	$(540 \pm 92) \times 10^3$	15	36.5	–
Humus on pine forest floor	$(230 \pm 220) \times 10^3$	16.1	58.1	–
Litter layer on mixed deciduous forest floor (2–5 cm thick)	$(30 \pm 30) \times 10^3$	–	–	–
Pine forest litter (6–7 cm thick)	$(9 \pm 5) \times 10^3$	28.6	38.9	–
Tall crops	4×10^4–5×10^4	–	–	0
Grass				
Sports field	1.4×10^6 [a] 5.1×10^5 [b] 9.9×10^5 [c]		0.24	−270
Lawn	1.2×10^5–7.5×10^5 [a] 1.2×10^5–3.1×10^5 [b] 3.9×10^4–3.7×10^5 [c]		0.34–1 [b]	−76 to 47
Long grass	4.7×10^4–1.8×10^5 [a] 3.7×10^4–8.5×10^4 [b] 1.4×10^5–8.0×10^5 [c]		0.38–1 [b]	−9 to 48
Grassland	7×10^4–8.5×10^5 1×10^5–2.4×10^5	–	0.3–0.7	0–250
Pasture	1.1×10^5–1.1×10^6 [a] 1.0×10^5–9.5×10^5 [b] 2.5×10^4–8.2×10^5 [c]	–	0.2–1 [b]	−400 to 80
Grass, rough pasture, airport, public buildings, etc.	15×10^4–30×10^4	–	–	–
Grass root layer in loamy sand	$(150 \pm 90) \times 10^3$	–	48 ± 4	–
Earth				
Earth, exposed and rain-packed	4×10^6–8×10^6	–	–	–
Bare earth	6.7×10^5 [a] 3.9×10^5 [b] 3.7×10^5 [c]		0.4 [b]	−49
Loose sand or dry cultivated soil	3×10^4–3.1×10^5	–	0.36–0.52	0–10
Sandy silt, hard packed by vehicles	8×10^5–2.5×10^6	–	–	–

(Continued)

Table 6.7 (Continued) Effective flow resistivity values for ground surfaces and other parameters

Surface	*Effective flow resistivity (rayl m⁻¹)*	*Water content (%)*	*Porosity*	*Porosity decay ζ (m⁻¹)*
Arable	9.3×10^4–1.9×10^6 [a]		0.15–1 [b]	−490 to 81
	7.5×10^4–1.6×10^6 [b]			
	2.2×10^4–1.7×10^6 [c]			
Cultivated soil	1×10^5–2×10^6		0.4–0.6	
Wet and compacted soil	4×10^7	–	–	−200 to 0
Heath	$1.7.3 \times 10^5$–3.2×10^5 [b]		0.71–1 [b]	17–33
	1.8×10^5–2.6×10^5 [b]			
	5.1×10^4–1.1×10^5 [c]			
Rock				
Roadside dirt, ill-defined, small rocks up to 0.1 m mesh	30×10^4–80×10^4	–	–	–
Clean limestone chops, thick layer (1–2.5 cm mesh)	1×10^6–4×10^6	–	–	–
Old dirt roadway, fine stones (5 cm mesh) interstices filled	2×10^6–4×10^6	–	–	–
Gravel	1.5×10^3–59×10^3	–	0.3–0.4	–
Other				
Quarry dust fine, very hard-packed by vehicles	5×10^6–20×10^6	–	–	–
Asphalt, sealed by dust and light use	$\approx 3 \times 10^7$	–	–	–
Porous asphalt, new	2×10^3–15×10^3	–	0.15–0.3	–
Urban	4.1×10^4–6.7×10^4 [a]	–	0.61–1 [b]	0–25
	3.5×10^4 [b]			
	2.1×10^4–6.7×10^4 [c]			
Sand				
Coarse sand, pore size 98 μm	5×10^4	0	–	−14
	1×10^5	11	–	−82
	4.7×10^5	51	–	−141
	9×10^4	95	–	1290
Fine sand	3.1×10^5	–	0.44	
Fine sand, pore size 65 μm	1.5×10^5	0	–	−28
	1.4×10^5	15	–	130
	1.5×10^5	48	–	620
	4.1×10^4	95	–	1130
Loamy sand	$(420 \pm 17) \times 10^3$	11.2	37.5	–
Bare sandy plain	$(370 \pm 110) \times 10^3$	9.3	26.9	–

Sources: Horoshenkov, K.V., Mohamed, M.H.A., *J. Acoust. Soc. Am.*, 120, 1910–1921, 2006; Attenborough, L., Bashir, I., Taherzadeh, S., *J. Acoust. Soc. Am.*, 129, 2806–2819, 2011; Attenborough, K., *J. Acoust. Soc. Am.*, 81, 93–102, 1982; Attenborough, K., *Acta Acust.*, 1, 213–226, 1993; Berryman, J.G., *Appl. Phys. Lett.*, 37, 382–384, 1980; Embleton, T.F.W., Piercy, J.E., Daigle, G.A., *J. Acoust. Soc. Am.*, 74, 1239–1244, 1983; Martens, M.J.M., van der Haijden, L.A.M., Walthaus, H.H.J., van Rens, W.J.J., *J. Acoust. Soc. Am.*, 78, 970–980, 1985; Datt, P., Kapil, J.C., Kumar, A., Srivastava, P.K., *Appl.Acoust.*, 101, 15–23, 2016.

[a] Fitted using Delany and Bazley model.
[b] Fitted using semi-phenomenological model.
[c] Fitted using variable porosity model.

works reasonably well above about 250 Hz, at lower frequencies, the characteristic impedance and wavenumber have significant errors, resulting in inaccurate predictions of excess attenuations. Modelling the porosity to vary with depth (see Section 6.6.3) or using a semi-phenomenological model (see Section 6.5.3) is usually more accurate.[75]

The problem with empirical models is that they do not readily give information about how the microscopic properties of the porous absorber, such as the pore size and orientation of the pores, affect the absorption produced. This means that it is difficult to use the empirical models to inform design, beyond finding the optimum flow resistivity. To improve materials, more detailed models of the propagation may well be required. In the next section, some of the parameters needed for more physically based porous absorber models are presented. Following on from this, some of these more complex semi-phenomenological models are presented.

6.5.2 Further material properties

Most theoretical models of porous absorbers assume that the material (or frame) of the absorbent is rigid. Then it is possible to apply some classical theories of sound propagation in small pores. There are a variety of analytical solutions for some geometries such as bundles of cylindrical pores; arrays of cylindrical fibres[63]; packing of identical grains[64,82]; and arrays of pores with slit, square, and triangular pores.[83] Alternatively, a semiempirical approach is often adopted, where a mixture of experiment and theory determines the key properties of the material. The necessary parameters are detailed in the following.

6.5.2.1 Viscous and thermal characteristic lengths

While the porosity and the flow resistivity are usually the most important parameters in determining the sound propagation in porous absorbents, other secondary parameters such as viscous and thermal characteristic lengths and tortuosity (see next section) can be important. The shape of the pores influences the sound propagation and, hence, the absorption. Altering the pore shapes changes the surface areas and hence changes the thermal and viscous behaviour. Analytically obtaining the characteristic lengths for most porous absorbents is impossible as they do not usually conform to simple geometric shapes. Consequently, the viscous and thermal characteristic lengths are usually empirically found by fitting acoustic measurements of the effective density and bulk modulus of the material. The characteristic lengths are therefore dependent on the model being used to predict the propagation within the absorbent. Later in this chapter, formulations for the effective density and bulk modulus are found, from which the characteristic impedance and wavenumber can be obtained. For now, the important characteristic lengths (Λ and Λ') used in the formulations are defined.

The viscous characteristic length Λ describes viscous effects at mid–high frequencies. It is defined as[84]

$$\Lambda = 2\frac{\int_{V_p} v_i^2 \, dV_p}{\int_{A_p} v_i^2 \, dA_p}, \qquad (6.13)$$

where v_i is the velocity of a nonviscous fluid, the integral on the numerator is carried out over the pore volume V_p, and for the denominator carried out over the pore surfaces A_p. It can be found for simple pore shapes using the following formulation:

$$\Lambda = \frac{1}{s}\sqrt{\frac{8\eta\alpha_\infty}{\phi\sigma}}, \tag{6.14}$$

where s is a constant. For most porous absorbers, s lies between 0.3 and 3, taking the following values 1, 1.07, and 1.14 for circular, square, and triangular pores, respectively, and 0.78 for slits. α_∞ is the tortuosity as defined in the following section, and η is the viscosity of air.

The thermal characteristic length Λ' is needed to account for the thermal boundary layer in the high-frequency analysis of heat conduction. It is a geometrical parameter defined by

$$\Lambda' = \frac{2V_p}{A_p}, \tag{6.15}$$

where A_p and V_p are the surface area and volume of the pores, respectively. This is the same ratio as used for Λ but without the weighting for microscopic velocity. Cylindrical pores are a special case where $\Lambda = \Lambda'$. In general,[85] $\Lambda' \geq \Lambda$, and to a first approximation, $\Lambda' = 2\Lambda$ and $s = 1$ can be used to derive simpler formulations for the sound propagation in rigid framed fibrous materials.[72]

Sections 4.5.6 and 4.5.7 detail the measurement of the characteristic lengths, and Table 6.8 shows some typical values. They can also be numerically calculated if the geometry is well described.[86] Complications arise when the material is anisotropic, which is common in many materials. For instance, mineral wool is often laid down in layers, in these cases, the characteristic lengths depend on the incident angle of the sound wave.

6.5.2.2 Tortuosity

The orientation of the pores relative to the incident sound field has an effect on the sound propagation. This effect[76] is represented by the parameter tortuosity, denoted α_∞. (Some older publications used the term *structural form factor*, and there were some differences in the definition.) How tortuous the propagation path is through the material affects the absorption. For simple cylindrical pores all aligned in the same direction, the tortuosity is simply related to the angle between the pores and the incident sound, i.e., $\alpha_\infty = 1/\cos^2(\psi)$. Formulations also exist for packed spheres. An empirical formulation is[77]

$$\alpha_\infty = \frac{1}{\sqrt{\phi}}. \tag{6.16}$$

An alternative expression is[78]

$$\alpha_\infty = 1 + \frac{1-\phi}{2\phi}. \tag{6.17}$$

Table 6.8 Various characteristic length values

Material	Characteristic length (μm) Viscous Λ	Thermal Λ'
Fibrous material		
Felt[40,41]	30–57	60–62
Fibreglass[41,45]	132–182	237–400
Polyester fibres[43]	73–86	133–161
Cellular material		
Cellular rubber[87]	9	15
Melamine foam[41,45]	81–240	255–470
Metal foam[46]	20	–
Plastic foam[40]	25, 207, 230	70 and 690
Poroelastic foam[65]	41–48	103–171
Polyimide foam[47]	39	–
Polylactide and polyethylene glycol foam[48]	5–12	75–167
PU, fully reticulated[12]	96, 200–450	280–600
PU, partially reticulated[12,47]	24–240	140–320
Granular materials		
2.1 mm lead shot[55]	280	490
4 mm lead shot[55]	500–550	730–830
9 mm gravel[55]	190, 290	–
Glass beads, 0.1 mm diameter	90	180
Perlite (expanded)[5]	5.1	15.4
Other		
Porous aluminium[55]	470, 770	–
Porous ceramic[88]	62	273
Snow[81]	49–156	131–582

And for granular materials[64]:

$$\alpha_\infty = \frac{1}{2}(3 - \phi)\,. \tag{6.18}$$

Real absorbents, however, are not normally that well ordered. Consequently, tortuosity needs to be measured. Techniques for doing this are discussed in Chapter 4. Some values are given in Table 6.9. Another approach is to calculate the tortuosity from the structure of the material at a microscopic scale.[23,82,86,89]

6.5.3 Semi-phenomenological models

Given the material properties (flow resistivity, porosity, tortuosity, viscous, and thermal characteristic lengths), it is possible to calculate the characteristic impedance and propagation wavenumber by considering the microscopic propagation within the pores. Many people have been involved in the development of the models, and Allard and Atalla[90] give a comprehensive summary of all the models and their variants. The absorber frame is assumed to be rigid.

Table 6.9 Example tortuosity values from the literature

Material	*Tortuosity*
Fibrous materials	
Felt[41]	1.01
Common fibrous absorbents, e.g., rock wool	1–1.06
Polyester[43]	1.03–1.05
Hemp[41]	1.01–1.05
Cellular materials	
Melamine foam[41,45]	1.01
Metal foam[46]	1.27
Plastic foam[65]	1.06 and 1.7
Poroelastic foam	1.24–4.45
Polyimide foam[47]	1.17
Polylactide and polyethylene glycol foam[48]	1.2–1.6
PU foam[41,47]	1.08–1.41
PU foam, fully reticulated[12]	1.04–1.06
PU foam, partially reticulated[12]	1.25–2.30
Granular materials	
Cellular rubber[87]	2.64
Consolidated foam granulates[50]	1.92
Coustone (Quietstone)[50]	1.66
Fused glass bead sample[91]	1.75–3.84
Lead shot[55]	1.46–1.54
Loose sand or dry cultivated soil	1.27–3.32
Glass beads, 0.1, 0.68, and 1.64 mm diameter[50]	1.46–1.87
Granular materials	1.1–1.8
Gravel[52,55]	1.5–1.8
Open porous asphalt[54,67]	2–3.3
Perlite[5]	2.04
Rubber crumb[53]	1.13–1.26, 1.38–1.56
Rubber, open cell synthetic[41]	2.64
Vermiculite[53]	1.48–1.58, 1.8–2.46
Other	
Asphalt[52]	1.8
Compacted soil[52]	1.4
Forest floor, top layer[52]	1.1
Nitrile foam granulate[53]	1.31, 1.49
Porous ceramic[88]	1.5
Porous aluminium[55]	1.07
Soft soil[52]	1.3
Snow[52,81]	1–1.6
Snow (new)	1.5–2.7
Snow (old crusted)	4

The Champoux-Allard model[70] is one of these semi-phenomenological models. The effective density of the porous material is given by

$$\rho_e = \frac{\alpha_\infty \rho_0}{\phi}\left[1 + \frac{\sigma\phi}{j\omega\rho_0\alpha_\infty}\sqrt{1 + \frac{4j\alpha_\infty^2\eta\rho_0\omega}{\sigma^2\Lambda^2\phi^2}}\right]. \tag{6.19}$$

The effective (or dynamic) bulk modulus of the air in the material is given by

$$K_e = \frac{\gamma P_0}{\phi}\left(\gamma - (\gamma - 1)\left(1 + \frac{8\eta}{j\Lambda'^2 N_p\omega\rho_0}\sqrt{1 + \frac{j\rho_0\omega N_p\Lambda'^2}{16\eta}}\right)^{-1}\right)^{-1}, \tag{6.20}$$

where γ is the ratio of the specific heat capacities ($\approx$1.4), P_0 is atmospheric pressure $\approx$101,320 Nm^{-2}, and N_p is the Prandtl number given by

$$N_p = \left(\frac{\delta_v}{\delta_h}\right)^2, \tag{6.21}$$

where δ_v and δ_h are the size of the viscous and thermal boundary layers. At 1 atmosphere and 20°C, the Prandtl number is about 0.77. The thickness of the viscous boundary layer is given by

$$\delta_v = \sqrt{\frac{2\eta}{\rho_0\omega}}. \tag{6.22}$$

Typically, the viscous boundary layer is submillimetre in size; for example, at 100 Hz, it is about 0.2 mm. The thickness of the thermal boundary layer is given by

$$\delta_h = \sqrt{\frac{2\kappa}{\rho_0 c_p\omega}}, \tag{6.23}$$

where $\kappa \approx 2.41 \times 10^{-2}$ WmK^{-1} is the thermal conductivity of air and $c_p \approx 1.01$ Jkg^{-1}K^{-1} is the specific heat capacity of air at constant pressure.

Once the effective density and bulk modulus have been determined from Equations 6.19 and 6.20, it is then possible to calculate the characteristic impedance and wavenumber for the porous materials. The characteristic impedance z_c is given by

$$z_c = \sqrt{K_e\rho_e}, \tag{6.24}$$

and the propagation wavenumber by

$$k = \omega\sqrt{\frac{\rho_e}{K_e}}. \tag{6.25}$$

The formulations give the correct high- and low-frequency asymptotic behaviour but are only approximately correct at mid frequencies for complicated pore geometries.

Figures 6.19 through 6.21 show the normalized characteristic impedance, propagation constant, and absorption coefficient for the model from Equations 6.19 and 6.20; these lines are labelled *semi-phenomenological*. These values are compared to the Delany and Bazley formulations of Equations 6.9 and 6.10. The following assumptions were made to implement the semi-phenomenological model: $\phi = 0.98$, $\alpha_\infty = 1$, $\Lambda' = \Lambda$, and $s = 1$. Both models give very similar results. As stated previously, the Delany and Bazley predictions are known to give inaccurate results at low frequency (the real part of the surface impedance of the porous absorber goes negative, implying that the absorbent increases the energy in the wave), but these are frequencies at which the absorption from the porous absorber is relatively small anyway.

The comparison in Figure 6.21 shows the absorption coefficient for the porous material on a rigid backing. How the absorption coefficient and surface impedance are calculated from the characteristic impedance and wavenumber is detailed in Section 6.6.

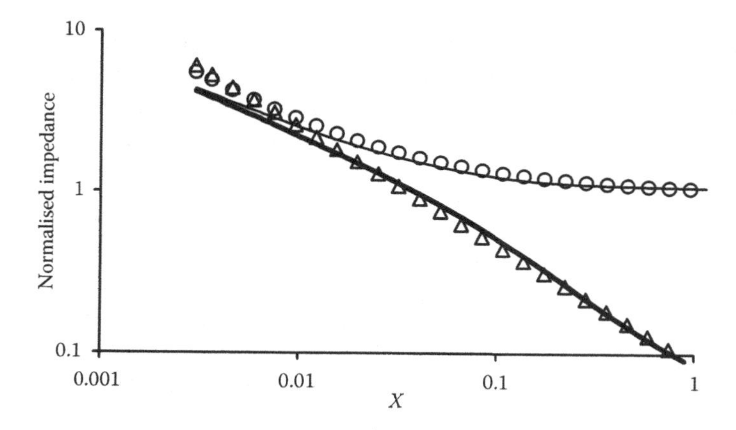

Figure 6.19 Two models for the normalized characteristic impedance ($z_c/\rho_0 c_0$) of a porous absorber. The x axis is $X = \rho_0 f/\sigma$. o Re (Delany and Bazley); —— Re (Champoux-Allard); Δ–Im (Delany and Bazley); and ▬▬ –Im (Champoux-Allard).

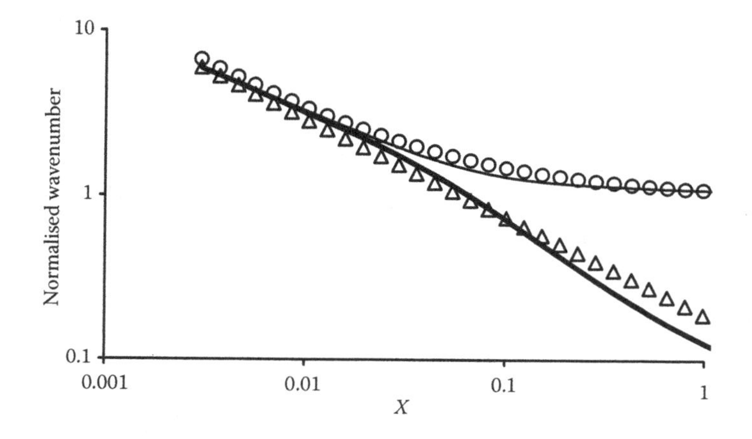

Figure 6.20 Two models for the normalized wavenumber (k/k_0, where k_0 is the wavenumber in air) for sound propagation through a porous absorber. The x axis is $X = \rho_0 f/\sigma$. o Re (Delany and Bazley); —— Re (Champoux-Allard); Δ–Im (Delany and Bazley); and ▬▬ –Im (Champoux-Allard).

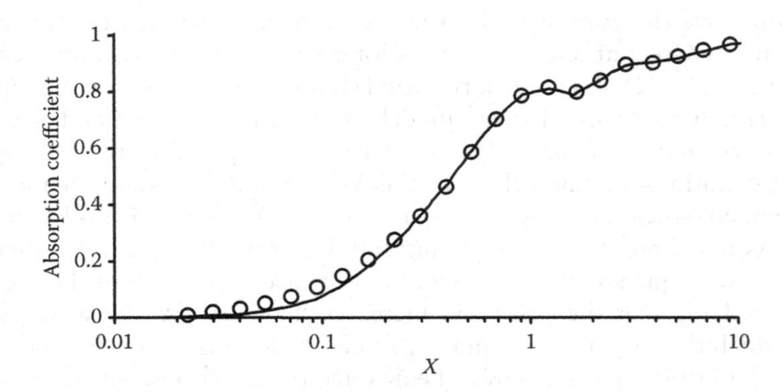

Figure 6.21 Two models for the absorption coefficient of a porous absorber. The *x* axis is $X = \rho_0 f/\sigma$: o Delany and Bazley and —— Champoux-Allard.

The Champoux-Allard model is known to sometimes produce errors in the estimation of the transition frequency between the isothermal and adiabatic regime.[90] This transition frequency is where thermal dissipation is at its greatest. This does not often cause too many problems when predicting surface impedance and absorption coefficients for conventional porous materials, however, because viscous effects normally dominate.

By measuring or predicting two additional parameters, the static thermal permeability, q_0', and the *static viscous permeability*, q_0, a more correct semi-phenomenological model can be developed: the JCAL (Johnson et al. Champoux-Allard, Lafarge et al.) model. A note on nomenclature: the subscript 0 indicates that these are the static values ($\lim \omega \to 0$), and the dash indicates it is a thermal value and the absence of the dash indicates a viscous value.

The JCAL model uses q, the dynamic viscous permeability given by[84]

$$q = q_0\left(j\frac{\omega}{\omega_v} + \sqrt{1 + \frac{j\omega}{\omega_v}\frac{M_v}{2}}\right)^{-1}, \tag{6.26}$$

where ω_v is the *viscous characteristic frequency* given by

$$\omega_v = \frac{\eta\phi}{\alpha_\infty q_0 \rho_0}, \tag{6.27}$$

and M_v is the *viscous shape factor* given by

$$M_v = \frac{8q_0\alpha_\infty}{\phi\Lambda^2}. \tag{6.28}$$

The only parameter needed to calculate q not previously defined is the static viscous permeability, q_0. By definition this is given by

$$q_0 = \eta/\sigma. \tag{6.29}$$

Consequently, the viscous effects in the JCAL model are accounted for with the same parameters that are used in the Champoux-Allard model. This is not the case for the formulations for the dynamic thermal permeability, q'.[56,92] The thermal formulations are

$$q'(\omega) = q_0'\left(j\frac{\omega}{\omega_t} + \sqrt{1 + \frac{j\omega}{\omega_t}\frac{M_t}{2}}\right)^{-1}, \tag{6.30}$$

where the *thermal characteristic frequency* is given by

$$\omega_t = \frac{\kappa\phi}{c_p\rho_0 q_0'}, \tag{6.31}$$

and the *thermal shape factor* is given by

$$M_t = \frac{8q_0'}{\phi\Lambda'^2}. \tag{6.32}$$

These formulations require the static thermal permeability, q_0'. This can be found from impedance tube measurements that yield the bulk modulus, with the frame of the porous material held static by needles so it cannot vibrate. Then, inverse analytical solutions derived from semi-phenomonological models can be used to get the permeability.[93,94] Measured values in the literature range from 1.2×10^{-10} to 2×10^{-8} m^2. (Note, the static thermal permeability is always greater than or equal to the static viscous permeability).

Once the thermal and viscous permeabilities are known (Equations 6.30 and 6.26), then the effective density and bulk modulus can be calculated from the following:

$$\rho_e = \frac{\eta}{j\omega q}, \tag{6.33}$$

and

$$K_e = \frac{\gamma P_0}{\phi}\left(\gamma - (\gamma - 1)j\omega\rho_0 c_p \frac{q'}{\kappa\phi}\right)^{-1}, \tag{6.34}$$

where the effective density (Equation 6.33) is identical to the Champoux-Allard model (Equation 6.19). Equations 6.24 and 6.25 can then be applied to get the characteristic impedance and wavenumber of the porous absorbent.

6.5.4 Relaxation model

Wilson[95] took an alternative approach to modelling the propagation through porous absorbents, viewing the thermal and viscous diffusion as relaxation processes. When sound propagates through porous materials, temperature perturbations result within the air inside the absorbent that then relax over time towards the equilibrium temperature. The characteristic time for this thermal relaxation process is denoted τ_t. Similarly, the sound wave sets up pressure gradients that induce changes in the flow velocity, which also relax towards the steady state. The characteristic time for this viscous relaxation process is denoted, τ_v. Writing the inverse of the bulk modulus and effective density in a relaxational form allows the characteristic impedance and wavenumber of the acoustic absorber to be derived in terms of the relaxation times and other characteristics of the absorber:

$$z_c = \frac{\rho_0 c_0 \sqrt{\alpha_\infty}}{\phi}\left[\left(1+\frac{\gamma-1}{\sqrt{1+j\omega\tau_t}}\right)\left(1+\frac{1}{\sqrt{1+j\omega\tau_v}}\right)\right]^{-0.5}, \tag{6.35}$$

and

$$k = \frac{\omega\sqrt{\alpha_s}}{c}\sqrt{\left(1+\frac{\gamma-1}{\sqrt{1+j\omega\tau_t}}\right)\Bigg/\left(1-\frac{1}{\sqrt{1+j\omega\tau_v}}\right)}. \tag{6.36}$$

With appropriate choice of relaxation times, these formulations give predictions that match the empirical and semi-phenomenological models discussed earlier in this chapter.[96] For instance, with $\tau_v = 2.54/\sigma$ and $\tau_t = 3.75/\sigma$, $\alpha_\infty = 1$, and $\varepsilon = 1$, the previous equations mimic the behaviour of the Delany and Bazley empirical formulations and have the advantage of not producing nonphysical values at low frequency.

The relaxation times are therefore tuned to fit the behaviour of the porous material in the frequency range of interest based on either measurement or other prediction models. So for frequency domain modelling, this relaxation method does not offer any great advantages over other models, except at low frequency.

6.6 PREDICTING THE SURFACE IMPEDANCE AND ABSORPTION COEFFICIENT

Once the characteristic impedance and wavenumber for the material are known, it is necessary to convert these to the surface impedance and absorption coefficient for a particular thickness of a porous material with known boundary conditions. The most flexible approach is the transfer matrix method. Consequently, this section starts by discussing the general case of propagation in one layer of a porous absorber. Then the specific case of a single layer with a rigid backing will be presented. The prediction model can be extended to multiple layers of absorbent if required.

Figure 6.22 shows the arrangement being considered. Only plane wave propagation will be presented, and for now, only normal incidence is considered. Section 2.6.1 showed how sound behaves when propagating from one medium to another. At each interface between the layers, continuity of pressure and particle velocity is assumed. This allows a relationship between the pressure and particle velocity at the top and bottom of a layer to be produced, which is compactly given in matrix format:

$$\begin{Bmatrix} p_{3b} \\ u_{3b} \end{Bmatrix} = \begin{Bmatrix} p_{2t} \\ u_{2t} \end{Bmatrix} = \begin{Bmatrix} \cos(k_2 d_2) & j\frac{\omega\rho_2}{k_2}\sin(k_2 d_2) \\ j\frac{k_2}{\omega\rho_2}\sin(k_2 d_2) & \cos(k_2 d_2) \end{Bmatrix} \begin{Bmatrix} p_{2b} \\ u_{2b} \end{Bmatrix}, \tag{6.37}$$

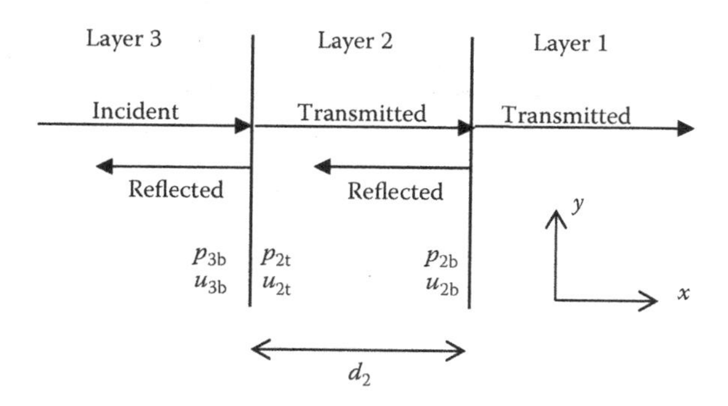

Figure 6.22 Geometry for propagation of sound through a layer of an acoustic medium.

where p_{2b} and u_{2b} are the pressure and particle velocity, respectively, at the bottom of layer 2; p_{3b} and u_{3b} are the pressure and particle velocity, respectively, at the bottom of the third layer; p_{2t} and u_{2t} are the pressure and particle velocity, respectively, at the top of layer 2; d_2 is the thickness of the second layer; ρ_2 the density of layer 2; and k_2 is the wavenumber for the second layer.

This formulation can be applied recursively to successive layers, and it is particularly powerful when calculating the surface impedance of an absorber, rather than simply a way of gaining values for pressure and velocity at the layer boundaries. If the top of the first layer has an impedance of z_{s1}, and layer 2 has a characteristic impedance z_{c2}, then the impedance at the top of the second layer is

$$z_{s2} = \frac{-jz_{s1}z_{c2}\cot(k_2d_2)+(z_{c2})^2}{z_{s1}-jz_{c2}\cot(k_2d_2)}. \tag{6.38}$$

This formulation can be applied repeatedly to calculate the surface impedance of a multilayered absorbent. Next a simple case is considered.

6.6.1 Single layer of a porous absorber with a rigid backing

Consider a single layer of absorbent with a rigid backing. The impedance at the bottom of the layer is infinite ($z_{s1} \to \infty$). Then, Equation 6.38 simplifies and gives the impedance on the surface of the absorbent as

$$z_{s2} = -jz_{c2}\cot(k_2d_2). \tag{6.39}$$

This can then be turned into absorption coefficients using Equations 2.21 and 2.24. Figure 6.23 shows the impedance changes as the thickness of the layer increases, using the Delany and Bazley empirical formulations for the porous material properties. As discussed previously, as the porous layer increases in thickness the absorption increases at low frequency as expected. MATLAB scripts can be downloaded from the URL in Reference 68 to demonstrate the use of these equations.

It is also possible to demonstrate the usefulness of air gaps in increasing absorption. Consider two cases: 2.5 cm of an absorbent mounted on a rigid backing and 1.25 cm of the

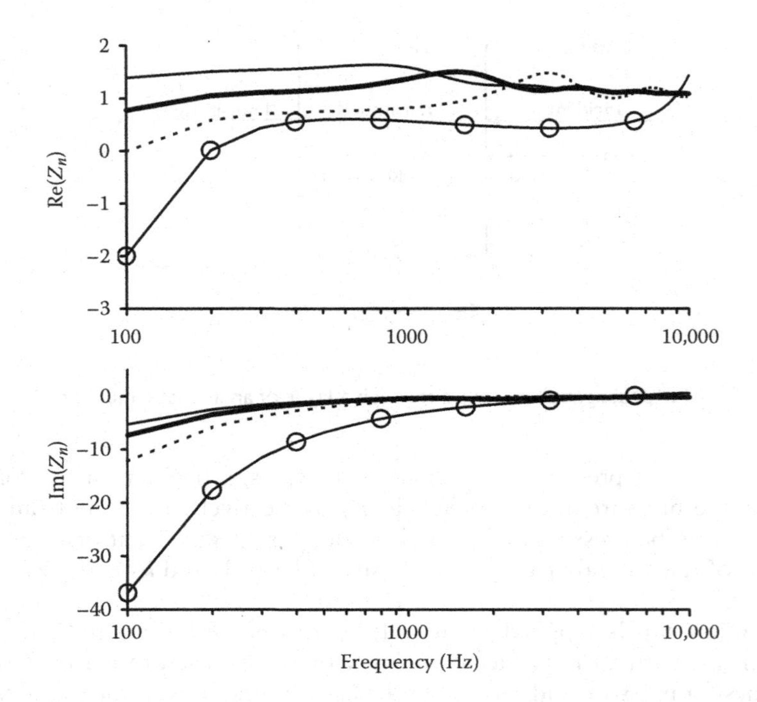

Figure 6.23 Effect of absorbent thickness: —— 88.9, ▬▬ 63.5, - - - - 38.1, and −⊖− 12.7 mm on the normalized acoustic impedance. Using Delany and Bazley model combined with the transfer matrix method.

same porous material mounted in front of a rigid backing with a 1.25 cm air gap. For the second case, two formulations are needed: for the air layer (layer 1),

$$z_{s1} = -j\rho_0 c_0 \cot(kd_1), \tag{6.40}$$

and for the porous absorbent layer (layer 2),

$$z_{s2} = \frac{-jz_{s1}z_{c2}\cot(k_2 d_2) + z_{c2}^2}{z_{s1} - jz_{c2}\cot(k_2 d_2)}. \tag{6.41}$$

Figure 6.24 compares the absorption coefficients for the two configurations. It shows that the absorbent with the air gap has very similar performance to a single thick layer of the porous material. This confirms the usefulness of air gaps, as discussed in Section 6.1 and elsewhere.

6.6.2 Modelling covers

Often, an absorbent is finished with a thin porous layer to make the treatment look better or more robust. Section 6.3 discussed how covers are used in detail. If the covering is not free to vibrate, then its resistance should be added to the impedance predicted from the previous formulations, such as Equation 6.39. If the covering material can vibrate, then the effect of

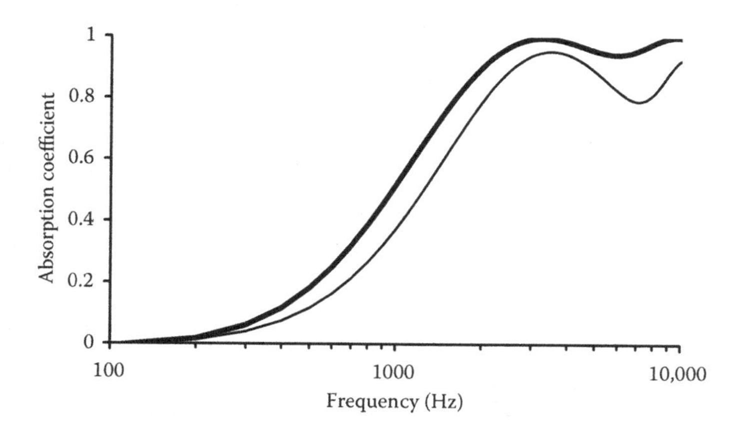

Figure 6.24 Comparison of the normal incidence absorption coefficient for two configurations of porous absorbent using transfer matrix methodology: —— 12.5 mm absorbent with 12.5 mm air gap and ▬ 25 mm absorbent. Both on a rigid backing.

the moving mass can become significant. The following impedance should be added to the surface impedance to allow for this:

$$z = \frac{j\omega\rho_s r_s}{j\omega\rho_s + r_s}, \tag{6.42}$$

where ρ_s is the mass per unit area and r_s is the resistance of the covering material. When the covering is not porous, and the resistance is large, this formulation simply adds the mass of the vibrating material.

6.6.3 Ground

A transfer matrix approach can be applied to the ground once the characteristic impedance and wavenumber have been determined. For surfaces such as grassland, assumptions about the layering of the soil and the backing impedance conditions need to be made—what is especially important are the characteristics of the first few centimetres. For many surfaces it is appropriate to assume a hard backing at some depth, but for others, such as forest floors, it is more appropriate to assume a nonhard backing.

More complex models, such as the semi-phenomenological model outlined earlier, can be used, but Attenborough[76] showed that in many cases, it is possible to use a simplified two-parameter model. The normalized surface impedance z_n of a single layer of ground of thickness d above an acoustically hard backing is approximately given by

$$z_n \approx z_c - \frac{jc_0}{4\pi d\gamma f\phi}, \tag{6.43}$$

where z_c is the normalized characteristic impedance of the layer, c_0 is the speed of sound in air, ϕ is the porosity of the layer, γ is the ratio of specific heats for air, and f is the frequency in Hz. If it is assumed that the upper layer has a relatively high effective flow resistivity, and the frequency is not too high, then the normalized characteristic impedance of the layer is given by

$$z_c \approx \sqrt{\frac{\sigma_e}{\pi\gamma\rho_0 f}}(1-j), \tag{6.44}$$

where p_0 is the density of air and σ_e is the effective flow resistivity given by

$$\sigma_e \approx \frac{4s_p^2\sigma}{\phi}, \tag{6.45}$$

where σ is the flow resistivity and s_p is a characteristic length, which can be taken to be 0.5 in many cases, meaning the effective flow resistivity is just the flow resistivity scaled by the porosity.

This model can also be used for the case where the deeper soil is more compacted, when the substrate has a lower porosity and higher flow resistivity. The change in soil properties is modelled as an exponential decrease in porosity with depth. The rate of change in porosity with depth is given by ζ, and some values for this parameter are given in Table 6.7. The normalized surface impedance becomes

$$z_n \approx z_c - \frac{jc\zeta}{8\gamma\pi f\phi}. \tag{6.46}$$

Figure 6.25 shows predicted and measured surface impedances for *institutional* grass. The single-parameter model is similar to one using the Delany and Bazley equations and shows inaccuracy at low frequencies, below about 300 Hz. The two-parameter model, based on the previous equations, is more accurate.

Horoshenkov and Mohamed[51] examined the effects of water saturation on sand in the laboratory. They found great changes in the surface impedance with the amount of moisture.

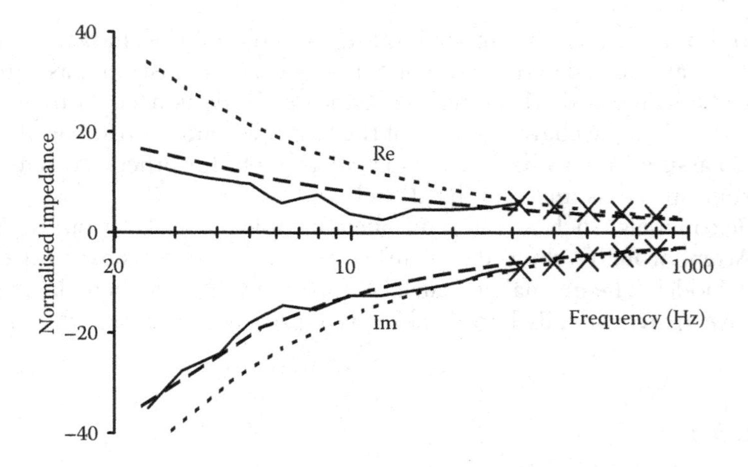

Figure 6.25 Comparison of two prediction models and measurements for grassland ground surface impedance: —— measurement, – – – one parameter model, and - - - - - two parameter model with σ_e = 35,000 rayls m^{-1} and k = 45 m^{-1}. (After Attenborough, K., *J. Acoust. Soc. Am.*, 81, 93–102, 1982.)

They also found that the two-parameter model of Attenborough was more suitable than one based on the Delany and Bazley formulations; the parameters for the model they deduced are shown in Table 6.7.

Porous road surfaces can reduce tyre and other noise. Both changing the road texture and introducing open voids in the asphalt are beneficial,[67] and reductions in A-weighted sound levels of 3 to 5 dB compared to a dense surface have been measured.[54] However, the pores will gradually become clogged as the surface is used, and then the noise reduction is lost. Asphalt consists of mineral grains bound together by a bituminous binder. Open porous asphalt uses an appropriate grain size distribution to form large voids that do not fill up with binder, although the porosity is rather small, typically around 20%. This material can be modelled as a porous absorber using three parameters: porosity, tortuosity, and flow resistivity, in a relaxation or semi-phenomenological model. Sarradj et al.[67] give empirical relationships for estimating the parameters from relevant technological parameters of the road material.

6.6.4 Multilayer porous absorbers

By using layers of different porous materials, it is possible to gain additional absorption. Ideally, a porous absorbent should offer an impedance that matches that of air to prevent reflections, while offering high internal acoustic attenuation. These two requirements are usually difficult to achieve in a thin layer of a single material. However, Mahasaranon et al.[97] showed that by controlling reaction time during the formation of an acoustic foam, a sample could be made where the pore size and distribution could be varied from the front to the rear of a sample. This stratified the acoustic properties of the foam, giving a 10%–20% increase in the absorption coefficient.

A similar effect can more easily be achieved by stacking different porous absorbents on top of each other, with the front material having the necessary impedance matching to air and the inner layers providing greater attenuation of the sound wave. This is achieved by the outside layer having a low flow resistivity and the inner layers offering more resistance. Ideally, the impedance should only change gradually between internal layers to minimize reflections. This arrangement can be modelled by repeated application of the transfer matrix equations.[46]

An alternative approach to achieve impedance matching is to use a metamaterial as a matching layer.[98] More simply, the front face of the porous absorber can be corrugated to form wedges. But this is appropriate only if the magnitude of the surface impedance of the porous material is greater than the characteristic impedance of air. In that case, cutting away material to form wedges can result in a better impedance match between air and the absorbent material. Foam sold for studios comes in a variety of shapes, including wedges. One study showed, however, that for most commercial samples, forming the foam into wedges reduces the absorption.[99] This happened because the samples had too little resistance. When foam is removed to form the wedges in this case, this does not improve the impedance matching; indeed, it makes matters worse by removing material that would otherwise usefully absorb sound.

An example where multilayer impedance matching is used is flat-wall linings for anechoic chambers. Mostly, anechoic chambers use wedges to achieve a gradual change in impedance from the air into the absorbent and so prevent strong reflections from the walls. The advantage of multilayer flat-walled linings is that they are simpler to make and install. Xu et al.[100] showed that it is possible to use a three-layered system with an overall thickness of about a sixth of a wavelength at the cutoff frequency. This is about 80% of the depth normally used for anechoic linings made from wedges. A process of trial and error, or an optimization algorithm, can be used to find the appropriate materials to use.

6.7 LOCAL AND EXTENDED REACTION

The propagation direction within many porous absorbers is normal to the surface, even for oblique incidence sound, because of refraction. Consequently, the reaction of the material at any point is independent of what happens elsewhere. In this case, the surface is termed *locally reacting*, as the surface impedance is independent of the nature of the incident wave. This is an extremely useful approximation. It means that in multilayered absorbents, the propagation can be assumed normal to the surface and are therefore much easier to evaluate. These assumptions will break down for large sound pressure levels as non-linear propagation is significant. Other common examples of locally reacting materials include resonant absorbers whose cavities are partitioned and massive walls made of materials like concrete, where the stiffness effect is small enough to be ignored in comparison with the effect of mass.

Fibrous materials such as mineral wool often behave as an extended reacting surface because they are anisotropic as the fibres are laid down in layers. The impedance produced is dependent on the incident wave type and angle of incidence. This is one reason why predictions from empirical models that are based on normal incidence impedance tube measurements are difficult to accurately translate into random incidence values comparable to measurements from the reverberation chamber.

Jeong[101] explored this by comparing five local and extended reaction prediction models. For a porous material with a rigid backing and a flow resistivity of 5000 Nm^{-3}s, assuming local reaction created errors of greater than 10% in the octave-band random incidence absorption coefficient, when the material was less than 12 cm thick. For a more resistive material, with a flow resistivity of 44,000 Nm^{-3}s, the minimum thickness for 10% error decreased to 4 cm. When there was an air gap behind the absorbent, significant oblique propagation can occur in this air layer. If the air gap is large compared to the material thickness, then this creates significant overprediction of the absorption coefficient at low frequencies if local reaction is assumed.

In such a case, a proper model of an extended reacting surface needs to deal with the entire wave field inside the medium. While there are formulations that allow for the anisotropic behaviour of mineral wool[90] (see the next section), these formulations produce similar results to the isotropic models because they do not account for the extended reaction.

6.8 OBLIQUE INCIDENCE

Consider a sound wave in air incident at an angle ψ to a finite layer of porous material with a rigid backing. The geometry is shown in Figure 6.26. The wavenumber in air is k_0, and the

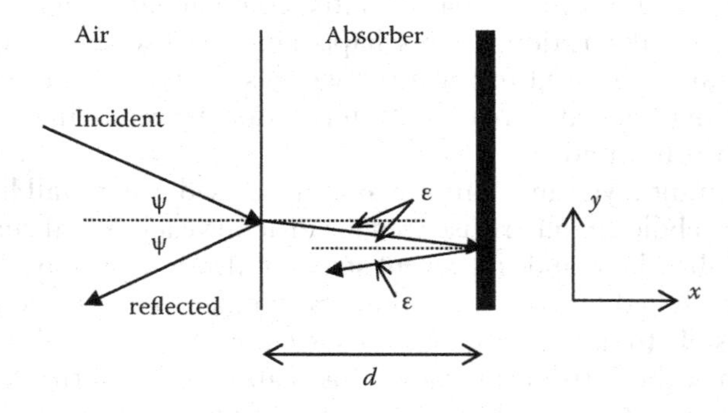

Figure 6.26 Geometry for propagation of sound through a finite layer of a rigid-backed porous absorber.

wavenumber in the material is k. In vector form, these are $k_0 = \{k_{0,x}, k_{0,y}, k_{0,z}\}$, $k = \{k_x, k_y, k_z\}$. For simplicity, it is assumed that $k_{0,z} = k_z = 0$. Snell's law relates the angles of propagation to the speed of sound in the material, as discussed in Section 2.5.1. In terms of wavenumber, this gives the following:

$$k_y = k_{0,y} = k_0 \sin(\psi) = k \sin(\varepsilon). \tag{6.47}$$

For many porous materials, the differences in the wavenumber between the air and the absorbent are so large that $\varepsilon \rightarrow 0$ and the previous derived normal incident formulae are sufficiently accurate. For other cases, a different formulation can be derived. As $k^2 = k_x^2 + k_y^2 + k_z^2$, this can be combined with Equation 6.47 to give

$$k_x = \sqrt{k^2 - k_y^2}. \tag{6.48}$$

The square root with the positive real part should be chosen. This can then be used to form an alternative form of the surface impedance for a rigidly backed absorbent for oblique incident sound:

$$z_{s1} = -jz_c \frac{k}{k_x} \cot(k_x d), \tag{6.49}$$

where z_c is the characteristic impedance of the porous absorbent.

Figure 6.4 shows how the attenuation varies with angle of incidence using the Delany and Bazley formulations. The surface impedance does not vary much with angle of incidence, but the reflection and absorption coefficients vary greatly as the pressure component perpendicular to absorbent drops off with a $1/\cos(\psi)$ relationship, where ψ is the angle of incidence (Equation 2.21).

Fibrous porous absorbents are anisotropic, in other words their acoustic properties vary depending on the angle of the wave to the fibre orientation. In this case, alternative forms for the surface impedance can be deduced. If the effective densities and bulk moduli are measured separately for propagation parallel and perpendicular to the fibres, Equation 6.25 can be applied to obtain the wavenumber parallel and perpendicular to the fibres. The wavenumber perpendicular to the fibres will be denoted k_N and parallel to the fibres k_p. The porous absorbent is placed so that the fibres are parallel to the rigid backing, as is most common. Then the component of the wavenumber in the x direction perpendicular to the backing and fibres is given by[90]

$$k_x = k_N \sqrt{1 - \frac{k_y^2}{k_p^2}}, \tag{6.50}$$

$$k_0 \sin(\psi) = k_y, \tag{6.51}$$

and

$$z_{s1} = -jz_N \frac{k_N}{k_x} \cot(k_x d), \tag{6.52}$$

where z_N is the characteristic impedance for propagation perpendicular to the fibres.

A method for obtaining the parallel and perpendicular propagation wavenumber and characteristic impedance is to use different flow resistivities in a model such as the one developed by Delany and Bazley. Typically, the flow resistivity perpendicular σ_N and parallel σ_p to the fibres are related by a constant factor[102,103]:

$$\sigma_N \approx A\sigma_p, \tag{6.53}$$

where A is a constant with a value of 0.6 or 0.5. Table 6.4 gave some expressions to allow the two flow resistivities to be evaluated from the fibre diameters. When this formulation is used, it makes some difference to the absorption coefficient and surface impedance, but the change is not that large.

6.9 BIOT THEORY FOR ELASTIC FRAMED MATERIAL

In the previous theories, the frame of the porous material was assumed to be rigid and waves only propagated in the air pores. In reality, porous absorbers have elastic frames that can support wave propagation. The consequence of this to the absorption properties of the material is often small. For instance, if the porous absorber is anchored to a rigid surface, for example, attached to the wall or resting on a floor, this will constrain the motion of the frame material. For this reason, the rigid frame models discussed previously are used much more often than models that allow for elastic motion of the frame.

Furthermore, as most of the previous models involve some form of empirical fitting, in many cases, this can partially compensate for inaccuracies introduced by not properly modelling the frame and frame–air resonances. If the frame of the porous absorber is not constrained, for example, if the material is hanging in free air, then resonances of the frame material can be seen in the characteristic impedance of the surface. In this case, a more complete model may be required, and most authors favour using Biot theory.[104,105] Other models of note are presented by Ingard[3] and Zwikker and Kosten.[105]

Biot theory is summarized in more detail in References 7 and 90, where the necessary formulations are given. The equations of motion for the displacement and strain tensors of the air in the pores and the frame are defined. These equations of motion include a set of coefficients that detail the coupling between the air and frame. These coefficients can be identified with physical properties such as the bulk modulus of the air in the pores and the elastic frame. The former, the bulk modulus of air, is taken from the rigid framed theories detailed earlier. Once these coefficients are determined, the equations of motion can be solved to give the surface impedance of porous layers.

There are now three waves to consider in the structure. There are two compressional waves. In most air saturated porous materials, the coupling between the frame and air is negligible and these waves can be identified as the frame-borne and airborne waves. Where there is weak coupling, the airborne wave remains mostly within the pores, but the frame-borne wave actually propagates through both the frame and pores. The third wave, the shear wave, is also frame-borne and, in most porous absorbents, is unaffected by the fluid air (for normal incidence, this is not excited and can be ignored).

Figure 6.27 shows a comparison of theory and experiment from Allard and Atalla.[90] For many frequencies, the rigid framed model is accurate, but deviations occur around 500 Hz. These inaccuracies are due to the resonances of the frame material, which, by definition, cannot be predicted by the rigid framed model. This is a dense fibreglass material, and so only the quarter wave resonance of the frame is seen. For these frequencies, Biot theory offers better predictions, although it should be noted that some fitting of prediction

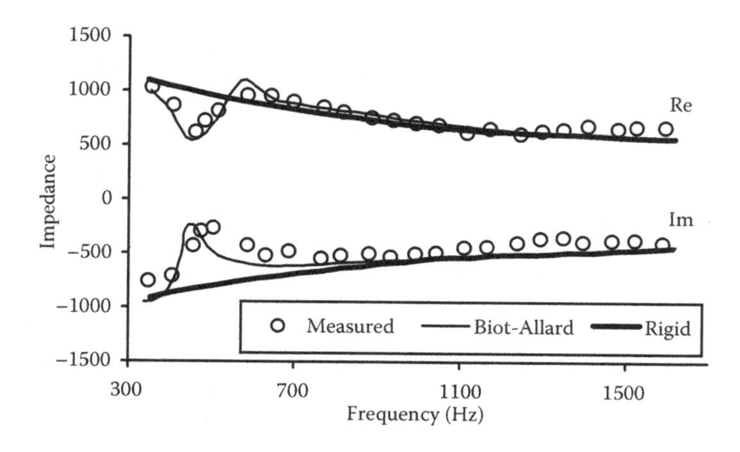

Figure 6.27 Comparison of two prediction models and measurements for a fibrous absorber with a distinct frame resonance. (After Allard, J.F., Atalla, N., *Propagation of Sound in Porous Media: Modelling Sound Absorbing Materials*, John Wiley & Sons, 2009.)

parameters had to be undertaken to gain this match. So even with Biot theory, the rather circular nature of model verification is still a problem.

6.10 TIME DOMAIN MODELS

The relaxation model described in Section 6.5.4 can allow a time domain representation of the propagation through the porous absorbent. Unfortunately, the model requires different thermal and viscous relaxation times at low and high frequencies, which constrains its usefulness. Alternatively, the Johnson-Allard model can be formulated in the time domain exactly, without any restrictions on frequency or pulse duration.[96] These approaches allow the modelling of porous absorbers through techniques such as finite difference time domain. A time domain formulation is much more computationally expensive than one in the frequency domain, however.

6.11 SUMMARY

This chapter has described porous absorbers both qualitatively and quantitatively. There are a large number of possible materials, and they form an important part of the acoustic palette for acoustic designers and noise control engineers. Being able to predict the absorption caused by porous materials, and understanding what causes it, is important to enable materials to be designed with high performance. In the next chapter, resonant absorbers will be discussed. Since many of these use porous absorbers within them, the understanding gained about porous absorber modelling from this chapter will be invaluable in the next.

REFERENCES

1. L. Cremer and H. A. Müller, *Principles and Applications of Room Acoustics Vol. 2*, Applied Science Publishers (translated by T. J. Schultz), London (1978).
2. J. Pfretzschner and E. M. Rodríguez, "Acoustical absorption and critical thickness", *Proc. 17th ICA* (2001).

3. U. Ingard, *Notes on Sound Absorption Technology*, Noise Control Foundation, Poughkeepsie, NY (1994).
4. P. D'Antonio, "Nestable sound absorbing foam with reduced area of attachment", US patent 5,665,943 (1997).
5. Siim Sepp, "Sand from Gobi Desert", commons.wikimedia.org/wiki/File:Sand_from_Gobi _Desert.jpg Licensed under CC BY-SA 3.0, accessed 3 May 2015.
6. R. Venegas and O. Umnova, "Acoustical properties of double porosity granular materials", *J. Acoust. Soc. Am.*, **130**, 2765–76 (2011).
7. F. P. Mechel, *Formulas of Acoustics*, Springer, Section G, Berlin (2002).
8. IARC Monographs on the Evaluation of Carcinogenic Risks to Humans, "Man-made Vitreous Fibres", **81** (2002).
9. R. A. Baan and Y. Grosse, "Man-made mineral (vitreous) fibres: Evaluations of cancer hazards by the IARC Monographs programme", *Mutation Research*, **553**, 43–58 (2004).
10. L. Lipworth, C. La Vecchia, C. Bosetti, and J. K. McLaughlin, "Occupational exposure to rock wool and glass wool and risk of cancers of the lung and the head and neck: A systematic review and meta-analysis", *J. Occup. Environ. Med.*, **51**(9), 1075–87 (2009).
11. F. Chevillotte, C. Perrot, and E. Guillon, "A direct link between microstructure and acoustical macro-behavior of real double porosity foams", *J. Acoust. Soc. Am.*, **134**(6), 4681–90 (2013).
12. O. Doutres, N. Atalla, and K. Dong, "Effect of the microstructure closed pore content on the acoustic behavior of polyurethane foams", *J. Appl. Phys.*, **110**(6), 064901 (2011).
13. T. Dupont, P. Leclaire, O. Sicot, X. L. Gong, and R. Panneton, "Acoustic properties of air-saturated porous materials containing dead-end porosity", *J. Appl. Phys.*, **110**, 094903 (2011).
14. P. Leclaire, O. Umnova, T. Dupont, and R. Panneton, "Acoustical properties of air-saturated porous material with periodically distributed dead-end pores", *J. Acoust. Soc. Am.*, **137**(4) 1772–82 (2015).
15. J. D. McRae, H. E. Naguib, and N. Atalla, "Mechanical and acoustic performance of compression-molded open-cell polypropylene foams", *J. Appl. Polym. Sci.*, **116**(2), 1106–15 (2010).
16. V. Desarnaulds, E. Costanzo, A. Carvalho, and B. Arlaud, "Sustainability of acoustic materials and acoustic characterization of sustainable materials", *Proc. 12th ICSV* (2005).
17. K. O. Ballagh, "Acoustical properties of wool", *Appl. Acoust.*, **48**(2), 101–20 (1996).
18. P. Glé, E. Gourdon, L. Arnaud, K. V. Horoshenkov, and A. Khan, "The effect of particle shape and size distribution on the acoustical properties of mixtures of hemp particles", *J. Acoust. Soc. Am.*, **134**(6), 4698–709 (2013).
19. J. P. Arenas and M. J. Crocker, "Recent trends in porous sound-absorbing materials", *Sound Vib.*, **44**(7), 12–8 (2010).
20. D. J. Oldham, C. A. Egan, and R. D. Cookson, "Sustainable acoustic absorbers from the biomass", *Appl. Acoust.*, **72**(6), 350–63 (2011).
21. J. Pfretzschner, "Rubber crumb as granular absorptive acoustic material", *Proc. Forum Acusticum*, MAT-01-005-IP (2002).
22. M. J. Swift and K. V. Horoshenkov, "Acoustic properties of recycled granular foams", *Proc. Euronoise* (2001).
23. R. Venegas, "Microstructure Influence on Acoustical Properties of Multi-Scale Porous Materials", PhD thesis, University of Salford, UK (2011).
24. C. M. Harris (ed), *Handbook of Noise Control*, 2nd edn, McGraw-Hill, New York (1991).
25. L. L. Beranek, *Acoustics*, McGraw-Hill, New York (1954).
26. F. A. Everest, *Master Handbook of Acoustics*, 4th edn, McGraw-Hill, New York (2001).
27. L. Forest, V. Gibiat, and A. Hooley, "Impedance matching and acoustic absorption in granular layers of silica aerogels", *J. Non-Cryst. Solids*, **285**(1–3), 230–5 (2001).
28. X. Olny and C. Boutin, "Acoustic wave propagation in double porosity media", *J. Acoust. Soc. Am.*, **114**(1), 73–89 (2003).
29. R. Venegas and O. Umnova, "Influence of sorption on sound propagation in granular activated carbon", submitted to *J. Acoust. Soc. Am.* (2015).
30. F. Bechwati, M. R. Avis, D. J. Bull, T. J. Cox, J. A. Hargreaves, D. Moser, and R. Venegas, "Low frequency sound propagation in activated carbon", *J. Acoust. Soc. Am.*, **132**, 239–48 (2012).

31. D. A. Bies and C. H. Hansen, *Engineering Noise Control: Theory and Practice*, 4th edn, CRC Press, Abingdon, UK (2009).
32. H. S. Yang, J. Kang, and C. Cheal, "Random-incidence absorption and scattering coefficients of vegetation", *Acta Acust. uw Acust.*, **99**(3), 379–88 (2013).
33. K. Attenborough, T. Waters-Fuller, K. M. Li, and J. A. Lines, "Acoustical properties of Farmland", *J. Agric. Eng. Res.*, **76**, 183–95 (2000).
34. S. Dance and R. Lorenzetto, "A new type of absorber for use by classical musicians in rehearsal rooms", *Proc. Euronoise* (2009).
35. F. Chevillotte, "Controlling sound absorption by an upstream resistive layer", *Appl. Acoust.*, **73**(1), 56–60 (2012).
36. O. Umnova, K. Attenborough, E. Standley, and A. Cummings, "Behavior of rigid-porous layers at high levels of continuous acoustic excitation: Theory and experiment", *J. Acoust. Soc. Am.*, **114**(3), 1346–56 (2003).
37. D. A. Bies and C. H. Hansen, "Flow resistance information for acoustical design", *Appl. Acoust.*, **13**, 357–91 (1980).
38. L. L. Beranek and I. L. Vér (eds), *Noise and Vibration Control Engineering*, John Wiley & Sons (1992).
39. M. H. Fouladi, M. Ayub, and M. J. M. Nor, "Analysis of coir fiber acoustical characteristics", *Appl. Acoust.*, **72**(1), 35–42 (2011).
40. F. Fohr, D. Parmentier, B. R. Castagnede, and M. Henry, "An alternative and industrial method using low frequency ultrasound enabling to measure quickly tortuosity and viscous characteristic length", *J. Acoust. Soc. Am.*, **123**(5), 3118 (2008).
41. P. Bonfiglio and F. Pompoli, "Inversion problems for determining physical parameters of porous materials: Overview and comparison between different methods", *Acta Acust. Acust.*, **99**(3), 341–51 (2013).
42. M. Garai and F. Pompoli, "A simple empirical model of polyester fibre materials for acoustical applications", *Appl. Acoust.*, **66**(12), 1383–98 (2005).
43. N. Kino and T. Ueno, "Evaluation of acoustical and non-acoustical properties of sound absorbing materials made of polyester fibres of various cross-sectional shapes", *Appl. Acoust.*, **69**(7) 575–82 (2008).
44. C. Wassilieff, "Sound absorption of wood-based materials", *Appl. Acoust.*, **48**(4), 339–56 (1996).
45. N. Kino and T. Ueno, "Comparisons between characteristic lengths and fibre equivalent diameters in glass fibre and melamine foam materials of similar flow resistivity", *Appl. Acoust.*, **69**(4), 325–31 (2008).
46. R. Panneton and X. Olny, "Acoustical determination of the parameters governing viscous dissipation in porous media", *J. Acoust. Soc. Am.*, **119**(4), 2027–40 (2006).
47. E. Lind-Nordgren and P. Göransson, "Optimising open porous foam for acoustical and vibrational performance", *J. Sound Vib.*, **329**(7), 753–67 (2010).
48. S. G. Mosanenzadeh, O. Doutres, H. E. Naguib, C. B. Park, and N. Atalla, "A numerical scheme for investigating the effect of bimodal structure on acoustic behavior of polylactide foams", *Appl. Acoust.*, **88**, 75–83 (2015).
49. J. P. Dunn and W. A. Davern, "Calculation of acoustic impedance of multi-layer absorbers", *Appl. Acoust.*, **19**, 321–34 (1986).
50. K. V. Horoshenkov and M. J. Swift, "The acoustic properties of granular materials with pore size distribution close to log-normal", *J. Acoust. Soc. Am.*, **110**(5), 2371–8 (2001).
51. K. V. Horoshenkov and M. H. A. Mohamed, "Experimental investigation of the effects of water saturation on the acoustic admittance of sandy soils", *J. Acoust. Soc. Am.*, **120**(4), 1910–21 (2006).
52. D. K. Wilson, V. E. Ostashev, S. L. Collier, N. P. Symons, D. F. Aldridge, and D. H. Marlin, "Time-domain calculations of sound interactions with outdoor ground surfaces", *Appl. Acoust.*, **68**, 173–200 (2007).
53. N. N. Voronina and K. V. Horoshenkov, "A new empirical model for the acoustic properties of loose granular media", *Appl. Acoust.*, **64**(4), 415–32 (2003).

54. M. C. Berengier, M. R. Stinson, G. A. Daigle, and J. F. Hamet, "Porous road pavements: Acoustical characterization and propagation effects", *J. Acoust. Soc. Am.*, **101**(1), 155–62 (1997).
55. U. Umnova, K. Attenborough, H.-C. Shin, and A. Cummings, "Deduction of tortuosity and porosity from acoustic reflection and transmission measurements on thick samples of rigid-porous materials", *Appl. Acoust.*, **66**, 607–24 (2005).
56. Y. Champoux and J. F. Allard, "Dynamic tortuosity and bulk modulus in air-saturated porous media", *J. Appl. Phys.*, **70**(4), 1975–9 (1991).
57. N. D. Yilmaz, P. Banks-Lee, N. B. Powell, and S. Michielsen, "Effects of porosity, fiber size, and layering sequence on sound absorption performance of needle-punched nonwovens", *J. Appl. Polym. Sci.*, **121**(5), 3056–69 (2011).
58. J. Alba, R. del Rey, J. Ramis, and J. Arenas, "An inverse method to obtain porosity, fibre diameter and density of fibrous sound absorbing materials", *Arch. Acoust.*, **36**(3), 561–74 (2011).
59. A. Putra, Y. Abdullah, H. Efendy, W. M. F. W. Mohamad, and N. L. Salleh, "Biomass from paddy waste fibers as sustainable acoustic material", *Adv. Acoust. Vib.*, 605932 (2013).
60. A. Putra, Y. Abdullah, H. Efendy, W. M. Farid, M. R. Ayob, and M. S. Py, "Utilizing sugarcane wasted fibers as a sustainable acoustic absorber", *Procedia Eng.*, **53**, 632–8 (2013).
61. N. Kino and T. Ueno, "Experimental determination of the micro and macrostructural parameters influencing the acoustical performance of fibrous media", *Appl. Acoust.*, **68**, 1439–58 (2007).
62. M. Vasina, D. C. Hughes, K. V. Horoshenkov, and L. Lapcik, "The acoustical properties of consolidated expanded clay granules", *Appl. Acoust.*, **67**(8), 787–96 (2006).
63. O. Umnova, D. Tsiklauri, and R. Venegas, "Effect of boundary slip on the acoustical properties of microfibrous materials", *J. Acoust. Soc. Am.*, **126**(4), 1850–61 (2009).
64. C. Boutin and C. Geindreau, "Estimates and bounds of dynamic permeability of granular media", *J. Acoust. Soc. Am.*, **124**(6), 3576–93 (2008).
65. M. T. Hoang, G. Bonnet, and C. Perrot, "Multi-scale acoustics of partially open cell poroelastic foams", *J. Acoust. Soc. Am.*, **133**(5), 3289 (2013).
66. R. T. Muehleisen, C. W. Beamer, and B. D. Tinianov, "Measurements and empirical model of the acoustic properties of reticulated vitreous carbon", *J. Acoust. Soc. Am.*, **117**(2), 536–44 (2005).
67. E. Sarradj, T. Lerch, and J. Hubelt, "Input parameters for the prediction of acoustical properties of open porous asphalt", *Acta Acust. Acust.*, **92**, 86–96 (2006).
68. MathWorks, uk.mathworks.com/matlabcentral/fileexchange/54004-modelling-of-acoustic-absorbers, accessed 8 December 2015.
69. M. E. Delany and E. N. Bazley, "Acoustical properties of fibrous absorbent materials", *Appl. Acoust.*, **3**, 105–16 (1970).
70. J. F. Allard and Y. Champoux, "New empirical equations for sound propagation in rigid frame fibrous materials", *J. Acoust. Soc. Am.*, **91**(6), 3346–53 (1992).
71. Q. L. Wu, "Empirical relations between acoustical properties and flow resistivity of porous plastic open-cell foam", *Appl. Acoust.*, **25**(3), 141–8 (1988).
72. A. Cummings and S. P. Beadle, "Acoustic properties of reticulated plastic foams", *J. Sound Vib.*, **1975**, 115–33 (1993).
73. Y. Miki, "Acoustical properties of porous materials. Modifications of Delany-Bazley models", *J. Acoust. Soc. Japan (E)*, **11**(1), 19–24 (1990).
74. R. Kirby, "On the modification of Delany and Bazley formulae", *Appl. Acoust.*, **86**, 47–9 (2014).
75. K. Attenborough, I. Bashir, and S. Taherzadeh, "Outdoor ground impedance models", *J. Acoust. Soc. Am.*, **129**(5), 2806–19 (2011).
76. K. Attenborough, "On the acoustic slow wave in air filled granular media", *J. Acoust. Soc. Am.*, **81**, 93–102 (1982).
77. K. Attenborough, "Models for the acoustical characteristics of air filled granular materials", *Acta Acust.*, **1**, 213–26 (1993).
78. J. G. Berryman, "Confirmation of Biot's theory", *Appl. Phys. Lett.*, **37**, 382–4 (1980).

79. T. F. W. Embleton, J. E. Piercy, and G. A. Daigle, "Effective flow resistivity of ground surfaces determined by acoustical measurements", *J. Acoust. Soc. Am.*, **74**(4), 1239–44 (1983).
80. M. J. M. Martens, L. A. M. van der Haijden, H. H. J. Walthaus, and W. J. J. van Rens, "Classification of soils based on acoustic impedance, air flow resistivity and other physical soil parameters", *J. Acoust. Soc. Am.*, **78**, 970–80 (1985).
81. P. Datt, J. C. Kapil, A. Kumar, and P. K. Srivastava, "Experimental measurements of acoustical properties of snow and inverse characterization of its geometrical parameters", *Appl. Acoust.*, **101**, 15–23 (2016).
82. C. Boutin and C. Geindreau, "Periodic homogenization and consistent estimates of transport parameters through sphere and polyhedron packings in the whole porosity range", *Phys. Rev. E.*, **82**, 036313 (2010).
83. M. R. Stinson and Y. Champoux, "Propagation of sound and the assignment of shape factors in model porous materials having simple pore geometries", *J. Acoust. Soc. Am.*, **91**(2), 685–95 (1992).
84. D. L. Johnson, J. Koplik, and R. Dashen, "Theory of dynamic permeability and tortuosity in fluid-saturated porous media", *J. Fluid Mech.*, **176**, 379–402 (1987).
85. D. Lafarge, "Porous and stratified porous media: Linear models of propagation", in M. Bruneau and C. Potel (Eds), *Materials and Acoustics Handbook*, John Wiley & Sons, Hoboken, NJ (2009).
86. C. Perrot, "Microstructure et macro-comportement acoustique: Approche par reconstruction d'une cellule élémentaire représentative", PhD thesis, University of Sheerbroke, Canada (2006).
87. P. Shravage, P. Bonfiglio, and F. Pompoli, "Hybrid Inversion technique for predicting geometrical parameters of Porous Materials", *J. Acoust. Soc. Am.*, **123**(5), 3284 (2008).
88. T. G. Zielinski, "Inverse identification and microscopic estimation of parameters for models of sound absorption in porous ceramics", *Proc. ISMA2012-USD2012, 95–108* (2012).
89. S. Gasser, "Etude des propriétés acoustiques et méchaniques d'un matériau métallique poreaux modèle à base de sphères creuses de nickel", PhD thesis, Institut National Polytechnique de Grenoble, France (2003).
90. J. F. Allard and N. Atalla, *Propagation of Sound in Porous Media: Modelling Sound Absorbing Materials*, John Wiley & Sons, Chichester, UK (2009).
91. D. L. Johnson, T. J. Plona, C. Scala, F. Pasierb and H. Kojima, "Tortuosity and acoustic slow waves", *Phys. Rev. Lett.*, **49**(25), 1840–4 (1982).
92. D. Lafarge, P. Lemarinier, J. F. Allard, and V. Tarnow, "Dynamic compressibility of air in porous structures at audible frequencies", *J. Acoust. Soc. Am.*, **102**(4), 1995–2006 (1997).
93. M. Sadouki, M. Fellah, Z. E. A Fellah, E. Ogam, N. Sebaa, F. G. Mitri, and C. Depollier, "Measuring static thermal permeability and inertial factor of rigid porous materials (L)", *J. Acoust. Soc. Am.*, **130**(5) 2627–30 (2011).
94. M. Henry and J. F. Allard, "Acoustical measurement of the trapping constant of foams with open cells", *CR. Acad. Sci. II B*, **325**(6), 331–8 (1997).
95. D. K. Wilson, "Simple, relaxational model for the acoustical properties of porous media", *Appl. Acoust.*, **50**(3), 171–88 (1997).
96. O. Umnova and D. Turo, "Time domain formulation of the equivalent fluid model for rigid porous media", *J. Acoust. Soc. Am.*, **125**(4), 1860–3 (2009).
97. S. Mahasaranon, K. V. Horoshenkov, A. Khan, and H. Benkreira, "The effect of continuous pore stratification on the acoustic absorption in open cell foams", *J. Appl. Phys.*, **111**(8), 084901 (2012).
98. A. S. Elliott, R. Venegas, J. P. Groby, and O. Umnova, "Omnidirectional acoustic absorber with a porous core and a metamaterial matching layer", *J. Appl. Phys.*, **115**(20), 204902 (2014).
99. T. J. Cox, "Choosing and Using Porous Absorbers", Sound on Sound (July 2015).
100. J. Xu, J. Nannariello, and F. R. Fricke, "Optimising flat-walled multi-layered anechoic linings using evolutionary algorithms", *Appl. Acoust.*, **65**(11), 1009–26 (2004).
101. C. H. Jeong, "Guideline for adopting the local reaction assumption for porous absorbers in terms of random incidence absorption coefficients", *Acta Acust. Acust.*, **97**(5), 779–90 (2011).
102. J. F. Allard, R. Bourdier, and A. L'Esperance, "Anisotropy effect in glass wool for normal impedance in oblique incidence", *J. Sound. Vib.*, **114**, 233–8 (1987).

103. V. Tarnow, “Measured anisotropic air flow resistivity and sound attenuation of glass wool”, *J. Acoust. Soc. Am.*, **111**(6), 2735–9 (2002).
104. M. A. Biot, “Theory of propagation of elastic waves in a fluid-saturated porous solid. II Higher frequency range”, *J. Acoust. Soc. Am.*, **28**(2), 179–91 (1956).
105. C. Zwikker and C. Kosten, *Sound Absorbing Materials*, Elsevier, New York (1949).

Chapter 7

Resonant absorbers

By exploiting resonance, it is possible to get absorption at low to mid frequencies. At these frequencies, it is inefficient to use porous absorbers because of the required thickness of the material. Furthermore, treatments are often placed at room boundaries where porous absorbers are inefficient as the particle velocity is low. For many resonant absorbers, placing the device at the boundaries will improve their effectiveness. The absorption characteristics of these resonant devices are a peak of absorption, as shown in the thick line in Figure 7.1. Unlike porous materials, wide band absorption is difficult to achieve in one device, and so one of the frequent challenges in the design of resonant structures is to extend the bandwidth.

There are two common forms of the device: the first is the Helmholtz absorber, which is named after the German physician and physicist Hermann von Helmholtz (1821–94), and the second is a membrane or panel absorber. The ideas and concepts of resonant absorption have been known for a very long time. More recently, some more specialist devices have been produced, for instance, clear absorbers, but these are still based on the same basic physics. While some devices, such as many basic Helmholtz absorbers, can be predicted with reasonable accuracy, others, such as membrane devices, are still designed by trial and error through experimentation. These treatments are commonly employed to treat low-frequency room modes and as parts of silencers within ventilation systems. Perhaps even more common, architectural acoustics often uses perforated wood backed with porous absorbent to reduce reverberation and noise at speech frequencies.

7.1 MECHANISMS

Resonant absorbers involve a mass vibrating against a spring, and the two most common types are illustrated in Figure 7.2. In the case of a Helmholtz absorber, the mass is a plug of air in the opening of a perforated sheet. The resonance is produced by the same mechanism that generates a note when you blow across a beer bottle. To make this into an absorber, losses are provided by damping to remove sound energy. This is often provided by a layer of mineral wool, but a notable exception is microperforated absorbers where damping is provided by viscous losses within the small perforations. For a membrane (or panel) absorber, the mass is a sheet of material that vibrates, such as rubber, mass loaded vinyl, or plywood. The spring in both cases is provided by air enclosed in the cavity. By changing the vibrating mass and the stiffness of the air spring, the resonant frequency of the device, where absorption is a maximum, can be tuned.

To achieve losses, damping is required. Often, this is best achieved by placing porous absorbent where the particle velocity is large—in the neck of the Helmholtz resonator or just behind the membrane in the panel absorber. In the latter case, normally, the absorbent

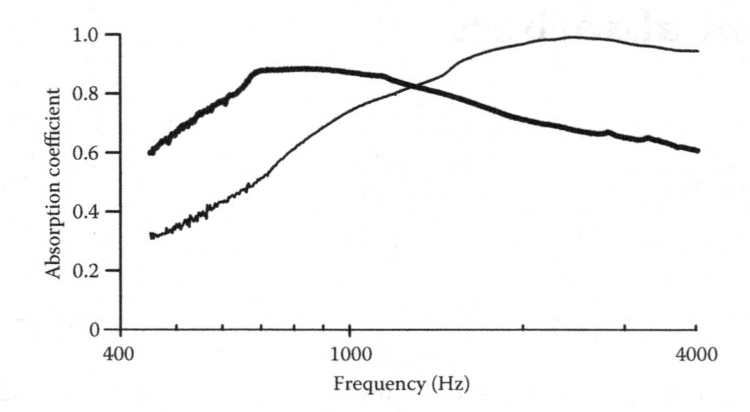

Figure 7.1 Normal incidence absorption coefficient measured in an impedance tube for: —— mineral wool and ▬▬ the same material covered with a perforated sheet to form a Helmholtz absorber.

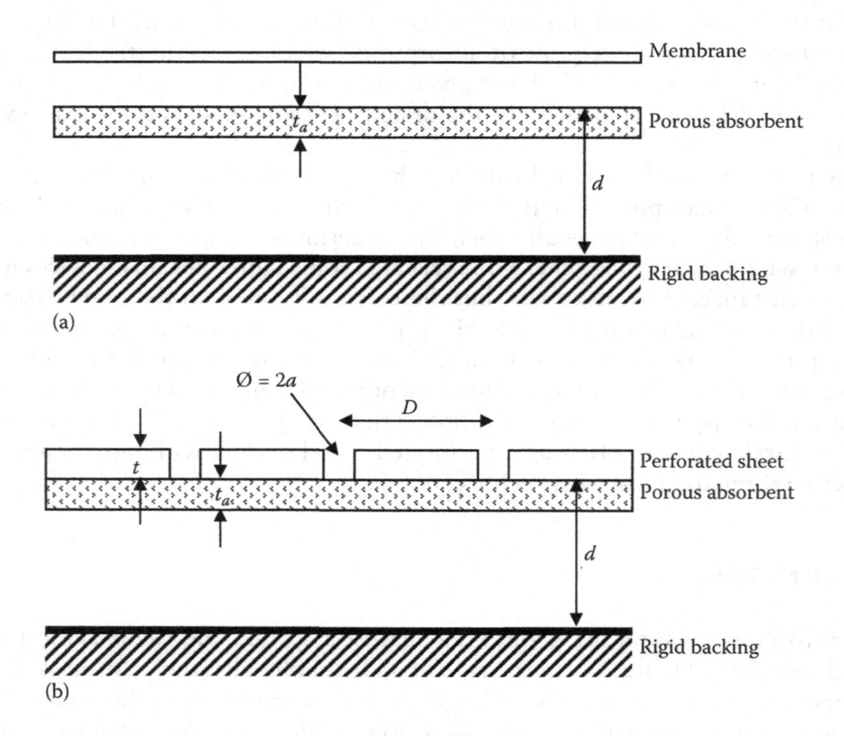

Figure 7.2 Typical constructions for (a) membrane and (b) Helmholtz absorbers.

should not be so close as to inhibit the movement of the membrane. Alternatively, for Helmholtz devices with small openings, viscous losses within the neck can be used to gain absorption; this leads to devices without porous materials, such as microperforated absorbers. For panel absorbers, there are also internal losses within the vibrating membrane, but these may be too small to give high absorption. More significant are the losses that come from the mounting between the membrane and the enclosure. Problems arise in predicting the performance of membrane devices, as the absorption from the boundaries is hard to characterize.

Before discussing the relevant design equations, some example constructions are given to provide a sense of what commercial devices are like. After the design equations, more complex and unusual constructions will be considered.

7.2 EXAMPLE CONSTRUCTIONS

7.2.1 Low-frequency membrane absorber

Small rooms often exhibit a poor low-frequency response with significant emphasis due to standing wave modes and deemphasis where modal excitation is small. With music, the most audible defect is the resonances ringing on after bass notes have ended.[1] This means that treatment such as splaying walls and choosing the right room ratio can only ever be a partial solution to room resonances. What is needed is absorbent to damp the modes and speed the decay of sound.

There is usually limited space within which to fit treatments. Porous absorbers are ineffective at these modal frequencies because the particle velocity near walls and in corners is essentially 0 for these long wavelengths and also the treatment would have to be made so deep that significant space within the room would be lost. Furthermore, if sufficient porous absorbent was used to treat the room modes, then the mid–high-frequency absorption would then be too high and the room very *dead*.

It has been shown that multiple subwoofers located at specific room locations with the right delays and gains can be used to reduce excitation of room modes.[2] But a more common and flexible solution is resonant absorption, such as a membrane design, which in audio engineering is often called *bass traps*. A membrane absorber converts the high sound pressure fluctuations typically found at wall surfaces and in corners into selective absorption in the modal frequency range.

Figure 7.2a shows a typical device. The membrane converts pressure fluctuations into air motion. As the membrane sympathetically vibrates over a selective low-frequency range, determined by its mass and the air spring stiffness, it pushes air through an internal porous layer producing low-frequency absorption. Simple relationships exist between the design frequency of these resonant systems and the membrane mass, stiffness, and cavity depth, and these will be outlined later. Figure 7.3 illustrates the absorption coefficient for this type of device and also shows an example application where the treatment is placed in the corner of a room where the pressure is a maximum for all standing wave modes. There is also a prediction of the performance using a transfer matrix approach, as discussed later. This prediction illustrates, however, that for membrane absorbers, the design equations are often inexact. There are many reasons for this; for instance, the physical mass of the membrane is often different from the vibrating acoustic mass due to mounting conditions. Finite element analysis might be used to help overcome these problems.

Consequently, experimental verification of the absorption is necessary, although at these low frequencies, this is not easy. The results shown in Figure 7.3 were measured in a very large impedance tube.

Specific problem frequencies can be addressed with individually tuned absorbers that minimize the decay time of a particular room mode.[3] If the absorber has a sharp resonant peak—a high *Q* factor—there is a risk of creating a notch at the wrong frequency should

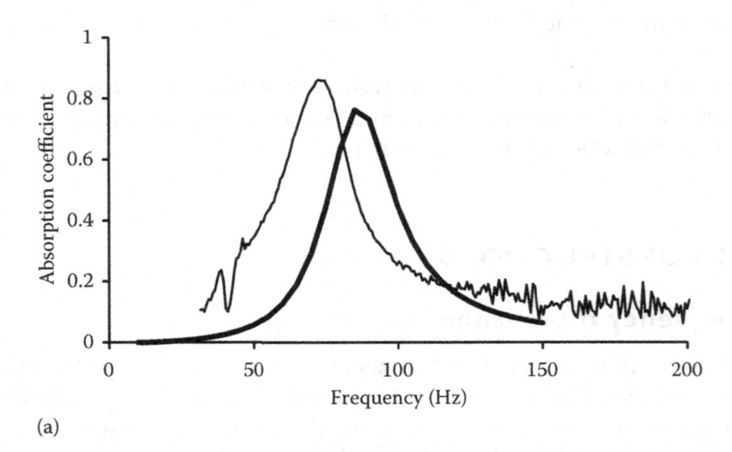

Figure 7.3 (a) —— Measured and ▬▬ predicted normal incidence absorption coefficient for a commercial membrane absorber (Modex™). (b) A typical installation in a small reproduction room; the resonant absorber is in the corner of the room. (Courtesy of RPG Diffusor Systems, Inc.)

room conditions change, thus aggravating rather than ameliorating the modal problems. Consequently, it is more common to broaden the absorption by introducing additional damping in the air cavity, although this often lowers the maximum absorption. A number of modules are used, each tuned to a different one-third octave band. A considerable amount of the room boundary must be covered with absorbent to get broadband low-frequency absorption.

In multipurpose spaces, bass absorbers, which can be turned on and off, are useful to alter the acoustic for different uses. It has been suggested that this can be achieved using inflatable absorbers like air mattresses.[4]

7.2.2 Absorbing wood

Wood is more often than not the preferred surface treatment. Perforated wood panels and planks of a wide variety of designs are widely used in architectural acoustics. The percentage open area, panel thickness, and cavity depth, typically including porous materials, can be used to design systems that absorb over a desired frequency range. Figure 7.4 illustrates such a system and an example application. By varying the groove (or hole) spacing, the groove width (or hole diameter), the rear air cavity depth, and the porous material, it is possible to obtain absorption over a wide variety of frequencies. The design equations are given later in this chapter and are much more successful than the equivalent formulations for membrane absorbers.

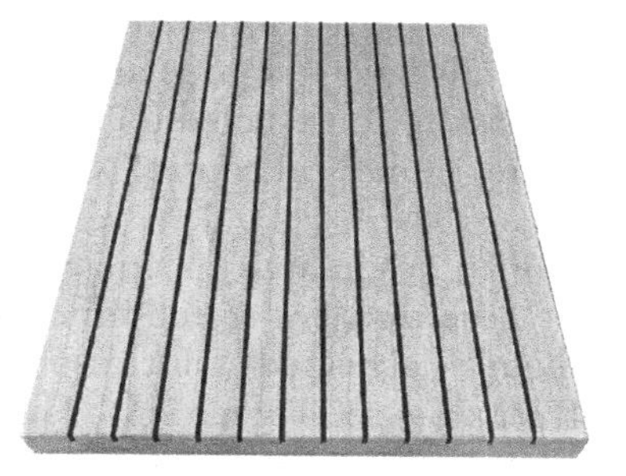

Figure 7.4 Perforated wood plate for a commercial Helmholtz absorber (Dado™ 14/2) and an application at the Cab Calloway Auditorium, Wilmington, DE. Acoustician: Acentech. (Courtesy of RPG Diffusor Systems, Inc.)

These Helmholtz absorbers are constructed from Class A medium-density fibreboard (MDF) cores either painted or surfaced with wood veneers or simulated wood grain melamine. The rear of the panel is covered with a black nonwoven glass matt to provide a resistive layer and also to conceal the contents behind the panel. In addition, a fibreglass panel is attached to the back of the glass matt to provide further resistance and losses.

Figure 7.5 shows typical absorption coefficients for different cavity depths and open areas of the perforated plate. As the open area increases, there is greater absorption at higher frequencies and a slight increase in resonant frequency. These trends can be predicted by the design equations given later, but they will not predict the absorption coefficients greater than one. These arise because of issues with reverberation chamber measurements, such as edge diffraction, which is not included in the prediction models (see Chapters 4 and 13 for further discussion of this). Increasing the cavity depth causes the stiffness of the spring to decrease, and consequently, the peak absorption decreases in frequency; again this is predictable.

The absorbers exploit mechanisms that have been understood for more than a century. One problem with this type of construction is getting the perforated sheet with the correct hole size and open area. Standard perforated board, such as peg board, has too small an open area, and most perforated metals have too large an open area. Consequently, the perforated sheet often has to be specifically constructed for acoustic purposes, which makes it more expensive. However, computer numerical control (CNC) fabrication offers an unlimited choice of design parameters.

Perforated absorbers will have an attenuated specular reflection within their absorption bandwidth but reflect all of the incident sound efficiently like a solid flat panel at other frequencies. A measurement of reflected polar responses, as is normally done for diffusers, can be used to illustrate this (see Figure 7.6). The device consists of a 19-mm-thick perforated wood panel. The face veneer has 0.8-mm-diameter holes, with a hole spacing of 4 mm. The MDF core contains 9-mm-diameter holes spaced 11 mm apart. For the polar response measurement, the rear of the device opens to free space and so mimics a very deep layer of porous absorbent. For the normal and random incidence absorption coefficient measurements, the perforated plate was backed by 50 mm of fibreglass.

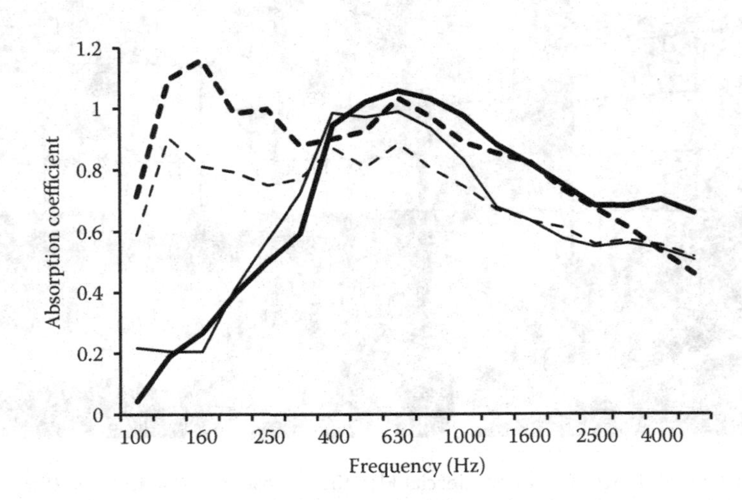

Figure 7.5 Random incidence absorption coefficient for different Helmholtz absorbers: —— small open area, shallow cavity; ▬▬ large open area, shallow cavity; - - - - small open area, deep cavity; and ▪▪▪▪ large hole, deep cavity.

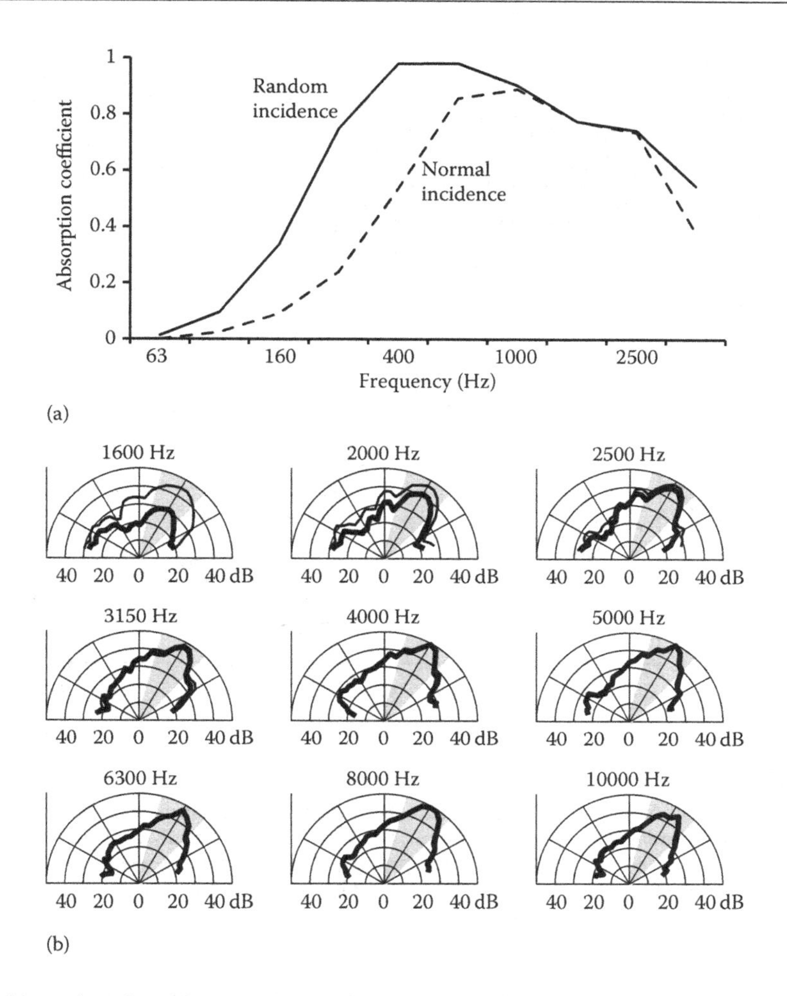

Figure 7.6 (a) Normal and random incidence absorption coefficients for a perforated wood absorber (Perfecto Micro). (b) Measured polar responses for 30° incident sound at various one-third octaves as indicated: —— flat rigid surface and ▬▬ perforated absorbent. For many frequencies, the two lines overlay each other.

The polar responses are examples for sound incidence at 30°. From roughly 2500 Hz and above, both perforated and unperforated panels reflect incident sound similarly. Below that, the perforated wood provides attenuated specular reflections within the absorption bandwidth, presumably down to about 500 Hz. (The polar response measurements are unfortunately limited to a lower frequency of 1600 Hz.) The polar responses illustrate that absorptive wood provides attenuated reflections within its absorption bandwidth and specular reflections outside this frequency band.

7.2.3 Absorption and diffusion

Figure 7.7 shows a Schroeder-style diffuser, which has rectangular holes to provide mid-frequency absorption via resonance. These are relatively shallow devices and so are only efficient diffusers above 3 kHz, unless modulation is used to extend the bandwidth, as discussed in Chapter 10. Below 3 kHz, absorption is provided by a resonant mechanism via

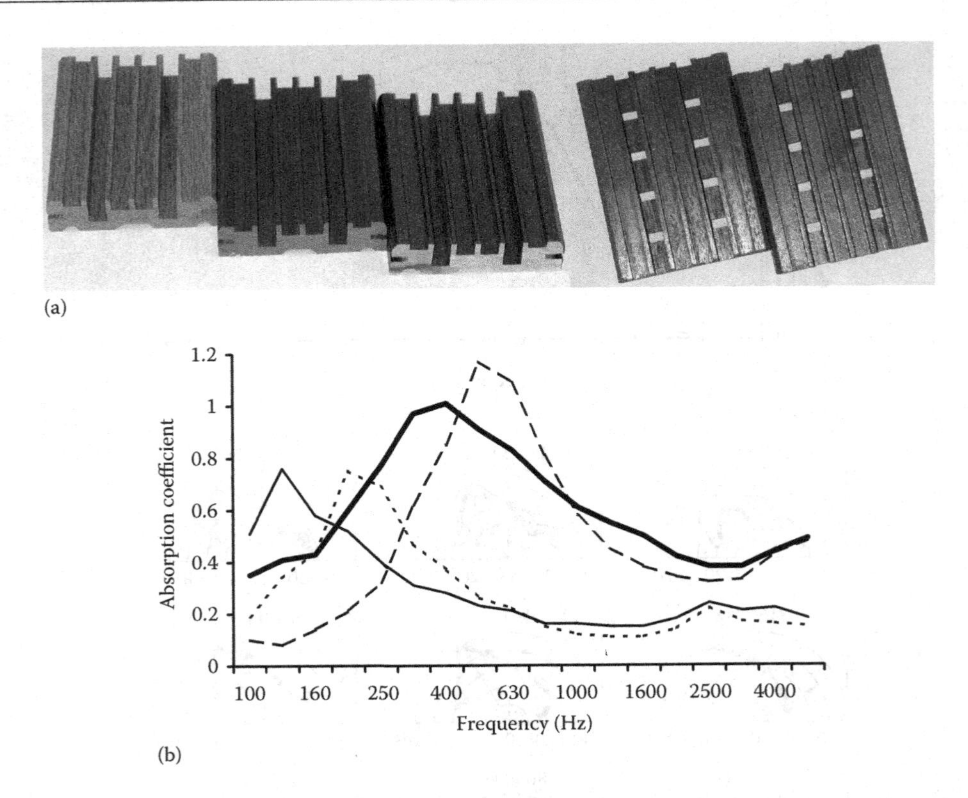

Figure 7.7 (a) A small-scale Schroeder diffuser, FlutterFree (left) and a perforated version with rectangular holes (right) forming a Helmholtz absorber. (b) Random incidence absorption coefficient for various systems: ---- no holes, Helmholtz mounts, shallow cavity; ——— square holes, normal mount, shallow cavity; —— no holes, Helmholtz mounts, deep cavity; and ▬▬ rectangular holes, normal mount, deep cavity.

the rectangular holes. The diffuser is mounted over porous absorbent with a cavity behind, similar to Figure 7.2b. Consequently, this is a hybrid device providing absorption and diffusion in different frequency ranges. As the frequency ranges are different, it is assumed that the absorption mechanism should not have too much effect on the diffusion performance and vice versa. The frequency of absorption can be varied by choosing the hole size, open area, and cavity depth, although in this case, the diffuser surface profile limits how much the geometry can be altered.

This surface is also provided without holes. In Figure 7.7 the absorption coefficients for two mounting types are shown. The panel can be mounted with a 1.6 mm gap between the laths (planks)—this is labelled *Helmholtz mounts*. This gives a small slit opening to the back cavity, and as this contains a porous absorbent, it generates additional bass absorption via resonance. This absorption can be useful in treating spaces with excessive bass reverberation. Similar absorption can also be achieved by using a tuned membrane device, but a Helmholtz mechanism is generally easier to achieve.

The rectangular holes shown in Figure 7.7 are responsible for the peaks in absorption around 500–600 Hz. The Helmholtz slits provide absorption at a few hundred Hertz or less. Again, the resonant frequency can be tuned by choosing appropriate cavity depth and plank spacing.

The Binary Amplitude Diffsorber (BAD™) panel[5] is a flat hybrid surface that provides absorption at mid frequencies and diffusion at high frequencies. Figure 7.8 shows some possible perforated patterns and an application. To fabricate these panels, a porous absorber, such as fibreglass, is faced with a complex perforated mask. The panel can then be fabric wrapped or the mask can be left exposed, as shown in Figure 7.8.

Figure 7.9 shows the random incidence absorption coefficient for such a hybrid surface compared to the mineral wool alone; the effect of changing the backing depth is also shown. The additional vibrating mass within the holes of the mask causes the absorption curve to shift down in frequency, generating additional low- to mid-frequency absorption. At high frequency, the hard parts of the mask reflect sound and hide some parts of the mineral wool; this causes a reduction in the absorption coefficient. It is at these high frequencies, where the absorption is reduced, that the surface needs to disperse the reflected sound.

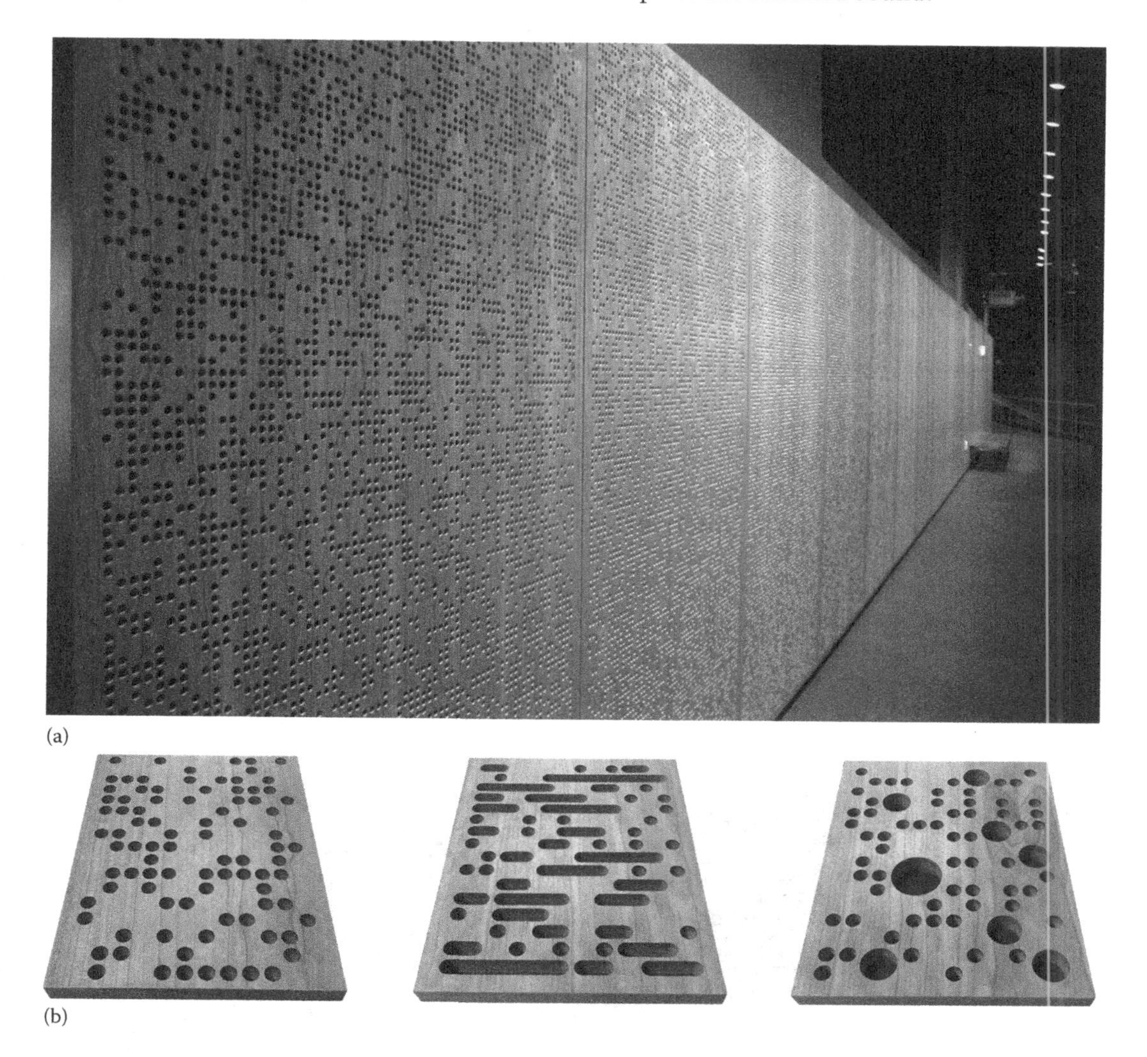

Figure 7.8 (a) Photo of the rear wall of a video theatre at the Newseum, Washington, DC. (b) At the left is the standard pattern mask (Expo™), shown in the photo above, alongside two examples of other aesthetic possibilities. (Courtesy of RPG Diffusor Systems, Inc.)

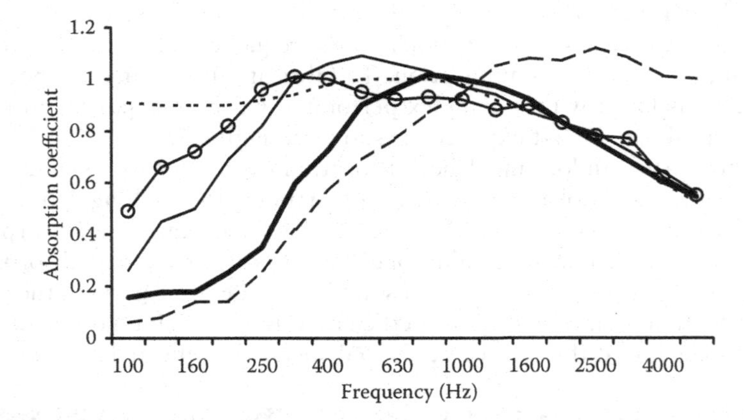

Figure 7.9 Random incidence absorption coefficient for a hybrid surface (BAD panel) compared to mineral wool. Four different backing depths for the panel are shown: ▬▬ 2.5 cm BAD, —— 5.1 cm BAD, —O— 7.6 cm BAD, ----- 10.2 cm BAD, and – – – 2.5 cm fibreglass.

To accomplish high-frequency dispersion, various hole patterns can be used, with a 2D pseudorandom binary sequence being a good choice. For example, an array of 31 × 33 absorptive and reflective areas might be used. The reflective areas map to the 1 bit and the absorptive areas map to the 0 bit in the binary pseudorandom number sequence. The distribution of these binary elements is based on an optimal binary sequence with a flat power spectrum as this maximizes dispersion. For example, maximum length sequences can be used.

The construction is simple and inexpensive. The mask can be made from a variety of materials provided that they are rigid and nonabsorbing, such as wood and metal, and a nonwoven matt can be used on the back surface to conceal the porous absorber. Furthermore, the acoustic function can be hidden, which can lessen the conflict between visual aesthetics and acoustic requirements. Alternatively, many architects are interested in seeing the mask due to its unique appearance and the fact that it offers an alternative to the traditional periodic perforated metal patterns. It is also possible to use the perforated holes to form an image, by mapping the grey scale of a picture to the hole size and density. Currently, this design is carried out using trial and error to find the right mapping from image to hole size and density to get the desired absorption coefficient. An example is shown in Figure 7.10.

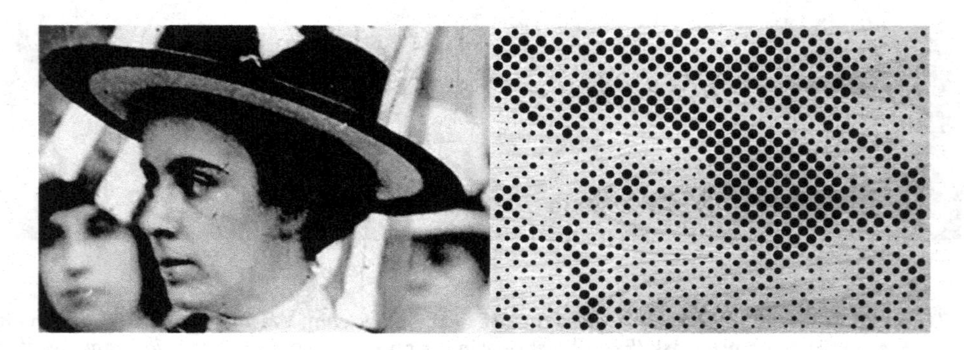

Figure 7.10 Youngest parader in New York City suffragist parade (Courtesy of the American Press Association) and a rendering as a perforated absorber (Courtesy of RPG Diffusor Systems, Inc.).

7.2.4 Microperforation

Microperforated devices are Helmholtz resonators with very small holes. They provide absorption through high viscous losses as air passes through the holes that are only a bit larger than the boundary layer. This inherent damping eliminates the need for fibreglass or other porous materials in the air cavity between the perforated sheet and the reflective surface behind it. This enables clear absorption (see next section). Microperforation is also useful even when transparency is not needed. Made from the right material, these absorbers can be more robust than porous absorbers in harsh environments. Indeed, the performance of these structures is largely independent of the material used (assuming the material does not significantly vibrate). They are also less susceptible to getting clogged up even in very dusty environments, probably due to the vibrating plug of air within the pores. Wood, metal, and other materials can be microperforated to gain absorption. Such absorbents might be used where fibreless materials are required, for instance, in situations where there are concerns about bacterial contamination in food and pharmaceutical industries. They might also be used where fibre contamination is a problem, such as in the microelectronic industries or ventilation system silencers.[6] They could also be used within double-leaf partitions where low weight is important, for instance, in aircraft and cars.

Later in this chapter, the design equations for microperforated devices will be outlined and shown to be accurate. The use of double layers hung in free space as resistive absorbers will also be considered.

Figure 7.11 shows typical absorption coefficients. For the thicker sheet material, the absorption is not as controllable as with Helmholtz devices with porous absorbent. The requirement for small holes restricts the frequency range over which the resonant absorption can be achieved. So the thicker materials are useful devices for treating troublesome low- to mid-frequency noise and reverberance in spaces such as atria.

Maa,[7] who developed the concepts of microperforation in the 1960s, showed that the sheet thickness and hole diameter should roughly be the same for high absorption. Consequently, to get absorption for frequencies important for speech, it is necessary to use a thinner material, say 0.1–0.2 mm thick, and to make the holes smaller and closer together.

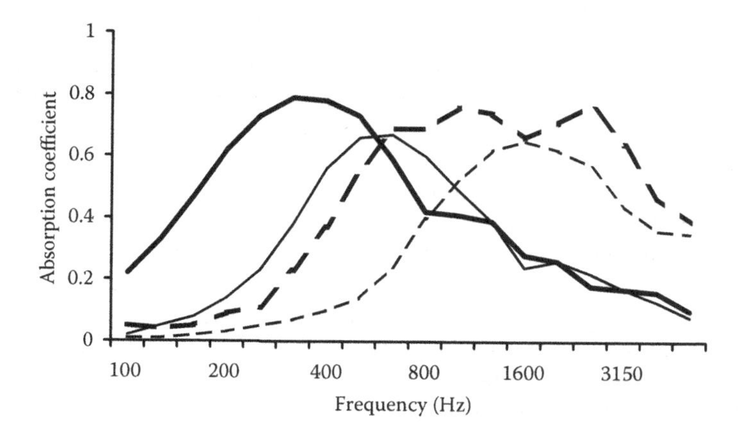

Figure 7.11 Measured random incidence absorption coefficient for four microperforated devices: ▬▬ 1 mm sheet, 200 mm backing depth, 0.5 mm holes spaced at 5 mm; —— 1 mm sheet, 50 mm backing depth, 0.5 mm holes spaced at 5 mm; ▬ ▬ two 0.1 mm foils, 50 mm backing depth, 0.2 mm holes spaced at 2 mm; and - - - 0.1 mm foil, 50 mm backing depth, 0.2 mm holes spaced at 2 mm.

Figure 7.12 A microperforated wood absorber. (a) Front view; (b) rear of panel; with (c) rear of panel nonwoven glass matt. The light coming through the five microperforations in each larger hole can be seen. The panel can be used with only the nonwoven substrate, but typically, fibreglass is used behind the microperforated panel. (Courtesy of RPG Diffusor Systems, Inc.)

The microperforated material might be fixed to another material with much larger holes to give a more structurally robust treatment.[8] Figure 7.12 shows such a device made from wood; the absorption coefficients are shown in Figure 7.1. Manufacturing the microperforations is easier in the thinner materials.

7.2.5 Clear absorbers

Acousticians have long sought a fully transparent absorbing finish to control reverberation while maintaining the view through glazing. Glazing is a popular building material, and there are considerable advantages in combining lighting and acoustic function into one device to save on materials and cost. This can be achieved by exploiting microperforation. Figure 7.13 shows a variety of clear microperforated materials. These are most often used as Helmholtz devices, but without the normal resistive material. For instance, the device might be like a double glazing unit, but with the first pane being a 5-mm-thick panel with submillimetre diameter holes spaced 5 mm apart, with the holes being drilled mechanically. Alternatively, narrow microslits cut into acrylic or polyethylene terephthalate glycol-modified (PETG) using lasers might be used, which is less expensive and gives the possibility of creating decorative micropatterns. These panels are typically suspended in front of glazing or integrated into the window mullions.

To augment the mid- to low-frequency absorption, the device can be curved, tilted, or shaped to provide redirection or diffusion at mid and high frequencies. The surfaces are transparent when looked at from straight on, but at oblique angles, the holes become more apparent, although the surface is still translucent. The optical transmittance of a single microperforated sheet at normal incidence is about 80%, so some loss of optical performance occurs.

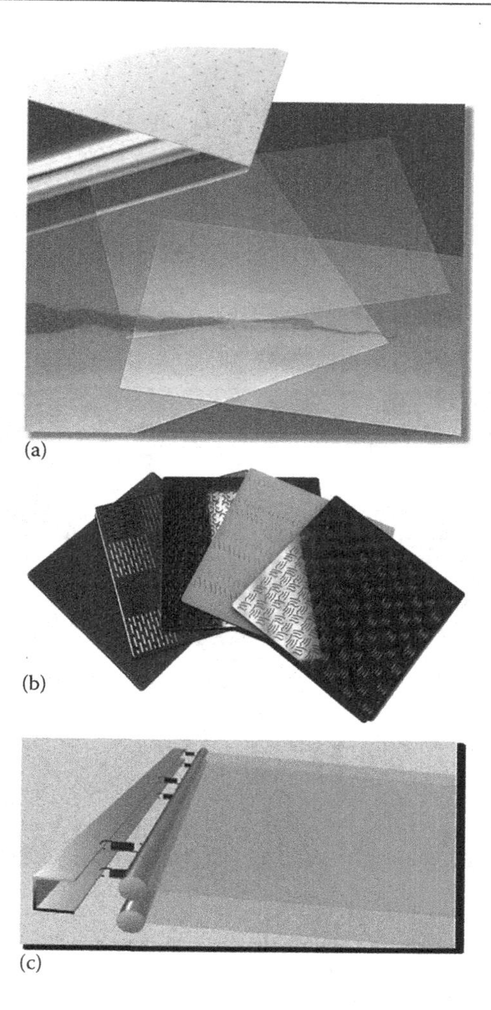

Figure 7.13 (a) Microperforated foil 150–180 μm, with exploded view showing 0.5 mm perforations; (b) microslit plastic panels with a variety of decorative micropatterns; and (c) tensioned hembar mounting illustrating the ability to use two foil layers to broaden the absorption bandwidth. (Courtesy of RPG Diffusor Systems, Inc.)

The thin foil might be made from a material such as polycarbonate or ethylene tetrafluoroethylene and stretched to provide a clear wrinkle-free finish. Double layers can also be used to broaden the absorption bandwidth. An example result for two layers of microperforated foils is shown in Figure 7.11. Another way to broaden performance is to add porous absorbent at the back of the air cavity.

One advantage of clear microperforated absorbers is that they can be hung some way from the backing surface without making the room visually smaller; the additional backing depth can help low-frequency performance.

Microperforated clear absorbers also have potential applications within double glazing units; the absorbent could help prevent the build up of reverberant sound between the glazing panes and alter the mass–spring–mass resonance. The absorbent can also be used in glazing units that have natural ventilation openings.[9] It has even been suggested that they can be used in transparent noise barriers.[10]

(a)

(b)

Figure 7.14 (a) CMU that uses two slotted Helmholtz absorbers to provide bass absorption, DiffusorBlox® (one slot is difficult to see). (b) Application on the upper wall at Zanesville United Methodist Church, Zanesville, IN. Acoustician: dBA Acoustics. (Courtesy of dBA Acoustics.)

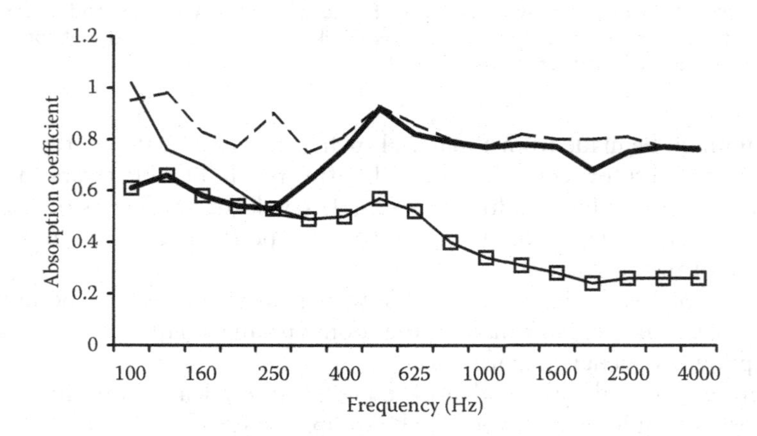

Figure 7.15 Random incidence absorption coefficient for a masonry unit. The slots provide absorption via a Helmholtz mechanism, producing low-frequency absorption: - - - - slotted, unpainted; ▬▬ not slotted, unpainted; —— slotted, painted; and —⊟— not slotted, painted.

7.2.6 Masonry devices

In 1917, Straub patented the CinderBlox, the first concrete masonry unit (CMU). In 1965, slotted blocks were introduced to provide low-frequency absorption. The slots produce a Helmholtz device to provide bass absorption. Figure 7.14 shows a modern equivalent. While the old blocks were useful for noise control, the flat or split face of these blocks can create reflection problems that degrade acoustics. Consequently, a Schroeder diffuser is used to break up the reflected sound wavefronts. The design utilizes two slotted Helmholtz resonator chambers, as well as the phase grating pressure gradient absorption mechanism (see Chapter 8). Typical absorption coefficients are shown in Figure 7.15. Painting reduces the high-frequency absorption as it seals the porous concrete surface but does not affect low-frequency absorption as would be expected. Good insulation against sound transmission is achieved because of the heavy construction.

7.2.7 Plate resonators

Most devices described in this chapter use the air within a cavity to act as a spring. But it is also possible to make a membrane absorber where the spring is the matrix material of the porous absorbent. A possible construction is sheet steel, PVC, or wood plate 0.5–3 mm in thickness, with a surface mass of 5–25 kg m^{-2}, which vibrates against a foam or polyester spring.[11] One advantage of the device is that less depth is required to get bass absorption; for instance, a 100-mm-thick device can provide absorption starting from about 50 Hz.[12]

Figure 7.16a illustrates the absorption mechanism. Sound strikes the plate, which then vibrates against the foam spring. The porous absorption also damps plate bending modes and absorbs higher frequencies, which diffract around the plate through the perforated metal frame. This arrangement provides low-frequency absorption as seen in the heavy line in Figure 7.17. To provide broad bandwidth absorption, part of the rear porous material can be moved to the facing side of the plate resonator. In Figure 7.16b,c the foam also absorbs high frequencies striking the face.

By exploiting all three mechanisms, it is possible to make an absorber that operates over about six octaves. Figure 7.17 shows the measured random incidence absorption coefficients for the two devices. Fuchs et al.[13] give formulations for calculating the key resonant frequencies of the absorber, but not the absorption coefficient.

7.3 DESIGN EQUATIONS: RESONANT FREQUENCY

Having discussed some example designs, this section outlines the most simple design equations. Consider a simple absorber formed by a cavity with a covering sheet. The sheet could either be perforated to form a Helmholtz design or solid but flexible to form a membrane absorber (Figure 7.2). It could even be a flexible perforated membrane: a combination of the two. In the first two cases, the impedance of the cavity given in Equation 6.39 will simply be altered by the addition of mass ($j\omega m$) and resistance (r_m) terms. These are the acoustic mass and resistance, respectively, due to the perforated sheet or membrane. The surface impedance of the resonant system is

$$z_{s1} = r_m + j[\omega m - \rho c \cot(kd)], \tag{7.1}$$

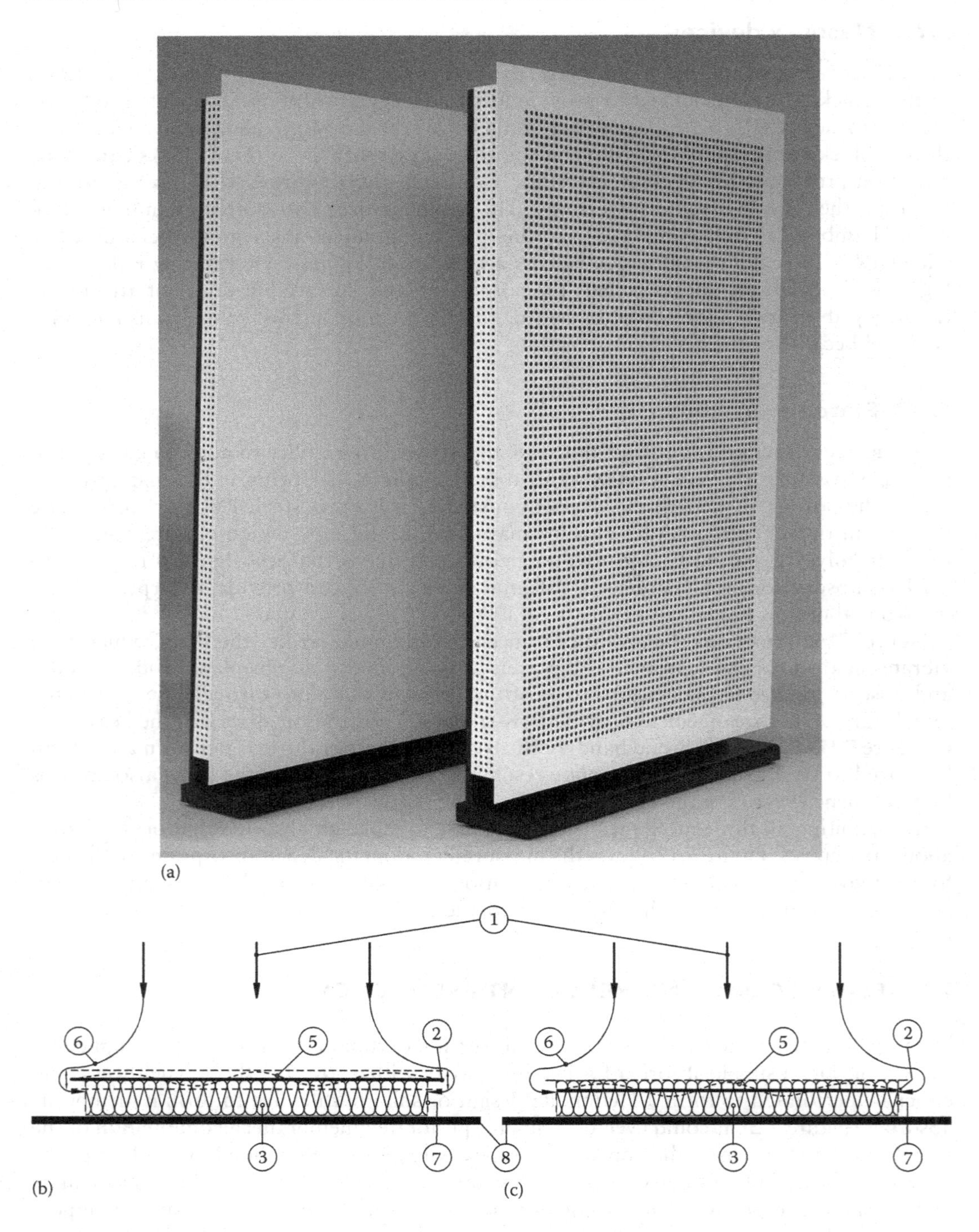

Figure 7.16 (a) Pictures of the two systems. (b and c) Illustrations show sound (1) striking two absorber systems. The steel plates (2) pistonically vibrate against the foam spring (3), mounted on a rigid backing (8). The porous absorption also damps plate bending modes (5) and absorbs higher frequencies, which diffract around the plate (6) through a perforated (7) metal frame. The right hand device has some of the porous absorbent in front of the steel plate, protected by a perforated sheet, which generates additional mid–high-frequency absorption.

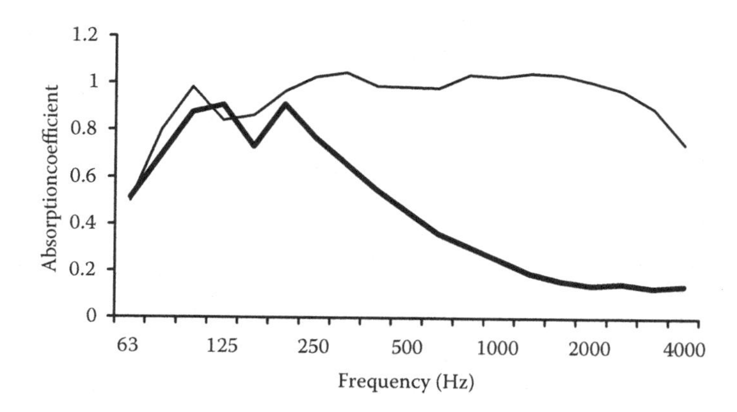

Figure 7.17 Random incidence absorption coefficients for two plate resonators: —— with front perforations to exploit porous absorption at mid–high frequencies and ▬▬ without perforations. (Data courtesy of Fraunhofer Institute for Building Physics, Stuttgart, Germany.)

where $k = 2\pi/\lambda$ is the wavenumber in air, d is the cavity depth; m is the acoustic mass per unit area of the panel, ω is the angular frequency, ρ is the density of air, and c is the speed of sound in air.

For now, the cavity is modelled without porous absorbent for simplicity. Systems resonate when the imaginary part of the impedance is 0, so to find the resonant frequencies, the imaginary part of Equation 7.1 is set equal to 0. Consider a case where the cavity size is much smaller than the acoustic wavelength, i.e., $kd \ll 1$ so $\cot(kd) \rightarrow 1/kd$, then the resonant frequency f is given by

$$f = \frac{c}{2\pi}\sqrt{\frac{\rho}{md}}. \tag{7.2}$$

This is a general formulation that can be applied to Helmholtz and membrane absorbers.

7.3.1 Helmholtz resonator

The perforated surface is divided into individual cells that are assumed to behave independently with a repeat distance D. D is defined in Figure 7.2, which shows a cross-section through the absorber. The absorber is assumed to be perforated in two directions, with the repeat length being the same in both directions. The individual cells will not be entirely independent at low frequency, and consequently, physical subdividing of the volume may be required as the wavelength becomes large. This is especially true if absorption at oblique incidence is required, as would be needed for good random incidence performance. In this case, lateral propagation within the cavity must be suppressed to maximize absorption. When a porous absorbent is placed in the cavity, sound propagation is generally normal to the surface, as discussed in Chapter 6, and so the need for subdividing is less critical, except at very low frequencies.

The hole spacing should be large compared to the hole diameter. The acoustic mass per unit area is then $m = pD^2t'/\pi a^2$, where t' is the thickness of the perforated sheet with the end corrections (end corrections allow for the radiation impedance of the orifices and are discussed later) and other variables are as defined in Figure 7.2. The sheet thickness t and hole radius a are assumed to be much smaller than the wavelength of sound in air. Under these assumptions, the resonant frequency is

$$f = \frac{c}{2\pi}\sqrt{\frac{S}{t'V}}, \tag{7.3}$$

where $S = \pi a^2$ is the area of a hole and V the volume = $D^2 d$ of each unit cell.

This is the same formulation as derived by other methods, such as lumped parameter, equivalent electrical circuits.[14] The transfer function approach is used here because it can more easily generalize to other cases, for example, when the cavity size is no longer shallow compared to wavelength. It is also consistent with the theories used elsewhere in this book.

An alternative, but entirely equivalent, formulation for the Helmholtz resonator uses the fraction of open area, ε, of the perforated sheet. This is often more convenient to work with when using perforated sheets and can be derived by considering the geometry in Figure 7.2 to revise Equation 7.3:

$$f = \frac{c}{2\pi}\sqrt{\frac{\varepsilon}{t'd}}, \tag{7.4}$$

$$\varepsilon = \frac{\pi a^2}{D^2}$$
$$\varepsilon = \frac{2\pi a^2}{\sqrt{3}D^2}, \tag{7.5}$$

where the first formulation of Equation 7.5 is for holes on a square lattice, and the second is for when the holes are spaced D apart on the apexes of equilateral triangles.

The vibrating plug of air within the perforations provides the mass of the device. The length of the plug of air is not just the perforated plate thickness, however. The effect of radiation impedance must be considered, including the mutual interaction between neighbouring holes. Consequently, the vibrating plug of air has a length given by the thickness of the panel plus end corrections to allow for the radiation impedance of the orifice. A full expression for the mass in Equation 7.2 is[15]

$$m = \frac{\rho}{\varepsilon}\left[t + 2\delta a + \sqrt{\frac{8\nu}{\omega}\left(1 + \frac{t}{2a}\right)}\right]. \tag{7.6}$$

The last term in the equation is due to the boundary layer effect, and $\nu = 15 \times 10^{-6}$ m^{2}s^{-1} is the kinemetric viscosity of air. This last term is often not significant unless the hole size is small, say, submillimetre in diameter. δ is the end correction factor, which, to a first approximation, is usually taken as 0.85 and derived by considering the radiation impedance of a baffled piston. A value of 0.85 does not, however, allow for the influence of neighbouring orifices because it is based on a calculation for a single piston. Consequently, other more accurate formulations exist and are summarised in Table 7.1.

In the limit of only one hole in an infinite plane, these formulations equal 0.85, as given earlier. For unusual shapes, the radiation impedance of the plug of air can be numerically

Table 7.1 Formulations for end correction of perforated sheet

Formulation	*Open area limit*	*Applicability*	*Source*
$\delta = 0.85(1 - 1.14\varepsilon^{1/2})$	$\varepsilon < 0.13$	Circular holes in a square pattern	Allard and Atalla[28]
$\delta = 0.85(1 - 1.25\varepsilon^{1/2})$	$\varepsilon < 0.16$	Square holes in a square pattern or circular holes in a circular pattern	Ingard[16]
$\delta = 0.85(1 - 1.47\varepsilon^{1/2} + 0.47\varepsilon^{3/2})$	None	Circular holes in a circular pattern	Nesterov[17]
$\delta = 0.85(1 - 1.13\varepsilon^{1/2} - 0.09\varepsilon + 0.27\varepsilon^{3/2})$	None	Circular holes in a square pattern	Jaouen and Bécot[18]
$\delta = 0.85(1 - 1.33\varepsilon^{1/2} - 0.07\varepsilon + 0.4\varepsilon^{3/2})$	None	Square holes in a square pattern	Jaouen and Chevillotte[17]

evaluated using boundary or finite element models, but the changes that this makes to the final resonant frequency are likely to be small. In theory, these formulations are for normal incidence, but they are also close approximations for oblique incidence cases as well.

An added complication with end corrections is that imperfections in constructions, such as burrs, may have an effect that will be ill-defined. For high-amplitude sound, turbulence will reduce the acoustic mass and so the resonant frequency will increase. Grazing mean air flow is generally observed to also decrease reactance and increase resistance of the perforated sheets.

Figures 7.18 and 7.19 illustrate the effect of changing the open area on the resonant frequency. When the open area is greater than 20%, the perforated sheet makes little difference to the acoustic performance of the porous absorbent behind it. The first figure shows the absorption coefficient, with the frequency at which peak absorption occurs decreasing as the open area reduces. The second figure shows the change in resonant frequency in terms of impedance where the frequency of the zero crossing of the imaginary part decreases as the open area reduces. The script effect_of_perforated_sheet_open_area.m downloaded from the URL in Reference 19 gives the code to generate the results. As the open area decreases, additional low-frequency absorption is generated mainly due to the decreased stiffness of the spring in the unit cell as the backing volume increases. The high-frequency absorption

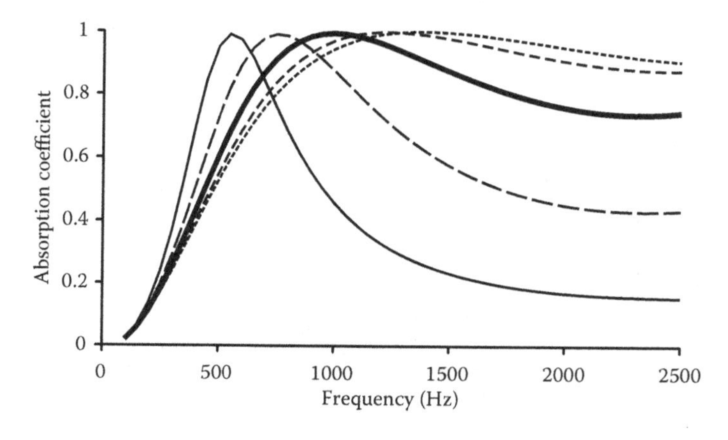

Figure 7.18 Absorption coefficient of a Helmholtz absorber showing the effect of open area. Hole radius, 2.5 mm; porous absorbent flow resistivity, 20,000 rayls m^{-1}; thickness, 2.5 cm; air layer thickness, 2.5 cm; and perforated sheet thickness, 6.3 mm. Open areas: ——— 6%, – – – 12.5%, ━━━ 25%, - - - - 50%, and ------- 100%.

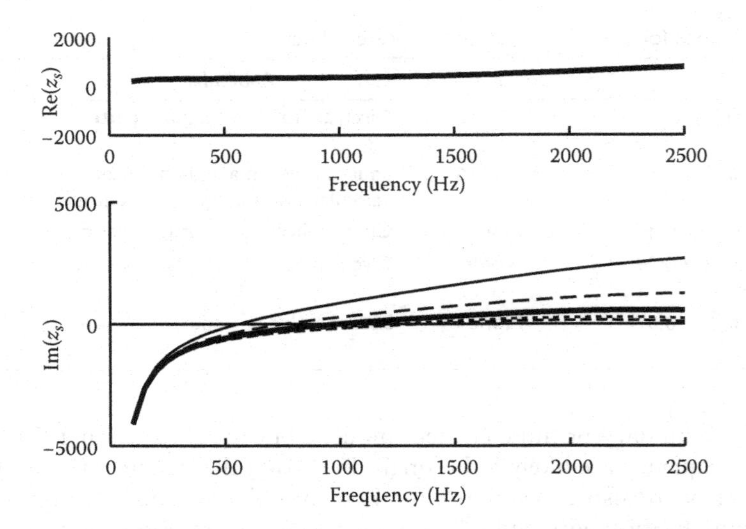

Figure 7.19 Real and imaginary surface impedance of some Helmholtz absorbers showing the effect of changing open area. Geometry same as Figure 7.18. The real lines overlay each other. Open areas: —— 6%, – – – 12.5%, ▬▬ 25%, ------- 50%, and ---- 100%.

decreases because the proportion of solid parts of the perforated sheet increases, and these parts reflect high-frequency sound. Similar results are seen in measurements.[20]

The maximum absorption decreases somewhat as the resonant frequency gets smaller. If these absorbers were tuned to a lower frequency, this decrease in absorption would be more marked. The reason for this is that the impedance of the porous material moves further from the characteristic impedance of air at low frequencies, making the absorbent less efficient. The peak absorption can be altered by changing the porous material's flow resistivity, as illustrated in Figure 7.20. In the case shown, when the flow resistivity is 25,000 $Nm^{-4}s$, the resistance is close to the characteristic impedance of air leading, to high absorption. An additional effect of changing the flow resistivity is to change the bandwidth over which absorption is effective by altering the Q of the resonance. In this case,

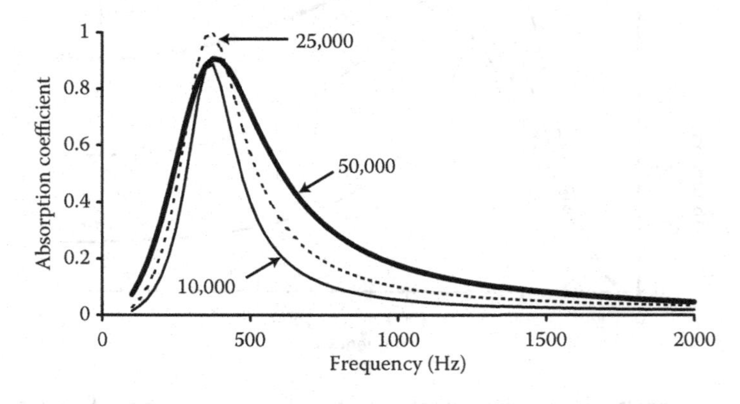

Figure 7.20 Effect of flow resistivity (labels on the lines are in rayls m^{-1}) on absorption of a Helmholtz resonator.

a higher flow resistivity would achieve a greater bandwidth but will reduce the maximum absorption, as the resistance exceeds the characteristic value. A lower flow resistivity leads to an impedance less than characteristic, which results in a reduction in bandwidth and maximum absorption.

Figure 7.21 shows the trade-off between cavity depth and perforated sheet thickness. The perforated sheet thickness has been varied while keeping the total thickness of the device, cavity plus perforated sheet, constant. Making the covering sheet thicker can generate additional bass absorption. But this is at the expense of reduced bandwidth, including decreased high-frequency absorption.

Another common geometry is a device where slots are used instead of holes, see, for example, the CMU shown in Figure 7.14. In some cases, slits can be easier to make than holes because the orifices can be formed by sawing slots in the board or by leaving spaces between parallel planks in wood claddings. For a lath or plank cladding, manufacturers can offer different mounting conditions, with and without spaces between the planks, which enables designers to choose the desired absorption characteristics (see Figure 7.7 for an example). For the end correction, Kristiansen and Vigran[21] used a formulation originally derived by Smits and Kosten[22]:

$$\delta = -\frac{1}{\pi}\ln\left[\sin\left(\frac{1}{2}\pi\varepsilon\right)\right]. \tag{7.7}$$

This then gives a mass term of

$$m = \frac{\rho}{\varepsilon}(t + 2\delta w), \tag{7.8}$$

where w is the width of the slots. As shall be shown later, this gives accurate results.

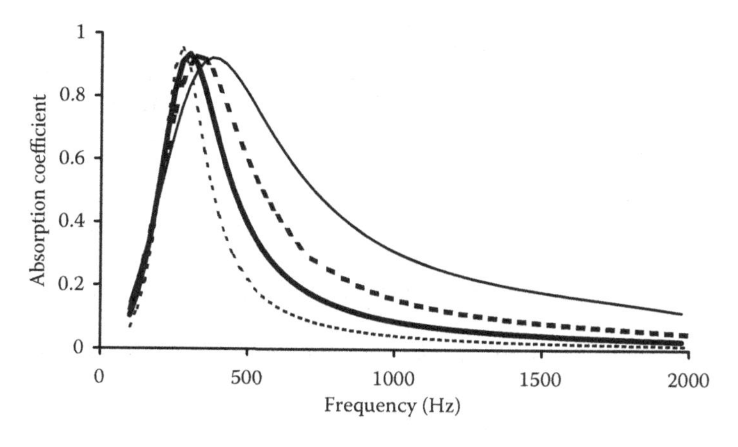

Figure 7.21 Effect of facing thickness on absorption of a Helmholtz resonator. Facing thicknesses: —— 0.8; - - - 6.3, ▬▬ 12.5, and ------ 21.9 mm. The total thickness of the device (cavity plus facing) is kept constant.

7.3.2 Membrane absorber

For a membrane absorber, m in Equation 7.2 is simply the mass per unit area of the panel. A common simplification of the formulation is derived after straightforward algebraic manipulation. This gives the resonant frequency as

$$f = \frac{60}{\sqrt{md}}. \tag{7.9}$$

This is correct when the cavity is filled with air. If the cavity is filled with porous absorbent, then the system is no longer necessarily adiabatic, and an isothermal case must be considered below about 500 Hz. In addition, the porosity of the porous absorber should be included, although this is a minor effect for more commonly used materials, as their porosity is close to 1. Under an isothermal assumption, Equation 7.9 becomes[23]

$$f = \frac{50}{\sqrt{md}}. \tag{7.10}$$

This formulation holds for oblique incidence when a porous absorbent is in the cavity because the porous absorbent enforces propagation normal to the front face. For an air cavity without partitions, a new formulation is required for oblique incidence:

$$f = \frac{60}{\cos(\psi)\sqrt{md}}, \tag{7.11}$$

where ψ is the angle of incidence.

Unfortunately, these simple formulations for membrane absorbers are often inaccurate. The membrane system is not as simple to model as a perforated absorber. Simple models assume that the membrane does not support higher-order modes within the bandwidth of interest. The mass of the membrane is being treated as a single mass, and therefore, the membrane should move as one, like a piston. The effect of bending stiffness is to increase the resonant frequency, but usually, Equation 7.2 is dominant. Problems can occur if the membrane is small, because the whole mass may not be able to vibrate freely if it is secured at the edges. In this case, the actual vibrating mass may be less than expected, and additional losses at the fixings may occur. One solution is to attach the edges of the membrane using resilient foam so that the whole membrane can vibrate, including the edges. If such a fixing is used, it is important that the cavity remains air tight. Alternatively, a surround from a loudspeaker can be used to increase the effective moving mass; this can reduce the resonant frequency of a membrane absorber by 60%–70%. As the angle of incidence increases, there is an increasing chance of bending waves being excited. Consequently, the simple formulations can break down. Unfortunately, the modelling of such bending wave problems is complex, as it is very dependent on the construction used.

More complex models for panel absorbers do exist,[24,25] but the techniques are not that useful for designing practical surfaces. It is possible to model the plate vibration and then use a mode matching approach to derive the power absorbed. This is complex, and many parameters concerning real surfaces, such as the mounting conditions of the panel, will not conform to simple conditions that the prediction models use. Consequently, predictions are unlikely to match measurements well. This has already been illustrated in Figure 7.3. For the prediction shown in the figure, a transfer function matrix method was used, as detailed later. There is a 10% error in the prediction of the peak frequency. In this case, the peak absorption frequency is somewhere between the values given by Equations 7.10 and 7.11. If a more accurate model is required, it is probably best to use a finite element analysis.[26]

7.3.3 Losses

So far, the previous formulations have only allowed a calculation of the resonant frequency. A proper design method must also allow the absorption coefficient and surface impedance to be determined. To do this, the losses within the device must be modelled. These are determined by the resistance term r_m in Equation 7.1. For a Helmholtz device with no additional porous absorbent, this can be calculated using the following[15]:

$$r_m = \frac{\rho}{\varepsilon}\sqrt{8\nu\omega}\left(1+\frac{t}{2a}\right). \tag{7.12}$$

This formulation assumes that the hole radius is not submillimetre in size, to ensure that it is larger than the boundary layer thickness. An alternative formulation for this resistive term derived by Ingard[16] is often used:

$$r_m = \frac{\sqrt{2\rho\eta\omega}}{2\varepsilon}, \tag{7.13}$$

where η is the viscosity of the air, with a value of 1.84×10^{-5} Nsm^{-2}. These theoretical equations do not allow for increased resistance that happens if burrs are present. Indeed, Ingard carried out empirical work to show that Equation 7.13 was approximately correct, but doubling the resistance enables predictions to match experimental results better. Equation 7.13 is more commonly quoted than Equation 7.12, but for most practical absorbers, both are negligible, as is the difference between them! The exception is with devices such as micro-perforated absorbers where the size of the resistance is critical. For most designs, the losses contributed by Equations 7.12 or 7.13 are very small, and in order to get good absorption, it is necessary to add porous materials.

The effect of the porous absorbent depends on where it is placed. Ideally, it should be placed where the particle velocity is a maximum. Porous absorption works primarily by viscous losses as sound penetrates the small pores. For this to be maximized, the air motion must be at its greatest, and this is achieved where the particle velocity is highest. For a Helmholtz resonator this means the absorbent being as close to the openings as possible, or even in the openings. A balance must be struck, however, as too much absorption in the neck might prevent resonance. The effect of placing an air gap between the perforated sheet and the porous absorbent is to reduce the resistance, and in most cases, this will result in a decrease in absorption.[20]

For a membrane absorber, the porous absorbent should be just behind, but not touching, the membrane. Without the porous absorbent, the primary losses are most likely to come from within the membrane or from friction at the fixings between the membrane and the supporting structure. If the porous absorbent behind the membrane does not provide sufficient absorption, perforating the membrane to allow easier access to the porous absorber behind can be done. This then creates a hybrid Helmholtz-membrane design. Then the design equations should be altered somewhat. The impedance of the membrane alone, z_{mem}, will be a combination of resistance and mass:

$$z_{mem} = r_{mem} + j\omega m_{mem}, \tag{7.14}$$

where r_{mem} and m_{mem} are the acoustic resistance and mass, respectively, of the membrane. Similarly, the impedance of the perforated sheet forming the Helmholtz device, z_{helm}, will be formed from acoustic resistance and mass:

$$z_{helm} = r_{helm} + j\omega m_{helm}, \tag{7.15}$$

where r_{helm} and m_{helm} are the acoustic resistance and mass, respectively, of the perforated sheet. It is necessary to make some assumptions about how these impedances interact. The simplest model is that the impedances act independently and in parallel; then, a combined impedance can be derived. The impedance of the device with an air cavity is given by

$$z_{s1} = -j\rho c \cot(kd) + \frac{z_{mem} + z_{helm}}{z_{mem} z_{helm}}, \tag{7.16}$$

where d is the cavity depth. To find the resonant frequency of Equation 7.16, the easiest technique is to plot z_{s1} versus frequency, and from the graph, find the zero crossing of the imaginary part or use a numerical root-finding algorithm. Later on, this type of formulation will be discussed in more detail for a microperforated thin membrane. The problem with applying this formulation is in properly defining the impedance of the membrane.

For a Helmholtz device with porous absorbent, the design equations to be used depend on where the porous layer is located. First, some relatively simple formulations are considered, and then a more complex treatment using transfer matrixes will be developed. The accuracy of these formulations will be demonstrated in Section 7.4.

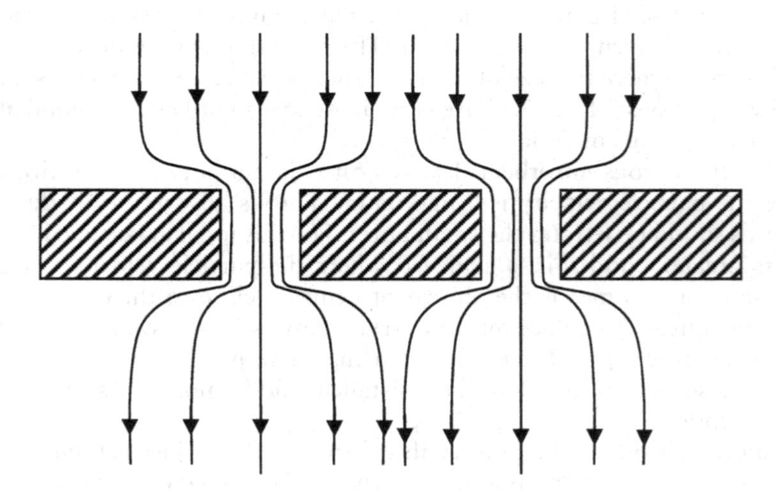

Figure 7.22 Flow through a perforated sheet.

7.3.3.1 Porous layer right behind perforations

When the porous layer is right in front of or behind the perforated plate, then the resistance behaves as though it is actually in the openings. This comes from a consideration of the flow through the device. As sound is squeezed through the holes, the particle velocity is increased. On the other side of the perforated sheet, the flux lines return to a free field case somewhat gradually; this is shown schematically in Figure 7.22. If the porous layer is within a hole diameter of the perforated sheet, it is assumed that the flux has not yet had time to return to a free field state. Consequently, the resistance added by the porous material, r_m, is altered by the fractional open area of the perforated plate, ε. The resistance is as follows:

$$r_m = \frac{\sigma t_a}{\varepsilon}, \tag{7.17}$$

where t_a is the thickness and σ is the flow resistivity of the resistive layer. It is assumed that the volume velocity is reduced by the open area of the plate, ε. The key in absorber design is to make this resistance in Equation 7.17 as close to the characteristic impedance of air as possible, as this maximizes absorption. If characteristic impedance is achieved at resonance, absorption will be complete at that frequency. Consequently, a balance between the open area, flow resistivity, and absorbent thickness must be struck, while remembering that the resonant frequency of the device is also dependent on the open area of the perforated sheet. In addition to changing the resistance, the presence of the porous material directly behind the perforations also increases the end correction. While the effect is smaller than the change in resistance, it can still vary by 30%–100%, depending on the open area of the perforated sheet,[27] with the biggest changes being for the most open sheets.

7.3.3.2 Porous layer in the middle of cavity with a perforated covering

It is assumed that the porous material is further than a hole diameter away from the perforated sheet; the materials are not too thick and are also away from the rigid backing. This is not a common situation as it is awkward to construct. As the bulk of the porous layer is away from the perforations, it is assumed that the velocity through the surface is the same as in free space. Consequently, the resistance term is given by

$$r_m = \sigma t_a. \tag{7.18}$$

A more exact formulation would use a full transfer matrix approach, as detailed in the following section.

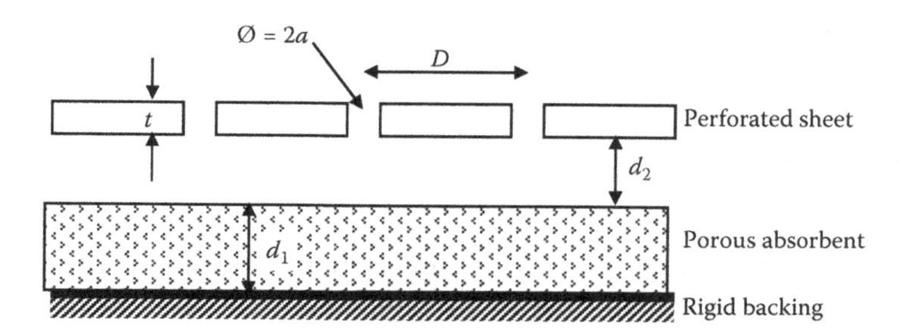

Figure 7.23 Construction for predictions around Equation 7.19.

7.3.3.3 More complete solution using transfer matrixes

A full multilayer solution first calculates the impedance just below the perforated sheet or membrane, and then the effect of the sheet is considered by adding on this impedance. This is a very flexible solution method as it can allow for many different combinations in design. The solution discussed here is split into two forms: the first is when air is immediately behind the perforated sheet and the second is when a porous absorber backs the perforated sheet.

The first case is shown in Figure 7.23. First, the impedance just behind the perforated sheet is calculated; this can be done to a first approximation using the equations set out in Section 6.6.1. Consider a simple case of a layer of absorbent of thickness d_1 and an air layer of thickness d_2. The surface impedance at the top of the absorbent is z_{s1}:

$$z_{s1} = -jz_{c1}\cot(k_1 d_1), \tag{7.19}$$

where z_{c1} is the characteristic impedance and k_1 is the wavenumber of the porous absorbent. The impedance at the top of the air layer and just below the perforation, z_{s2}, can be found by using a transfer matrix (Equation 6.38):

$$z_{s2} = \frac{-z_{s1} j\rho c\cot(kd_2) + \rho^2 c^2}{z_{s1} - j\rho c\cot(kd_2)}, \tag{7.20}$$

where ρ and c are the density and speed of sound in air, respectively. The impedance of the Helmholtz absorber, z_{s3}, is given by using Equation 7.6 and 7.12[28]:

$$z_{s3} = \frac{\rho}{\varepsilon}\left(\frac{t}{2a}+1\right)\sqrt{8\nu\omega} + (2\delta a + t)\frac{j\omega\rho}{\varepsilon} + z_{s2}, \tag{7.21}$$

where the additional viscous term in Equation 7.6 is ignored as it is generally small.

The second case, with the porous layer next to the perforated sheet, is shown in Figure 7.2b. For simplicity, it is assumed that the entire cavity is filled with porous absorbent, and the cavity depth is d. This is a common construction and is simple to make. Two solution methods can be attempted. The most simple is to consider that only plane waves propagating normally to the perforated sheet are present in the porous layer. Then the impedance immediately below the perforated sheet is given by

$$z_{s1} = -jz_{c1}\cot(k_1 d), \tag{7.22}$$

where z_{c1} is the characteristic impedance and k_1 is the wavenumber of the porous absorbent. Then the mass effect of the perforations can be added, and the effect of open area is taken into account to give the surface impedance of the absorber, z_{s2}, as

$$z_{s2} = \frac{1}{\varepsilon}(2\delta a + t)j\omega\rho + z_{s1}. \tag{7.23}$$

The mass and resistance terms due to viscous forces in and around the perforated plate are ignored because they will usually be negligible compared to the z_{s1} term.

A more complex solution to this problem allows for the multiple waves propagating in the porous media. The surface is considered in a series of elementary cells of size D by D, each containing one hole. The velocity at the cell boundaries parallel to the perforated sheet is assumed to be 0. Unless the cavity is actually partitioned, this is only an approximation. This enables the pressure within the cells to be decomposed into a sum of modes within the cell in an analogous way to solving the modes in a room. The impedance below the perforated sheet, z_{s1}, is then given as a sum over modal terms[28]:

$$z_{s1} = \frac{z_{0,0}\varepsilon}{\phi} + \frac{4}{\pi}\sum_{m}\sum_{n,(n\neq 0 \& m\neq 0)} \frac{v_{m,n} z_{m,n} J_1^2\left(\frac{2\pi a}{D}\sqrt{m^2+n^2}\right)}{\phi(m^2+n^2)}, \tag{7.24}$$

$$z_{m,n} = -jz_{c1}\frac{k_1}{k_{m,n}}\cot(k_{m,n}d), \tag{7.25}$$

$$k_{m,n} = \sqrt{k_1^2 - \frac{4m^2\pi^2}{D^2} - \frac{4n^2\pi^2}{D^2}}, \tag{7.26}$$

and

$$v_{m,n} = \begin{cases} 0.5; & m=0 \text{ or } n=0 \\ 1 & \text{otherwise,} \end{cases} \tag{7.27}$$

where k_1 and z_{c1} are the wavenumber and characteristic impedance, respectively, of the porous absorber; ϕ is the porosity of the porous absorber; and J_1 is the Bessel function of the first kind and first order. The sum is carried out over all combinations of n and m when both are not equal to 0.

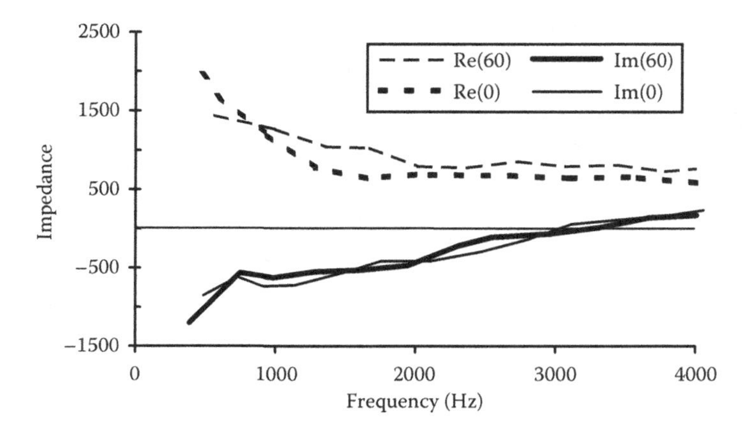

Figure 7.24 Measured impedance for two angles of incidence for a Helmholtz absorber. Angles are indicated in legend in degrees. (Data from Guignouard, P., Meisser, M., Allard, J.F., Rebillard, P., Depollier, C., *Noise Control Eng. J.*, 36, 129–135, 1991.)

The sum converges as the contributions from higher modes reduce. In fact, in many cases, only the plane wave term ($n = 0$ and $m = 1$) need be considered as the dominant propagation mode in a porous medium will be perpendicular to the perforations due to refraction. This is especially true when high-flow-resistivity materials are used. When only the first term is considered, Equations 7.24 through 7.27 give similar results to the more simple formulation given in Equation 7.22. Once the impedance immediately below the perforated sheet is known, z_{s1}, Equation 7.23 can be applied to get the surface impedance of the whole system, including the perforated sheet.

7.3.3.4 Oblique incidence

For oblique incidence, it can be assumed to a first approximation that the impedance of the Helmholtz absorber will be very similar to the normal incidence value, provided that there is a significant amount of porous absorbent in the cavity and/or the cavity is partitioned. In these cases, the dominant propagation direction will be normal to the front face due to refraction. Lateral propagation could change the impedance of the device at oblique incidences, but this is not normally significant. Figure 7.24 shows the measured impedance for a sample at normal and 60° incidence,[29] confirming this assertion. At low frequencies, without partitions within the cavity, this may become less true as lateral propagation modes become more significant. Any lateral propagation would be expected to decrease the absorption achieved for most angles of incidence.

There is a more complex and complete prediction model for oblique incidence.[28] As the surface is periodic, it is possible to solve the problem with a Fourier decomposition. This method can only produce a solution when the wavelength in air, λ, projected onto the surface is an integer multiple of the spacing between the perforations, i.e.,

$$ND = \lambda/\sin(\psi), \tag{7.28}$$

where N is a positive integer and ψ is the angle of incidence. With this principle, it is possible to carry out a Fourier decomposition into a series of modes within the porous material. Consider the case of a Helmholtz device where there the cavity is filled with porous material. The impedance just below the perforated sheet is given by[28,29]

$$z_{s1} = \frac{2}{\pi\phi}\sum_{m=0}^{\infty}\sum_{n=-1,-1\pm N,-1\pm 2N\ldots} v_{m,n} z_{m,n} \frac{J_1^2\left\{2\pi a\sqrt{\dfrac{m^2}{D^2}+\dfrac{n^2}{N^2D^2}}\right\}}{m^2+\dfrac{n^2}{N^2}}, \tag{7.29}$$

$$z_{m,n} = -j\rho c\frac{k_1}{k_{m,n}}\cot(k_{m,n}d), \tag{7.30}$$

$$k_{m,n} = \sqrt{k_1^2 - \chi}, \tag{7.31}$$

and

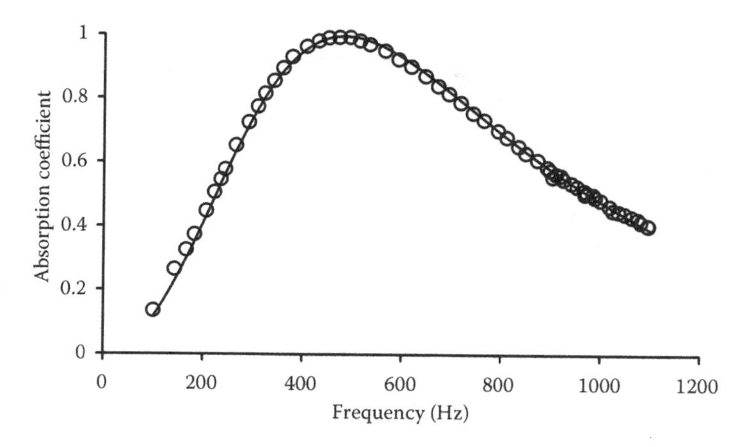

Figure 7.25 —— Predicted and O measured normal incidence absorption coefficient for a slotted Helmholtz absorber. (Measured data from Kristiansen, U.R., Vigran, T.E., *Appl. Acoust.*, 43, 39–48, 1994.)

$$\chi^2 = \left(\frac{2\pi m}{D}\right)^2 + \left(\frac{2\pi n}{ND}\right)^2, \tag{7.32}$$

where k_1 is the wavenumber of the porous layer. The porous material has a depth d and porosity ϕ. $v_{m,n}$ is defined in Equation 7.27. Once z_{s1} has been evaluated, Equation 7.23 can be applied to get the impedance of the surface above the perforated sheet at the front face of the absorber.

7.4 EXAMPLE CALCULATIONS

7.4.1 Slotted Helmholtz absorber

Kristiansen and Vigran[21] carried out impedance tube measurements on a slotted absorber, which allows the accuracy of the previous formulations to be illustrated. The absorber had an open area of about 24%, the slots were 15 mm deep and 10 mm wide, the cavity depth

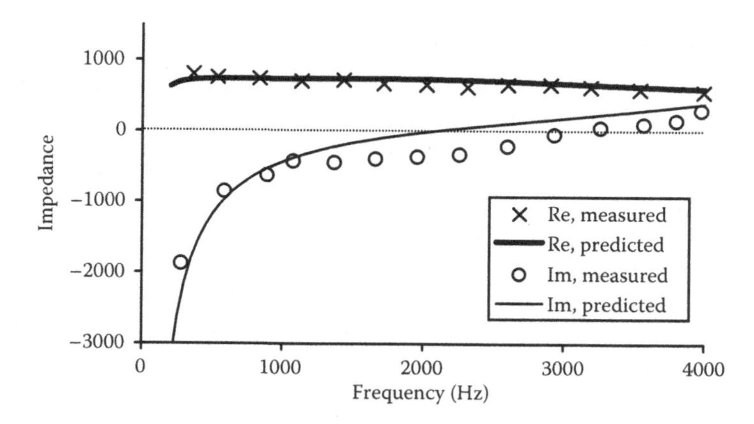

Figure 7.26 Impedance predicted and measured for a Helmholtz absorber. (Measured data from Guignouard, P., Meisser, M., Allard, J.F., Rebillard, P., Depollier, C., *Noise Control Eng. J.*, 36, 129–135, 1991.)

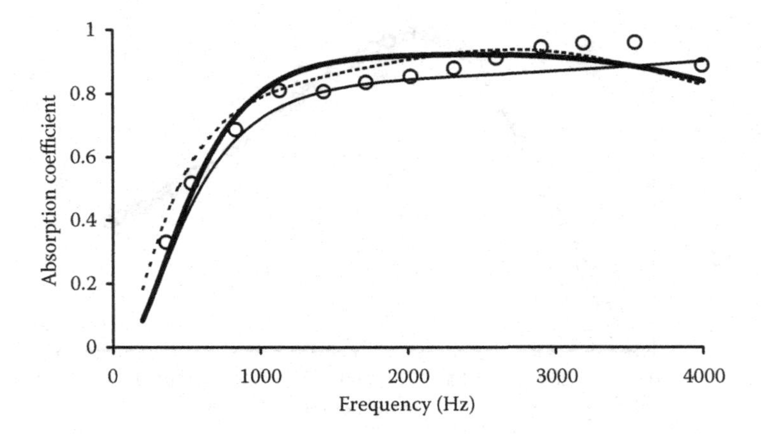

Figure 7.27 Absorption coefficient predicted for three absorbers, plus measurement for one case. (From Guignouard, P., Meisser, M., Allard, J.F., Rebillard, P., Depollier, C., *Noise Control Eng. J.*, 36, 129–135, 1991.) Some interpolation of the measured impedance data was used to obtain the measured absorption coefficient: —— porous absorbent only, predicted; ▬▬ porous absorbent with perforated facing, predicted; - - - - porous absorbent with perforated facing and air gap, predicted; and O porous absorbent with perforated sheet, measured.

was 100 mm, and a material with an air flow resistance of 86 Pa s m^{-1} was attached to the bottom of the slotted plate. Script slotted_absorber.m available for download from Reference 19 predicts the scattering from the slotted absorber, and it is compared with the experimental data in Figure 7.25.[21] Using the transfer matrix method with Equations 7.7, 7.8, 7.17, and 7.1 gives accurate results as shown. Adding the resistance term, Equation 7.12 or 7.13 has a negligible effect and changes the absorption coefficient only by less than a hundredth.

A simple calculation of the peak of absorption using Equations 7.4 and 7.7 yields a predicted resonant frequency about 100 Hz greater than measured. This shows the power and usefulness of the transfer function matrix procedure for Helmholtz absorbers. Similarly, accurate results were also found by Ingard[16] when he examined circular perforations with a thin resistive layer behind the perforated sheet.

7.4.2 Porous absorbent filling the cavity

Figure 7.26 compares the predicted and measured normal incidence impedance for a Helmholtz absorber. These include measurements by Guignouard et al.[29] using a two-microphone free field method to obtain both normal and 60° incidence results. The two-microphone technique is described in Section 4.2. Figure 7.27 shows the predicted absorption coefficients for three arrangements: a porous absorber and a porous absorbent faced with a porous sheet with and without an air gap between the porous absorber and the rigid backing. In addition, a measurement for one of the configurations is given. The predictions for normal incidence used the transfer function matrix technique given in Equations 7.19 through 7.21. The more complex modal decomposition model is unnecessary because the simple model gives satisfactory results. The porous absorber had a flow resistivity of 70,000 rayls m^{-1} and was 3 cm thick. For the perforated sheet, the holes had a radius of 2.5 mm, the open area was 17.5%, and the thickness was 0.75 mm. The prediction model gives reasonable accuracy.

7.4.3 Bass Helmholtz absorber

Room acoustic books often include design charts to calculate the desired dimensions of a Helmholtz resonator. Designing an absorber with charts using a simplified model for a Helmholtz resonator risks poor performance in practice, yielding the wrong resonant frequency with an inappropriate absorption bandwidth.

To explore this further, the simple model of Equation 7.4 was used to design 600 bass absorbers with various hole sizes and spacing, backing depth, and plate thickness and a further thousand bass absorbers made using slots between planks rather than round holes. The end correction factor was taken to be $\delta = 0.85$ as this is the most commonly quoted figure in texts. The absorbers were designed to have a peak absorption somewhere between 30 and 160 Hz. These devices were then properly modelled in a transfer matrix model to get the normal incidence absorption coefficient. As Section 7.4.1 shows, this transfer matrix is an accurate model for these devices. A backing layer was modelled that included a 25 mm layer of mineral wool with a flow resistivity of 20,000 MKS rayls m^{-1} immediately behind the perforations. For this, the Wilson model was used (Section 6.5.4).[30]

Only 15% of the round-hole designs and 23% of the slotted designs, created using the simple model, met all the criteria for a good bass absorber. For the round perforations, of the 91 devices that met the design criteria, the simple theory tended to underpredict the resonant frequency. In a third of these cases, the frequency of peak absorption predicted by the simple theory was wrong by more than a third-octave bandwidth. For the slotted designs, the tendency was for the simple theory to overpredict the resonant frequency, with nearly every design being wrong by over a third-octave bandwidth.

Most designs failed to meet the design specification because the bandwidth of absorption was wrong. For the round perforations, in 24% of cases the bandwidth was significantly narrower than a one-third octave, and for the slotted case this was true of 42% of designs. Given the uncertainty in predicting the peak absorption frequency, this risks creating a device that fails to absorb over the right bandwidth. Most other designs had a bandwidth significantly greater than a one-third octave and so risk attenuating neighbouring frequencies that do not require treatment.

7.5 OTHER CONSTRUCTIONS AND INNOVATIONS

Having described the most common designs, more complex constructions and others innovations will now be discussed. Much can be gained from application of the simple resonator design as discussed previously, but more complicated systems do exist. In trying to decide whether to use more complex constructions, the trade-off lies mainly between acoustic performance and cost of manufacture and installation. After all, a piece of porous material covered with a perforated sheet is relatively inexpensive to produce. Once more complex designs are considered, like complex neck plates, the cost of the device will naturally increase. Consequently, the designs discussed in this section are most often used where space is a particular premium or where special requirements, such as transparency or fibreless absorption, need to be achieved.

7.5.1 Shaped holes and slots

Vigran[31] examined the effects of making conical rather than cylindrical holes for Helmholtz absorbers where the front plate is at least 1 cm thick. The conical holes are such that the smallest opening faces the source. In comparison to cylindrical holes, conical holes broaden the bandwidth of absorption; the frequency of peak absorption increases somewhat. The conical holes also have high-order absorption modes that are broader, resulting in increased absorption at mid–high frequencies. While it may be complex to form conical holes, similar devices can be made from shaped slots very easily.

7.5.2 Double resonators

The problem with resonant absorbers is that they have a relatively limited bandwidth. It is common to want a bigger bandwidth than can be achieved by a single resonator alone. One possibility is to use a device that has multiple absorption mechanisms, such as the plate resonators described in Section 7.2.7. Another possibility is to stack a high-frequency Helmholtz absorber in front of a low-frequency device. The disadvantage of this is that the surfaces become very deep, and depth is often restricted by nonacoustic constraints. This double system can be most easily modelled with the transfer matrix approach. Such a double design was a standard construction used by the British Broadcasting Corporation for many decades.

7.5.3 Microperforation

If the holes of a Helmholtz resonator are made small enough, then losses will occur due to viscous boundary layer effects in the perforations. To achieve this, the perforations must be submillimetre in diameter so that they are comparable to the boundary layer thickness. Then it is possible to achieve absorption without using a porous material. If the perforated sheet is made from transparent acrylic or glass, a clear absorber can be made. Commercial realizations of this were discussed in Sections 7.2.4 and 7.2.5. From a modeller's viewpoint, this is a neat device because the physics of the system is simple and accurate predictions are readily achieved. A microperforated device was reported by Cremer and Müller, was reported in Ref. 23 (pp. 204–206) where a multilayer system originally devised by Rschevkin is briefly outlined. It is the work of Maa,[7] however, that has led to many researchers investigating the concept.

Formulations for the impedance of microperforated devices can be derived by considering the sound propagation within a cylindrical hole. This problem is well established and is the theoretical foundation of much work on microscopic propagation in porous absorbents. For a tube that is short compared to wavelength, it can be shown that the specific acoustic impedance of the tube is given by[7]

$$z_1 = j\omega\rho t\left(1-\frac{2J_1\left(k'\sqrt{-j}\right)}{k'\sqrt{-j}J_0\left(k\sqrt{-j}\right)}\right)^{-1}, \tag{7.33}$$

and

$$k' = a\sqrt{\frac{\rho\omega}{\eta}}, \tag{7.34}$$

where J_0 and J_1 are the Bessel functions of the first kind, of zero and first order, respectively; t is the tube length; and a is the tube radius.

To get the specific acoustic impedance of the perforated sheet, Equation 7.33 must be divided by the plate open area, ε. Maa details approximate solutions to the previous equation, but with the advent of modern numerical tools on computers, it is as easy to implement

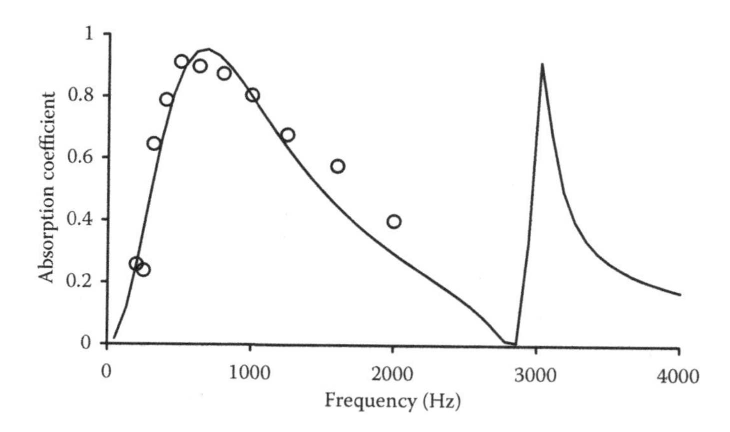

Figure 7.28 —— Predicted and O measured absorption coefficient for a microperforated Helmholtz absorber. (Measurement data from Maa, D.-Y., *Noise Control Eng. J.*, 29, 77–84, 1984.)

Equation 7.33 directly as to use an asymptotic solution. To model the Helmholtz resonator, a transfer matrix can be used to get the surface impedance, z_h:

$$z_h = \frac{z_1}{\varepsilon} - j\rho c \cot(kd) + \frac{\sqrt{2\omega\rho\eta}}{2\varepsilon} + \frac{j1.7\omega\rho a}{\varepsilon} . \tag{7.35}$$

The second term is the impedance of the air cavity, which is assumed to be d deep. The final term is the end correction to allow for the radiation reactance of the tube. The penultimate term is the radiation resistance for an orifice. Maa uses the formulation from Ingard[16] given in Equation 7.13 for the radiation impedance. Once the impedance is known, the normal incidence absorption coefficient can be readily obtained. These equations are most applicable for common sound intensities. For large intensities, the impedance will change due to non-linear effects; flow also affects the impedance, and formulations are given in Reference 32.

Figure 7.28 compares the prediction according to Equation 7.35 to measurements presented by Maa.[7] The hole separation is 2.5 mm, the hole radius 0.1 mm, the plate thickness is 0.2 mm, and the cavity depth is 6 cm. Reasonable agreement between measurement and prediction is achieved, although not as good as given in the paper. Script microperforated_absorber.m downloadable from Reference 19 gives the code for the predictions. The prediction shows a sharp peak due to a second-order resonance. Such resonances are relatively narrow in frequency, and so if the results are summed in one-third octave bands, the second-order peak appears less significant.

With microperforated treatments, it can be difficult to get broadband absorption as the devices rely on resonance. To extend the bandwidth, Maa and others have shown that multiple layers can be used. Each layer is then tuned to a different frequency range. This can then be predicted using a transfer matrix solution, taking each layer in turn. The problem with double-layer devices is they increase the depth and cost of the device, both of which are usually under strict restrictions by nonacousticians. Another solution to increasing the absorption at low frequency is to use a construction such as that shown in Figure 7.23, where porous absorbent is attached to the rigid backing.

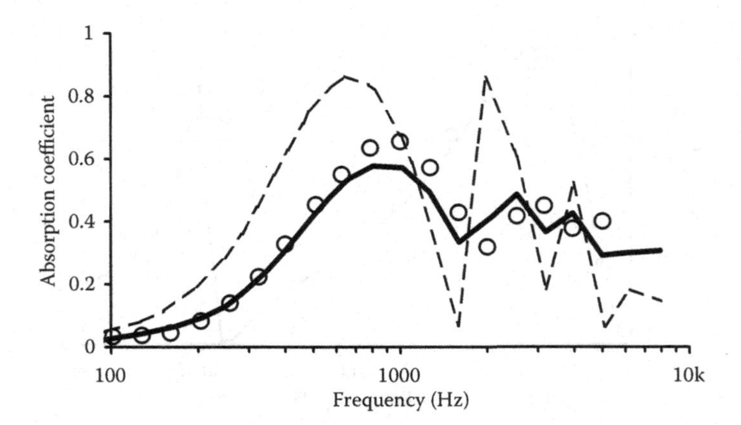

Figure 7.29 Measured and predicted absorption coefficient for a microperforated membrane for different incident sound conditions: O measured, random incidence; ▬▬ predicted, random incidence; and - - - predicted, normal incidence. (After Kang, J., Fuchs, H.V., *J. Sound Vib.*, 220, 905–920, 1999.)

For oblique incidence, the sound in the cavity travels at an angle to the normal, which is the angle of incidence. Consequently, Equation 7.35 should be altered to

$$z_b = \left[\frac{z_1}{\varepsilon} - j\rho c\cot(kd) + \frac{\sqrt{2\omega\rho\eta}}{2\varepsilon} + \frac{j1.7\omega\rho a}{\varepsilon}\right]\cos(\psi), \tag{7.36}$$

where ψ is the angle of incidence. The effect of this is to increase the resonant frequency. For large angles of incidence, however, the lateral coupling between adjacent holes within the cavity will become significant. This might be expected to lower the absorption for most if not all frequencies. Consequently, in a diffuse field, the absorption would be expected to be broader, but with lower maximum absorption.

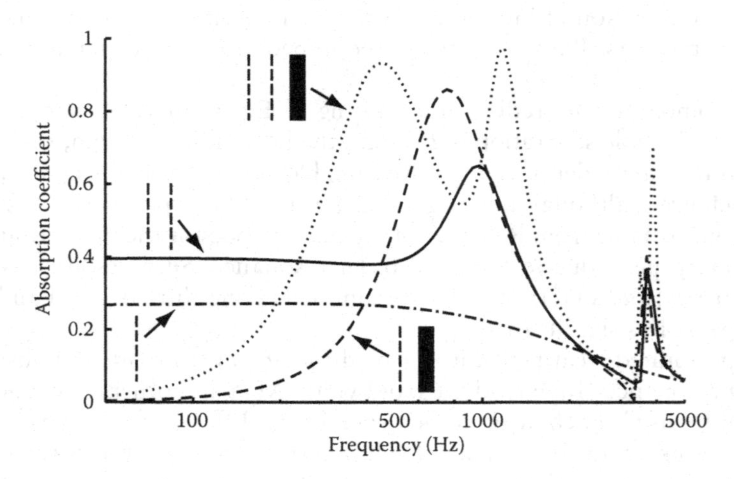

Figure 7.30 Predicted absorption coefficient for various microperforated devices: -·-·- single microperforated sheet in free space, - - - single microperforated sheet in front of rigid backing, ········ double microperforated sheet in front of rigid backing, and ——— double microperforated sheet in free space. (Adapted from Sakagami, K., Morimoto, M., Koike, W., *Appl. Acoust.*, 67, 609–619, 2006.)

As discussed previously, it is possible to combine membrane and Helmholtz mechanisms in a single device. Kang and Fuchs[33] tested a microperforated plastic membrane; the theory was also applied to glass fibre textiles. This treats the membrane and Helmholtz effects in parallel as discussed previously around Equation 7.16. Good agreement was found between impedance tube and reverberation chamber measurement, and the transfer matrix theory. An example result is shown in Figure 7.29, where random and normal incidence absorption coefficients are compared. The device had a mass per unit area m = 0.14 kg m^{-2}; thickness t = 0.11 mm; hole radius a = 0.1 mm; hole spacing D = 2 mm; and a cavity depth of d = 10 cm. As with the previous microperforated systems, the random incidence is less than the normal incidence absorption, and a shift in the frequency of maximum absorption is also seen.

If a microperforated material is used without a rigid backing, it can be used as a suspended absorber, attenuating sound incident from both sides. Figure 7.30 compares the attenuation generated by two microperforated materials with a rigid backing with that generated by a single- and double-layer device suspended in free space. (In the last two cases, the absorption is purely that which is dissipated in the device; the transmitted sound is not considered to be absorbed.) The single and double microperforated sheets suspended in free space provide low-frequency absorption via the device's flow resistance only, and this is maximized when the flow resistance of the structure is $2\rho c$.[34] The additional low-frequency absorption for the double-layer material suspended in free space is, however, at the expense of some of the mid-frequency attenuation.

It is also possible to make microperforated devices using narrow slots instead of cylindrical holes.[35,36] This makes manufacture easier and allows aesthetically pleasing etchings to be made. For an infinitely long slit of width w in a plate of thickness t the impedance can be written as

$$z_h = \frac{j\rho_e \omega t}{\varepsilon} - j\rho c \cot(kd) - \frac{2j\rho\omega w}{\pi\varepsilon} \ln\left[\sin\left(\frac{1}{2}\right)\pi\varepsilon\right], \tag{7.37}$$

where ρ_e is the effective air density in the slit due to viscosity effects and is given by

$$\rho_e = \frac{\rho}{1 - \frac{2}{k'w}\tan\left(\frac{k'w}{2}\right)}, \tag{7.38}$$

where ρ is the density of air and $k' = \sqrt{-j\omega\rho/\eta}$.

Figure 7.31 A Helmholtz absorber that uses lateral space between two perforated sheets as part of the neck of the device.

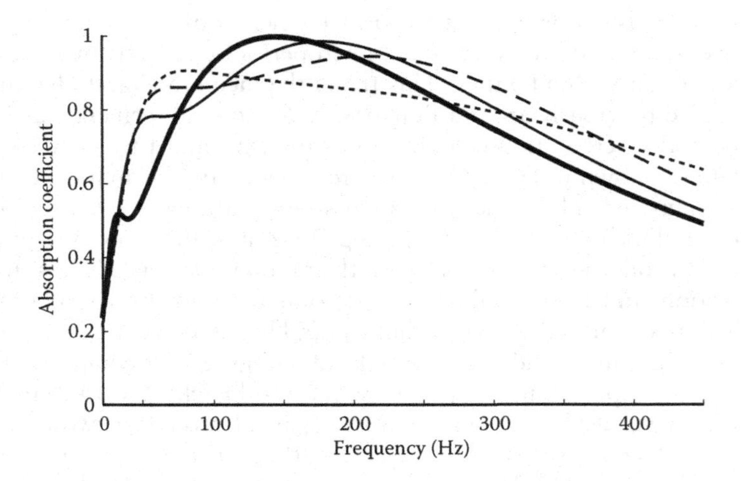

Figure 7.32 Predicted absorption coefficients for a loudspeaker with an inductor in series with a 110 μF capacitor as a passive load: ------- 1, – – – 10, —— 25, and ▬▬ 100 H.

7.5.4 Lateral orifices

Another way to get clear absorption is to elongate the neck of the absorber laterally. Again, the principle is to exploit viscous boundary layer losses in narrow openings and so remove the need for resistive material. Randeberg developed such a technique using both perforations[37] and slots.[38] Figure 7.31 shows such a device. The front and rear plates are perforated with relatively large perforations (1–3 mm in diameter); the viscous losses occur in the propagation parallel to and between the plates. Strict control of the plate spacing is required, which must be of the order of the boundary layer thickness, about 0.2 mm. This spacing must be achieved to a high precision, as the results by Randeberg demonstrate that a change in spacing of 0.05 mm makes significant difference to the absorption obtained. Predicting the absorption of the system is complicated and requires a finite difference solution of the Navier-Stokes Equation. A simple solution using a calculation of vibrating mass based on the volume of the openings and the elongated orifice does not work.

The device gives very similar performance to the microperforated systems discussed previously and as such offers a different construction rather than improved acoustic performance. The absorption is limited to low to mid frequencies.

7.5.5 Passive electroacoustic absorption

A sealed loudspeaker within an enclosure can be used as a membrane absorber: the diaphragm offers the mass and the compliance of the air in the enclosure is the restoring spring. Any porous material within the enclosure provides additional damping. The loudspeaker is not being driven—this is not an active impedance system—it is just responding to the incident sound with passive electrical components across the input terminals.

Using a loudspeaker as a membrane absorber has several advantages: loudspeakers are cheap, readily available, and already configured for rudimentary acoustic treatment. In addition, the electrical properties of the loudspeaker can be manipulated to change how the system responds to incident sound. Consequently, the bandwidth and quantity of absorption can be tuned, within certain limits, by altering the electrical coefficients of the loudspeaker.[39]

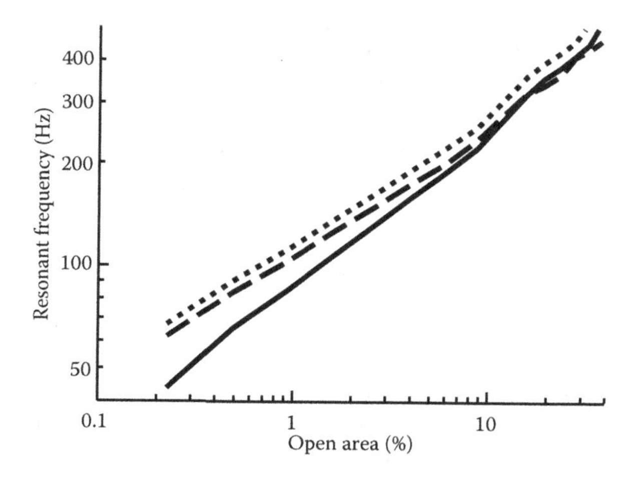

Figure 7.33 Resonant frequencies for Helmholtz resonators vs. open area of the top plate with three backing conditions: – – – empty, —— activated carbon, and ······· sand. (After Bechwati, F. Avis, M.R., Bull, D.J., Cox, T.J., Hargreaves, J.A., Moser, D., Venegas, R., *J. Acoust. Soc. Am.*, 132, 239–248, 2012.)

By applying resistive loads across the terminal, the absorption of an underdamped moving coil loudspeaker can be considerably increased.[40] Without the load, there is an impedance mismatch that limits the absorption. Adding resistance can also decrease the Q of the resonant absorption peak. Measurements show that the absorption bandwidth can be altered by up to 50%. Applying capacitive loads across the terminals alters the resonant frequency by up to 30%, but it also affects the damping in the system somewhat. Predictions indicate that using inductors and capacitors together allows even greater changes in resonant frequency and broader absorption over two and a half octaves, as Figure 7.32 shows.

It is also possible to sense the pressure on the surface of the loudspeaker or the velocity of the cone and to feed these signals back into the electrical circuit of the loudspeaker. Lissek et al.[40] derive formulations that show how these feedback signals change the absorption coefficient of a loudspeaker. Impedance tube measurements demonstrate that this can extend the absorption to cover five octaves. This requires very precise selection of electrical components to work, however.

A similar use of electromechanical systems has also been suggested for Helmholtz absorbers[41] and for reducing cavity resonance in pipes.[42] Tao et al.[43] showed that by using a shunt loudspeaker at the back of a cavity behind a microperforated absorber, additional bass absorption could be obtained.

7.5.6 Activated carbon

Experiments have shown that activated carbon can be used to enhance the compliance of acoustic enclosures and can improve the low-frequency performance of loudspeakers.[44] As discussed in detail in Section 6.2.8, activated carbon provides high sound attenuations at low frequency due to changes in surface reactance. It appears that this happens because it is both a double porosity material and is an adsorbent. Adsorption is a process whereby gas molecules adhere to a surface due to the Van der Waals potential. At low frequency, during compression cycles in a sound wave, gas molecules such as water vapour adhere to the surface of the activated carbon, and during rarefractions, the gas is lost from the surface. Activated carbon causes the effective volume of an acoustic enclosure experienced by sound waves to be larger than its physical volume. This allows membrane and Helmholtz absorbers to be made with smaller backing volumes.

Bechwati et al.[45] explored how the compliance of Helmholtz resonators could be altered by activated carbon. Figure 7.33 shows the resonant frequencies of 16 Helmholtz resonators with different backing materials: air, activated carbon, and sand. The activated carbon lowers the resonant frequency, with the effect being strongest at the lowest frequencies.

7.6 SUMMARY

This chapter has outlined some design principles, typical applications, and theoretical models for resonant absorbers. These devices play a crucial role in improving acoustic conditions. They are exploited to control modes and reverberation within rooms and enclosures and for reducing sound levels in many noise control applications. The next chapter details some absorbers and diffusers that did not fit neatly into other chapters, including seating, metamaterials, and vegetation.

REFERENCES

1. M. Wankling, B. Fazenda, and W. J. Davies, "The assessment of low-frequency room acoustic parameters using descriptive analysis", *J. Audio Eng. Soc.*, **60**, 325–37 (2012).
2. A. Celestinos and S. B. Nielsen, "Controlled acoustic bass system (CABS): A method to achieve uniform sound field distribution at low frequencies in rectangular rooms", *J. Audio Eng. Soc.*, **56**, 915–31 (2008).
3. J. Klaus, I. Bork, M. Graf, and G. P. Ostermeyer, "On the adjustment of Helmholtz resonators", *Appl. Acoust.*, 77, 37–41 (2014).
4. N. W. Larsen, E. R. Thomson, and A. C. Gade, "Variable low-frequency absorber for multipurpose concert halls", *Proc. Forum Acusticum*, 616 (2005).
5. J. A. S. Angus, "Sound diffusers using reactive absorption grating", *Proc. 98th Convention Audio Eng. Soc.*, **3953** (1995).
6. M. Q. Wu, "Micro-perforated panels for duct silencing", *Noise Control Eng. J.*, **45**(2), 69–77 (1997).
7. D.-Y. Maa, "Microperforated-panel wideband absorbers", *Noise Control Eng. J.*, **29**(3), 77–84 (1984).
8. J. Pfretzschner, P. Cobo, F. Simon, M. Cuesta, and A. Fernandez, "Microperforated insertion units: An alternative strategy to design microperforated panels", *Appl. Acoust.*, **67**(1), 62–73 (2006).
9. J. Kang and M. W. Brocklesby, "Feasibility of applying micro-perforated absorbers in acoustic window systems", *Appl. Acoust.*, **66**(6), 669–89 (2005).
10. F. Asdrubali and G. Pispola, "Properties of transparent sound-absorbing panels for use in noise barriers", *J. Acoust. Soc. Am.*, **121**(1), 214–21 (2007).
11. H. V. Fuchs, "Alternative fibreless absorbers—New tools and materials for noise control and acoustic comfort", *Acustica*, **87**, 414–22 (2001).
12. H. V. Fuchs, *Applied Acoustics: Concepts, Absorbers, and Silencers for Acoustical Comfort and Noise Control: Alternative Solutions-Innovative Tools-Practical Examples*, Springer Science & Business Media, Berlin (2013).
13. H. V. Fuchs, X. Zha, X. Zhou, and H. Drotleff, "Creating low-noise environments in communication rooms", *Appl. Acoust.*, **62**, 1375–96 (2001).
14. L. E. Kinsler, A. R. Frey, A. B. Coppens, and J. V. Sanders, *Fundamentals of Acoustics*, 4th edn, John Wiley & Sons, Oxon, UK (2000).

15. A. W. Guess, "Result of impedance tube measurements on the acoustic resistance and reactance", *J. Sound Vib.*, **40**, 119–37 (1975).
16. U. Ingard, "On the theory and design of acoustic resonators", *J. Acoust. Soc. Am.*, **25**(6), 1037–61 (1953).
17. L. Jaouen and F. Chevillotte, "Clarification of the expressions for the end corrections of duct and perforation for linear acoustics and large wavelengths", personal communication (2015).
18. L. Jaouen and F. X. Bécot, "Acoustical characterization of perforated facings", *J. Acoust. Soc. Am.*, **129**(3), 1400–6 (2011).
19. MathWorks, uk.mathworks.com/matlabcentral/fileexchange/54004-modelling-of-acoustic-absorbers, accessed 15 November 2015.
20. W. A. Davern, "Perforated facings backed with porous materials as sound absorbers--An experimental study", *Appl. Acoust.*, **10**, 85–112 (1977).
21. U. R. Kristiansen and T. E. Vigran, "On the design of resonant absorbers using a slotted plate", *Appl. Acoust.*, **43**(1), 39–48 (1994).
22. J. M. H. Smits and C. W. Kosten, "Sound absorption by slit resonators", *Acustica*, **1**, 114–122 (1951).
23. L. Cremer and H. A. Müller, *Principles and Applications of Room Acoustics vol. II*, Applied Science Publishers (translated by T. J. Schultz), Barking, UK (1978), p. 191.
24. R. D. Ford and M. A. McCormick, "Panel sound absorbers", *J. Sound Vib.*, **10**, 411–23 (1969).
25. F. P. Mechel, "Panel absorber", *J. Sound. Vib.*, **248**(10), 43–70 (2001).
26. W. H. Chen, F. C. Lee, and D. M. Chiang, "On the acoustic absorption of porous materials with different surface shapes and perforated plates", *J. Sound Vib.*, **237**, 337–55 (2000).
27. I. Lee, A. Selamet, and N. T. Huff, "Acoustic impedance of perforations in contact with fibrous material", *J. Acoust. Soc. Am.*, **119**(5), 2785–97 (2006).
28. J. F. Allard and N. Atalla, *Propagation of Sound in Porous Media: Modelling Sound Absorbing Materials*, John Wiley & Sons, Chichester, UK (2009).
29. P. Guignouard, M. Meisser, J. F. Allard, P. Rebillard, and C. Depollier, "Prediction and measurement of the acoustical impedance and absorption coefficient at oblique incidence of porous layers with perforated facings", *Noise Control Eng. J.*, **36**(3), 129–35 (1991).
30. D. K. Wilson, "Simple, relaxational models for the acoustical properties of porous media", *Appl. Acoust.*, **50**, 171–88 (1997).
31. T. E. Vigran, "Conical apertures in panels; sound transmission and enhanced absorption in resonator systems", *Acta Acust. Acust.*, **90**, 1170–7 (2004).
32. L. L. Beranek and I. L. Vér (eds), *Noise and Vibration Control Engineering*, John Wiley & Sons, New York (1992).
33. J. Kang and H. V. Fuchs, "Predicting the absorption of open weave textiles and micro-perforated membranes backed by an air space", *J. Sound Vib.*, **220**, 905–20 (1999).
34. K. Sakagami, M. Morimoto and W. Koike, "A numerical study of double-leaf microperforated panel absorbers", *Appl. Acoust.*, **67**, 609–19 (2006).
35. D. X. Mao and Z. M. Wang, "Theory and analogous design of microslitted-panel absorbers", *J. Tongji Univ.* (in Chinese), **28**, 316–9 (2000).
36. T. E. Vigran and O. K. Ø. Pettersen, "The absorption of slotted panels revisited", *Proc. Forum Acusticum*, Budapest, Hungary, 2037–40 (2005).
37. R. T. Randeberg, "A Helmholtz resonator with a lateral elongated orifice", *Acustica*, **86**, 77–82 (2000).
38. R. T. Randeberg, "Adjustable slitted panel absorber", *Acta Acust. Acust.*, **88**, 507–12 (2002).
39. R. G. Oldfield and T. J. Cox, "Passive tuned loudspeakers as absorbers for room acoustics", *Proc. 19th ICA* RBA-16-008 (2007).
40. H. Lissek, R. Boulandet, and R. Fleury, "Electroacoustic absorbers: Bridging the gap between shunt loudspeakers and active sound absorption", *J. Acoust. Soc. Am.*, **129**, 2968–78 (2011).
41. F. Liu, S. Horowitz, T. Nishida, L. Cattafesta, and M. Sheplak, "A multiple degree of freedom electromechanical Helmholtz", *J. Acoust. Soc. Am.*, **122**, 291–301 (2007).

42. A. J. Fleming, D. Niederberger, S. O. R. Moheimani, and M. Morari, "Control of resonant acoustic sound fields by electrical shunting of a loudspeaker", *IEEE Trans. Control Syst. Technol.*, **15**(4), 689–703 (2007).
43. J. Tao, R. Jing, and X. Qiu, "Sound absorption of a finite micro-perforated panel backed by a shunted loudspeaker", *J. Acoust. Soc. Am.*, **135**, 231–8 (2014).
44. J. R. Wright, "The virtual loudspeaker cabinet", *J. Audio Eng. Soc.*, 51, 244–7 (2003).
45. F. Bechwati, M. R. Avis, D. J. Bull, T. J. Cox, J. A. Hargreaves, D. Moser, and R. Venegas, "Low frequency sound propagation in activated carbon", *J. Acoust. Soc. Am.*, **132**, 239–48 (2012).

Chapter 8

Other absorbers and diffusers

This chapter deals with some absorbers and diffusers that do not easily fit into categories but are nevertheless important to airborne acoustics. The first subject is seating and audience absorption. In many auditoria, the audience form the main absorption in the room, and consequently, being able to correctly measure and predict the absorption coefficient of the audience area is very important. The second subject is how to make efficient absorbers from Schroeder diffusers (phase grating devices). Researchers started by looking into why the absorption from badly constructed Schroeder diffusers could be large, and ended up inventing a new style of absorber.

There has been considerable interest in *sonic crystals* that absorb sound. It could be argued that phase grating surfaces designed to absorb sound are sonic crystals, but the concern here is with volumetric devices analogous to *photonic crystals*, which can produce high attenuations—but only over limited bandwidths. The simplest sonic crystals are made from a periodic arrangement of shapes such as spheres or cylinders. If a random arrangement of the elements is made, then instead of attenuation, dispersion is created. These volumetric diffusers are also discussed.

Over the years, there has been disagreement about the absorbing ability of vegetation and ground. This chapter ends by examining the mechanisms and ability of achieving noise control through natural materials, whether that is a green wall or a belt of trees.

8.1 AUDIENCE AND SEATING

The reverberation time in performance spaces is often dominated by the absorption of the seating and audience. It is essential that these are measured or predicted accurately for correct design. Section 4.3.1 discussed how the absorption of seating should be measured, and so this discussion concerns the actual values of absorption coefficients that are available in literature and what they mean.

Beranek[1] and Kosten[2] have both produced data for the absorption coefficients of occupied and unoccupied seating. The data were averaged from measurements in many halls and are useful for estimating reverberation time in the early stages of design. The use of average data is not reliable for later design work, however, as there is too much variation in the construction of seating, and consequently, absorption coefficients can vary greatly. Figure 8.1 shows the spread and mean of the absorption coefficients measured by Davies et al.[3] for nine seating types. Also shown are the average values from Beranek,[1] which are in common use. At higher frequencies, as Bradley[4] explains, Beranek's absorption data are possibly affected by differences in air absorption between the many halls measured. Discrepancies may also arise because modern theatre seating has slightly more padding than the older ones

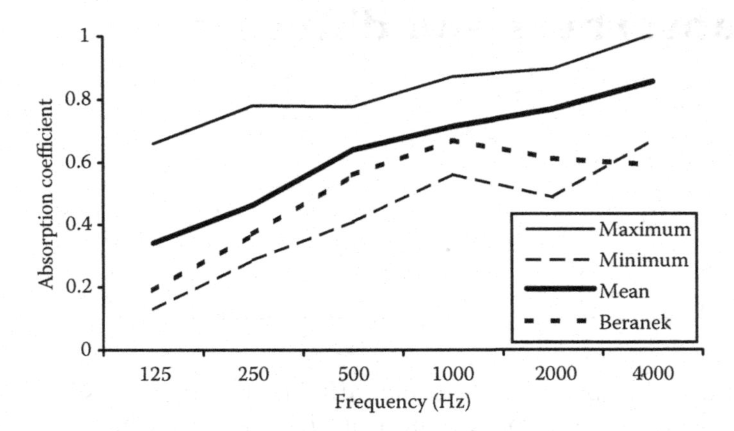

Figure 8.1 The minimum, maximum, and mean absorption coefficients for theatre audience seating. (Modified from Davies, W.W., Orlowski, R.J., Lam, Y.W., *J. Acoust. Soc. Am.*, 96, 879–888, 1994.) Also shown are the mean values from Beranek. (Modified from Beranek, L.L., *J. Acoust. Soc. Am.*, 45, 13–19, 1969.)

forming the bulk of Beranek's data. Bradley et al.[5] also presented absorption coefficients for eight different seat types, again demonstrating how the absorption coefficient varies greatly depending on the seat construction. (The absorption coefficients were generally larger than those shown in Figure 8.1. The data shown in Figure 8.1 exclude absorption by the edges of the seating. This means that in practice, the seating will offer more absorption than shown in the graph.)

For occupied seating, measurements by different authors are much more similar.[4,6] It appears that the absorption of occupied upholstered seats is dominated by the absorption of the occupants and does not vary much over different seat types. Consequently, it is possible to predict the absorption of occupied audience seating with more confidence during design. The absorption coefficient of a block of seating, α, is given by

$$\alpha = \alpha_\infty + \beta \frac{P}{A}, \tag{8.1}$$

where α_∞ is the absorption of an infinite block of seating, P is the perimeter, A is the surface area of the block of seating, and β is an experimentally determined constant. A value of P/A of 0.5 m^{-1} is typical for larger blocks of chairs in auditoria. Table 8.1 gives values for the constants.

For churchgoers in pews or standing audiences, absorption per person is a better measure as the perimeter effect is weaker. Martellotta et al.[7] derived expressions to enable the

Table 8.1 Constant values used to evaluate occupied audience absorption coefficients based on the average of three chair types

	125	*250*	*500*	*1000*	*2000*	*4000*
α_∞	0.69	0.89	1.05	0.97	0.97	0.97
β	0.05	0.14	0.23	0.35	0.37	0.37

Source: After Bradley, J.S., Choi, Y.J., and Jeong, D.U., *Appl. Acoust.*, 74, 1060–1068, 2013.

absorption coefficient to be derived from the occupation density. The expressions are frequency dependent and are shown in Table 8.2 for people wearing heavy winter clothing. For lighter summer clothing, the resultant absorption coefficient would usually be considerably lower at mid–high frequency. As a rough guide, for the octave bands 125 to 4000 Hz, the anticipated reduction would be 0.1, 0.2, 0.4, 0.6, 0.4, and 0.2, respectively. Appendix A gives some alternative absorption coefficients for standing audiences.

Figure 8.2 shows how the absorption coefficient of seating varies with row spacing over the small range commonly found in auditoria. Increasing the row spacing decreases the absorption coefficient. Figure 8.3 shows the effect of carpet on the absorption coefficient. The addition of carpet, even below the seating, significantly increases the absorption and so is generally avoided in large concert venues designed for classical music.

Figure 8.4 compares occupied and unoccupied absorption coefficients. Although it is normal practice to try and make the absorption of seating the same whether occupied or not, this is not entirely successful. On a related issue, Hidaka et al.[8] have suggested that draping the seats with felt can simulate occupied conditions. This appears to be successful at mid frequencies, but at low frequencies (100–200 Hz), it does not always work because the felt alters the seat dip effect.[9,10]

Table 8.2 Formulations to calculate the absorption coefficient of a block of standing audience from the occupation density (d) in persons per metre (heavy winter clothing)

Frequency (Hz)	*Absorption coefficient*
125	$\alpha = 0.142d$
250	$\alpha = 0.239d$
500	$\alpha = -0.082d^2 + 0.797d - 0.146$
1000	$\alpha = -0.086d^2 + 0.986d + 0.081$
2000	$\alpha = -0.115d_2 + 1.109d$
4000	$\alpha = -0.114d^2 + 1.125d - 0.017$

Source: After Martellotta, F., D'alba, M., Della Crociata, S., *Appl. Acoust.*, 72, 341–349, 2011.

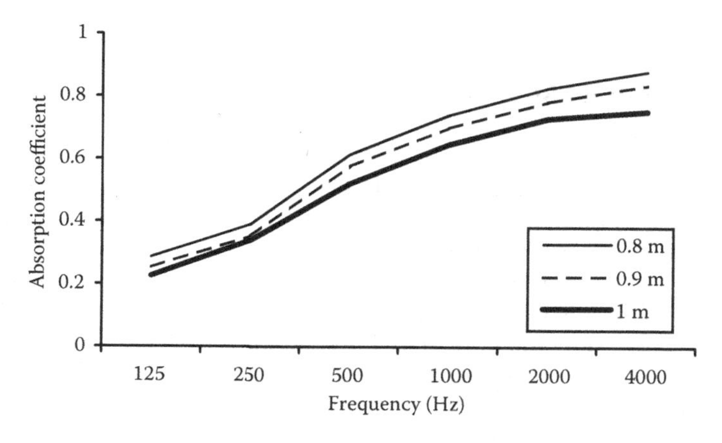

Figure 8.2 Effect of row spacing on the absorption coefficient for a large block of seating. (After Davies, W.J., "The Effects of Seating on the Acoustics of Auditoria", PhD thesis, University of Salford, 1992.)

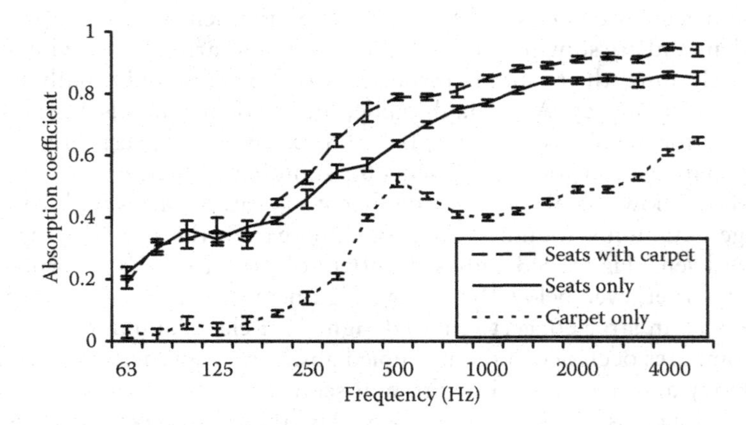

Figure 8.3 The absorption coefficient for seating with and without carpet. (Data from Davies, W.J., "The Effects of Seating on the Acoustics of Auditoria", PhD thesis, University of Salford, 1992.)

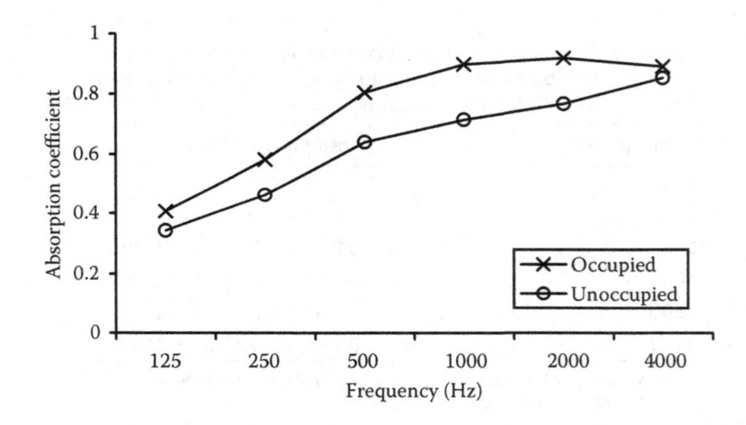

Figure 8.4 Absorption coefficients for occupied and average unoccupied seating. (Data adapted from Davies, W.W., Orlowski, R.J., Lam, Y.W., *J. Acoust. Soc. Am.*, 96, 879–888, 1994.)

8.2 ABSORBERS FROM SCHROEDER DIFFUSERS

Numerous pictures and sketches of Schroeder diffusers can be found in Chapter 10, for example, Figure 10.1. The Schroeder diffuser was designed to diffuse rather than absorb sound, but with changes in geometry and design, it is possible to change the device to make an absorber. Absorption is of great concern for diffuser installation, as it is very easy to accidentally make a highly absorbing surface through bad workmanship. Section 10.8 discusses some general principles to achieve low absorption from Schroeder diffusers. What is discussed here is the mechanism of the absorption and how this can be exploited to form a good absorber.

When Marshall and Hyde[11] implemented their revolutionary use of Schroeder diffusers in the Michael Fowler Centre (see Figure 3.25), they used shallow and wide wells. This was borne out not only of a desire to achieve moderate diffusion but also of a concern that the devices could cause excess absorption. Dramatic levels of absorption from Schroeder diffusers were measured by Fujiwara and Miyajima[12] in 1992, with the absorption coefficient ranging from 0.3 to 1, and at the time, this could not be explained. Fujiwara and Miyajima[13] later reported that the quality of construction was to blame for some of the excess absorption; achieving low absorption requires good workmanship. Commins et al.[14] experimentally investigated the absorption characteristics of a Schroeder diffuser and found values peaking at 0.5. They showed that by sloping the bottom of the diffuser wells, the absorption could be reduced. In 1983, D'Antonio made the first absorption measurements of a commercial quadratic residue diffuser with seven 86.4-mm-wide wells, with a maximum depth of 196.9 mm. The average absorption coefficient was 0.24 between 125 and 4000 Hz, with a maximum value of 0.35 at 500 Hz.

Although workmanship can explain the excess absorption in many cases, even diffusers constructed to a high standard can have absorption coefficients higher than expected. Resonant absorption occurs due to the one-quarter wave resonances in the wells, but the absorption measured is too high to be explained by this phenomenon alone. It was Kuttruff[15] who first postulated energy flow between the wells as a probable cause for the excess absorption, although his theoretical model could not predict the high absorption measured by others. Mechel[16] thoroughly discussed the theoretical basis for the absorption effect, and although his studies lacked direct experimental verification, the prediction model developed was shown by others to be accurate. Wu et al.[17] then brought together measurement and Mechel's prediction model to provide evidence that the energy flow or strong coupling between the wells was indeed responsible for the high absorption.

8.2.1 Energy flow mechanism

Consider a pure tone wave incident on two neighbouring wells of a Schroeder diffuser. Consider a frequency where one well is in resonance and the neighbouring well is not, as illustrated in Figure 8.5. The energy at the mouth of the resonating well will be much greater than that of the nonresonating well. This means that there will be energy flow from the resonating well to the well that is not resonating. Consequently, around the entrances to the wells, there is high particle velocity. Indeed, Fujiwara et al.[18] showed that the particle velocity is up to 14 times greater at the mouth of the wells compared to the incident field. As sound moves around the front of the fins, from one well to the next, excess absorption occurs. This is the source of the additional absorption in Schroeder diffusers and occurs even in properly constructed structures.

Knowing that the front face of the diffuser is a region of high particle velocity, placing resistive material at the well entrance can create an absorber.[16] Figures 8.6 and 8.7 show the absorption coefficient and surface impedance for a profiled structure with and without a resistive layer. Two different resistive layers of different flow resistance are illustrated. The effect of the resistive layer is to broaden the resonant peaks, thereby generating absorption over a greater bandwidth. It also increases the impedance closer to the characteristic impedance for air and thereby gains more absorption. The resistive layer can be made from wire mesh, cloth, or any material with an appropriate acoustic resistance. The advantage of using wire mesh is that the absorber is then washable and durable, which can be useful in some applications.

To maximize absorption, the resistance of the covering must be such that the total resistance of the wells are close to the characteristic impedance of air. Too large a resistance

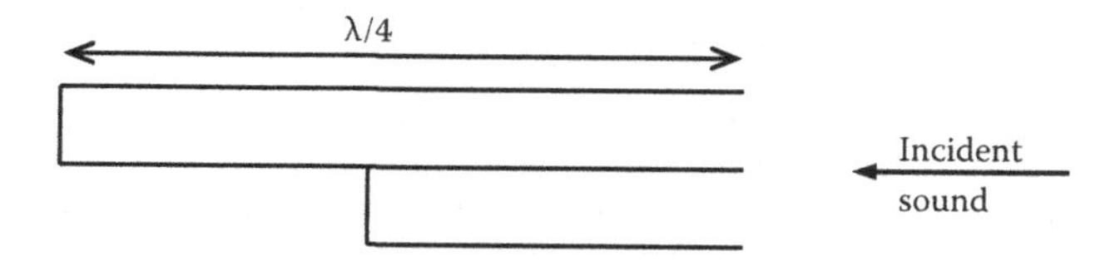

Figure 8.5 Two wells of a Schroeder diffuser.

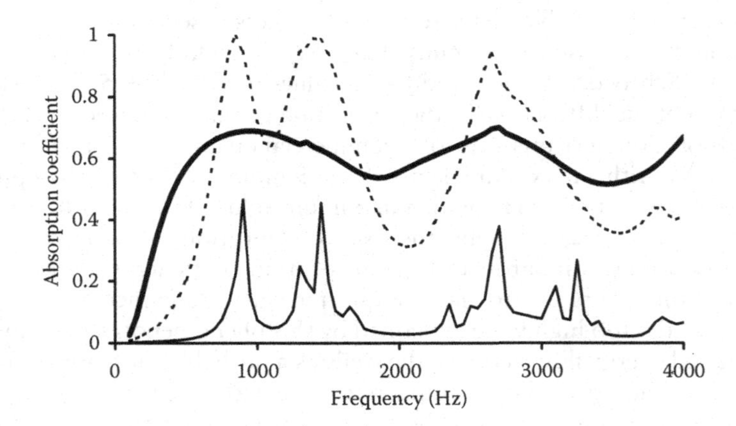

Figure 8.6 Normal incidence absorption coefficient for a quadratic residue diffuser with narrow wells, showing dependence on whether there is a covering at well entrance and what the flow resistance of the covering is: —— no covering, - - - - covering of 65 rayls, and ▬▬ covering of 550 rayls.

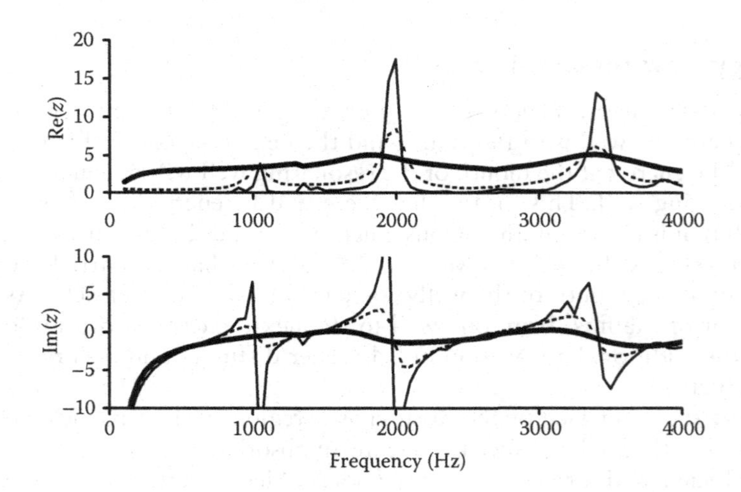

Figure 8.7 Normal incidence surface impedance for a quadratic residue diffuser with narrow wells showing dependence on whether there is a covering at well entrance and what the flow resistance of the covering is: —— no covering, - - - - covering of 65 rayls, and ▬▬ covering of 550 rayls.

leads to an overly damped system and the peaks of absorption are significantly lowered. This is illustrated in Figures 8.6 and 8.7. Too little resistance (no covering) leads to an uneven performance; too much resistance (550 rayls) leads to over damping, whereas 65 rayls gives the highest peak absorption.

8.2.2 Boundary layer absorption

When the wells of a Schroeder surface become narrow, the losses at the well walls due to viscous boundary layer effects can become significant. This can be exploited to produce greater absorption, but the role of the resistive layer must not be forgotten. The key to obtaining a high absorption is that the combination of the covering material and the losses at the well walls should approach the characteristic impedance of air. If the wells are wide, a higher resistance will be needed from the covering material to compensate for the lack of boundary layer absorption. Similarly, if the walls of the wells are rough, then there will be more boundary layer absorption than with a smooth material, and this must be allowed for in the design.

8.2.3 Absorption or diffusion

It is possible to construct Schroeder surfaces to maximize the absorption or maximize the diffusion. Although the surface used to produce high absorption has the same ancestry as those used to disperse sound, crucial design differences result in radically different absorption properties. The two different design remits are contrasted in Table 8.3.

As indicated in Table 8.3, measurements show that 2D surfaces usually absorb more than 1D surfaces, as shown in Figure 10.39. The reason for this is probably twofold:

1. There are often a greater number of different well depths in a 2D compared to a 1D device. This means that there are more quarter wave resonances in the 2D surface, leading to more frequencies at which resonance is occurring. This in turn means that the absorption due to quarter wave resonance is significant for more frequencies, and the energy flow between the wells is also greater, leading to more losses.
2. There is a greater surface area of well boundaries in a 2D compared to a 1D device. It is at these boundaries that viscous boundary layer losses occur. Consequently, it is expected that the greater the boundary area, the greater the absorption (unless a high flow resistivity covering is used).

Some of the other features summarized in Table 8.3 are discussed in the following sections.

Table 8.3 Construction differences between Schroeder diffusers and absorbers

	Absorber	*Diffuser with little absorption*
Well width	Usually narrow to exploit viscous boundary layer losses.	Usually >2.5 cm to minimize boundary layer losses.
Covering	Key to good absorption. Covering should be chosen so surface resistance is $\approx\rho_0 c$ when added to well resistance to maximize absorption.	Should not be covered. If covering is unavoidable, use low-flow-resistivity material away from well entrances.
1D vs. 2D	2D surface often gives more absorption.	2D surface gives hemispherical dispersion, 1D surface diffuses in a single plane.
Number of different depth wells, *N*	Determined by the need to have a sufficient number of quarter wave resonances in absorption bandwidth.	A larger *N* usually makes a better diffuser.
Depth sequence	Well depths should be chosen to evenly distribute well resonances across absorption bandwidth, best done using numerical optimization.	Chosen to maximize dispersion, best done using numerical optimization. Narrow period widths should be avoided.
Deepest well depth	Determines low frequency limit of absorption.	Determines low frequency limit of diffusion, except when period width is small.
Construction	Well sealed, no slits.	Well sealed, no slits.
Mass elements (addition of perforated sheets or membranes)	Can be used to lower bandwidth of absorption.	Can be used to lower bandwidth of diffusion.
Well sides	Can be rough.	Should be smooth.

8.2.4 Depth sequence

An optimum depth sequence for diffuse reflections does not necessarily produce the best absorber. For absorption, the sequence should produce a set of well depths with many resonant frequencies, distributed evenly in frequency and optimally arranging to maximize energy flow between the wells. Mechel[16] was the first to suggest this; he discussed how a primitive root sequence could result in a better absorber than the more common quadratic residue sequence can. This is because the primitive root sequence generates more different well depths than a quadratic residue sequence does. The simple procedure outlined in the following to determine well depths works even better, however, as the primitive root sequence does not evenly space resonant modes in frequency. Another possibility is to use a numerical optimization to find the best well depth sequence. This can follow the principles outlined in Chapters 10 and 11 for diffuse reflection optimization. The optimization can be tasked with maximizing the average absorption coefficient across the bandwidth of interest. As with diffuse reflection optimization, this is a slow and moderately complex procedure.

A simple procedure to determine the well depths is based on determining the resonant frequencies of the wells. To a first approximation, neglecting viscous boundary layer losses in the well, each well is a quarter wave resonator with resonant frequencies, f, given by

$$f = \frac{(2m-1)c}{4d_n};m = 1,2,3\ldots, \tag{8.2}$$

where d_n is the depth of the nth well and c is the speed of sound.

To maximize the absorption, it is necessary to evenly space these resonant frequencies over the design bandwidth avoiding degenerate modes—modes with similar resonant frequencies. This can be achieved using a trial-and-error process and a spreadsheet. Once the depths are determined, it is necessary to order them to maximize the losses due to energy flow between the wells. To achieve this, wells causing adjacent in-frequency resonances should not be physically next to each other. This can be done quickly by rearranging the well order by hand.

Figure 8.8 compares the performance of an absorber made following this simple design method to that of an absorber produced using a numerical optimization.[17] The performance

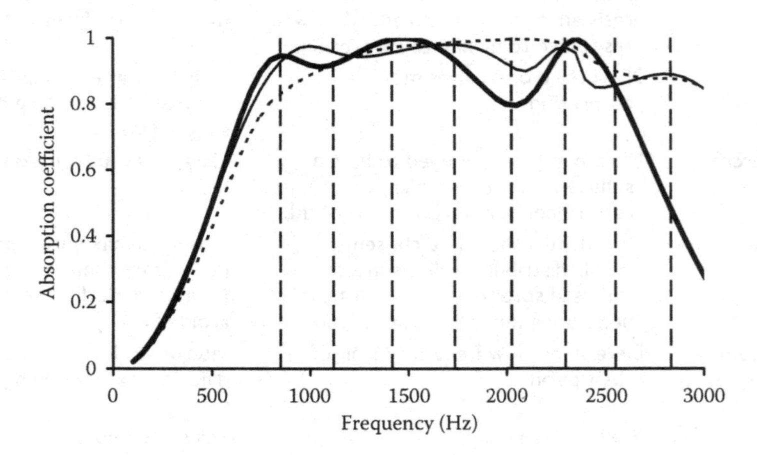

Figure 8.8 Absorption coefficient for profiled absorbers using different methods to determine the depth sequence: ---- simple design method, —— designed using numerical optimization, and ▬ simple design method, ascending well order. Vertical lines indicate resonant frequencies for simple design method. (Data from Wu, T. Cox, T.J., Lam, Y.W., *J. Acoust. Soc. Am.*, 108, 643–650, 2000.)

of the absorber using the simpler design procedure is good. As might be expected, the optimization gives slightly better results, but that design involves considerably more computation and encoding effort. The resonant frequencies used during the simple design are also marked as vertical dashed lines. The drop at high frequencies >2.5 kHz occurs due to lack of resonances in the region above 3 kHz (beyond the frequency range shown in Figure 8.8). To illustrate that the ordering of the wells is important, Figure 8.8 also shows the results when the absorber is designed using the simple design method, but with the wells stacked in ascending size order. There is considerably less absorption at some frequencies.

8.2.5 Use of mass elements

High absorption at low frequencies is often the most difficult to achieve. Using perforated plates in some wells can significantly extend the absorption range towards lower frequencies by adding mass to the system and so lowering the resonant frequencies. Another possibility would be to use membranes to act as limp mass elements. A typical construction is shown in Figure 8.9.

Fujiwara et al.[18] first published measurement results on a structure with Helmholtz resonators in the wells. Wu et al.[19] took this work further by producing a prediction model validated against measurement and some basic design methodologies. The simple concept of spacing resonant frequencies, as discussed in Section 8.2.4, can be used again, although predicting the resonant frequencies is more awkward with mass elements. In addition, multiple resonances from each of the wells need to be considered.

Wu et al.[19] found that wells with and without perforations are needed to get a wide enough range of resonant frequencies. Both well types are shown in Figure 8.9. The added mass within the perforations makes it difficult to keep the reactance of the impedance small at high frequencies and so too many wells with perforations make it difficult to achieve high-frequency absorption. The holes of the perforations must be carefully chosen. If they offer significant resistance, it may be necessary to lower the resistance of the covering material to achieve good absorption. These devices can produce greater absorption than a set of standard Helmholtz resonators stacked next to each other because of the multiple resonances within some of the wells, but they are more expensive to construct.

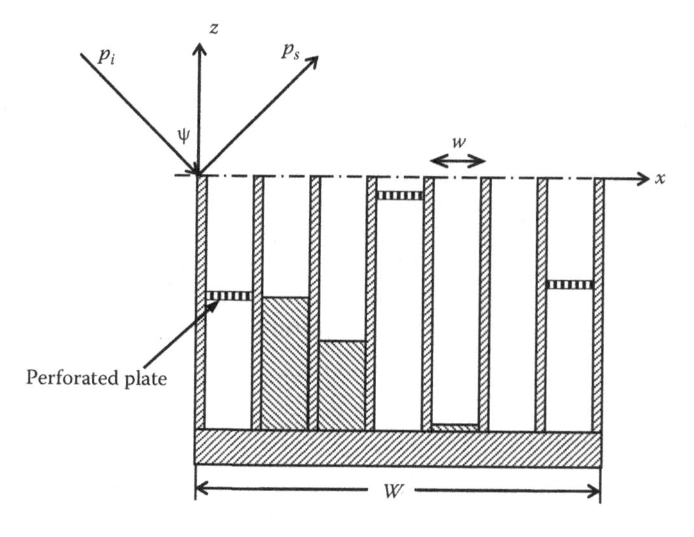

Figure 8.9 One period of a profiled sound absorber with perforated plates. (Adapted from Wu, T., Cox, T.J., Lam, Y.W., *J. Acoust. Soc. Am.*, 108, 643–650, 2000.)

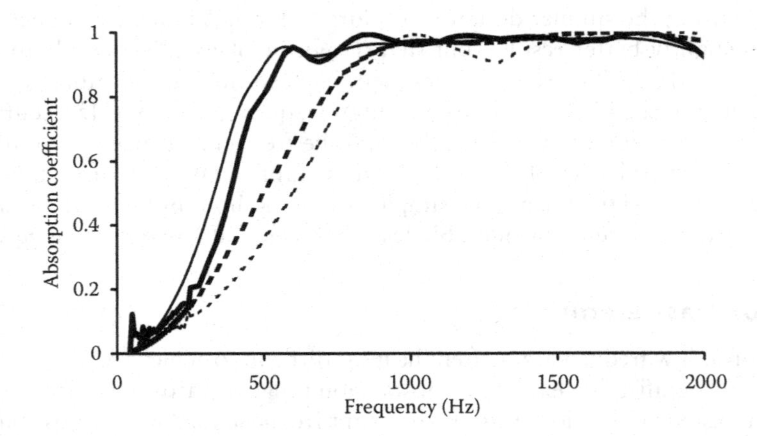

Figure 8.10 Measured and predicted absorption coefficient for two different profiled structures: —— with perforated sheets, predicted; ▬▬ with perforated sheets, measured; ▪▪▪▪▪▪▪ no perforated sheets, predicted; and - - - - - - - no perforated sheets, measured. (After Wu, T., Cox, T.J., Lam, Y.W., *J. Acoust. Soc. Am.*, 110, 3064–3070, 2001.)

Figure 8.10 shows a typical result for two optimized designs, one with perforated sheets and one without. Measurements from the impedance tube and predictions are shown. This demonstrates that adding mass elements can extend the low-frequency performance of these devices.

8.2.6 Number of wells

For a narrow bandwidth, only a few different well depths are needed, simplifying the design and reducing manufacturing costs. Wu et al.[19] designed a diffuser to work up to 3 kHz that needed only three different well depths. With this small number of wells, however, the density of resonances is insignificant above 3 kHz, and so the absorption decreases at higher frequencies. The choice of the correct flow resistance for the covering material is even more important in this case.

8.2.7 Theoretical model

Boundary element methods (BEMs) can be used, but it is also possible to construct a prediction model using a Fourier decomposition of the infinite periodic surface. This later model could also be used for profiled diffuser scattering. It is applicable to periodic structures and is almost as accurate as a BEM but requires considerably less computation time. Both theoretical approaches divide into two parts: first, the admittance of the individual wells must be calculated, and then the absorption should be calculated from these well admittances.

8.2.7.1 Admittance of wells

The approach follows similar lines to the transfer matrix approaches used for porous and resonant absorbers in Chapters 6 and 7 and also described in Chapter 2. The admittance (or impedance) is needed at the entrance of the wells. The well width of a profiled absorber is often narrow; therefore, the energy losses caused by viscous and thermal conduction in the wells cannot always be neglected. Consider the case where the well width $w << \lambda/2$, where λ is the wavelength of the sound, so that only fundamental modes are considered to propagate in each well. The wavenumber in the wells, k_t, is[20]

$$k_t \approx k + \frac{k}{2w}(1-j)[d_v + (\gamma - 1)d_h], \tag{8.3}$$

where k is the wavenumber in air, γ is the ratio of the specific heat ≈7/5 for air, and d_v and d_h are the thickness of the viscous and thermal boundary layers, respectively.

The thickness of the viscous boundary layer can be found from

$$d_v = \sqrt{\frac{2\eta}{\rho\omega}} \approx \frac{0.0021}{\sqrt{f}}, \tag{8.4}$$

where η is the coefficient of viscosity for air, ω is the angular frequency, ρ is the density of air, and f is the frequency. The formulation for the thickness of the thermal boundary layer is

$$d_h = \sqrt{\frac{2K}{\rho\omega c_p}} \approx \frac{0.0025}{\sqrt{f}}, \tag{8.5}$$

where K is the thermal conductivity and c_p is the heat capacity per unit mass of air at constant pressure.

With no perforated sheet, the impedance at the top of the well of depth, l_n, is

$$z_1 = r_m - \rho_e c \frac{k}{k_t} \cot(k_t l_n), \tag{8.6}$$

where r_m is the resistance of the covering material and p_e is the effective density of air in the slit.[21] The effective density can be calculated using

$$\rho_e = \rho[1 + (1 - j)d_v/w]. \tag{8.7}$$

For a well with a perforated sheet a distance d_n from the bottom, the impedance at the top of the perforated plate is given by

$$z_p = r_p + j\left(\omega m_p - \rho_e c \frac{k}{k_t} \cot(k_t d_n)\right), \tag{8.8}$$

where r_p and m_p are the added resistance and mass, respectively, due to the perforated sheet. These can be calculated from Equations 7.6 and 7.12 or 7.13. The impedance at the top of the well is

$$z_1 = \frac{-j\rho c z_p \cot(k_t l_n) + (\rho c)^2}{z_p - j\rho c \cot(k_t l_n)} + r_m, \tag{8.9}$$

where l_n is the distance from the perforated plate to the top of the well.

8.2.7.2 From well impedance to absorption: BEM

Once the well impedances are known, a method for gaining the absorption coefficient is needed. One possibility is to apply one of the BEMs described in Chapter 9. The absorber is treated as a box with an impedance distribution on the front face. A source is placed in the far field and irradiates the absorber. An array of receivers on a sphere measures the far field scattered energy, which is then integrated to give the sound power reflected, P_a. A box with zero admittance on the front face of the same dimensions is placed in the same set-up. The sound power reflected is calculated and in this case gives the incident power, P_i. From these two powers, the absorption coefficient of the surface can be calculated:

$$\alpha = 1 - \frac{P_a}{P_i}. \tag{8.10}$$

Figure 8.11 compares a prediction of the absorption using a boundary element model and a multimicrophone free field measurement.[22] Reasonable agreement is achieved. The low frequency discrepancies are as likely to be due to measurement inaccuracies as the BEM model. The BEM is rather laborious, and consequently, a different method can be used exploiting periodicity; this is detailed in the next section.

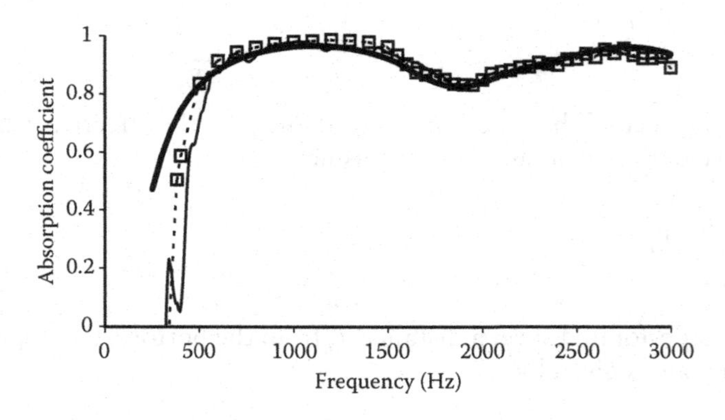

Figure 8.11 Measurement and prediction of absorption for a periodic profiled absorber: —— multi-microphone measurement, ▬▬ infinite sample prediction, and —□— BEM prediction. (Adapted from Wu, T., Lam, Y.W., Cox, T.J., *Proc. 17th ICA, Rome*, 2001.)

8.2.7.3 From well impedance to absorption: wave decomposition

As the surface is periodic, it is possible to decompose the scattered wave according to the periodicity of the surface. This greatly reduces the computation burden compared to a BEM. There is an assumption that the surface is infinitely wide, and consequently, the prediction accuracy will be compromised at low frequencies for finite samples. The analysis here closely follows the method used by Mechel[16] and Wu et al.[17] The sound field in front of the absorber, shown in Figure 8.9, is decomposed into the incident plane wave, $p_i(x,z)$, and scattered field, $p_s(x,z)$, which is made up of propagating and nonpropagating evanescent waves:

$$\begin{aligned} p(x,z) &= p_i(x,z) + p_s(x,z) \\ p_i(x,z) &= P_i e^{j(-xk_x+zk_z)} \\ p_s(x,z) &= \sum A_n e^{j(-x\beta_n - z\gamma_n)}, \end{aligned} \tag{8.11}$$

where $W = Nw$ is the width of one period. For simplicity, variation in the y direction is ignored.

To use this set of equations, the amplitudes A_n need to be obtained for the nonevanescent waves. (Evanescent waves decay rapidly as they propagate away from the surface and so do not reach the far field.) These coefficients represent the magnitude of the grating lobes, and consequently, one coefficient needs to be obtained for every grating lobe. The corresponding radiating harmonics indices, n, which can propagate to the far field, must satisfy the following relationship:

$$\left(\sin(\psi) + n\frac{\lambda}{W}\right)^2 \leq 1. \tag{8.12}$$

The outward particle velocity along the positive z direction and the pressure can be related to the surface impedance, as discussed in Section 2.5.2. For this theory, it is more convenient to work with admittance. The relationship between particle velocity, u_z, and pressure, p, is thus

$$\rho c u_z(x, 0) = -\beta(x) p(x, 0), \tag{8.13}$$

where β is the surface admittance, which can be calculated using the transfer matrix approach outlined in Section 8.2.7.1. The relations in Equation 8.11 are differentiated to give the particle velocity, and these are then related to the pressure relations in Equation 8.11 using the admittance relationship in Equation 8.13. This is imposing the boundary condition of the surface admittance onto the system of equations. This gives the following:

$$\cos(\psi)p_i - \sum_{n=-\infty}^{\infty} \frac{\gamma_n}{k} A_n e^{-j2\pi x n/W} = \beta(x)\left[p_i + \sum_{n=-\infty}^{\infty} A_n e^{-j2\pi x n/W}\right]. \tag{8.14}$$

As the surface admittance is periodic, a Fourier analysis can be applied. Since the period is W, this gives the admittance as

$$\beta(x) = \sum_{n=-\infty}^{\infty} B_n e^{-j2\pi xn/W}, \tag{8.15}$$

and

$$B_n = \frac{1}{W}\int_0^W \beta(x) e^{j2\pi xn/W}\, dx. \tag{8.16}$$

Equations 8.14 through 8.16 are combined to impose the periodicity of the boundary conditions. After multiplication by $e^{j2\pi mx/W}$ and integration over W, this gives the following:

$$\sum_{n=-\infty}^{\infty} A_n \left[B_{m-n} + \delta_{m,n}\left(\frac{\gamma_n}{k}\right)\right] = P_i(\delta_{m,0}\cos(\psi) - B_m) \tag{8.17}$$

$$m = -\infty, \ldots, +\infty,$$

and

$$\delta_{m,n} = \begin{cases} 1 & m = n \\ 0 & m \neq n. \end{cases} \tag{8.18}$$

The infinite sum in m can be terminated by monitoring convergence as more terms are added. On the samples tested, the index limits were taken as $|m| \le 2N$, where N is the number of wells in one period.

Equation 8.17 gives a set of simultaneous equations relating the coefficients of the nonevanescent waves, A_n, to the surface admittance and other known factors of geometry, such as incident angle. These simultaneous equations can be solved using standard solution techniques to get the unknown amplitudes.

By considering the energy in the scattered and incident waves shown in Equation 8.11, it is possible to derive an equation for the absorption coefficient. This is given by

$$\alpha = 1 - \left|\frac{A_0}{P_i}\right|^2 - \frac{1}{\cos(\psi)}\sum \left|\frac{A_n}{P_i}\right|^2 \sqrt{1-(\sin(\psi)+n\lambda/W)^2}, \tag{8.19}$$

where the summation runs over radiating spatial harmonics only. The middle term on the right-hand side is the specularly reflected energy, and the rightmost term, the scattered energy. For a small period width, W, the specular reflection is the only nonevanescent reflection. In this case, a normalized impedance on the surface of the structure, z_n, can be derived from

$$z_n = \frac{1 + A_0/P_i}{1 - A_0/P_i}. \tag{8.20}$$

Figure 8.11 compares predictions from this Fourier model with the BEM described in the previous section. Free field measurements are also shown. Good agreement is obtained between the prediction models and measurements. Some discrepancies at low frequencies occur between the Fourier and BEM models because the Fourier model assumes an infinite sample and the BEM model does not.

Figure 8.10 compared the Fourier theory and impedance tube measurements for two different samples. Again, good accuracy is obtained. To get good comparisons between theory and measurement, high-quality samples are required. Even apparently small imperfections in the samples can lead to large measurement errors.

This Fourier model can also be applied to periodic diffusers designed for scattering rather than absorption. Although this approach has not been verified, it is assumed that the predictions would be accurate. The advantage in using this method over a BEM is the reduction in computation time and storage requirements.

8.3 VOLUMETRIC DIFFUSERS

Modern diffuser design has focussed on surface diffusers attached to the boundaries of a room. But there is a different approach to achieving scattering. The idea is to place the diffuser in the volume of the room rather than on surfaces. Doing this means the scattering elements have the possibility of influencing 4π space, whereas surface diffusers can only work on 2π space, and so volume diffusers have the potential to be more efficient. Surface diffusers often have limited bass response because of depth restrictions, and depending on the layout of a room, this problem can be overcome using volumetric designs.

The problem with applying these structures is that they might reduce sight lines and get in the way of a room's functionality. They will also inherently reduce the reverberation time by reducing the mean free path and by making a space more diffuse. Consequently, applying large-volume diffusers in a concert hall designed for classical music may be problematic. For spaces where speech is important, however, such as large reverberant railway stations, using volumetric devices along paths causing long delayed echoes is one way to improve the acoustic.

Volumetric devices are commonly used in reverberation chambers (see Figure 3.24), and there have been a few examples of diffusing objects hung from the ceiling of auditoria. Overhead stage canopies could also be viewed as volumetric diffusers. But no one has tested the effectiveness, except in the case of reverberation chambers or stage canopies.

Sonic crystals are regular arrangements of objects, say cylinders or spheres, which, when placed in the path of sound waves, result in frequencies that are not readily transmitted. These are known as *band gaps*. Sonic crystals also produce acoustic iridescence, where different frequencies of sound are scattered into different directions.[23] Both of these features arise from the periodicity of the structure, and neither behaviour is desirable for a good diffuser. This means a random or pseudorandom arrangement of scattering elements in a multilayered or 3D array is needed; an example is shown in Figure 8.12.

The density of the device, i.e., the ratio of the volume of scattering elements to total volume of the diffuser, is important. With insufficient density, too much sound will pass through the array unaltered. If the density is too high, then too much sound will reflect from the outer layers without penetrating the device, and this will reduce performance. Consequently, the ratio of the power transmitted compared to that back-scattered is an important design specification. The device shown in Figure 8.12 is designed so that half the energy is back-scattered. The low-frequency power limit is determined by the size of the elements. At high frequency, the back-scatter criterion requires there to be roughly a 50%

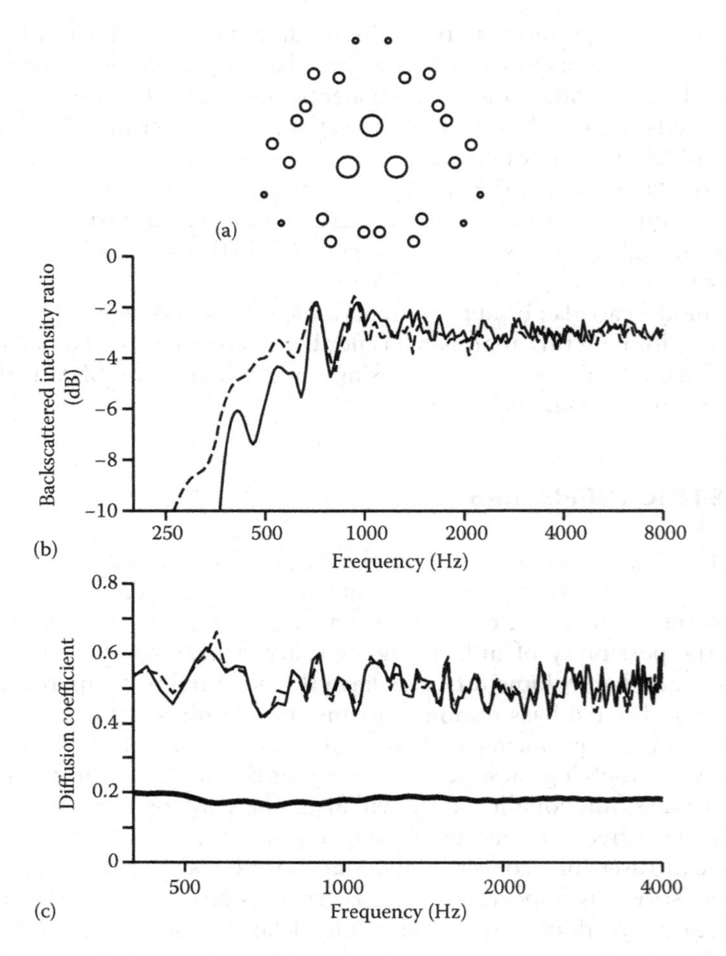

Figure 8.12 (a) Plan view of a volumetric diffuser made from cylinders; (b) the back scattered power; and (c) the diffusion coefficient evaluated over the whole circle: —— BEM prediction and - - - - measurement. ▬▬ Flat plate, bottom graph only. (After Hughes, R.J., Angus, J.A., Cox, T.J., Umnova, O., Gehring, G.A., Pogson, M., Whittaker, D.M., *J. Acoust. Soc. Am.*, 128, 2847–2856, 2010.)

line of sight, meaning that looking through the array, half of the view should be blocked by objects.

Hughes et al.[24] examined a number of array construction techniques based on applying number theory to reduce periodicity. This is done because the dispersion at high frequency is dictated by the spatial similarity of the array. The design in Figure 8.12 uses a Costas array, a 2D sequence whose separation vectors are unique and so does not suffer from periodicity. First, a square $N \times N$ grid is formed, where each row or column only contains one cylinder. Then, a shear transformation is applied to form a hexagonal grid that is then more isotropic.

The cylinders that make up the volumetric diffuser need to have a variety of sizes to scatter the different wavelengths of sound. The inner elements need to be large and the outer elements smaller to facilitate penetration into the diffuser. Figure 8.12 also shows the diffusion coefficient measured and predicted in all directions, i.e., around a complete circle, demonstrating the acoustic performance of the device.

8.4 METAMATERIALS AND ABSORBING SONIC CRYSTALS

When a wave passes through a periodic structure, interesting effects happen. For instance, when x-rays pass through a crystal, scattered energy is concentrated into particular directions to form grating lobes. The diffraction directions depend on the wavelength and the lattice spacing(s) in the unit cell. In optics, similar effects happen in what are called *photonic crystals*. In these periodic nanostructures of regularly repeating internal regions of high and low dielectric constant, photons propagate through the structure or not, depending on their wavelength. Wavelengths that pass through are known as modes, and disallowed bands of wavelengths are called *photonic band gaps*.

The acoustic equivalent is a *phononic crystal*, which is a material that exhibits stop bands for phonons, preventing sounds at certain frequencies from being transmitted through the material. A key factor for acoustic band gap engineering is impedance mismatch between periodic elements comprising the crystal and the surrounding medium. If sound is incident on a set of periodically arranged cylinders, as shown in Figure 8.13, then there will be certain frequencies that will not pass through the structure. Consequently, these sonic crystals offer the chance of reducing the transmission and/or absorbing particular frequencies. As described, such a structure is also known as a sonic crystal.

To demonstrate why gaps appear, an analysis on a periodic waveguide will be used, which is a 1D sonic crystal. This is done because it simplifies the explanation, and the findings can be qualitatively generalized to 2D and 3D structures.

At low frequency, the crystal in Figure 8.13 has a periodic disturbance of the impedance because sound cannot enter the cylinders. An analogous 1D structure would be a corrugated tube, as shown in Figure 8.14. This structure is best analyzed through a transfer matrix approach. A volume velocity rather than a particle velocity is used in the formulations to account for the change in cross-sectional area of the tube. The pressure, p_n, and volume

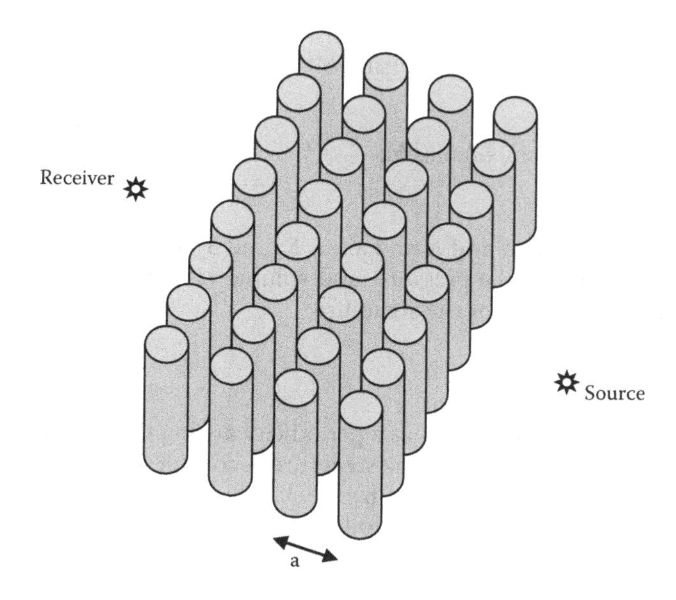

Figure 8.13 A 2D sonic crystal.

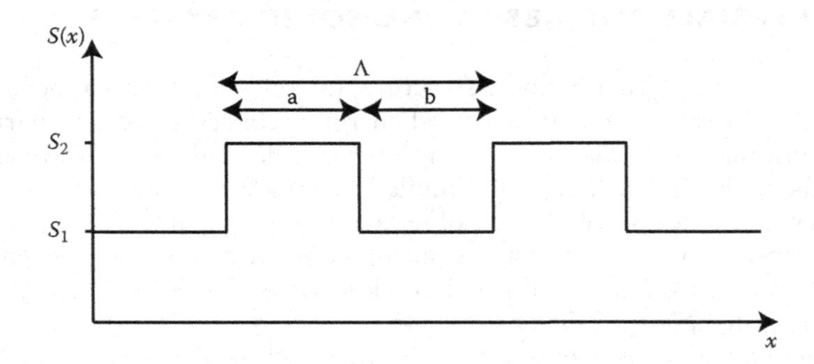

Figure 8.14 A 1D sonic crystal waveguide.

velocity, V_n, in the nth unit cell can be related to the pressure, p_{n+1}, and volume velocity, V_{n+1}, in the $(n + 1)^{th}$ cell via an adapted form of Equation 2.29:

$$\begin{Bmatrix} p_n \\ V_n \end{Bmatrix} = \mathbf{P} \begin{Bmatrix} p_{n+1} \\ V_{n+1} \end{Bmatrix} = \begin{Bmatrix} P_{11} & P_{12} \\ P_{21} & P_{22} \end{Bmatrix} \begin{Bmatrix} p_{n+1} \\ V_{n+1} \end{Bmatrix}, \tag{8.21}$$

where

$$\begin{aligned} P_{11} &= \cos(ka)\cos(kb) - \frac{S_1}{S_2}\sin(ka)\sin(kb) \\ P_{12} &= j\rho_0 c\left(\frac{1}{S_1}\cos(ka)\sin(kb) + \frac{1}{S_2}\sin(ka)\cos(kb)\right) \\ P_{21} &= \frac{j}{\rho_0 c}(S_2\sin(ka)\cos(kb) + S_1\cos(ka)\sin(kb)) \\ P_{22} &= \cos(ka)\cos(kb) - \frac{S_2}{S_1}\sin(ka)\sin(kb), \end{aligned} \tag{8.22}$$

where $b = \Lambda - a$; the distances a and Λ and areas S_1 and S_2 are defined in Figure 8.14.

As the structure is periodic, the pressure (and volume velocity) must be a periodic disturbance with the same periodicity of the structure[25]:

$$p(x) = e^{\pm jK\Lambda} f(x), \tag{8.23}$$

where $f(x)$ is the Bloch function, which has a periodicity arising from the lattice, i.e., $f(x + \Lambda) = f(x)$, and K is the Bloch wave vector. If the structure is considered to be infinitely long, both the right and left propagating waves must be Bloch waves. The pressure and volume velocities between two periods can be stated as follows:

$$\begin{Bmatrix} p_n \\ V_n \end{Bmatrix} = e^{\pm jK\Lambda} \begin{Bmatrix} p_{n+1} \\ V_{n+1} \end{Bmatrix}. \tag{8.24}$$

Comparing Equations 8.21 and 8.24 shows that $e^{\pm jK\Lambda}$ are the eigenvalues of **P**. The determinant of **P** is 1, and therefore, the eigenvalue solutions of **P** are

$$e^{\pm jK\Lambda} = \frac{1}{2}(P_{11} + P_{22}) \pm \sqrt{\frac{1}{2}(P_{11} + P_{22}) - 1}. \tag{8.25}$$

Adding the two Bloch wave solutions and rearranging yields the Bloch wave vector,

$$K = \frac{1}{\Lambda}\cos^{-1}\left(\frac{1}{2}(P_{11} + P_{22})\right). \tag{8.26}$$

Consequently, if $1/2(P_{11} + P_{22}) \leq 1$, then waves can propagate through the structure. But if $1/2(P_{11} + P_{22}) > 1$, then K is complex and the Bloch waves are evanescent, and band gaps arise. Figure 8.15 shows the measured and predicted transmitted intensity through a corrugated waveguide. In the case shown, $a = b$ and $2s_1 \approx s_2$. Using these values in Equations 8.22 and 8.25 yields $1/2(P_{11} + P_{22}) = 1 - 9\sin^2(ka)/4$. So waves are not propagated when the frequency, f, is $775 \leq f \leq 1200$ Hz and $2750 \leq f \leq 3180$ Hz.

Close inspection of Figure 8.15 shows that the band gaps measured are slightly lower in frequency than predicted. Furthermore, there is an additional dip at around 2 kHz that is not expected by the previous analysis. These effects are due to the radiation impedance as the size of the waveguide changes.[26] This is something that needs considering for accurate predictions.

So a periodic crystal, whether 1D, 2D, or 3D, has the potential to have band gaps—frequencies that are not transmitted. All the different periodicities need to be considered. For a square lattice structure as shown in Figure 8.13, the smallest repeat distance, a, is along the side of the lattice. However, there is also periodicity diagonally across the crystal with a repeat distance of $\sqrt{2}a$. Consequently, the band gaps occur at different frequencies proportional to $1/a$ and $1/\sqrt{2}a$. If these band gaps overlap, then any wave is reflected completely from this periodic structure in the overlapping frequency range.

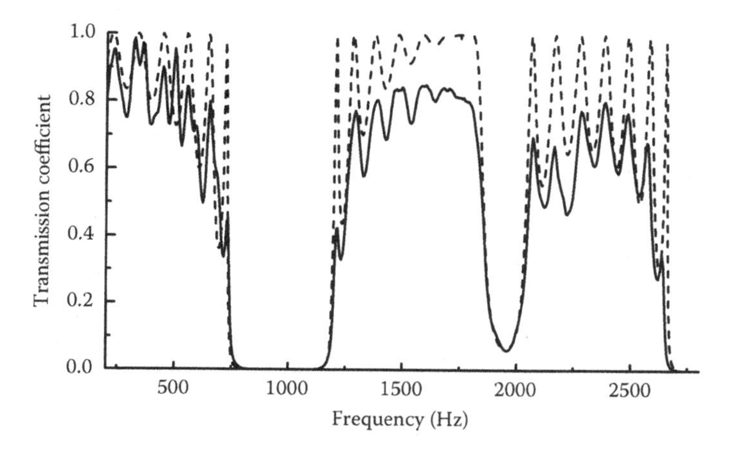

Figure 8.15 The transmission coefficient for eight periods of a 1D sonic crystal with $a = b = 9.6$ cm formed in a waveguide with areas of $s_1 = 0.0542$ m^2 and $s_2 = 0.038$ m^2: —— measurement and ---- prediction. (Data from King, P.D.C., Cox, T.J., *J. Appl. Phys.* 102, 014902–014902-8, 2007.)

The filling ratio, the fractional area occupied by the scatterers, is an important variable in determining behaviour, and dense arrays are needed to gain wide band gaps. Umnova et al.[27] measured sets of cylinders at model scale with and without an absorbent covering. The cylinders had a filling ratio of 33%, meaning that the cylinders occupied a third of the volume of the array and were therefore quite closely packed together. Figure 8.16 shows the insertion loss, which is the change in sound level between the measurement with the cylinders and a measurement without the cylinders. A large insertion loss means more sound is prevented from being transmitted. With hard cylinders, the reduction in sound level at the receiver is uneven with frequency and shows evidence of band gaps. For most noise control applications, such an uneven response with frequency is not useful, and consequently, the structure needs altering.

Adding defects to the crystals, removing some of the cylinders, or introducing local variations in the lattice spacings can help generate additional pseudo-band gaps. However, the response is still uneven. The addition of absorbent material provides energy dissipation and so improves the attenuation and has potential to make the response more even with frequency. It is important to use the right amount of absorbent, however. Too much, and the structure will not resonate and so the band gaps will disappear.

These sonic crystals need large lattice spacing for audio frequencies, and this makes them impractical devices in many situations. By replacing the cylinders with resonators, say by using hollow cylinders with slits to form a set of Helmholtz resonators, it is possible to get band gaps at lower frequencies.[28] This is an example of a *locally resonant acoustic metamaterial*.[29] Metamaterials are structures made up of arrays of subwavelength components, which allow the global acoustic properties of a material to be altered by changing the components. By appropriate choice of components, these materials can appear to defy the laws of nature, producing negative densities and other strange properties. For example, negative effective bulk modulus implies that the applied pressure and resultant volume change are out of phase: increased pressure results in a bigger volume!

Mostly, metamaterials are of most use to those working in seismology, underwater acoustics, or ultrasonics because the wavelength of audible frequencies in air makes devices too large to be practical. One notable exception are devices where air-filled cylindrical Helmholtz resonators are embedded into a porous material.[30,31] Two additional peaks of absorption are seen, one at the resonant frequency of the Helmholtz devices and the other from a trapped mode due to the presence of the inclusions in the porous material created by the outer shell

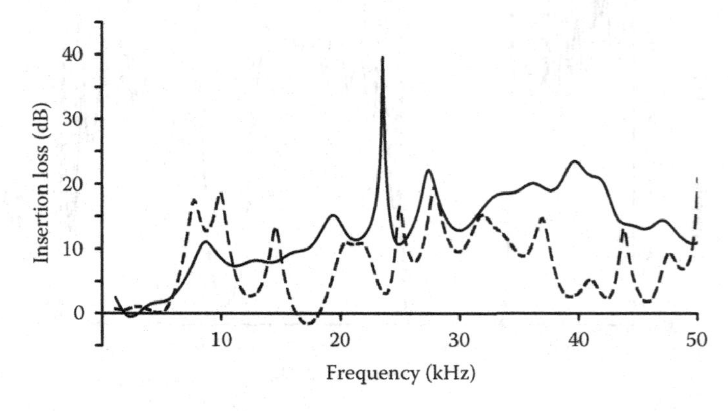

Figure 8.16 Insertion loss for two model-scale sonic crystals made from ---- rigid cylinders and —— cylinders covered with felt. (Data from Umnova, O., Attenborough, K., Linton, C.M., *J. Acoust. Soc. Am.*, 119, 278–284, 2006.)

of the Helmholtz resonators. Constructions using weighted membranes to resist transmission of sound are promising areas for the application of metamaterials in air-borne acoustics at audio frequencies.[32] But methods to allow efficient and cost-effective manufacture of metamaterials need developing. Other devices, such as acoustic cloaks, waveguides, lenses, and black holes, are of more interest to researchers than to practicing acousticians.

8.5 NATURAL ABSORBERS

8.5.1 Tree belts, hedges, shrubs, and crops

For many years, it was believed that trees have no practical part to play in noise control. Certainly, a single row of trees is not going to reduce noise significantly. There is evidence, however, that a tree belt 30 m or more in width can cause significant sound reduction compared with open grassland.[33–36]

Fang and Ling[37] investigated 35 monoculture, evergreen tree belts in a field study. They found that visibility distance is a good guide to how much attenuation is achieved, with depth of the belt being the next most important factor. The highest attenuation of over 6 dBA was achieved for shrubs where the visibility was less than 5 m. Tree belts are less sensitive to changes in performance caused by changes in weather conditions.[38] Whereas temperature inversion can lead to higher sound levels over grassland, the attenuation provided by woodland is largely unaffected.[39]

The way a mature forest affects sound propagation divides into different frequency ranges.[40] Figure 8.17 shows how the sound from a typical traffic noise spectrum is attenuated by the presence of either a 100 m belt of pine trees or by open pasture. The results shown are for the excess attenuation, which is the measured level (corrected for air absorption) minus the free field level. The free field level is that which would exist with no ground, obstacles, and neutral metrology. The pine forest attenuates the broadband A-weighted sound pressure level by 10 dB more than the pasture.

At low frequency, the ground effect produces additional attenuation compared to the free field. This is where the sound reflected from the ground destructively interferes with the direct sound between source and receiver. (The scattering from trunks and branches is small and the absorption from the foliage itself is negligible at these low frequencies.) The

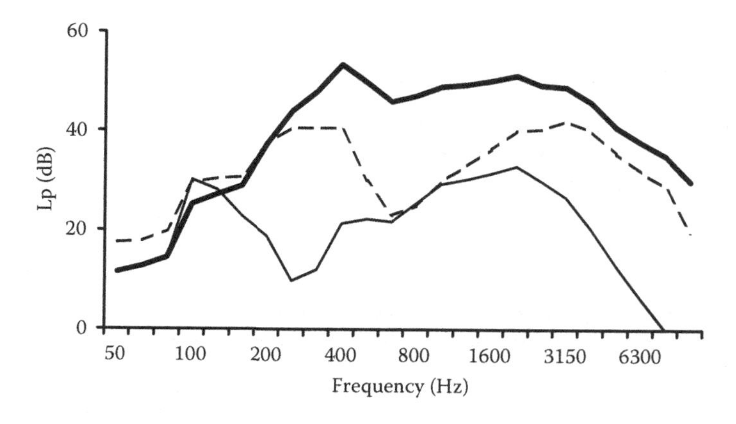

Figure 8.17 Sound pressure level from traffic after propagating 100 m: ▬▬ in free field, - - - - over pasture, and —— through a pine forest. A-weighted corrections applied to spectrum. (After Huisman, W.H.T., Attenborough, K., *J. Acoust. Soc. Am.*, 90, 2664–2667, 1991.)

dip created by the ground reflection in the forest is usually around a few hundred Hertz, although this is geometry dependent. In a mature forest, a thick litter layer of partially decomposing vegetation usually lies on the floor. In this case, the ground surface consists of a highly porous layer with relatively low flow resistivity. The ground needs to be modelled as a complex layered structure. Reported effective flow resistivity values range from 1×10^4 to 8×10^4 rayls m^{-1}.[41] This means that the ground effect occurs at a lower frequency for the forest in comparison to the pasture. Whether this shift to a lower frequency is useful or not depends on the frequency spectrum of the noise to be attenuated. For many environmental noise sources such as traffic, this shift to a lower frequency is undesirable.

At mid frequencies, say around 1 kHz, the trunks and large branches begin to scatter the sound out of the direct path between the source and receiver. There is little difference in the attenuation between the pasture and forest at these frequencies, however. Tree belts with higher biomass, i.e., more or thicker trunks, are better. This poses challenges because trees cannot survive if they are too densely planted. When using tree belts to reduce traffic noise, planting densely parallel to a road and more sparsely perpendicular to the road can be done to allow the trees to thrive.[42]

At high frequency, typically above 1 kHz, scattering is still important, and in addition, the foliage attenuates the sound by viscous friction and thermal dissipation into the leaves. Achieving most absorption requires trees where the foliage has a large surface area to maximize viscous losses. In addition, the canopy should extend as low to the ground as possible, to ensure that there are no paths through the forest that bypass the foliage. This helpfully also increases the total surface area of the foliage. This implies that trees with high biomass are required. Evergreen trees are also important if the sound attenuation is to be maintained during winter.

Predictions of mid to high frequencies might use multiple scattering theory for an idealized random infinite array of identical parallel impedance-covered cylinders, with the foliage represented by arrays of much smaller cylindrical scatterers than the trunks.[43] Huisman and Attenborough[40] used a stochastic particle-bounce method. In conjunction with a two-parameter impedance model and an assumption about the dependence of incoherent scattering on distance and frequency, good agreement was obtained with measurements in a monoculture of 29-year-old Austrian pines.

Shrubs, tall crops, and low hedging are not as effective as tree belts for traffic noise. Van Renterghem et al.[44] measured insertion losses of only 1.1 to 3.6 dBA from thick (1.3–2.5 m) and dense hedges.

8.5.2 Green walls, roofs, and barriers

Green walls are vertical gardens where soil in the structure allows vegetation to grow. Figure 2.15 illustrated how applying absorption to the inside of roadside barriers can reduce noise. One possible way of providing this absorption is to use a green wall. A prediction study has indicated that 4–8 dBA of attenuation can be achieved if two parallel roadside barriers are replaced by green walls for flat terrain.[45]

The absorption of green walls comes from both the vegetation and the soil. Figure 8.18[46,47] shows the absorption coefficients from three different green walls from the literature. As might be expected, the exact absorption achieved depends on the construction of the green wall. The depth of soil and any air gap behind can have a significant influence on the absorption at bass frequencies. The species will affect the density of the foliage, and this influences the amount of absorption. Also important is the orientation of the leaves: Horoshenkov et al.[48] found that species with larger leaf surface area and with leaves parallel to the soil created greater absorption.

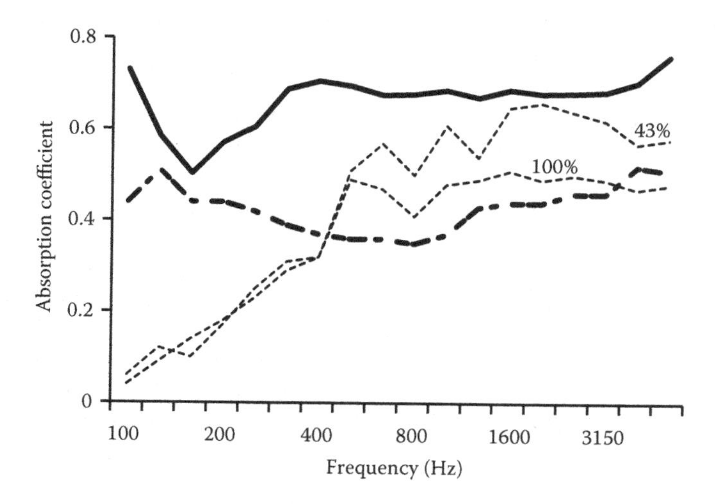

Figure 8.18 Absorption coefficient for three green wall systems. The two thinner lines are the same soil with different percentage foliage coverage as indicated. (Data from Azkorra, Z., Pérez, G., Coma, J., Cabeza, L.F., Bures, S., Álvaro, J.E., Erkoreka, A., Urrestarazu, M., *Appl. Acoust.*, 89, 46–56, 2015; Wong, N.H., Tan, A.Y.K., Tan, P.Y., Chiang, K., Wong, N.C., *Build. Environ.*, 45, 411–420, 2010; and Yang, H.S., Kang, J., Cheal, C., *Acta Acust. Acust.*, 99, 379–388, 2013.)

The effect of foliage coverage is illustrated by the reverberation chamber measurements made by Yang et al.[49] in Figure 8.19. They show that more foliage can increase the absorption through viscous losses and the vibration of the leaves at low and mid frequency. But at high frequency, the more foliage decreased absorption, presumably because the outer leaves reflect some of the sound, preventing it from reaching the soil.

Courtyards behind buildings can suffer from noise due to nearby roads. Appropriate application of absorption can help, with green walls and roofs being one option.[50] One study showed that green roofs can reduce the noise in the courtyard by 2.6–7.5 dB by reducing the noise propagating over the buildings, but attenuation depends on the roof shape.

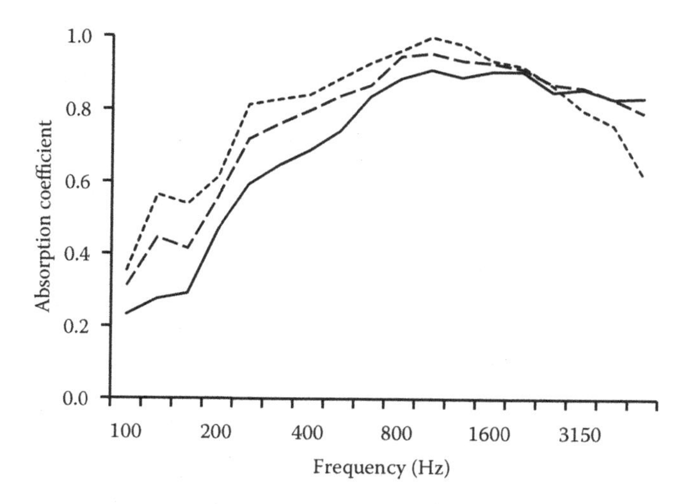

Figure 8.19 Absorption coefficient for top soil with different amounts of vegetative cover: —— 0%, – – – – 40%, and - - - - - - - 100%. (Data from Yang, H.S., Kang, J., Cheal, C., *Acta Acust. Acust.*, 99, 379–388, 2013.)

The effectiveness of green walls in reducing noise in the courtyard is highly dependent on the geometry and where the absorption is applied, with attenuations being between 1 and 4 dBA. In street canyons and city squares, where the listener and noise sources are in the same space, green walls provide only marginal acoustic benefits.

The foliage of a green wall will provide some scattering. Yang et al.[49] measured how the random incidence scattering coefficient varied with species type and density of planting. The scattering is relatively weak and confined to high frequencies (2000 Hz and above). Appendix D includes the measured values.

Vegetation can also help with the degradation of barrier performance that can be caused by wind.[45] When the wind speed varies with height, this can cause downward refraction of sound due to the changes in the speed of sound with height. This can allow noise to more efficiently bend over the top of barriers. One case study for a 4 m barrier near a highway found that wind caused barrier insertion loss to reduce by 3.5–4 dBA. Tall, dense tree canopies such as conifers can be used to decrease the wind speed gradient and so reduce this loss of performance.

8.6 SUMMARY

This chapter discussed the absorption of seating, how to turn a Schroeder diffuser into an absorber, sonic crystals for absorption and diffusion, metamaterials, and the absorption of trees and green walls. Accurate estimations of seating and audience absorption are vital to good room design, especially large concert halls for orchestral music. The absorption of Schroeder diffusers is a more esoteric subject. While the concept of a Schroeder absorber is interesting, and good absorption can be obtained, the cost of construction is high and this limits the commercial exploitation of these concepts. A similar comment could be made about sonic crystals and metamaterials. The absorption of trees and green walls is useful because vegetation is regularly planted, and the evidence is, with appropriate choice of plants and management, that this can provide useful amounts of absorption.

REFERENCES

1. L. L. Beranek, "Audience and chair absorption in large halls. II", *J. Acoust. Soc. Am.*, **45**, 13–9 (1969).
2. C. W. Kosten, "New method for the calculation of the reverberation time of halls for public assembly", *Acustica*, **16**, 325–30 (1965).
3. W. J. Davies, R. J. Orlowski, and Y. W. Lam, "Measuring auditorium seat absorption", *J. Acoust. Soc. Am.*, **96**, 879–88 (1994).
4. J. S. Bradley, "Predicting theater chair absorption from reverberation chamber measurements", *J. Acoust. Soc. Am.*, **91**, 1514–24 (1992).
5. J. S. Bradley, Y. J. Choi, and D. U. Jeong, "Understanding chair absorption characteristics using the perimeter-to-area method", *Appl. Acoust.*, **74**(9), 1060–8 (2013).
6. W. J. Davies, "The Effects of Seating on the Acoustics of Auditoria", PhD thesis, University of Salford (1992).
7. F. Martellotta, M. D'alba, and S. Della Crociata, "Laboratory measurement of sound absorption of occupied pews and standing audiences", *Appl. Acoust.*, **72**(6), 341–9 (2011).
8. T. Hidaka, N. Nishihara, and L. L. Beranek, "Relation of acoustical parameters with and without audiences in concert halls and a simple method for simulating the occupied state", *J. Acoust. Soc. Am.*, **109**(3), 1028–42 (2001).

9. T. J. Schultz and B. G. Watters, "Propagation of sound across audience seating", *J. Acoust. Soc. Am.*, **36**, 855–96 (1964).
10. G. M. Sessler and J. E. West, "Sound transmission over theatre seats", *J. Acoust. Soc. Am.*, **36**, 1725–32 (1964).
11. A. H. Marshall, J. R. Hyde, and M. F. E. Barron, "The acoustical design of Wellington Town Hall: Design development, implementation and modelling results", *Proc. IoA(UK)*, Edinburgh (1982).
12. K. Fujiwara and T. Miyajima, "Absorption characteristics of a practically constructed Schroeder diffuser of quadratic-residue type", *Appl. Acoust.*, **35**, 149–52 (1992).
13. K. Fujiwara and T. Miyajima, "A study of the sound absorption of a quadratic-residue type diffuser", *Acustica*, **81**, 370–8 (1995).
14. D. E. Commins, N. Auletta, and B. Suner, "Diffusion and absorption of quadratic residue diffusers", *Proc. IoA(UK)*, **10**(2), 223–32 (1988).
15. H. Kuttruff, "Sound absorption by pseudostochastic diffusers (Schroeder diffusers)", *Appl. Acoust.*, **42**, 215–31 (1994).
16. F. P. Mechel, "The wide-angle diffuser—A wide-angle absorber?", *Acustica*, **81**, 379–401 (1995).
17. T. Wu, T. J. Cox, and Y. W. Lam, "From a profiled diffuser to an optimised absorber", *J. Acoust. Soc. Am.*, **108**(2), 643–50 (2000).
18. K. Fujiwara, K. Nakai, and H. Torihara, "Visualisation of the sound field around a Schroeder diffuser", *Appl. Acoust.*, **60**(2), 225–36 (2000).
19. T. Wu, T. J. Cox, and Y. W. Lam, "A profiled structure with improved low frequency absorption", *J. Acoust. Soc. Am.*, **110**(6), 3064–70 (2001).
20. P. M. Morse and K. Ingard, *Theoretical Acoustics*, McGraw-Hill, New York, Chapter 6, 285–91 (1968).
21. J. F. Allard and N. Atalla, *Propagation of Sound in Porous Media: Modelling Sound Absorbing Materials*, John Wiley & Sons, Chichester, UK (2009).
22. T. Wu, Y. W. Lam, and T. J. Cox, "Measurement of non-uniform impedance surface by the two microphone method", *Proc. 17th ICA*, Rome (2001).
23. T. J. Cox, "Acoustic iridescence", *J. Acoust. Soc. Am.*, **129**(3), 1165–72 (2011).
24. R. J. Hughes, J. A. Angus, T. J. Cox, O. Umnova, G. A. Gehring, M. Pogson, and D. M. Whittaker, "Volumetric diffusers: Pseudorandom cylinder arrays on a periodic lattice", *J. Acoust. Soc. Am.*, **128**(5), 2847–56 (2010).
25. F. Bloch, "Uber die Quantenmechanik der Elektronen in Kristallgittern", *Z. Phys.*, **52**, 555–600 (1928).
26. P. D. C. King and T. J. Cox, "Acoustic band gaps in periodically and quasiperiodically modulated waveguides", *J. Appl. Phys.* **102**(1), 014902–014902-8 (2007).
27. O. Umnova, K. Attenborough, and C. M. Linton, "Effect of porous covering on sound attenuation by periodic arrays of cylinders", *J. Acoust. Soc. Am.*, **119**, 278–84 (2006).
28. V. Romero-García, A. Krynkin, L. M. Garcia-Raffi, O. Umnova, and J. V. Sánchez-Pérez, "Multi-resonant scatterers in sonic crystals: Locally multi-resonant acoustic metamaterial", *J. Sound Vib.*, **332**(1), 184–98 (2013).
29. Z. Liu, X. Zhang, Y. Mao, Y. Y. Zhu, Z. Yang, C. T. Chan, and P. Sheng, "Locally resonant sonic materials", *Science*, **289**(5485), 1734–6 (2000).
30. J. P. Groby, B. Nennig, C. Lagarrigue, B. Brouard, O. Dazel, O. Umnova, and V. Tournat, "Design of sound absorbing metamaterials by periodically embedding 3D inclusions in rigidly backed porous plate", *Proc. EURODYN* 3423–8 (2014).
31. C. Boutin and F. X. Becot, "Theory and experiments on poro-acoustics with inner resonators", *Wave Motion*, **54**, 76–99 (2015).
32. Z. Yang, H. M. Dai, N. H. Chan, G. C. Ma, and P. Sheng, "Acoustic metamaterial panels for sound attenuation in the 50-1000 Hz regime", *Appl. Phys. Lett.*, **96**, 041906 (2010).
33. G. A. Parry, J. R. Pyke, and C. Robinson, "The excess attenuation of environmental noise sources through densely planted forest", *Proc. IoA(UK)*, **15**(3), 1057–65 (1993).

34. J. Kragh, *Road Traffic Noise Attenuation by Belts of Trees and Bushes*, Danish Acoustical Laboratory Report no. 31 (1982).
35. G. M. Heisler, O. H. McDaniel, K. K. Hodgdon, J. J. Portelli, and S. B. Glesson, "Highway noise abatement in two forests", *Proc. Noise-Con*, 87 (1987).
36. L. R. Huddart, *The Use of Vegetation for Traffic Noise Screening*, TRRL Research Report 238 (1990).
37. C. F. Fang and D. L. Ling, "Investigation of the noise reduction provided by tree belts", *Landscape Urban Plan.*, **63**(4), 187–95 (2003).
38. N. Barrière, "Theoretical and Experimental Study of Traffic Noise Propagation through Forest", PhD thesis, Ecole Centrale de Lyon (1999).
39. W. H. T. Huisman, "Sound Propagation Over Vegetation-Covered Ground", PhD thesis, University of Nijmegen, The Netherlands (1990).
40. W. H. T. Huisman and K. Attenborough, "Reverberation and attenuation in a pine forest", *J. Acoust. Soc. Am.*, **90**(5), 2664–7 (1991).
41. D. G. Albert, *Past Research on Sound Propagation through Forests*, ERDC/CRREL TR-04-18, Hanover, NH (2004).
42. T. Van Renterghem, "Guidelines for optimizing road traffic noise shielding by non-deep tree belts", *Ecol. Eng.*, **69**, 276–86 (2014).
43. M. A. Price, K. Attenborough, and N. W. Heap, "Sound attenuation through trees: Measurements and models", *J. Acoust. Soc. Am.*, **84**(5), 1836–44 (1988).
44. T. Van Renterghem, K. Attenborough, M. Maennel, J. Defrance, K. Horoshenkov, J. Kang, I. Bashir, S. Taherzadeh, B. Altreuther, A. Khan, Y. Smyrnova, and H-S. Yang, "Measured light vehicle noise reduction by hedges", *Appl. Acoust.* **78**, 19–27 (2014).
45. T. Van Renterghem, J. Forssén, K. Attenborough, P. Jean, J. Defrance, M. Hornikx, and J. Kang, "Using natural means to reduce surface transport noise during propagation outdoors", *Appl. Acoust.*, **92**, 86–101 (2015).
46. Z. Azkorra, G. Pérez, J. Coma, L. F. Cabeza, S. Bures, J. E. Álvaro, A. Erkoreka, and M. Urrestarazu, "Evaluation of green walls as a passive acoustic insulation system for buildings", *Appl. Acoust.*, **89**, 46–56 (2015).
47. N. H. Wong, A. Y. K. Tan, P. Y. Tan, K. Chiang, and N. C. Wong, "Acoustics evaluation of vertical greenery systems for building walls", *Build. Environ.*, **45**(2), 411–20 (2010).
48. K. V. Horoshenkov, A. Khan, and H. Benkreira, "Acoustic properties of low growing plants", *J. Acoust. Soc. Am.*, **133**, 2554–65 (2013).
49. H. S. Yang, J. Kang, and C. Cheal, "Random-incidence absorption and scattering coefficients of vegetation", *Acta Acust. Acust.*, **99**(3), 379–88 (2013).
50. T. Van Renterghem, M. Hornikx, J. Forssen, and D. Botteldooren, "The potential of building envelope greening to achieve quietness", *Build. Environ.*, **61**, 34–44 (2013).

Chapter 9

Prediction of reflection including diffraction

Being able to predict the reflected pressure from a surface, whether that is a reflector, a diffuser, or a noise barrier, enables efficient evaluation, design, and characterization of the performance. Currently, for reflectors and diffusers, this is usually done by considering the surface in the free field, in isolation from other objects and boundaries. The prediction techniques outlined in this chapter could, in theory, be used to predict the sound for whole rooms or outdoor environments such as street canyons (if the effects of weather are ignored). At the moment, however, long computation times and storage limitations mean that algorithms dealing with large spaces are forced to use cruder representations of reflection and diffraction. Consequently, when predicting the responses in rooms and semi-enclosed spaces, such as pavilions, it is more common to use geometric models. The issue of modelling scattering in geometric room acoustic models (GRAMs) is discussed in Chapter 13.

Therefore, the issue for this chapter is predicting the reflection from isolated surfaces using models that treat sound as a wave. There is a range of models, from the numerically exact but computationally slow, to the more approximate but faster. The techniques can also be differentiated as either time- or frequency-domain methods. In diffuser design, frequency-domain methods have dominated. For this reason, this chapter starts with these methods. Time-domain modelling, especially finite difference time domain (FDTD), is becoming more popular, and so that is also featured. Table 9.1 summarizes the frequency- and time-domain prediction models that will be considered in this chapter, along with their key characteristics.

The next section will start with the most accurate frequency-domain model, a boundary element method (BEM) based on the Helmholtz–Kirchhoff integral equation. Then, more approximate models will be derived from this integral equation, and the relative merits and limitations of the techniques will be discussed.

9.1 BOUNDARY ELEMENT METHODS

When BEMs are applied to diffusers, remarkable accuracy is achieved. The precision is much better than most engineers are used to achieving from an acoustics theory. Acousticians are used to using empirical fixes to make measurements match predictions, but that is not often needed when BEMs are used to predict diffuser scattering. The disadvantages of BEMs are that the method is prone to human error in meshing the surface and, most importantly, it is slow for high frequencies and large surfaces. Some have attempted to apply the prediction methods to whole rooms for low frequencies, but this is very computationally intense, requiring super-computing facilities or a considerable amount of patience while waiting for results.

Table 9.1 Key characteristics of various prediction models for reflectors and diffusers

Model	*Accuracy*	*Computing time*	*Notes*
Frequency-domain models			
Standard BEM	Highest	Slowest	Exact, provided surfaces are locally reacting and viscous boundary layer losses are small. Slow, especially for large surfaces or high frequencies.
Thin-panel BEM	↑	↑	An efficient method for thin surfaces, approximately halving the number of elements compared to a standard BEM model.
Kirchhoff			Uses the Kirchhoff boundary conditions to approximate surface pressures and so is much faster. Less accurate for oblique sources and receivers, low frequencies, complex surface impedance profiles, and surfaces with steep gradients.
Fresnel	↓	↓	Replaces the numerical integration of Kirchhoff model by quicker-to-compute Fresnel integrals. Requires scattering across width and along length of surface to be orthogonal. Some useful simplifications available for flat and curved surfaces.
Fraunhofer or Fourier	Lowest	Quickest	Simplifies numerical integral of the Kirchhoff method but is useable only in the far field. Allows simpler Fourier principles to be applied. Good for understanding physical processes and designs.
Time-domain models			
FDTD	Highest	Slowest	Slow, especially for large surfaces or high frequencies due to volume mesh and time stepping solution.
Kirchhoff (time domain)	↕	↕	Uses the Kirchhoff boundary conditions to solve problem with a quick, simple summation. Less accurate for oblique sources and receivers, low frequencies, complex surface impedance profiles, and surfaces with steep gradients.
GRAMs	Lowest	Quickest	Ray tracing and image source models cannot model wave effects properly, so they are not a precise method for single objects. They are, however, very useful and widely used for whole room predictions.

Note: The accuracy and computational efficiency columns are indicative; the rank ordering of the top four prediction models might vary depending on the surface type being considered and the particular implementation of the algorithm.

9.1.1 The Helmholtz–Kirchhoff integral equation

The Helmholtz–Kirchhoff integral equation forms the core of many of the prediction models. It is defined below and the following sections will then discuss how it is solved. For steady-state, constant frequency motion, the acoustic wave equation reduces to the Helmholtz equation:

$$\nabla^2 p(\mathbf{r}) + kp(\mathbf{r}) = 0, \tag{9.1}$$

where p is the acoustic pressure, $\mathbf{r}$ is a point in space, and k is the wavenumber. Green's first and second theorems are used to transform the differential equations based on Equation 9.1, which involve volume integrals, to an integral equation that is evaluated using integrals over the reflecting surface. This Helmholtz–Kirchhoff integral equation formulates the pressure at a point, as a combination of the pressure direct from the sources (the incident sound) and a surface integral of the pressure and its derivative over the reflecting surfaces (the scattered sound). The single frequency form of the integral equation gives the pressure p as[1]

$$\left.\begin{array}{ll} \mathbf{r}\in E & p(\mathbf{r}) \\ \mathbf{r}\in s & \frac{1}{2}p(\mathbf{r}) \\ \mathbf{r}\in D & 0 \end{array}\right\} = p_i(\mathbf{r}|\mathbf{r}_0) + \int_s p(\mathbf{r}_s)\frac{\partial G(\mathbf{r}|\mathbf{r}_s)}{\partial n(\mathbf{r}_s)} - G(\mathbf{r}|\mathbf{r}_s)\frac{\partial p(\mathbf{r}_s)}{\partial n(\mathbf{r}_s)}ds, \tag{9.2}$$

where $\mathbf{r} = \{x, y, z\}$ is the vector describing the receiver location; $\mathbf{r}_0 = \{x_0, y_0, z_0\}$ is the vector describing the source location; $\mathbf{r}_s = \{x_s, y_s, z_s\}$ is the vector for a point on the surface; $p_i(\mathbf{r}|\mathbf{r}_0)$ is the direct pressure radiated from the source at $\mathbf{r}_0$ to the receiver at $\mathbf{r}$; G is the Green's function; n is the normal to the surface pointing out of the surface; E is the external region; s is the surface; and D is the interior of the surface.

The geometry is illustrated in Figure 9.1. $p_i(\mathbf{r}|\mathbf{r}_0)$ is the pressure direct from the source at $\mathbf{r}_0$ to the receiver point at $\mathbf{r}$, and so the first term on the right-hand side represents the direct pressure. The integral is carried out over the surface, with $\mathbf{r}_s$ being a point on the surface and n a normal to the surface pointing out of the surface, so the integral gives the contribution of the reflected energy to the pressure at $\mathbf{r}$. By single frequency, it is meant that the system is in steady-state conditions so that the time variation $\exp(j\omega t)$ can be neglected. G is the Green's function, which gives how the pressure and its derivative propagate from one point in space to another. Consequently, in 3D the Green's function is simply a point source radiation equation:

$$G(\mathbf{r}|\mathbf{r}_0) = \frac{e^{-jkr}}{4\pi r}, \tag{9.3}$$

where $r = |\mathbf{r} - \mathbf{r}_0|$. Carrying out the solution in two dimensions is extremely useful for some reflectors and diffusers as it can greatly decrease the computational burden in terms

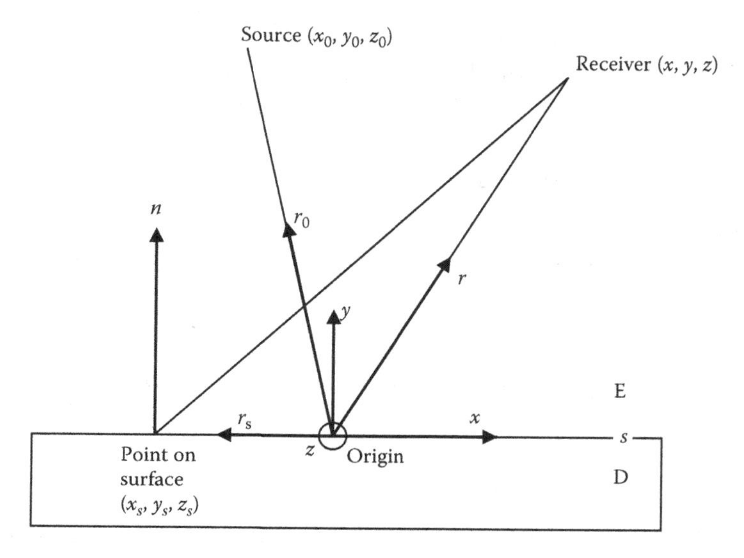

Figure 9.1 Geometry for prediction models.

of storage and calculation time. In that case, the Green's function is given by the Hankel function:

$$G(\mathbf{r}|\mathbf{r}_0) = \frac{-j}{4} H_0^{(2)}(kr), \tag{9.4}$$

where $H_0^{(2)}$ is the Hankel function of the second kind of order 0. The asymptotic version of the Hankel function when kr is large is

$$\underset{kr \gg 1}{G(\mathbf{r}|\mathbf{r}_0)} = \frac{Ae^{-jkr}}{\sqrt{kr}}, \tag{9.5}$$

where A is a constant. So this is a line source radiation equation as would be expected in a 2D world. The Hankel function is most efficiently evaluated by using polynomial expansions[2] when kr is small and by using the asymptotic form (Equation 9.5) when kr is large.

There are three possible equations shown in Equation 9.2. The top case is when the point **r** is external to the scattering surface ($r \in E$), the middle case has **r** on the surface ($r \in s$), and the bottom case is when **r** is internal to the scattering surface ($r \in D$).

The integral has two terms: one involving the surface pressure $p(\mathbf{r}_s)$ and one involving the surface pressure derivative $\partial p(\mathbf{r}_s)/\partial n(\mathbf{r}_s)$. If the surface is taken to be *local reacting*, the derivative of the surface pressure will be related to the surface pressure by the surface admittance:

$$jkp(\mathbf{r}_s)\beta'(\mathbf{r}_s) = \frac{\partial p(\mathbf{r}_s)}{\partial n(\mathbf{r}_s)}, \tag{9.6}$$

where β' is the surface admittance. In BEM modelling, it is normal to define quantities in terms of an outward pointing normal. Surface admittances would normally be defined with an inward pointing normal. The prime is used to signify this difference, where $\beta' = -\beta$, where β is the more usual surface admittance. This definition of an outward pointing normal also affects the interrelations between admittance and surface reflection coefficient given in Section 2.5.2. The assumption of local reaction means the surface admittance is independent of the incident and reflected pressure waves. For low absorption surfaces, where $\beta' \to 0$, the term involving the pressure derivative can be neglected.

9.1.2 General solution method

Having defined the terms and the nomenclature for the integral equation, the general solution technique for a BEM will be presented. The BEM involves the application of Equation 9.2 twice:

1. The surface pressures, $p(\mathbf{r}_s)$, on the reflecting surface(s) are found.
2. A numerical integral is carried out over the surface to determine the pressures at the desired external points.

9.1.2.1 Determining surface pressures

Obtaining the surface pressures is usually the rate-determining step of a BEM. The surface pressures depend not only on the incident sound field but also on each other. This is a statement that there are mutual interactions across the surface, as might be expected. To model this, the usual solution method is to discretize the surface into a number of surface (boundary) elements, across which it is approximated that the pressure and its derivative are constant. The elements must have sufficiently small dimensions to prevent errors in representing the continuous pressure variation by a set of discrete values. This is usually achieved by making elements smaller than an eighth of a wavelength in size for the highest frequency being modelled. Breaking the surface geometry into a series of elements—meshing the surface—can be a difficult process for complicated diffusers, but it can be greatly simplified by using specialist meshing programs. This is where human error is most likely to occur. Two-dimensional BEMs not only have computational speed advantages, they also require far simpler 2D meshes.

Once the surface has been discretized, a set of simultaneous equations can be set up with one equation for each boundary element. The formulations will be for the surface pressures, with **r** being taken for positions on the surface in the middle of each of the elements. In matrix form, Equation 9.2 can be rewritten as follows

$$\left(\frac{1}{2}\delta + \mathbf{A}\right)\mathbf{P} = \mathbf{P}_i, \tag{9.7}$$

$$\begin{aligned} \delta_{nm} &= 1; m = n \\ \delta_{nm} &= 0; m \neq n, \end{aligned} \tag{9.8}$$

and

$$\mathbf{A}_{mn} = \int_{s_m} \frac{\partial G(\mathbf{r}_n|\mathbf{r}_s)}{\partial n_m(\mathbf{r}_s)} - G(\mathbf{r}_n|\mathbf{r}_s) jk\beta'_m \, ds_m, \tag{9.9}$$

where **P** is a $(1 \times N)$ matrix of surface pressures; $\mathbf{P}_i$ is a $(1 \times N)$ matrix of incident pressures direct from the source(s) to the surface; N is the number of elements; the subscripts n and m refer to the $(n, m)^{\text{th}}$ element of the matrix, i.e., the contribution from the mth element to the nth element surface pressure; and s_m is the surface of the mth element.

The calculation of the matrix **A** is an important rate-determining step in the BEM. It is roughly an N^2 process, where N is the number of elements. For each element, it is relatively slower for the 2D formulations compared to 3D. This is because the Hankel function is slower to evaluate unless it is in-built and optimized for speed by the computing language used to code the numerical model. But then there are great time savings to be had in a 2D model, as Equation 9.9 is only a line integral rather than being a surface integration and the number of elements is much smaller.

The integration of Equation 9.9 can use various algorithms and efficient numerical techniques[3] that can make significant time savings. It is also possible to use more approximate integration for elements that are far away from each other when the mutual interactions are less strong. Another technique to speed the method when evaluating the response for many frequencies is to use interpolation to estimate the interaction matrixes rather than do an exact numerical integration at every frequency. Such approximations have to be done with care, however, because they risk compromising the accuracy of the solution.

Evaluating Equation 9.9 when $m = n$, in other words evaluating the influence of an element on itself, requires special consideration. The reason for this is that the integral includes a singularity. The singularity is relatively weak in this case, and so provided that care is taken to avoid evaluating the case where $\mathbf{r}_n = \mathbf{r}_s$ within the numerical integration, it will converge to a correct value.

Once the simultaneous equations are constructed, they can be solved using standard matrix solution techniques. The BEM forms full matrixes so sparse matrix solvers used in techniques such as finite element analysis (FEA) are not directly applicable.

If the simultaneous equations are solved, it is possible to get non-unique solutions at certain frequencies. These equate to eigensolutions of the interior of the surface being modelled. There are various methods for overcoming this problem. One solution is to form an over-determined set of equations; this is the combined Helmholtz integral equation formulation (CHIEF) method.[4] By placing some receivers in the body of the diffuser, where the pressure must be 0, it is possible to add additional equations that help to ensure that the correct solution is found. In choosing the receivers inside the body of the surface, often referred to as internal points, it must be remembered that if these internal points are at a node of an incorrect eigensolution, then they do not help. Consequently, several internal points should be used, avoiding lines of symmetry and simple integer relationships between the coordinates of internal point locations. Another remedy to ensure a unique solution is to combine Equation 9.2 with its derivative; this is the Burton Miller approach.[1]

In reality, most diffuser geometries are such that non-unique solutions are not usually a problem. Non-unique solutions are most common when the wavelength is small compared to the geometry, but cases with such small wavelength to structural size are not often attempted in diffuser calculations because the computational burden becomes too large.

A significant reduction in computation burden can be achieved if there are planes of symmetry in the surface, and the source lies on the planes of symmetry. In this case, a simple image source construction can be used to take the place of identical parts of the surface and so reduce the number of elements. This is shown in Figure 9.2. The pressure on identical parts of the surface will be the same, and consequently, it is possible to construct a solution using about half the number of elements that would be required to mesh the whole surface. This does not reduce time in setting up the simultaneous equations, but it greatly speeds up the solution of the equations and decreases memory requirements. The easiest method to exploit the image source construct is to modify the Green's function by an additional term:

$$G(\mathbf{r}|\mathbf{r}_0) = \frac{e^{-jkr}}{4\pi r} + \frac{e^{-jkr'}}{4\pi r'}, \tag{9.10}$$

where r' is the distance from an image source which is reflected in the plane of symmetry. Multiple symmetry planes may exist, and so multiple image sources may need to be considered. A similar process can also be applied to the 2D Green's function.

It is also possible to assume non-uniform pressure variation across the elements.[5] For instance, it is possible to define the matrices in terms of the pressures at the element boundaries and assume a linear relationship between these. This reduces the number of elements needed to correctly represent the pressure variation on the surface compared to constant pressure elements. This has potential to make a faster prediction model, but at the cost of a more complex implementation.

9.1.2.2 Determining external point pressures

Once the surface pressures are known, Equation 9.2 becomes a more straightforward numerical integration, which must be solved. This process is relatively quick. The use of

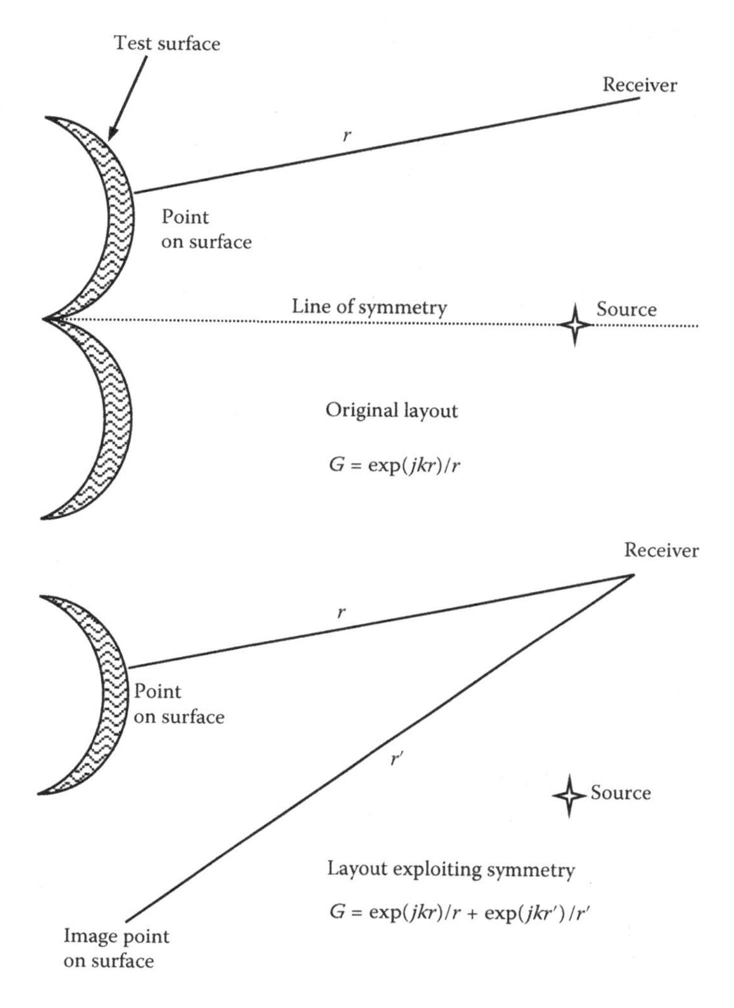

Figure 9.2 Illustration showing the use of an image source for a diffuser made of two arcs. The top illustration shows the original configuration where the source lies on a plane of symmetry. The bottom shows the exploitation of mirror symmetry to halve the number of elements required in a BEM.

efficient numerical integration algorithms and asymptotic solutions for the Green's function, when the receivers are far from the surface, can greatly speed up this process.

9.1.2.3 2D versus 3D

Cox[6] examined whether it is possible to predict the scattering from 3D diffusers using a 2D BEM model. The surfaces he tested were single-plane surfaces, which were extruded in one direction, like 1D Schroeder diffusers and cylinders. This meant that the scattering was roughly orthogonal across the width and along the extruded length. In these cases, 2D models provided accurate predictions except at low frequencies (for the surfaces tested, this meant below 500 Hz). Cox derived formulations that allow the overall scattered sound power level for the 2D polar response to be adjusted, to match the real (3D) case.

9.1.3 Thin-panel solution

When a surface becomes very thin, the previous solution method will not work. The front and back of the surface have elements that are too close to each other, and the solution method often becomes inaccurate. It is possible, however, to provide a formulation in terms of the pressure difference and sum across the panel. Not only does this regularize the equations to make them solvable, it also approximately halves the number of elements required and so speeds solution times and reduces storage requirements.

The solution method requires both Equation 9.2 and its derivative. Terai[7] showed that with correct regard for the jump relations, the integral equation and its derivative can be given by

$$\frac{1}{2}\{p(\mathbf{r}_1)+p(\mathbf{r}_2)\}=p_i(\mathbf{r}_0|\mathbf{r}_1)$$
$$+\iint_s \{p(\mathbf{r}_{s,1})-p(\mathbf{r}_{s,2})\}\frac{\partial G(\mathbf{r}|\mathbf{r}_{s,1})}{\partial n(\mathbf{r}_{s,1})}-\left\{\frac{\partial p(\mathbf{r}_{s,1})}{\partial n(\mathbf{r}_{s,1})}-\frac{\partial p(\mathbf{r}_{s,2})}{\partial n(\mathbf{r}_{s,1})}\right\}G(\mathbf{r}|\mathbf{r}_{s,1})\,ds, \tag{9.11}$$

and

$$\frac{1}{2}\left\{\frac{\partial p(\mathbf{r}_1)}{\partial n(\mathbf{r}_1)}+\frac{\partial p(\mathbf{r}_2)}{\partial n(\mathbf{r}_1)}\right\}=\frac{\partial p_i(\mathbf{r}_0|\mathbf{r}_{s,1})}{\partial n(\mathbf{r}_{s,1})}$$
$$+\iint_s \{p(\mathbf{r}_{s,1})-p(\mathbf{r}_{s,2})\}\frac{\partial^2 G(\mathbf{r}|\mathbf{r}_{s,1})}{\partial n(\mathbf{r}_1)\partial n(\mathbf{r}_{s,1})}-\left\{\frac{\partial p(\mathbf{r}_{s,1})}{\partial n(\mathbf{r}_{s,1})}-\frac{\partial p(\mathbf{r}_{s,2})}{\partial n(\mathbf{r}_{s,1})}\right\}\frac{\partial G(\mathbf{r}|\mathbf{r}_{s,1})}{\partial n(\mathbf{r}_1)}\,ds, \tag{9.12}$$

where the 1 and 2 in the subscripts refer to the front and the back, respectively, of an infinitesimally thick panel. These are the equations for points on the surface ($\mathbf{r}_1, \mathbf{r}_2 \in s$) and should be used to set up the simultaneous equations, which can then yield the surface pressures. If the desire is to achieve a reduction in computational burden, further simplifications can be obtained if more assumptions are made. Two approaches will be considered: first, the case of a non-absorbing surface and, second, the situation of a flat surface with non-zero surface admittance.

9.1.3.1 Non-absorbing surface

The surface is assumed to be non-absorbing and thin, then the differentials in the pressures on the front and rear surface are 0 as the surface velocity is 0. Under this assumption, Equation 9.12 can be simplified to yield a single equation in terms of the pressure difference across the panel:

$$0=\frac{\partial p_i(\mathbf{r}_0|\mathbf{r}_{s,1})}{\partial n(\mathbf{r}_{s,1})}+\iint_s \{p(\mathbf{r}_{s,1})-p(\mathbf{r}_{s,2})\}\frac{\partial^2 G(\mathbf{r}|\mathbf{r}_{s,1})}{\partial n(\mathbf{r}_1)\partial n(\mathbf{r}_{s,1})}\,ds. \tag{9.13}$$

Using this equation, it is possible to discretize the front surface into a set of elements across which the pressure is assumed constant and to set up simultaneous equations in the pressure difference between the front and rear of the panel $p(\mathbf{r}_{s,1}) - p(\mathbf{r}_{s,2})$. These simultaneous equations can then be solved to give the pressure difference for each element.

Once the pressure difference for each element is known, then the following equation is used to calculate the pressure at external points:

$$p(\mathbf{r}) = p_i(\mathbf{r}_0|\mathbf{r}_1) + \iint_s \{p(\mathbf{r}_{s,1}) - p(\mathbf{r}_{s,2})\} \frac{\partial G(\mathbf{r}|\mathbf{r}_{s,1})}{\partial n(\mathbf{r}_{s,1})} ds. \tag{9.14}$$

Incidentally, for a planar surface, it is simple to get the pressures on the front and the rear of the panel if these are wanted (they are not explicitly needed to get the external point pressures). The sum of the pressures on either side of the surface is equal to twice the incident pressure $p(\mathbf{r}_{s,1}) + p(\mathbf{r}_{s,2}) = 2p_i(\mathbf{r}_{s,1})$. This fact can be used with the values for the pressure difference between the front and rear of the panel, $p(\mathbf{r}_{s,1}) - p(\mathbf{r}_{s,2})$, to give the actual surface pressures on either side of the panel.

The matrix form of the integral Equation 9.13 is highly singular when the interaction of an element with itself is considered. To overcome this difficulty, Terai suggested using an asymptotic solution for calculating the contribution of an element's radiation to its own surface pressure. For the 3D case, this yields a line integral around the perimeter of the element:

$$\lim_{\mathbf{r} \to \mathbf{r}_s} \iint_s \frac{\partial^2 G(\mathbf{r}|\mathbf{r}_s)}{\partial n^2} ds = \frac{-1}{4\pi} \left\{ \oint \frac{e^{-jk\mathbf{r}(\theta)}}{\mathbf{r}(\theta)} d\theta + 2\pi jk \right\}, \tag{9.15}$$

where θ is defined in Figure 9.3.

For the 2D case, the corresponding equation is

$$\lim_{a \to 0} \int_0^a \frac{\partial^2 G(\mathbf{r}|\mathbf{r}_s)}{\partial n^2} ds = \frac{1}{2\pi a}. \tag{9.16}$$

In many cases, it is possible to set the distance a as being the length of the element, and so the previous factor can be used to directly calculate the contribution of the element to itself. For some problems with complex geometries, however, more accurate results are obtained if a smaller value of a is used and the rest of the element is integrated using normal numerical integration procedures.

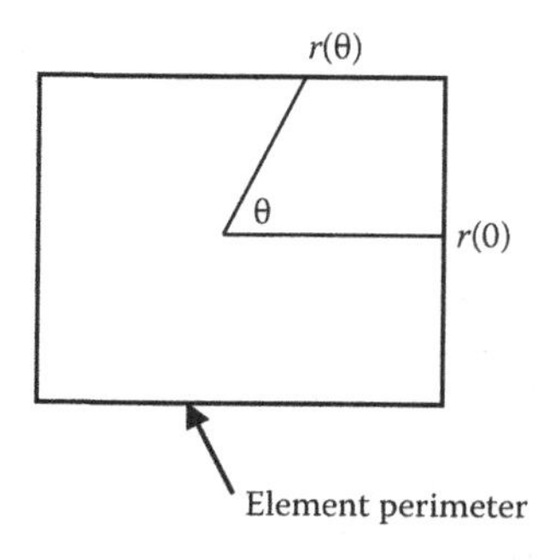

Figure 9.3 Definition of θ used for the asymptotic form of 3D thin-panel BEM.

The thin-panel BEM provides unique solutions, provided no enclosed volumes are created out of the elements. Consequently, there is no need to use an overdetermined system in many cases.

Apart from these details, the solution method proceeds in exactly the same way as for the standard BEM. The thin-panel formulation is very useful for many scattering surfaces. An overhead canopy above a stage can be treated as a thin rigid surface, and this formulation allows a faster solution than the standard BEM does. The accuracy is compromised close to grazing angles, as it does not model the scattering from the edges that exist with real surfaces. More details on the accuracy of this approach are given later.

The method can also be applied to Schroeder diffusers that have thin fins as part of the construction. Schroeder diffusers are discussed in Chapter 10, and several pictures are shown in Figure 10.1. The thin-panel BEM allows the geometry to be exactly modelled and the reflection predicted for any frequency. This is not true of most prediction methods, which are generally frequency limited as they assume plane wave propagation within the wells. An even neater solution for a Schroeder diffuser is to use a BEM that uses a combination of both normal and thin elements.[8]

9.1.3.2 *Planar, thin surface with non-zero admittance*

By assuming that the surface is planar, some of the terms in Equations 9.11 and 9.12 can be simplified and solutions can be obtained faster. One way of reducing the number of elements is to assume that the admittances on the front and rear of the panel at any point are the same, i.e., $\beta(\mathbf{r}_{s,1}) = \beta(\mathbf{r}_{s,2})$. For room acoustic diffusers, this is not going to introduce large errors into the calculation because it is the scattering on the bright side, which is of primary importance. For all but low frequencies, the pressure is low on the rear of the panel, and consequently, what admittance is assumed on the rear is not very important. This admittance assumption might not be accurate in all cases, however. For example, if a diffusing roadside barrier is being considered, the scattering in the shadow zone at low frequencies is of great interest.

Assuming $\beta(\mathbf{r}_{s,1}) = \beta(\mathbf{r}_{s,2})$ and the surface is planar, Equations 9.11 and 9.12 can be rewritten as

$$\frac{1}{2}\{p(\mathbf{r}_1) + p(\mathbf{r}_2)\} = p_i(\mathbf{r}_0|\mathbf{r}_1) - jk\iint_s \beta'(\mathbf{r}_{s,1})[p(\mathbf{r}_{s,1}) + p(\mathbf{r}_{s,2})]G(\mathbf{r}|\mathbf{r}_{s,1})\,ds, \tag{9.17}$$

and

$$\frac{1}{2}\beta'(\mathbf{r}_1)\{p(\mathbf{r}_1) - p(\mathbf{r}_2)\} = \frac{\partial p_i(\mathbf{r}_0|\mathbf{r}_{s,1})}{\partial n(\mathbf{r}_{s,1})} + \iint_s \{p(\mathbf{r}_{s,1}) - p(\mathbf{r}_{s,2})\}\frac{\partial^2 G(\mathbf{r}|\mathbf{r}_{s,1})}{\partial n(\mathbf{r}_1)\partial n(\mathbf{r}_{s,1})}\,ds. \tag{9.18}$$

The surface is again discretized into elements that are small compared to the wavelength. Then, two sets of simultaneous equations can be constructed from Equations 9.17 and 9.18. The first set of equations is in the sum of the pressures (Equation 9.17), and the second is in the pressure difference across the panel (Equation 9.18). These sets of equations are then separately solved. As two sets of simultaneous equations are being used, with half the number of elements when compared to a standard BEM, then the solution will be quicker by a factor of four to eight times, depending on the implementation.

The propagation from surface pressures to external receivers is carried out using the following equation:

$$p(\mathbf{r}) = p_i(\mathbf{r}_0|\mathbf{r}_1) + \iint_s \{p(\mathbf{r}_{s,1}) - p(\mathbf{r}_{s,2})\}\frac{\partial G(\mathbf{r}|\mathbf{r}_{s,1})}{\partial n(\mathbf{r}_{s,1})} - ik\beta'(\mathbf{r}_{s,1})[p(\mathbf{r}_{s,1}) + p(\mathbf{r}_{s,2})]G(\mathbf{r}|\mathbf{r}_{s,1})\,ds. \quad (9.19)$$

The accuracy of this technique will be presented later in this chapter.

9.1.4 Acceleration schemes

When large areas of diffusers are used, BEMs become too slow and storage requirements for the full matrixes can also cause problems. For these reasons, various schemes to speed the solution have been developed. One approach is the fast multipole methods (FMMs) that have been developed for BEM.[9] FMMs collect the boundary elements into clusters employing a classic BEM approach to elements in the near field and applying a more efficient FMM procedure for those in the far field. The computational effort for the conventional FMM increases for N elements by $N^{3/2}$ compared to the conventional BEM, which increases by N^2. Storage requirements are also reduced. The implementation of this technique is complex, however.

In addition to FMM-BEM, there are also various matrix acceleration libraries, which have the advantage that the integral equation formulation and discretization part of the algorithm are unchanged. BEM++ is an open source application for Mac OS and Linux that uses the AHMED adaptive cross approximation (ACA) algorithm and an H-matrix algebra library.[10]

Another approach to improving BEM predictions is to use information concerning the physical nature of the surface boundary conditions to reduce computational burden. In the case of periodic surfaces, Lam[11] showed that considerable improvements in prediction times can be achieved by representing the periodicity in the Green's function. This technique is useful because diffusers are often applied in a periodic formation.

9.1.5 BEM accuracy: thin rigid reflectors

The next two sections will examine the accuracy of BEMs. Consider first thin, rigid, planar and curved surfaces. These commonly occur in indoor and outdoor spaces and are relatively straightforward to mesh and predict using BEMs. It is also relatively easy to construct and measure the scattering from such surfaces and so enable the accuracy of the prediction methods to be directly compared to scale model measurements (see Chapter 5 for the measurement techniques used).

The full BEM solution based on Equation 9.2 produces accurate predictions over a wide range of frequencies for plane and curved surfaces.[12] To illustrate this, Figures 9.4 and 9.5 compare predicted and measured results for the total and scattered pressure for a flat thin panel.[13] The thin-panel model is based on Equations 9.13 and 9.14 and gives very similar results to the full BEM solution (for many angles, the lines overlay each other), and it is also faster. The thin-panel and full BEM model only deviate for grazing angles and high frequencies. The deviations occur because the thin-panel model does not properly represent the finite thickness of the surface, which becomes more critical at grazing angles when the wavelength is not large compared to the panel thickness. Inaccuracies arise because there are no edge elements in the thin-panel BEM and so edge scattering is not properly modelled.

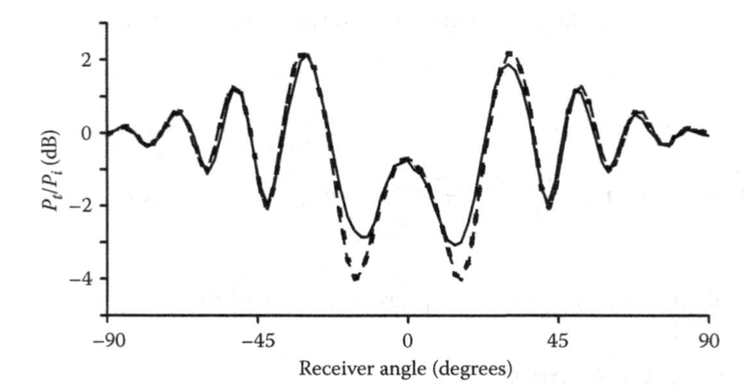

Figure 9.4 Comparison of total field (incident plus scattered) for a plane surface measured and predicted by two boundary element models. Normalized to incident sound at receiver: —— measured, – – – thin-panel BEM, and - - - - 3D BEM. (Data from Cox, T.J., "Objective and Subjective Evaluation of Reflection and Diffusing Surfaces in Auditoria", PhD thesis, University of Salford, UK, 1992.)

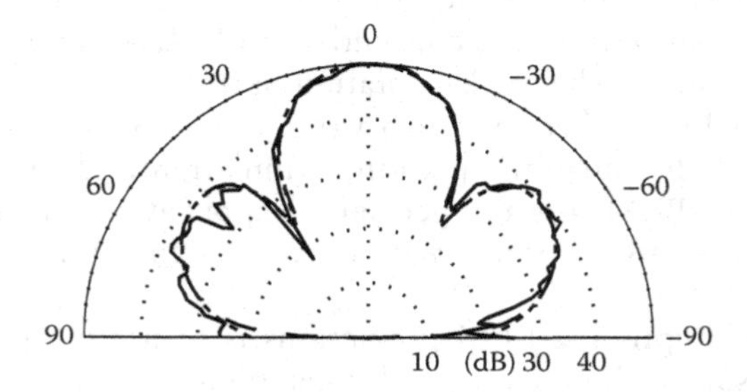

Figure 9.5 Comparison of scattered pressure from a plane surface: —— measured, – – – thin-panel BEM, and - - - - 3D BEM. Normalized to incident sound at each receiver and offset by 57.5 dB. (Data from Cox, T.J., "Objective and Subjective Evaluation of Reflection and Diffusing Surfaces in Auditoria", PhD thesis, University of Salford, UK, 1992.)

9.1.6 BEM accuracy: Schroeder diffusers

Figure 10.1 shows pictures of Schroeder diffusers and Chapter 10 discusses the design of these devices in great detail. The diffuser consists of a series of wells of the same width but different depths. There are two approaches to modelling this surface. The first uses the thin-panel solution and allows the diffuser shape to be exactly modelled; the second uses an approximate model of the surface as a box with a spatially varying front face admittance (or impedance).

The thin-panel solution allows explicit representation of the diffuser shape. The complete diffuser can be covered with thin-panel elements. Figure 9.6 shows a typical example. A drawback of this method is that it uses a very large number of elements. The complete enclosure of the diffuser by thin-panel elements forces the interior to have 0 pressure provided no critical frequencies are found. Two other problems could arise from this representation: (1) a large number of thin-panel elements with different sizes have to be sealed together, and the technique is therefore prone to meshing errors, and (2) the thin-panel solution for a plane panel showed small inaccuracies for scattering at grazing angles, especially at high frequencies. This could be a problem for Schroeder diffusers with fins, as these are presented edge-on to sources normal to the surface.

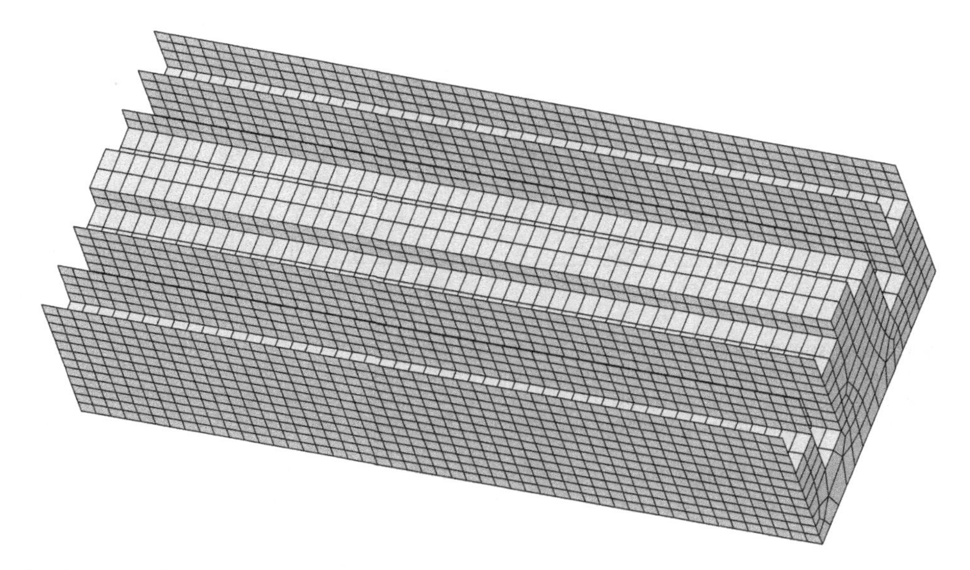

Figure 9.6 An example of a Schroeder diffuser meshed for prediction using a thin-panel model.

The approximate model, using a spatially varying admittance on the front face of a box, has been used by most authors investigating Schroeder diffusers. The admittance on the front face at the entrance of each well is derived from the phase change of plane waves propagating up and down the wells. This representation is expected to work under certain conditions: (1) the frequency must be below the cut-off frequency of the well so plane wave propagation in the wells dominates, and (2) the impedance at the opening of the wells must be locally reacting, which means that the radiation coupling between the wells has to be small. Also, unless the radiation impedance of each well is small, it should be represented in the admittance at the well entrances.

Comparisons with measurement show that the thin-panel predictions of Schroeder diffuser scattering are accurate. Figure 9.7 shows an example for the scattered pressure at normal incidence.[13] Similar accuracy is also achieved for the total field. The BEM model based on Equation 9.2, using the box representation with a spatially varying front face admittance, is also successful. This demonstrates that the simple phase change local reacting admittance assumption is reasonable—this is discussed in more detail in Section 10.9.

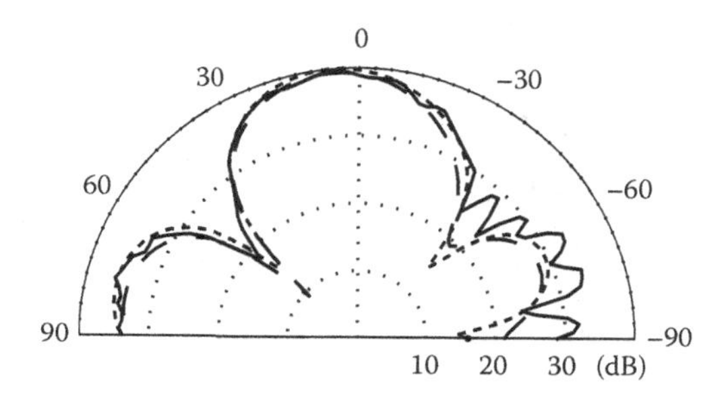

Figure 9.7 Comparison of scattered pressure from a Schroeder diffuser: —— measured, – – – thin-panel BEM, and - - - - BEM with box model. Normalized to incident sound at each receiver and offset by 50 dB. (Data from Cox, T.J., "Objective and Subjective Evaluation of Reflection and Diffusing Surfaces in Auditoria", PhD thesis, University of Salford, UK, 1992.)

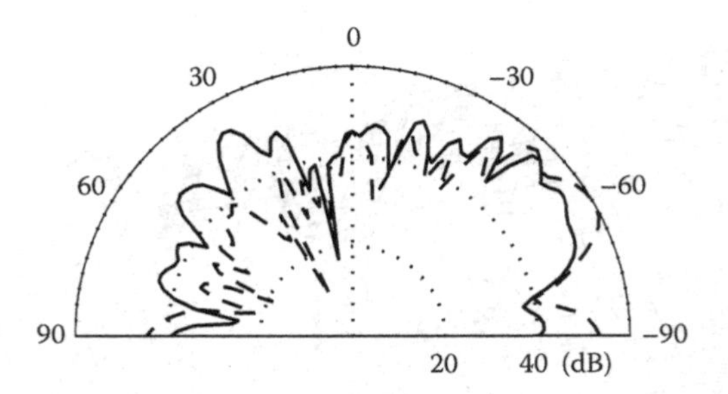

Figure 9.8 Comparison of two BEM models for the scattered pressure for oblique sound incident on a Schroeder diffuser: —— thin-panel and - - - box model. Normalized to incident pressure at each receiver and offset by 66 dB. (Data from Cox, T.J., "Objective and Subjective Evaluation of Reflection and Diffusing Surfaces in Auditoria", PhD thesis, University of Salford, UK, 1992.)

For a source that irradiates the surface from an oblique angle, there is greater interaction across the front face of the diffuser, and the box representation with a spatially varying front face admittance will be less accurate. Figure 9.8 shows an example for a source at 60°, where the thin-panel model, which is assumed to be correct, is compared to the BEM using the box with the spatially varying admittance.[13] The BEM using the box model is most accurately close to the specular reflection angle and becomes more inaccurate for receivers further from that angle.

9.1.7 BEM accuracy: hybrid surfaces

Hybrid surfaces use a mixture of absorbent and hard surfaces to generate a combination of absorption and dispersion. To test the accuracy of a BEM for such a surface, a single-plane surface shown in Figure 9.9 was constructed.[14] This was based on a fourth-order maximum length sequence; i.e., there were 15 patches of either absorption or reflection. Each of the

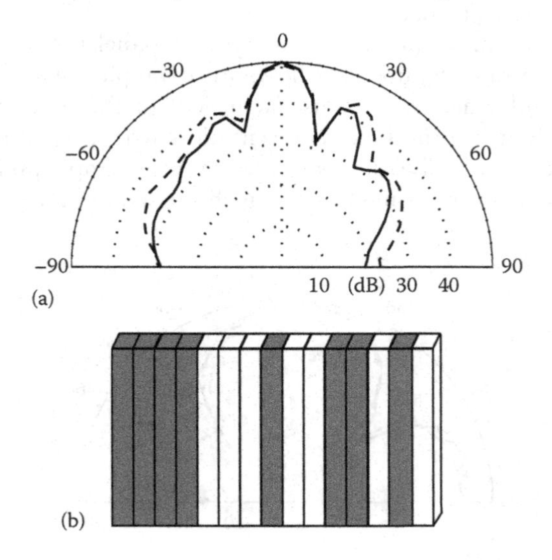

Figure 9.9 (a) Scattered pressure for normal incident sound and (b) the hybrid surface tested. The shaded sections are constructed of medium density fibreboard (MDF) and the unshaded sections of mineral wool —— BEM prediction and - - - measured. ([a] After Xiao, L.J., Cox, T.J., Avis, M.R., *J. Sound Vib.*, 285, 321–339, 2005.)

15 patches was 127 mm wide. Absorption was provided by mineral wool 76 mm thick, and reflection, by varnished wood 25 mm thick. The BEM was a conventional solution using Equation 9.2 and a 2D Green's function for speed. For the patches of absorbent, the impedance was modelled using the Delany and Bazley empirical formulation (see Section 6.5.1) with a flow resistivity of 50,000 rayl m^{-1} and a porosity of 0.98. Figure 9.9 shows an example result at 1.25 kHz. The BEM model agrees well with the measurement.

9.2 KIRCHHOFF

The rate-determining steps in carrying out BEM predictions are setting up and solving the simultaneous equations to determine the surface pressures. Consequently, faster methods for estimating the surface pressures have been derived, based on theories first applied to light. Photonics uses the Kirchhoff approximation to determine the propagation of light through an aperture. The Kirchhoff approximation gives the wave function and its derivative across the aperture as unaltered from the incident wave. On the surround defining the aperture, both the wave function and its derivative are assumed 0. Adapted for scattering in acoustics, this approximation can be used to obtain the surface pressures and their derivatives and yield reasonably accurate results for far field scattering. There are cases, however, when the method is not accurate, and so the approximation should be applied with care. An example MATLAB® code for this prediction model is available from the URL given in Reference 15.

Consider a large planar surface, with uniform surface impedance across the whole surface. By considering the definition of pressure reflection coefficient, Equation 2.20, it would be anticipated that the pressure p on the surface at $\mathbf{r}_s$ would be given by

$$p(\mathbf{r}_s) = [1 + R(\mathbf{r}_s)] \cdot p_i(\mathbf{r}_s|\mathbf{r}_0), \tag{9.20}$$

where R is the reflection coefficient of the surface. This formulation is sometimes referred to as the *Kirchhoff boundary condition*. If the surface is completely non-absorbing, $R = 1$, then the surface pressure is simply double the incident pressure. When the surface is completely absorbing, $R = 0$, then the surface pressure is just the incident pressure. It is necessary to assume that the diffuser is thin, so that the pressure from the sides of the surface can be neglected. It is also assumed that the surface is large compared to wavelength, so that the pressure on the rear of the panel can be assumed to be 0. Then substitution of Equation 9.20 into Equation 9.2 leaves a straightforward numerical integration over the front face that can be rapidly and readily evaluated.

Problems arise when applying the Kirchhoff boundary condition when the surface has significant thickness, is small compared to wavelength, or has a surface impedance that varies rapidly (spatially). Problems also arise for oblique sources and receivers. In the following paragraphs, these problems are highlighted and discussed.

When a surface becomes very deep, it is possible for second-order reflections to occur. These second-order reflections are not modelled by the Kirchhoff boundary condition. This is illustrated in Figure 9.10 for a couple of triangles. A model using the Kirchhoff boundary condition would predict the reflection from the first triangle, but not the second-order reflection from the neighbouring triangle. This would incorrectly result in a prediction of a significant reflection into grazing angles. In reality, however, second-order reflections mean that this scattered energy returns back towards the source. To prevent this problem, the Kirchhoff boundary conditions should be applied only to surfaces whose surface gradients are shallow. It is generally assumed that when the surface is steeper than about 30°–40°, the Kirchhoff boundary conditions are likely to become inaccurate.

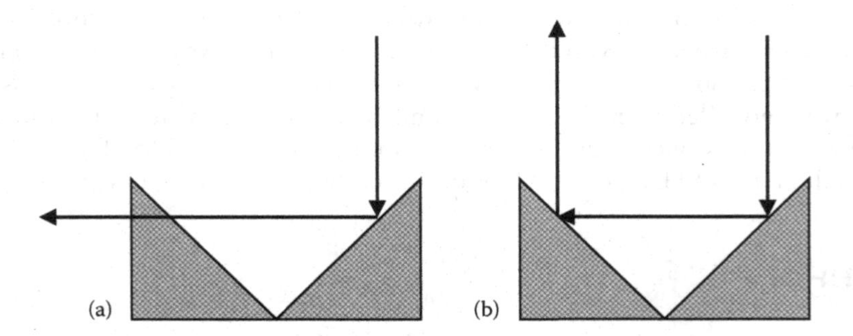

Figure 9.10 The Kirchhoff boundary conditions only model first-order reflections (a), leading to incorrect modelling of the true situation where second-order reflections dominate (b).

When the surface becomes small compared to wavelength, the surface pressures on the rear of the panel become significant, and the assumption of zero pressure on the rear can be inaccurate. Assuming zero pressure on the rear of the panel can also cause the predictions to become inaccurate if the angle of incidence or reflection becomes too large for finite-sized surfaces. Furthermore, inaccuracies can occur because the scattering from the edges is not modelled, as the pressure on the edges is also assumed to be zero. Incomplete modelling of edge diffraction will also be more problematical for oblique sources and receivers.

A surface that has a non-uniform surface impedance, where the spatial variation in impedance is rapid, can also cause problems. Consider the hybrid diffuser shown as an insert in Figure 9.11. The dark patches are absorbent ($R = 0$) and the light patches are reflective

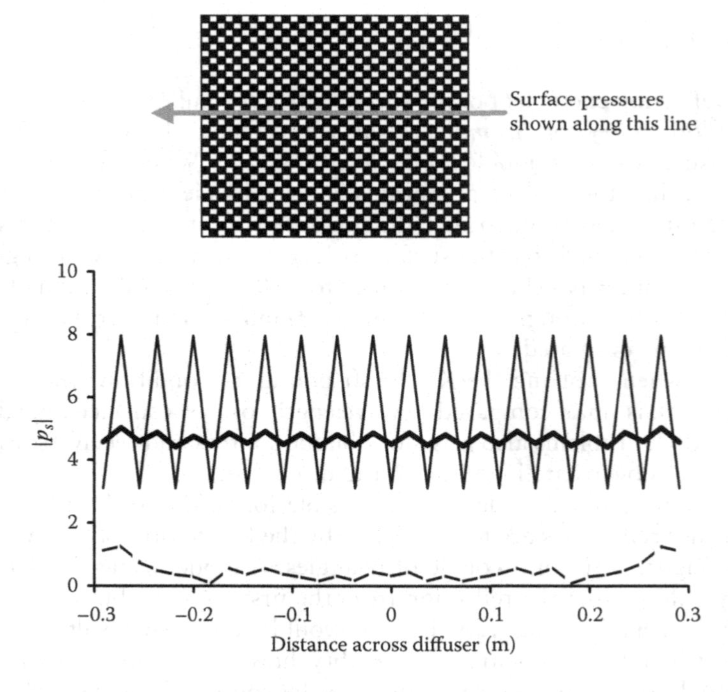

Figure 9.11 Surface pressures for a BEM compared to the Kirchhoff boundary conditions. Hybrid surface made up of a periodic arrangement of hard (white) and soft (black) patches as shown. The pressure distribution is shown for a line through the middle of the surface as indicated: —— Kirchhoff boundary conditions; ▬▬ front pressure, thin-panel BEM; and - - - - rear pressure, thin-panel BEM.

(R = 1). Figure 9.11 also shows the surface pressures predicted by a BEM, and these are compared to the Kirchhoff boundary conditions. The Kirchhoff boundary conditions predict a rapidly fluctuating pressure distribution due to the arrangement of hard and soft patches. The more accurate BEM shows that mutual interactions across the surface significantly alter the pressure distribution, smoothing out the spatial variation. If the far field polar response is considered, the Kirchhoff model in this case is inaccurate because of the incorrect surface pressures.

The surface pressures are inaccurate because there is an assumption that the surface is large in extent. Consequently, it is necessary to have a surface where the surface impedance spatial variation is large compared to wavelength; in other words, the patches of different impedances should be larger than half a wavelength. This would also appear to rule out the use of the Kirchhoff boundary conditions for Schroeder diffusers, but fortunately, the case of most Schroeder diffusers is less severe than that shown in Figure 9.11—in particular, the well width is wider for commercial implementations, and good prediction accuracy can be achieved.

Given all the previous reservations, the Kirchhoff solution is surprisingly good and useful for many acoustic diffusers, the exception being hybrid surfaces. For a flat surface, accurate results are achieved because the surface pressures are close to those given by the Kirchhoff boundary conditions. Figure 9.12 shows a typical example.[13] Close to the specular reflection direction, the accuracy of the Kirchhoff solution improves as the frequency increases. As the frequency increases, the pressure on the rear of the panel decreases, as does the significance of edge diffraction and mutual interactions across the surface. For single curved surfaces, better accuracy is obtained, although there is a tendency for the Kirchhoff solution to incorrectly smooth out local minima in the polar distribution.

To use the Kirchhoff solution for Schroeder diffusers, the model of a box with a spatially varying front face admittance must be used as described in Section 9.1.6. As the Kirchhoff boundary conditions do not allow for mutual interactions across the surface, it is not completely successful in predicting the sound field. This is most obvious at low frequencies. Figures 9.13 and 9.14 contrast the prediction accuracy achieved at low and mid–high frequencies.[13] Again, the accuracy is best near the specular reflection angle. The Schroeder diffuser tends to scatter more sound energy to the side than a plane surface, and this tends to mask the decreasing accuracy with angle that is normally found with the Kirchhoff model. For oblique receivers, the predictions become less accurate, as the Kirchhoff boundary condition breaks down.

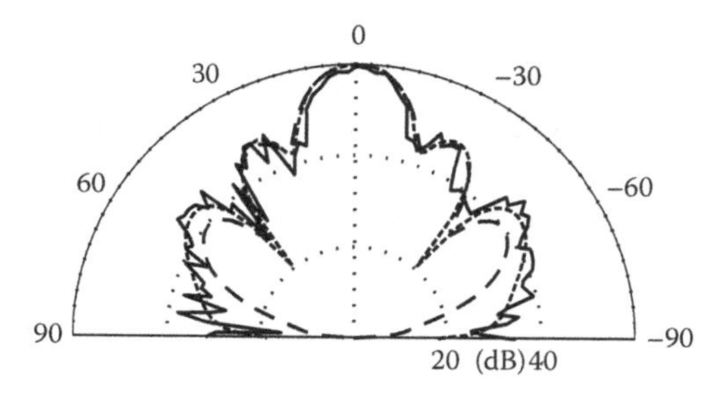

Figure 9.12 Pressure scattered from a plane surface. Comparing the accuracy of the – – – Kirchhoff solution to - - - - BEM and —— experiment. Normalized to incident pressure at each receiver and offset by 65.5 dB for plotting. (Data from Cox, T.J., "Objective and Subjective Evaluation of Reflection and Diffusing Surfaces in Auditoria", PhD thesis, University of Salford, UK, 1992.)

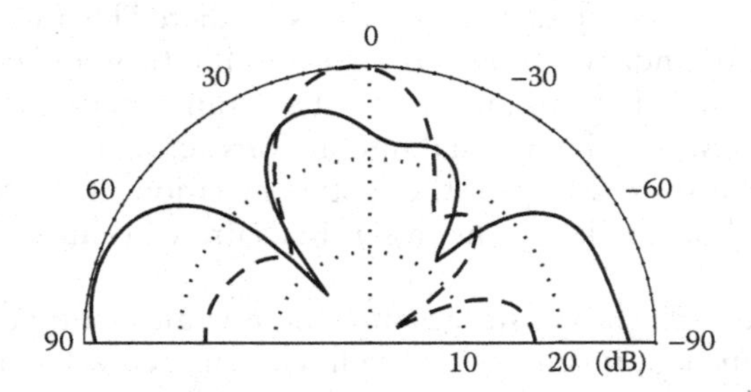

Figure 9.13 Pressure scattered from a Schroeder diffuser at a low frequency using two different prediction models: —— thin-panel BEM and - - - Kirchhoff solution. Normalized to incident pressure at each receiver and offset by 49 dB for plotting. (Data from Cox, T.J., "Objective and Subjective Evaluation of Reflection and Diffusing Surfaces in Auditoria", PhD thesis, University of Salford, UK, 1992.)

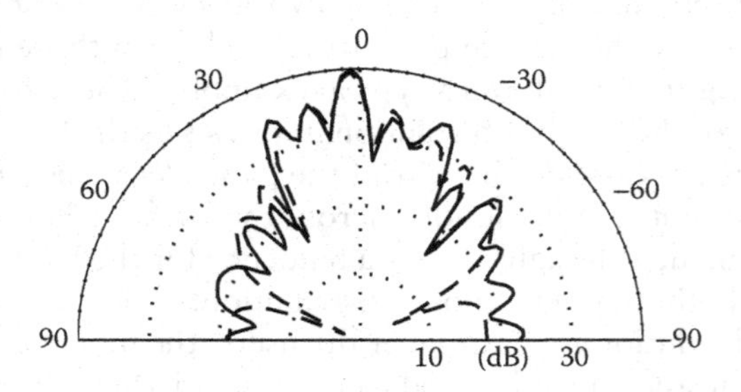

Figure 9.14 Pressure scattered from a Schroeder diffuser at a mid–high frequency using two different prediction models: —— thin-panel BEM and - - - Kirchhoff solution. Normalized to incident pressure at each receiver and offset by 48 dB for plotting. (Data from Cox, T.J., "Objective and Subjective Evaluation of Reflection and Diffusing Surfaces in Auditoria", PhD thesis, University of Salford, UK, 1992.)

9.3 FRESNEL

Once the Kirchhoff boundary conditions have been assumed, the resulting numerical integration can be simplified further. This can be done either to facilitate faster computation or to enable the derivation and understanding of simple design principles. The Kirchhoff boundary condition (Equation 9.20) is substituted into the Helmholtz–Kirchhoff integral equation (Equation 9.2). The usual relationship between surface admittance and pressure reflection coefficient, as given in Section 2.5.2, is also used. (Remembering that in this case, the normal to the surface is pointing outwards, as is usually the case with BEM models, whereas surface admittance is usually defined with an inwardly pointing normal.) Combining these equations gives the following formulation:

$$p(\mathbf{r}) = p_i(\mathbf{r}|\mathbf{r}_0) + \int_s p_i(\mathbf{r}_s|\mathbf{r}_s)\left[(R(\mathbf{r}_s)+1)\frac{\partial G(\mathbf{r}|\mathbf{r}_s)}{\partial n(\mathbf{r}_s)} - jk\cos(\psi)G(\mathbf{r},\mathbf{r}_s)(R(\mathbf{r}_s)-1)\right]ds, \tag{9.21}$$

where ψ is the angle of incidence. It is assumed that the receiver is sufficiently far from the surface that the differential of the Green's function can be approximately given by

$$\frac{\partial G(\mathbf{r}|\mathbf{r}_s)}{\partial n(\mathbf{r}_s)} \approx -jkG(\mathbf{r}|\mathbf{r}_s)\cos(\theta), \tag{9.22}$$

where θ is the angle of reflection (this relation is true for both the 2D and 3D Green's functions). Combining Equations 9.21 and 9.22 gives

$$p(\mathbf{r}) = p_i(\mathbf{r}|\mathbf{r}_0) - jk\int_s p_i(\mathbf{r}_s|\mathbf{r}_0)\,G(\mathbf{r}|\mathbf{r}_s)[(R(\mathbf{r}_s)+1)\cos(\theta) + \cos(\psi)(R(\mathbf{r}_s)-1)]\,ds. \tag{9.23}$$

Fresnel diffraction is a method normally designed to work with non-absorbing panels, i.e., $R = 1$. In that case, Equation 9.23 simplifies further:

$$p(\mathbf{r}) = p_i(\mathbf{r}|\mathbf{r}_0) - 2jk\int_s p_i(\mathbf{r}_s|\mathbf{r}_0)\,G(\mathbf{r}|\mathbf{r}_s)\cos(\theta)\,ds. \tag{9.24}$$

For conciseness, just the 3D case will be considered, although the findings here are readily translated into a 2D model. Consider a point source some way from a flat reflector, so the incident pressure p_i is given by the Green's function. It is necessary to come up with approximations for the distances $|\mathbf{r}_s - \mathbf{r}_0|$ and $|\mathbf{r} - \mathbf{r}_s|$. By considering the geometry shown in Figure 9.1 with the diffuser in the $y = 0$ plane, and a simple binomial expansion, it is possible to show that these distances are given by

$$\left|\mathbf{r}_s - \mathbf{r}_0\right| \approx \left|\mathbf{r}_0\right| - \frac{x_0x_s + z_0z_s}{\left|\mathbf{r}_0\right|} + \frac{x_s^2 + z_s^2}{\left|\mathbf{r}_0\right|}, \tag{9.25}$$

and

$$\left|\mathbf{r} - \mathbf{r}_s\right| \approx \left|\mathbf{r}\right| - \frac{xx_s + zz_s}{\left|\mathbf{r}\right|} + \frac{x_s^2 + z_s^2}{\left|\mathbf{r}\right|}. \tag{9.26}$$

If Equations 9.25 and 9.26 are substituted into Equation 9.24, the following expression is obtained for the scattered pressure:

$$p_s(\mathbf{r}) \approx -\frac{jk}{8\pi^2}\frac{e^{-jk(r_0+r)}}{rr_0}\cos(\theta)\iint e^{jk\frac{xx_s+zz_s+x_s^2+z_s^2}{r}}\,e^{jk\frac{x_0x_s+z_0z_s+x_s^2+z_s^2}{r_0}}\,dx_s dz_s. \tag{9.27}$$

It has been assumed that the variation in $|\mathbf{r}_s - \mathbf{r}_0|$ and $|\mathbf{r} - \mathbf{r}_s|$ in the denominator of the integral is negligible compared to the variation in the phase of the complex exponential; an assumption often applied in optics. This enables the denominator of the Green's function to be moved outside the integral. Similar arguments allow the cos(θ) factor to be moved outside the integration as well.

The phase terms of the complex exponentials are quadratic in x_s and z_s and so it is not possible to provide an analytical solution to this integration. In the past, this was overcome by using tables of Fresnel integrals. Nowadays, however, there is little point in using Fresnel integrals as computer power has increased to such an extent that the Kirchhoff approximation might as well be used. There are, however, some neat and simple shortcuts to calculating the above integration suggested by Rindel,[16] which could be used if speed is at a premium. The Fresnel solution does, however, lead to far field prediction models, which are key to understanding diffuser design.

9.4 FRAUNHOFER OR FOURIER SOLUTION

This solution is valid only in the far field, when both source and receiver are some distance from the surface. Then it is possible to neglect the quadratic terms in the exponentials in Equation 9.27 and obtain

$$p_s(\mathbf{r}) \approx -\frac{jk}{8\pi^2}\frac{e^{-jk(r_0+r)}}{rr_0}\cos(\theta)\iint e^{jk\frac{xx_s+zz_s}{r}} e^{jk\frac{x_0x_s+z_0z_s}{r_0}} dx_s dz_s. \tag{9.28}$$

For a planar surface, this then gives an analytical equation that can be solved. Assuming the panel is $2a$ long in the x direction, and $2b$ long in the z direction, the scattering is given by two sinc functions:

$$p_s(\mathbf{r}) \approx -\frac{jkab}{2\pi^2}\frac{e^{-jk(r_0+r)}}{rr_0}\cos(\theta)\cdot \operatorname{sinc}\left[k\left(\frac{x}{r}+\frac{x_0}{r_0}\right)a\right]\cdot \operatorname{sinc}\left[k\left(\frac{z}{r}+\frac{z_0}{r_0}\right)b\right], \tag{9.29}$$

where sinc(θ) = sin(θ)/θ. (Caution, in some texts, definitions of sinc() also include a factor of π.) This is a result familiar from optics and signal processing. The Fourier transform of a rectangular or top hat function gives a sinc() response. Equation 9.29 enables very quick estimations of the far field reflection from a rigid flat surface.

While this case is interesting, the Fraunhofer solution is arguably going to be most useful in analyzing surfaces that do not have uniform reflection coefficient. The most obvious example is the Schroeder diffuser, which can be modelled as having a spatially varying admittance on the front surface of a box. To carry out this analysis, it is necessary to return to Equation 9.23 and to apply the distance approximations outlined previously. For conciseness, consider just the scattered pressure:

$$\begin{aligned} p_s(\mathbf{r}) = &-\frac{jk}{16\pi^2}e^{-jk(r+r_0)}\iint e^{jk\frac{x_0x_s+z_0z_s}{r_0}} e^{jk\frac{xx_s+zz_s}{r}} \\ &\times[(R(\mathbf{r}_s)+1)\cos(\theta)+\cos(\psi)(R(\mathbf{r}_s)-1)]dx_s dz_s. \end{aligned} \tag{9.30}$$

To simplify the analysis, just normal incidence sound will be considered. Furthermore, it will be assumed that the surface admittance variation is in the x direction only. With these simplifications, it can be shown that the scattering is given by

$$p_s(\mathbf{r}) = -\frac{jk}{8\pi^2} e^{-jk(r+r_0)} \mathrm{sinc}\left(\frac{\mathrm{kb}}{\mathrm{r}}\right) \times \left\{ \int_{-a}^{a} R(\mathbf{r}_s) e^{jkx_s \sin(\theta)} [\cos(\theta)+1] dx_s \right\} + \left\{ \int_{-a}^{a} e^{jkx_s \sin(\theta)} [\cos(\theta)-1)] dx_s \right\}. \tag{9.31}$$

The term with [cos(θ) – 1] is usually less than the term with [cos(θ) + 1], especially away from grazing angles. Consequently, it can often be ignored. This leads to a scattered pressure of

$$p_s(\mathbf{r}) = -\frac{jk}{8\pi^2} e^{-jk(r+r_0)} \mathrm{sinc}\left(\frac{\mathrm{kb}}{\mathrm{r}}\right) [\cos(\theta)+1] \int_{-a}^{a} R(\mathbf{r}_s) e^{jkx_s \sin(\theta)} dx. \tag{9.32}$$

This is essentially the equation used by Schroeder in the design of phase grating diffusers, although he derived his formulations following a different philosophy. Furthermore, there are some additional factors outside the integral. Several authors neglect the [cos(θ) + 1], and this simplified form is often called a *Fourier theory* because the integration is essentially a Fourier transform. MATLAB code illustrating the Fraunhofer solution applied to various diffusers is available from the URL given in Reference 15.

9.4.1 Near and far field

As the analysis is now considering a far field prediction model, it is expedient to define what the near and far field mean. Unfortunately, with diffusers, the location of the near and far field is not as clear cut as for simple pistonic radiators, which is the case most often cited in textbooks. The *far field* is defined as the region where the difference between minimum and maximum path lengths from the panel to the receiver is small compared to the wavelength. In this region, all points on the panel are effectively at the same distance from the receiver.[17] This is illustrated in Figure 9.15. For diffusers, there is an added complication that both the source and receiver need to be considered, but to simplify discussions, it will be assumed that the source is at infinity. There is also a second requirement for the far field, which is that the receiver distance should be large compared to wavelength. This is, however, not usually the critical requirement for the geometries that occur with diffusers. Frequencies where the wavelength is large enough for this to be a consideration are usually below the lower frequency limit at which surface roughness effects are important.

In the far field, the polar response is independent of the receiver distance from the surface—this makes it a useful place to test diffusers. By considering the geometry in Figure 9.15 for a planar surface, it is possible to derive the following formulation for an on-axis receiver to be in the far field:

$$r > \frac{a^2}{\lambda}. \tag{9.33}$$

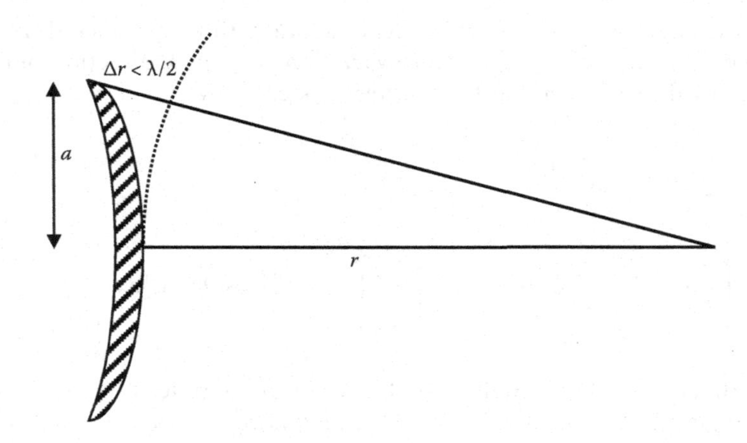

Figure 9.15 Common construction for determining near field extent.

Unfortunately, this far field formulation is not applicable to the case of oblique sources and receivers. As Figure 5.17 demonstrated, the true far field is achieved for many diffusers only when the receiver radius is many hundreds of metres! For further discussion of this issue, see Section 5.2.1.1.

9.4.2 Fraunhofer theory accuracy

Provided that sources and receivers are in the far field, for plane and Schroeder surfaces, the accuracy of the Fraunhofer theory is similar to that of the Kirchhoff solution. So if the results in Figure 9.12 were predicted with the Fraunhofer solution, similar results to the Kirchhoff solution would be obtained. Consequently, in the far field, the limiting factor for accuracy is the Kirchhoff boundary conditions. Where the Kirchhoff model fails, at low frequencies and for oblique sources and receivers, so does the Fraunhofer method.

Differences between the Kirchhoff and Fraunhofer solutions occur when the receiver is in the near field. Figure 9.16 shows the scattering from an N = 11 Schroeder diffuser in the near field.[13] It is assumed the BEM is accurate, and consequently, this shows that the inaccuracies in the Fraunhofer model are significant at this distance. Figure 9.17 shows the same situation, but now the receiver is 50 m from the diffuser, which is in the far field.[13] At these distances the Fraunhofer solution is as accurate as the Kirchhoff model. It is not often,

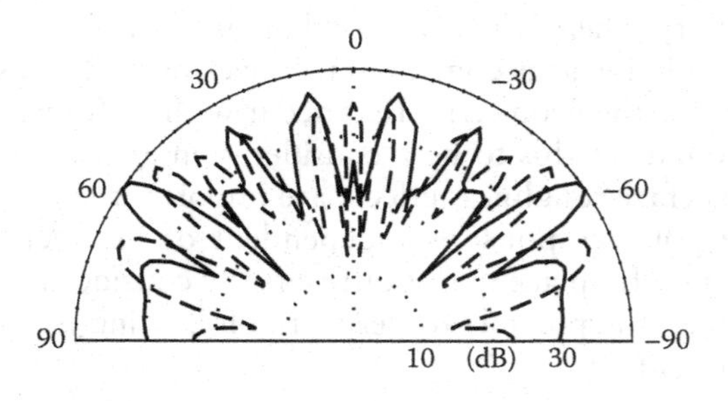

Figure 9.16 Scattered pressure from a surface. Comparison between —— BEM and – – – Fraunhofer solutions in the near field. Normalized to incident pressure at each receiver and offset by 49.6 dB for plotting. (Data from Cox, T.J., "Objective and Subjective Evaluation of Reflection and Diffusing Surfaces in Auditoria", PhD thesis, University of Salford, UK, 1992.)

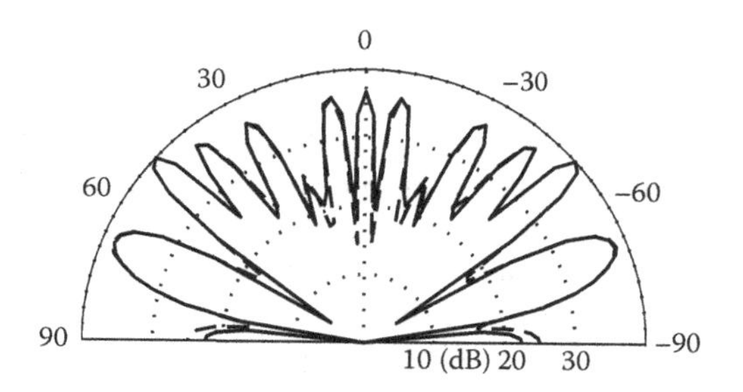

Figure 9.17 Scattered pressure from a surface. Comparison between —— Kirchhoff and - - - Fraunhofer solutions in the far field. Normalized to incident pressure at each receiver and offset by 77.6 dB for plotting. (Data from Cox, T.J., "Objective and Subjective Evaluation of Reflection and Diffusing Surfaces in Auditoria", PhD thesis, University of Salford, UK, 1992.)

however, that application realistic sources and receivers are this far from the panel. In diffuser design, however, the usual assumption is that a good far field diffuser will also work in the near field.

9.5 FINITE DIFFERENCE TIME DOMAIN

FDTD is a widely used simulation technique in electromagnetism and is becoming popular in acoustics. Both easy to understand and implement, it can cover a wide frequency range within a single prediction. It uses volumetric rather than surface meshes and can give accurate predictions of scattering.[18] (Chapter 3 used FDTD to generate pictures of wavefronts reflecting from various surfaces.) Figure 9.18 shows the scattering from a Schroeder diffuser predicted using three different models and illustrates that, especially near the specular reflection angle, FDTD gives accurate results. Giving the impulse response directly, it is natural to evaluate the temporal dispersion that diffusers generate using this method; however, the interpretation of this data is not yet well established. Furthermore, one of the biggest strengths of FDTD, the simulation of systems that are time-variant, still remains largely unexplored.

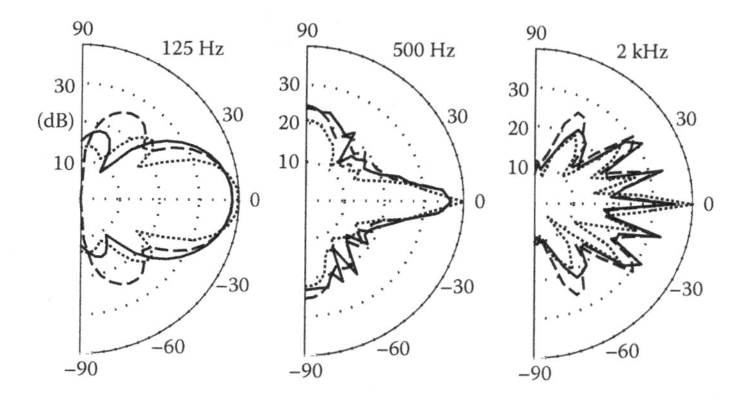

Figure 9.18 Normal incidence scattered pressure level from a quadratic residue diffuser in the far field for three prediction models: —— FDTD, - - - BEM, and • • • • FEA. Three one-third octave bands as labelled on charts.

One of the strengths of FDTD is that it is an extremely intuitive technique, so users can easily write their own codes. Its main weakness is that the entire computational domain must be meshed, and the spacing of the mesh must be small compared to the smallest wavelength and the physical features being modelled. Consequently, a very large system of equations must be considered, which results in very long solution times. Near to far field transformations combined with absorbing boundary conditions can help to overcome this problem in some particular cases. The first technique (near to far field transformation) requires some postprocessing, and the second one (absorbing boundary conditions) can complicate the code, losing the simplicity of the prediction model. Code execution times can be greatly reduced by exploiting graphics processor units (GPUs). Recent studies have demonstrated an improvement in parallelized FDTD code of up to 80 times over traditional implementations.[19–21]

For sound, the conservation of momentum and continuity equations that form the basis of the wave equation can be transformed to central-difference equations. This yields update formulations for the sound pressure and particle velocity. The equations are solved in a leapfrog manner: the sound pressure is solved at a given instant in time, then the particle velocity field is found at the next instant in time, and the process is repeated over and over again. Unless knowledge of both particle velocity and pressure is required throughout the entire volume being modelled, one may employ a faster numerical scheme based on a second-order wave equation, in which sound pressure is the only field variable. Simulation results are similar for the two cases.[22]

In 1966, Yee[23] described the first space-grid time-domain numerical technique in electromagnetism. Meloney and Cummings[24] adapted the method for acoustics based on the conservation of momentum and conservation of mass equations:

$$\frac{\partial p}{\partial t} = -\rho_0 c^2 \nabla \cdot \mathbf{u}, \tag{9.34}$$

and

$$\rho_0 \frac{\partial \mathbf{u}}{\partial t} = -\nabla p, \tag{9.35}$$

where $\mathbf{u} \equiv \mathbf{u}(x, y, z, t)$ denotes particle velocity, $p \equiv p(x, y, z, t)$ denotes acoustic pressure, ρ_0 is the density of air, and c is the speed of sound. The next step is to develop a finite difference scheme, to simulate the sound field governed by these mass and momentum equations. For this purpose, the mass and momentum equations are discretized in space and time such that

$$p\big|_{l,m,i}^{n} = p(x,y,z,t)\big|_{x=lX,y=mX,z=iX,t=nT}, \tag{9.36}$$

and

$$\mathbf{u}\big|_{l,m,i}^{n} = \mathbf{u}(x,y,z,t)\big|_{x=lX,y=mX,z=iX,t=nT}. \tag{9.37}$$

The superscripts denote the time index, n, and the subscripts denote the space indices, l, m, and i. At this point, we consider only discretization on a rectilinear grid, as shown in Figure 9.19. The parameter X represents the spatial sampling period (i.e., the distance

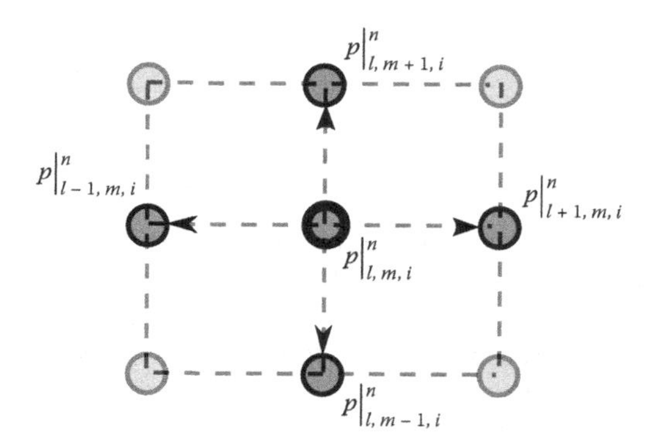

Figure 9.19 2D pressure grid for the FDTD model.

between any two neighbouring nodes on the grid), and $T = 1/f_s$ represents the temporal sampling period, where f_s is the sampling frequency. For the sake of simplicity, at this point, only 2D schemes are considered, in which case Equations 9.34 and 9.35 can be rearranged:

$$\frac{\partial p}{\partial t} = -\rho_0 c^2 \left(\frac{\partial u_x}{\partial x} + \frac{\partial u_y}{\partial y} \right), \tag{9.38}$$

$$\frac{\partial u_x}{\partial t} = -\frac{1}{\rho_0} \frac{\partial p}{\partial x}, \tag{9.39}$$

and

$$\frac{\partial u_y}{\partial t} = -\frac{1}{\rho_0} \frac{\partial p}{\partial y}. \tag{9.40}$$

To derive a numerical scheme for a rectilinear grid shown in Figure 9.19, the partial derivatives in the governing equations are represented using finite differences. For instance, the pressure derivative with respect to x is approximated by

$$\left. \frac{\partial p}{\partial x} \right|_{x=x_0} \approx \frac{p\left(x_0 + \frac{X}{2}, y, z, t \right) - p\left(x_0 - \frac{X}{2}, y, z, t \right)}{X}. \tag{9.41}$$

Using this approach, Equations 9.38 through 9.40 can be transformed into a set of update formulations that are used to obtain pressure and particle velocity in a time-stepping manner:

$$p\Big|_{l,m}^{n+\frac{1}{2}} = p\Big|_{l,m}^{n-\frac{1}{2}} - z_0 s \left(u_x\Big|_{l+\frac{1}{2},m}^{n} - u_x\Big|_{l-\frac{1}{2},m}^{n} + u_y\Big|_{l,m+\frac{1}{2}}^{n} - u_y\Big|_{l,m-\frac{1}{2}}^{n} \right), \tag{9.42}$$

$$u_x\Big|_{l+\frac{1}{2},m}^{n+\frac{1}{2}} = u_x\Big|_{l+\frac{1}{2},m}^{n} - \frac{s}{z_0} \left(p\Big|_{l+1,m}^{n+\frac{1}{2}} - p\Big|_{l,m}^{n+\frac{1}{2}} \right), \tag{9.43}$$

and

$$u_y\Big|_{l,m+\frac{1}{2}}^{n+\frac{1}{2}} = u_x\Big|_{l,m+\frac{1}{2}}^{n} - \frac{s}{z_0}\left(p\Big|_{l,m+1}^{n+\frac{1}{2}} - p\Big|_{l,m}^{n+\frac{1}{2}}\right), \tag{9.44}$$

where $z_0 = \rho_0 c$ is the specific acoustic impedance of the medium. The parameter $s = cT/X$ is called the *Courant number* and, as will be discussed later, is strongly linked to numerical propagation velocity and to the stability of the finite difference scheme.

The spatial grids are staggered in space to minimize the errors from higher-order terms when discretizing spatial and time derivatives. For instance, the mesh for the x component of the particle velocity is shifted a distance of $X/2$ with respect to the pressure mesh. The same goes for the time mesh; the particle velocity meshes are shifted $T/2$ in time with respect to the pressure mesh. This arrangement is termed a leap-frog scheme. All particle velocity computations are calculated and stored in memory for a particular time point using the previously stored values of the pressure. Then all the pressure computations are calculated and stored in memory using the previously calculated values of particle velocity. The cycle can be repeated as many times as needed to explore the changing sound field over time.

If only sound pressure is sought, then by substituting Equation 9.35 into 9.34, a second-order wave equation is obtained:

$$\frac{1}{c^2}\frac{\partial^2 p}{\partial t^2} - \nabla^2 p = 0. \tag{9.45}$$

In this case, second-order derivatives need to be used. For example, the second-order spatial derivative for pressure in the x direction is approximated by

$$\left.\frac{\partial^2 p}{\partial x^2}\right|_{x=x_0} \approx \frac{p(x+X,y,z,t) - 2p(x,y,z,t) + p(x-X,y,z,t)}{X^2}. \tag{9.46}$$

Using the same procedure as for the mass and momentum formulations, an explicit update rule for the second-order wave equation can be derived. For the case of a 3D rectilinear grid, as shown in Figure 9.20, this update equation is explicitly given by[25]

$$\begin{aligned} p\Big|_{l,m,i}^{n+1} &= s^2\left(p\Big|_{l+1,m,i}^{n} + p\Big|_{l-1,m,i}^{n} + p\Big|_{l,m+1,i}^{n} + p\Big|_{l,m-1,i}^{n} + p\Big|_{l,m,i+1}^{n} + p\Big|_{l,m,i-1}^{n}\right) \\ &\quad + 2(1-3s^2)\,p\Big|_{l,m,i}^{n} - p\Big|_{l,m,i}^{n-1}. \end{aligned} \tag{9.47}$$

In contrast to leap-frog schemes, the sound pressure across the domain needs to be stored in memory for three consecutive time-steps. However, in practice, one can overwrite values stored in data structures designated to $n - 1$ with those designated to $n + 1$, without violating any read-after-write dependencies.

Whilst rectilinear schemes are the most easy to derive and implement, they are largely inefficient in balancing computational efficiency with numerical errors. Savioja and Välimäki[26] proposed using spatially interpolated schemes to reduce dispersion errors, inherent in the FDTD method, and Kowalczyk and van Walstijn[27] introduced a family of compact explicit

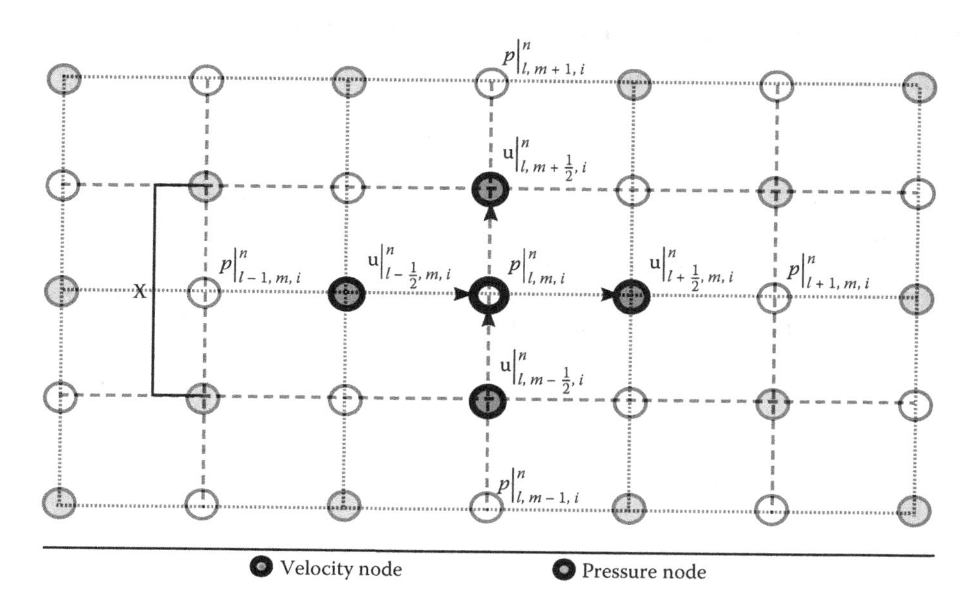

Figure 9.20 Position of pressure and velocity nodes in an FDTD grid for the scalar wave equation.

schemes, providing a wide range of design choices that maximize the spectral bandwidth and isotropy of the simulation.

9.5.1 Stability: spatial and time steps

According to the Lax-Richtmyer Equivalence Theorem,[28] in order to ensure numerical convergence, the finite difference scheme needs to be both consistent and stable. Assuming that the scheme is used to solve a well-posed initial value problem, consistency with the governing partial differential equations (PDEs) can be proved by showing that at the limit, as $(T, X \rightarrow 0)$, the finite difference scheme approaches the values of the PDEs that it approximates. For a numerical scheme to be considered stable, it should not allow any growing solutions to occur. Testing for stability can be done in a number of ways, most noticeably by using a von Neumann analysis,[28] effectively revealing for which values of s a growing solution will occur. The resulting stability condition is dependent on the specifics of the numerical scheme chosen. For the family of rectilinear schemes with uniform spatial sampling periods, the stability conditions are given by[25]

$$s \leq \sqrt{\frac{1}{2}}, \quad \text{2D,} \tag{9.48}$$

and

$$s \leq \sqrt{\frac{1}{3}}, \quad \text{3D.} \tag{9.49}$$

Typically, the Courant number is chosen at its upper stability bound. Stability conditions for other numerical schemes can be found in the literature.[25,29]

9.5.2 Numerical dispersion and simulation bandwidth

When sound waves propagate in a non-dispersive medium such as air, their propagation velocity is isotropic and independent of frequency. In an FDTD simulation, however, the phase velocities of the propagated waves deviate from this ideal. This artefact, referred to as numerical dispersion,[29] is rooted in the nature of finite difference approximations and is both frequency and direction dependent. In general, numerical dispersion becomes more dominant with increasing frequency and, as such, imposes a limitation on the usable bandwidth of the simulation.

Dispersion relations can be derived by transforming update equations to wavenumber space and rearranging to show the relationship between temporal and spatial frequencies. For the case of a 3D rectilinear scheme, when the Courant number is chosen at its upper stability bound, numerical dispersion constrains the spectral bandwidth of the simulation to a frequency of $f/f_s = 0.075$, when a maximum of 2% accuracy and isotropy is aimed for.[25] This translates to over 13 times temporal oversampling, highlighting the significant computational expense of the rectilinear FDTD. Other schemes, such as the Interpolated Wideband scheme,[25] require only about five times oversampling to achieve an equivalent degree of accuracy.

9.5.3 Boundary modelling and including objects in domain

Modelling scattering problems requires two types of boundary models. The first type is used to model wave interaction with a scattering object and can be realized by imposing a boundary condition inside the domain. The simplest form of such boundary condition is when the object offers no absorption and permits no transmission. If particle velocity is explicitly solved for, then one can simply impose a value of 0 for the particle velocity component normal to the boundary surface. When pressure is used as the field variable, the following locally reacting boundary condition must be satisfied:

$$\frac{\partial p}{\partial t} + \left(\frac{c}{\beta}\right)\mathbf{n}\cdot\nabla p = 0, \tag{9.50}$$

where β is the specific acoustic admittance of the surface, and $\mathbf{n}$ is the unit normal to the boundary. Discretizing this formulation will result in a set of boundary update equations that are specific to the type (face, edge, corner) and orientation (front–back, left–right, up–down) of the boundary nodes.[25] Therefore, embedding a scattering object in the grid requires that the boundary geometry is first voxellised (sampled in space into cubic elements), and then the resulting boundary nodes are classified according to the orientation of their respective surfaces. To simplify this process, and to facilitate an efficient implementation on a GPU, a generalized, uniform boundary update equation has been proposed in[20,30]

$$\begin{aligned} p\Big|_{l,m,i}^{n+1}(1 + B_K) &= (2 - Ks^2)\,p\Big|_{l,m,i}^{n} + (B_K - 1)\,p\Big|_{l,m,i}^{n-1} \\ &+ s^2\left(p\Big|_{l+1,m,i}^{n} + p\Big|_{l-1,m,i}^{n} + p\Big|_{l,m+1,i}^{n} + p\Big|_{l,m-1,i}^{n} + p\Big|_{l,m,i+1}^{n} + p\Big|_{l,m,i-1}^{n}\right), \end{aligned} \tag{9.51}$$

where

$$B_K = (6 - K)\lambda\frac{\beta}{2}, \tag{9.52}$$

and $K = 5$ for surface nodes, $K = 4$ for edge nodes, and $K = 3$ for corner nodes. Setting $K = 6$, reduces the formulations to the regular update equation for the medium, and hence, Equation 9.51 can be used as a uniform update equation for the entire computational domain. To facilitate the use of a frequency-dependent boundary impedance, the admittance terms need to be reformulated as digital impedance filters.[27] In addition, since the model geometry is voxellised on cubic grid cells, these boundary conditions are perfectly accurate only for surface geometries that consist of right angles, aligned along the principal Cartesian axes. While many diffusers (e.g., Schroeder diffusers) can be represented in such form, other types of scatterers, such as polycylindrical surfaces, may require a boundary formulation that can accurately model oblique and curved surfaces. If the wavelength is not very small compared to the discretization step, X (i.e., at low frequencies, or when the grid is highly oversampled), then such non-conformal boundary conditions might be sufficient. However, if high accuracy is required at short wavelengths, then one must employ a conformal boundary model. Recently, Bilbao[31] proposed such a model by considering boundary regions as finite volume elements.

When a locally reacting impedance cannot be assumed, one option is to treat objects as having extended reaction. In such cases, there are several approaches depending on the different models used to simulate the propagation of vibrations in solids.[32] For example, Drumm[33] coupled acoustic FDTD and FEA models to predict surface reflections, although the use of FEA significantly increases computation time.

The second type of boundary is used to terminate the simulation at the edges of the domain, without introducing reflections back into the grid. In order to do that, one can define the end of the integration area with an impedance equivalent to the characteristic impedance of air. In doing so, and due to inevitable small errors due to discretization, the absorption of the terminations will not be complete. Reducing the reflected waves further requires the use of perfectly matched layers (PMLs).[34] This technique defines a lossy medium in locations proximate to the boundaries, which in turn implies the modification of Equations 9.38 through 9.40 to include attenuation factors for each dimension considered (γ_x and γ_y), i.e.,

$$\frac{\partial p_x}{\partial t} + \gamma_x p_x + \rho_0 c^2 \left(\frac{\partial u_x}{\partial x} \right) = 0, \tag{9.53}$$

$$\frac{\partial p_y}{\partial t} + \gamma_y p_y + \rho_0 c^2 \left(\frac{\partial u_y}{\partial y} \right) = 0, \tag{9.54}$$

$$\frac{\partial p}{\partial x} + \rho \left(\frac{\partial u_x}{\partial t} + \gamma_x u_x \right) = 0, \tag{9.55}$$

and

$$\frac{\partial p}{\partial y} + \rho \left(\frac{\partial u_y}{\partial t} + \gamma_y u_y \right) = 0, \tag{9.56}$$

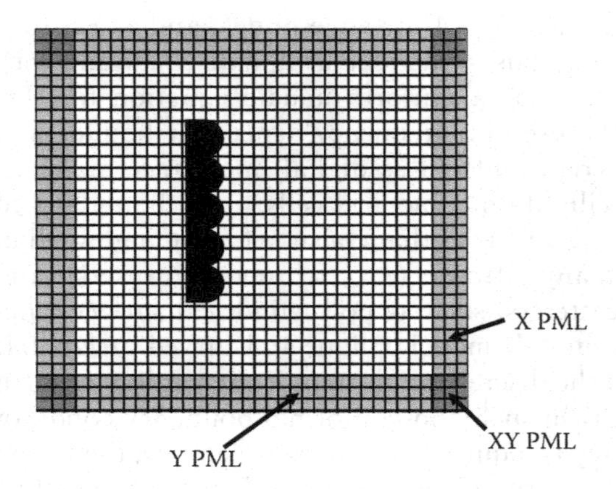

Figure 9.21 Structure of a 2D PML with a test specimen in the middle.

where the sound pressure has been split into two additive components p_x and p_y that have no physical sense. The attenuation factors are 0 inside the integration area and are gradually increased in areas near the boundaries using the following expression:

$$\gamma_x = \gamma_{x\,\max} \left| \frac{x - x_0}{x_{\max} - x_0} \right|^n, \tag{9.57}$$

where x_0 is the initial point in the PML, $x_{\max}$ is the last point of the PML, n is a number between 2 and 3, and $y_{x\max}$ is the maximum value of the attenuation factor. All these parameters should be optimized with some previous calculations.

Figure 9.21 illustrates the disposition of the different PMLs used in 2D simulations. At both the left and right sides of the grid, there are x direction PMLs ($y_x \neq 0$, $y_y = 0$). At both the upper and lower sides of the grid, there are y direction PMLs ($y_x = 0$, $y_y \neq 0$). At the four corners of the grid, there is an overlap of two perpendicular PMLs. This scheme ensures that the reflected sound is attenuated by at least 60 dB relative to the incident sound, provided that the PML parameters are chosen correctly. For three-dimensional problems, PMLs can impose a significant computational burden, especially when the modelled domain is large. When computational efficiency or implementation simplicity is desired, other choices for absorbing boundaries, such as Engquist-Majda[35] or Taylor-based[36] boundary conditions, can be used.

9.5.4 Excitation

There are several ways to introduce sources into the mesh. In the first papers on electromagnetism, the excitation was introduced as an initial condition in the whole integration area. Researchers focussed on the scattering caused by perfect conductors, equivalent to rigid walls in acoustics, and so they took a plane wave as an initial condition and observed how it was transformed by the particular objects included in the simulation domain. A similar approach was taken in few acoustics-related studies, e.g., see Reference 37. A more popular approach is to inject a time-dependent pressure or particle-velocity at a single or at multiple nodes in the grid. The most straightforward way to do this is to impose a pressure function at a node; this is referred to as a *Hard Source* formulation:

$$p\Big|_{l',m',i'}^{n+1} = F_p\Big|^{n+1}, \tag{9.58}$$

where (l', m', i') denote the discrete indices of the source position and F_p is the excitation function. Such excitation imposes a Dirichlet boundary condition at the source node, which results in artificial scattering from the source. This scattering artefact can be overcome by superimposing the excitation function, also known as a *Soft Source* excitation:

$$p\Big|_{l',m',i'}^{n+1} = \left\{p\Big|_{l',m',i'}^{n+1}\right\} + F_p\Big|^{n+1}, \tag{9.59}$$

where $\left\{p\Big|_{l',m',i'}^{n+1}\right\}$ represents the result of updating the source node with the regular update formulation for the medium. It should be noted that, due to the derivatives in the wave equation, such superposition causes the excitation function to be differentiated in time, resulting in a severe roll-off at low frequencies.[38]

The excitation function may take on many forms, depending on the desired temporal and spectral properties of the simulation. To prevent exciting the grid at frequencies prone to strong numerical dispersion errors, the excitation function can be bandlimited. Alternatively, in most cases, it is possible to excite the grid using a Kronecker delta function and perform filtering in postprocessing.[39] One of the most commonly used bandlimited excitation functions is the Gaussian pulse, given by

$$F_p\Big|^{n} = A_p \exp\left(\frac{-T^2[n-n_0]^2}{2\sigma^2}\right), \tag{9.60}$$

where n_0 is the number of samples by which the pulse is shifted, A_p is the pulse's amplitude, and σ is the pulse's variance, which is related to its half-power cut-off frequency by

$$f_c = \frac{\sqrt{2\ln 2}}{2\pi\sigma}. \tag{9.61}$$

While the Gaussian pulse has a relatively flat bandwidth down to zero frequency, it may excite frequencies near-DC that can resonate with the grid.[40] In such cases, the Gaussian pulse can be passed through a DC blocking filter. It is also possible to use excitation functions that are band-passed by nature, for example, a Ricker wavelet, which is a normalized second-derivative of a Gaussian function:

$$F_p\Big|^{n} = (1-2\gamma[n-n_0]^2)\exp(-\gamma[n-n_0]^2), \tag{9.62}$$

where $\gamma = (\pi f_c T)^2$. Other choices for excitation are a Blackman-Harris window, a differentiated Gaussian pulse, a sine-modulated Gaussian pulse, and an impulse response of a maximally flat finite impulse response filter. Figure 9.22 shows the frequency spectra of a number

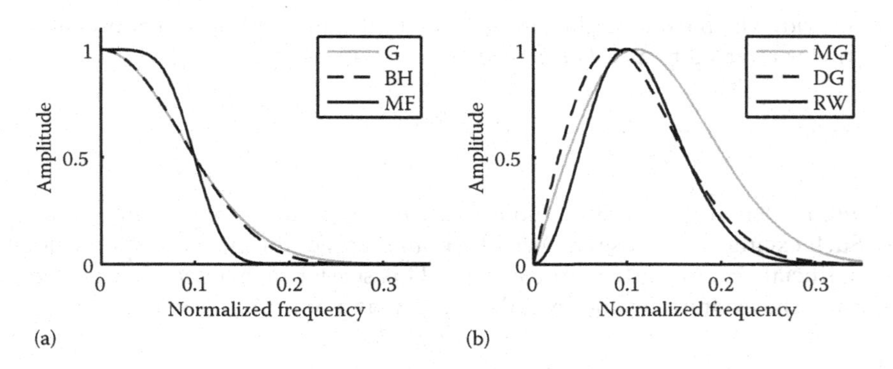

Figure 9.22 Frequency spectra of FDTD excitation functions with a normalized cut-off frequency of 0.1. (a) G: Gaussian; BH: Blackman-Harris; MF: maximally flat finite impulse response; (b) MG: sine-modulated Gaussian; DG: differentiated Gaussian; and RW: Ricker wavelet. (After Sheaffer, M., van Walstijn, M., Fazenda, B., *J. Acoust. Soc. Am.*, 135, 251–261, 2014.)

of possible excitation functions. For a detailed discussion on FDTD excitation signals and source injection techniques, see Sheaffer et al.[38]

9.5.5 Near to far field transformation

One of the main weaknesses of FDTD is that the entire computational domain must be meshed. If the aim of a simulation is to obtain the sound pressure reflected by an object in the far field, very large domains must be considered increasing the computational cost. To overcome this problem, near field to far field transformations (NFTFT) can be used. The standard transformation is based on the contour equivalence theorem. In this approach, the scattered pressure field along a closed virtual contour surrounding the structure of interest is computed via FDTD, and this is then integrated over the entire contour to provide the far field response. Since the virtual surface is independent of the geometry of the scatterer which it encloses, it is usually chosen as a rectangular shape to conform to the standard Cartesian FDTD grid. As there are no sources outside the transformation contour, the pressure at a far point can be computed via the Helmholtz–Kirchhoff integral, Equation 9.2; for further details, see Taflove.[41]

9.5.6 Realization

The FDTD method is straightforward to implement using a numerical computation package (see example MATLAB code available from the URL in Reference 42). However, when complex geometries, frequency-dependent boundaries, processing on GPU, and directional source and receivers are required, the implementation becomes much more complicated. There are a number of open-source packages available for simulating complex and large-scale problems using FDTD. Specific to linear acoustics problems, readers may wish to examine ParallelFDTD[43] and WaveCloud.[44]

9.6 TIME-DOMAIN BOUNDARY INTEGRAL METHODS

It is possible to formulate transient boundary integral models that operate in the time domain. Like FDTD, these methods have potential advantages over the constant frequency methods in that the impulse response can be generated more quickly. There is no need to separately calculate the pressure at many different frequencies and apply an inverse Fourier Transform to get the impulse response.

9.6.1 Kirchhoff and Fraunhofer solutions

These formulations apply the Kirchhoff boundary conditions to allow rapid prediction of the scattered impulse responses. FDTD models have a computational cost that scales with the maximum frequency f at $O(f^4)$, due to the combination of using a volumetric mesh and iterative time stepping. In contrast, the following time-domain models use a summation over a surface mesh with computation cost scaling at $O(f^2)$.

Like the frequency-domain Kirchhoff solution, inaccuracies will arise for surfaces with significant corrugation and small surfaces. In addition, for Schroeder diffusers, the radiation coupling and radiation impedance of the wells must be small. Results in Reference 45 show that the following time-domain models yield identical results to their frequency-domain equivalents. Example MATLAB codes for the time- and frequency-domain solutions are available from the URL given in Reference 15, demonstrating that the models yield the same results.

The sound field at receiver point $\boldsymbol{r}$ and time t in the vicinity of a scatterer s is represented by the pressure $p_t(\boldsymbol{r}, t) = p_i + p_s$, where p_i represents the sound travelling directly from the source to the receiver along vector $\boldsymbol{r}_d = \boldsymbol{r} - \boldsymbol{r}_0$, and p_s, the sound scattered off the surface. The pressure can be found by applying the Kirchhoff integral equation[46]:

$$p_t(r,t) = p_i(r,t) + \iint_s \left[p_t(r_s,t) * \hat{n} \cdot \nabla g\left(r|r_s,t\right) - g\left(r|r_s,t\right) * \hat{n} \cdot \nabla p_t(r_s,t) \right] ds, \tag{9.63}$$

where $\boldsymbol{r}_s$ is a point on that surface, $\boldsymbol{n}$ is the normal to surface, and * denotes convolution. See Figure 9.1 for the geometry. $g()$ is the time-domain Green's function that describes how the pressure created by an instantaneous point source at time $t = 0$ travels to another location. It is given by

$$g(r_a|r_b,t) = \frac{\delta\left(t - R_{ab}/c\right)}{4\pi R_{ab}}, \tag{9.64}$$

where $R_{ab} = r_a - r_b$, $\delta()$ is a Dirac delta function, and c is the speed of sound.

To evaluate the integration in Equation 9.63, the surface is discretized into N elements that are small compared to wavelength, so the pressure can be approximated to be constant across the elements and the integration is turned into a summation. For non-absorbing materials, the Kirchhoff boundary conditions simply states that there will be a doubling of the incident pressure at the surface, i.e., $p_t = 2p_i$ and $\hat{\boldsymbol{n}} \cdot \nabla p_t(r_s,t) = 0$.

The incident field is assumed to come from a monopole with time-dependent amplitude, $F(t)$:

$$p_i(r, t) = F(t) * g(\boldsymbol{r}|\boldsymbol{r}_0, t) = F(t - r_d/c). \tag{9.65}$$

A Gaussian function is a good choice for an excitation function:

$$F(t) = \frac{1}{\sqrt{2\pi}\sigma} e^{-\frac{t^2}{2\sigma^2}}. \tag{9.66}$$

The width of the pulse is governed by σ. This must be chosen to limit the bandwidth of the input energy to low frequencies where the surface discretization is valid.

Combining the previous formulations and simplifying assuming a far field solution yield[45]

$$p_t(r,t) = \frac{1}{4\pi r_d} F(t - r_d/c) - \sum_{n=1}^{N} \frac{\cos(\theta)\Delta s}{8\pi^2 r_1 r_2 \sigma^2 c}\left(t - \frac{r_1 + r_2}{c}\right) F\left(t - \frac{r_1 + r_2}{c}\right), \tag{9.67}$$

where θ is the angle of reflection relative to the surface normal; each element is assumed to have an area Δs; r_1 is the vector from the source to the point on the surface; and r_2 is the vector from the point on the surface to the receiver. The accurate modelling of arrival times of the Gaussian pulses is vital and so oversampling is used. Equation 9.67 is calculated with a sampling frequency at least 10 times the highest frequency of interest. Then the samples of each Gaussian pulse are moved to the nearest sampling point.

This is the time-domain equivalent to the *Fraunhofer models* outlined in Section 9.4. The diffusers are modelled as thin surfaces. The front face is discretized into elements that are smaller than an eighth of a wavelength for the highest frequency of interest. To model the effects of the wells in a Schroeder diffuser, the reflections need to be delayed by the time it takes the sound waves to propagate within the wells. If the nth element is at the mouth of a well of depth d_n, then the scattered pressure in Equation 9.67 is modified to

$$p_s(r,t) = -\sum_{n=1}^{N} \frac{\cos(\theta)\Delta s}{8\pi^2 r_1 r_2 \sigma^2 c}\left(t - \frac{r_1 + r_2 + d_n}{c}\right) F\left(t - \frac{r_1 + r_2 + d_n}{c}\right). \tag{9.68}$$

A more precise Kirchhoff model for Schroeder diffusers, a time-domain equivalent of Equation 9.23, would better represent the surface pressures at the mouth of the wells. The pressure at the well entrance can be approximated to be $p_t(r, t) = p_i(r, t) + p_i(r, t - 2d_n/c)$. The first term represents an incident wave travelling with an angle of $-\psi$ to the normal. The second term is the wave reflected from the surface at an angle of ψ' to the normal. The formulation in this case is

$$p_s(r,t) = \frac{-\Delta s}{4\sqrt{2\pi^2} r_1 r_2 c\sigma^2} \times \sum \left\{ \begin{aligned} &\left(t - \frac{r_1 + r_2}{c}\right) F\left(t - \frac{r_1 + r_2}{c}\right)(\cos(\theta) - \cos(\psi)) \\ &+ \left(t - \frac{r_1 + r_2 + 2d_n}{c}\right) F\left(t - \frac{r_1 + r_2 + d_n}{c}\right)(\cos(\theta) + \cos(\psi')) \end{aligned} \right\}. \tag{9.69}$$

Following the normal rules of reflection, for many surfaces, $\psi' = \psi$ would be an appropriate assumption. But when the wells of a Schroeder diffuser are much smaller than wavelength, a better approximation might be that the waves reradiate from the wells travelling parallel to the surface normal, $\psi' = 0$.

9.6.2 BEM solution

For a time-domain BEM, the surface quantities are discretized into elements in a similar way to frequency-domain BEM, and in addition, time is also discretized into steps. It can be shown from the time-dependent integral equation[47–49] that the velocity potential and its derivative on the surface can then be represented by the incident velocity potential from the source, plus contributions from the other elements at the current and previous times. Consequently, if initial silence is assumed, it is possible to formulate a system of equations for the velocity potential and its derivative that is marched on in time to obtain the full-time history of the pressure on each element. Once the surface pressures are known, these can be propagated to find the pressure versus time response at external receiver points.

The marching on in time is an iterative process and therefore has potential for divergence. Although the actual physical system being modelled is stable, numerical inaccuracies in the discretized equations can, and often do, result in instability. The interaction matrixes have to be integrated to a high accuracy, particularly for complex surfaces. As with the constant frequency BEM, the underlying integral equation is known to possess non-unique solutions at certain frequencies. These happen for fully enclosed surfaces and are physically interpreted as cavity resonances. Despite the fact that these should be precluded by the initial conditions, they can be excited by numerical error and diverge exponentially, corrupting the external solution. This behaviour can be overcome by using the combined field integral equation,[48] but it is still possible for structures with lightly damped behaviour to slide into instability.[49]

The instabilities and high computational cost have, to date, prohibited widespread use of the time-domain BEM, although a commercial implementation is now available.[50] In addition, a new approach,[51] *Convolution Quadrature*, has recently become popular, which does not solve the problem using time-marching and therefore eradicates the possibility of such instabilities. The algorithm has significant efficiency gains and has much in common with the notion of computing many separate frequencies and then performing an inverse transform. It undertakes this in a structured way utilizing complex (damped) frequencies.

9.7 OTHER METHODS

9.7.1 Finite element analysis

FEA[52] uses volumetric rather than surface meshes and can give accurate predictions of scattering.[18] Figure 9.18 shows the scattering from a Schroeder diffuser predicted using FEA and illustrates that, especially near the specular reflection angle, the model gives accurate results. As a mesh is formed throughout the space, a large system of sparse equations results. However, by applying an NFTFT (see Section 9.5.5), it is possible to mesh only the volume close to the surface.

FEA is much slower than a BEM when dealing with exterior domain acoustic problems such as the scattering from diffusers. Unlike FDTD, FEA does not produce a simple set of equations that compensate for the computational burden of using a volumetric mesh. Where FEA is useful is when there is fluid and structural motion. For example, it can model the behaviour of a non-rigid surface that exhibits structural vibration.

9.7.2 Edge diffraction models

Edge diffraction models can be used to produce the scattering from wedges and simple shapes. For a plane rigid surface, the total field can be seen as a sum of the direct sound, specular reflections, and edge diffraction components.[53–55] Consequently, it is possible to solve the scattering problem by integrating over the edges present in a diffuser. This type of method becomes rather slow if high orders of edge diffraction need be considered, as would be the case for complex surfaces. It does readily yield the sampled impulse response and consequently is particularly useful where broadband time-domain scattering needs to be calculated or if the results are to be integrated into GRAMs.[56]

9.7.3 Wave decomposition and mode matching approaches

It is possible to carry out a decomposition of the acoustic wave knowing the spatial distribution of the diffuser. Strube[57–59] used this approach to solve the scattering from Schroeder diffusers. In Section 8.2.7, an example of this type of theory was used to explain the absorption from Schroeder diffusers. In this theory, it is assumed that the diffuser structure is periodic and then it is possible to decompose the scattered wave into the different diffraction lobes using a Fourier decomposition. Then, it is possible to set up and solve simultaneous equations into the diffraction lobe scattered amplitudes. These methods offer an alternative approach to BEMs, but BEMs are considerably more useful as they can be applied to arbitrary, non-periodic surfaces. The modal decomposition models are particularly powerful for large arrays of periodic structures.

9.7.4 Random roughness

In the theories used so far, a deterministic approach has been taken, with the surface geometry and impedance properties being modelled exactly. For large-scale surfaces with small roughness, this can turn out to be a very inefficient approach for carrying out predictions. In that case, it may be advantageous to use a statistical approach, whereby the surface is only determined by some shape statistics, such as the mean square surface height and the slope probability function.[60] In diffuser design, these theories are not often useful because the size of the surface roughness is large and the sample of roughness that might be used is small in width, and so a few shape statistics are not sufficient to accurately predict the scattering from the surface. This is illustrated in Figure 9.23. The top shape is meant to generically represent deliberately designed diffusers, where the scale of the roughness is deep and relatively slow varying. In this case, the small number of bumps on the surface will dominate the scattering in such a way that a statistical approach is not applicable. In contrast, the bottom line of Figure 9.23 represents a more randomly rough surface, for which the theories based on a few shape statistics may be more applicable. To put it another way, there needs to be a sufficiently wide sample of the surface roughness for the shape statistics to be properly representative of the surface.

For accidental surface roughness, this statistical approach is more useful, especially as it may be impossible to get the exact geometry of all shapes in existing structures. Random rough theories are probably most commonly used in underwater acoustics, although both Cox and D'Antonio[61] and Embrechts et al.[62] have applied statistical approaches

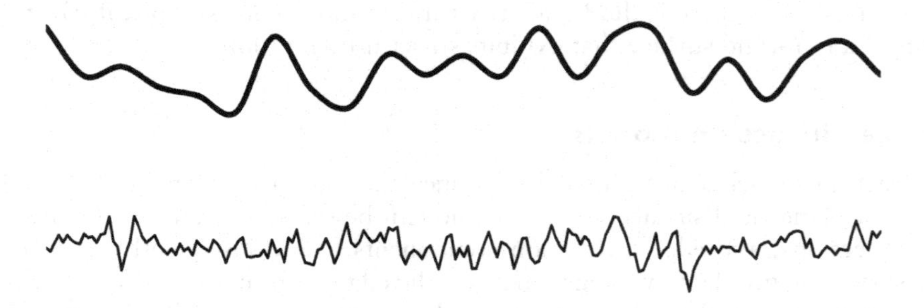

Figure 9.23 Two different randomly generated surfaces.

to diffuser reflections. These approaches usually assume the Kirchhoff boundary conditions, and so an additional limitation is that the surface gradients must not be too steep; otherwise, second- and high-order reflections become important, and the model becomes inaccurate.

9.7.5 Boss models

Boss models are hybrid approaches.[60] They use a deterministic solution for the scattering from a single element—examples include cylinders and hemispheres—and then model the distribution of the elements in a statistical manner. Twersky[63] developed the best known approach. This theory allows high-order scattering, across all frequencies, both in 2D and 3D, to be considered. Up-to-date versions of the theory also enable scattering from different sized bosses. One of the problems with applying this model is representing complex surfaces by a series of regular sized bosses. Torres et al.[64] have applied a boss model to predict scattering by hemispherical surface elements in auditoria.

9.8 SUMMARY

In this chapter, some commonly used prediction models for scattering have been outlined, and the necessary equations, developed. These theories will be drawn upon in the design of diffusers, which is the subject of the next three chapters.

REFERENCES

1. A. J. Burton, The Solution of Helmholtz Equation in Exterior Domains Using Integral Equations, National Physical Laboratory Report NAC30 (1973).
2. M. Ambramowitz and I. A. Stegun (Eds), *Handbook of Mathematical Functions*, Dover Publications Inc., New York (1965).
3. N. Atalla and F. Sgard, *Finite Element and Boundary Methods in Structural Acoustics and Vibration*, CRC Press, Boca Raton, FL (2015).
4. H. A. Schenck, “Improved integral formulation for acoustic radiation problems”, *J. Acoust. Soc. Am.*, **55**, 41–58 (1968).
5. E. Perrey-Debain, O. Laghrouche, P. Bettess, and J. Trevelyan, “Plane-wave basis finite elements and boundary elements for three-dimensional wave scattering”, *Philos. T. Roy. Soc. A.*, **362**(1816), 561–77 (2004).
6. T. J. Cox, “Predicting the scattering from reflectors and diffusers using 2D boundary element methods”, *J. Acoust. Soc. Am.*, **96**(2), 874–8 (1994).
7. T. Terai, “On the calculation of fields around three-dimensional objects by integral equation method”, *J. Sound Vib.*, **68**, 71–100 (1980).
8. T. W. Wu, “A direct boundary element method for acoustic radiation and scattering from mixed regular and thin bodies”, *J. Acoust. Soc. Am.*, **97**, 84–91 (1995).
9. N. Gumerov and R. Duraiswami, *Fast Multipole Methods for the Helmholtz Equation in Three Dimensions*, Elsevier Science Ltd., The Netherlands (2004).
10. T. Betcke, S. Arridge, J. Phillips, W. Smigaj, and M. Schweiger, “Solving boundary integral problems with BEM++”, *ACM Trans. Math. Softw.*, **41**(2), article 6 (2015), http://www.bempp.org/, accessed 16 September 2015.
11. Y. W. Lam, “A boundary integral formulation for the prediction of acoustic scattering from periodic structures”, *J. Acoust. Soc. Am.*, **105**(2), 762–9 (1999).
12. T. J. Cox and Y. W. Lam, “Evaluation of methods for predicting the scattering from simple rigid panels”, *Appl. Acoust.*, **40**, 123–40 (1993).

13. T. J. Cox, "Objective and Subjective Evaluation of Reflection and Diffusing Surfaces in Auditoria", PhD thesis, University of Salford, UK (1992).
14. L. J. Xiao, T. J. Cox, and M. R. Avis, "Active diffusers: Some prototypes and 2D measurements", *J. Sound Vib.*, **285**(1–2), 321–39 (2005).
15. MathWorks, http://uk.mathworks.com/matlabcentral/fileexchange/49561-acoustic-modelling-of-surface-scattering-from-reflectors-and-diffusers, accessed 18 September 2015.
16. J. H. Rindel, "Attenuation of sound reflections due to diffraction", *Proc. Nordic Acoustical Meeting*, 257–60 (1986).
17. L. E. Kinsler, A. R. Frey, A. B. Coppens, and J. V. Sanders, *Fundamentals of Acoustics*, 4th edn, John Wiley & Sons, New York (2000).
18. J. Redondo, R. Pico, B. Roig, and M. R. Avis, "Time domain simulation of sound diffusers using finite-difference schemes", *Acta Acust. Acust.*, **93**(4), 611–22 (2007).
19. J. Sheaffer and B. M. Fazenda, "FDTD/K-DWM simulation of 3D room acoustics on general purpose graphics hardware using compute unified device architecture (CUDA)", *Proc. IoA(UK)*, **32**(5) (2010).
20. C. J. Webb and S. Bilbao, "Computing room acoustics with CUDA-3D FDTD schemes with boundary losses and viscosity", in *Acoustics, Speech and Signal Processing (ICASSP), IEEE International Conference*, 317–20 (2011).
21. L. Savioja, "Real-time 3D finite-difference time-domain simulation of low-and mid-frequency room acoustics", *Proc. 13th DAFX*, **1** (2010).
22. J. Botts and L. Savioja, "Integrating finite difference schemes for scalar and vector wave equations", *Proc. ICASSP, IEEE International Conference*, 171–5 (2013).
23. K. S. Yee, "Numerical solution of initial boundary value problems involving Maxwell's equations in isotropic media", *IEEE T. Antenn. Propag.*, **14**, 302–7 (1966).
24. J. G. Meloney and K. E. Cummings, "Adaptation of FDTD techniques to acoustic modelling", *11th Annu. Rev. Prog. Applied Computational Electromagnetics*, **2**, 724 (1995).
25. K. Kowalczyk and M. van Walstijn, "Formulation of locally reacting surfaces in FDTD/K–DWM modelling of acoustic spaces", *Acta Acust. Acust.*, **94**(6), 891–906 (2008).
26. L. Savioja and V. Välimäki, "Reducing the dispersion error in the digital waveguide mesh using interpolation and frequency-warping techniques", *IEEE T. Acoust. Speech*, **8**(2), 184–94 (2000).
27. K. Kowalczyk and M. van Walstijn, "Room acoustics simulation using 3-D compact explicit FDTD schemes", *IEEE T. Acoust. Speech*, **19**(1), 34–46 (2011).
28. J. C. Strikwerda, *Finite Difference Schemes and Partial Differential Equations*, Siam, Philadelphia (2004).
29. S. Bilbao, *Wave and Scattering Methods for Numerical Simulation*, John Wiley & Sons (2004).
30. J. Botts and L. Savioja, "Spectral and pseudospectral properties of finite difference models used in audio and room acoustics", *IEEE/ACM T Audio Speech Lang. Process.*, **22**(9), 1403–12 (2014).
31. S. Bilbao, "Modeling of complex geometries and boundary conditions in finite difference/finite volume time domain room acoustics simulation", *IEEE/ACM T Audio Speech Lang. Process.*, **21**(7), 1524–33 (2013).
32. P. Fellinger, R. Marklein, K. J. Langenberg, and S. Klaholz, "Numerical modelling of elastic wave propagation and scattering with EFIT—Elastodynamic finite integration technique", *Wave Motion*, **21**, 47–66 (1995).
33. I. A. Drumm, "Hybrid finite element/finite difference time domain technique for modelling the acoustics of surfaces within a medium", *Acta Acust. Acust.*, **93**, 804–9 (2007).
34. X. Yuan, D. Borup, J. W. Wiskin, M. Berggren, R. Eidens, and S. Johnson, "Formulation and validation of Berenger's PML absorbing boundary for the FDTD simulation of acoustic scattering", *IEEE Trans. Ultrason. Ferroelect. Freq. Control*, **44**(4), 816–22 (1997).
35. B. Engquist and A. Majda, "Absorbing boundary conditions for numerical simulation of waves", *Proc. Natl. Acad. Sci. U.S.A.*, **74**(5), 1765–6 (1977).
36. D. T. Murphy and J. Mullen, "Digital waveguide mesh modelling of room acoustics: Improved anechoic boundaries", *Proc. DAFX*, **2**, 163–8 (2002).

37. T. Yokota, S. Sakamoto, and H. Tachibana, "Visualization of sound propagation and scattering in rooms", *Acoust. Sci. Technol.*, **23**(1), 40–6 (2002).
38. J. Sheaffer, M. van Walstijn, and B. Fazenda, "Physical and numerical constraints in source modeling for finite difference simulation of room acoustics", *J. Acoust. Soc. Am.*, **135**(1), 251–61 (2014).
39. T. Murphy, A. Southern, and L. Savioja, "Source excitation strategies for obtaining impulse responses in finite difference time domain room acoustics simulation", *Appl. Acoust.*, **82**, 6–14 (2014).
40. J. Botts and L. Savioja, "Effects of sources on time-domain finite difference models", *J. Acoust. Soc. Am.*, **136**(1), 242–7 (2014).
41. A. Taflove, *Computational Electrodyamics, The Finite-Difference Time-Domain Method*, Artech House, Norwood, MA (2005).
42. MathWorks, http://uk.mathworks.com/matlabcentral/fileexchange/53070-acoustic-fdtd-example, accessed 18 September 2015.
43. J. Saarelma and L. Savioja, "An open source finite difference time-domain solver for room acoustics using graphics processing units", *Proc. Forum Acusticum* (2014). Code available from https://github.com/juuli/ParallelFDTD.
44. J. Sheaffer and B. M. Fazenda, "WaveCloud: An open source room acoustics simulator using the finite difference time domain method", *Proc. Forum Acusticum* (2014). Code can be downloaded from https://github.com/jonessy/wavecloudm.
45. T. J. Cox, "Fast time domain modeling of surface scattering from reflectors and diffusers", *J. Acoust. Soc. Am.*, **137**(6), EL483–9 (2015).
46. J. A. Hargreaves and T. J. Cox, "A transient boundary element method model of Schroeder diffuser scattering using well mouth impedance", *J. Acoust. Soc. Am.*, **124**, 2942–51 (2008).
47. Y. Kawai and T. Terai, "A numerical method for the calculation of transient acoustic scattering from thin rigid plates", *J. Sound. Vib.*, **141**, 83–96 (1990).
48. A. A. Ergin, B. Shanker, and E. Michielssen, "Analysis of transient wave scattering from rigid bodies using a Burton–Miller approach", *J. Acoust. Soc. Am.*, **106**(5), 2396–404 (1999).
49. J. A. Hargreaves and T. J. Cox, "A transient boundary element method for acoustic scattering from mixed regular and thin rigid bodies", *Acta Acust. Acust.*, **95**(4), 678–89 (2009).
50. IMACS, http://imacs.polytechnique.fr/SONATE.htm, accessed 21 July 2015.
51. L. Banjai and M. Schanz, "Wave propagation problems treated with convolution quadrature and BEM", in U. Langer et al. (Ed.), *Fast Boundary Element Methods in Engineering and Industrial Applications*, Springer, Berlin, Heidelberg, 145–84 (2012).
52. D. G. Crighton, A. P. Dowling, J. E. Ffowcs Williams, M. Heckl, and F. G. Leppington, *Modern Methods in Analytical Acoustics, Lecture Notes*, Springer-Verlag, Berlin (1992).
53. U. P. Svensson, R. I. Fred, and J. Vanderkooy, "An analytical time domain model of edge diffraction", *J. Acoust. Soc. Am.*, **106**(5), 2331–44 (1999).
54. U. P. Svensson and P. T. Calamia, "Edge-diffraction impulse responses near specular-zone and shadow-zone boundaries", *Acta Acust. Acust.*, **92**(4), 501–12 (2006).
55. A. Asheim and U. P. Svensson, "Efficient evaluation of edge diffraction integrals using the numerical method of steepest descent", *J. Acoust. Soc. Am.*, **128**(4), 1590–7 (2010).
56. P. T. Calamia and U. P. Svensson, "Fast time–domain edge–diffraction calculations for interactive acoustic simulations", *EURASIP J. Adv. Signal Process.*, 63560 (2007).
57. H. W. Strube, "Scattering of a plane wave by a Schroeder diffuser: A mode matching approach", *J. Acoust. Soc. Am.*, **67**(2), 453–9 (1980).
58. H. W. Strube, "Diffraction by a planar, local reacting, scattering surface", *J. Acoust. Soc. Am.*, **67**(2), 460–9 (1980).
59. H. W. Strube, "More on the diffraction theory of Schroeder diffusers", *J. Acoust. Soc. Am.*, **70**(2), 633–5 (1981).
60. D. Chu and T. K. Stanton, "Application of Twersky's boss scattering theory to laboratory measurements of sound scattered by a rough surface", *J. Acoust. Soc. Am.*, **87**(4), 1557–68 (1990).
61. T. J. Cox and P. D'Antonio, "Fractal sound diffusers", *Proc. 103th Convention of the Audio Eng. Soc.*, preprint 4578, Paper K-7 (1997).

62. J. J. Embrechts, D. Archambeau, and G. B. Stan, "Determination of the scattering coefficient of random rough diffusing surfaces for room acoustics applications", *Acta Acust. Acust.*, **87**, 482–94 (2001).
63. V. Twersky, "On scattering and reflection of sound by rough surfaces", *J. Acoust. Soc. Am.*, **29**, 209–25 (1957).
64. R. R. Torres, G. Natsiopoulos, and M. Kleiner, "Room acoustics auralization with boss–model scattering, compared with a Lambert diffusion model", *Symposium on Surface Diffusion in Room Acoustics*, Liverpool (2000).

Chapter 10

Schroeder diffusers

One of the most significant occurrences in diffuser design, if not the most important event, was the invention of the *phase grating diffuser* by Schroeder.[1,2] Apart from very simple constructions, previous diffusers had not dispersed sound in a predictable manner. The Schroeder diffuser offered the possibility of producing *optimum* diffusion and also required only a small number of simple design equations. D'Antonio and Konnert[3] presented one of the most readable reviews examining the far field diffraction theory underpinning Schroeder's number theoretic surfaces; they experimentally measured performance and described applications in critical listening environments. Most crucially, they commercialized Schroeder diffusers and so made them widely available. During the four decades since the diffusers were invented, there have been many new developments, which are documented in this chapter. Some of this chapter was featured as a review article in the journal *Building Acoustics*.[4]

This chapter will start by outlining a largely qualitative view of the diffuser, how it works, and the basic design principles. Following this, a more detailed quantitative and theoretical analysis will be given. In these descriptions, the ingenuity of the original design concept should become apparent. In addition, more recent developments will be presented, illustrating weaknesses in the original design that can be overcome by modifying the design procedure. Finally, it will be shown that better phase gratings can be made using optimization.

10.1 BASIC PRINCIPLES AND CONSTRUCTION

Figure 10.1 shows the evolution of single-plane or 1D Schroeder diffusers. The simplest quadratic residue diffusor (QRD®) design was commercially introduced in 1983, is shown at the bottom of Figure 10.1, and consists of a series of wells of the same width and different depths. The wells are separated by thin dividers or fins. The depths of the wells are determined by a quadratic residue sequence based on the prime 7. In all of the photos, the zero-depth well is divided in half and placed far left and right. The diffusion coefficient response is shown to the right. Above the QRD is a photo of a diffusing fractal, referred to as a Diffractal, which was introduced in 1995. The flat wells of the QRD are replaced by scaled replicas of the original shape. This extends the diffusion response to higher frequencies, as shown in the graph. Next is a photo of a modulated, optimized, non-number theoretic diffuser, which was introduced in 2004, called a Modffusor. It has eight non-integer-related well depths, aimed at reducing the flat plate frequency where all of the wells reflect in phase. The diffusion response is compared with the QRD at the right. Finally, at the top is a photo of a modulated optimized fractal in which the flat wells of the Modffusor are replaced with a scaled replica. This diffusor was introduced in 2009 and is called a Modffractal. The Modffractal represents the highest performance reflection phase grating to date.

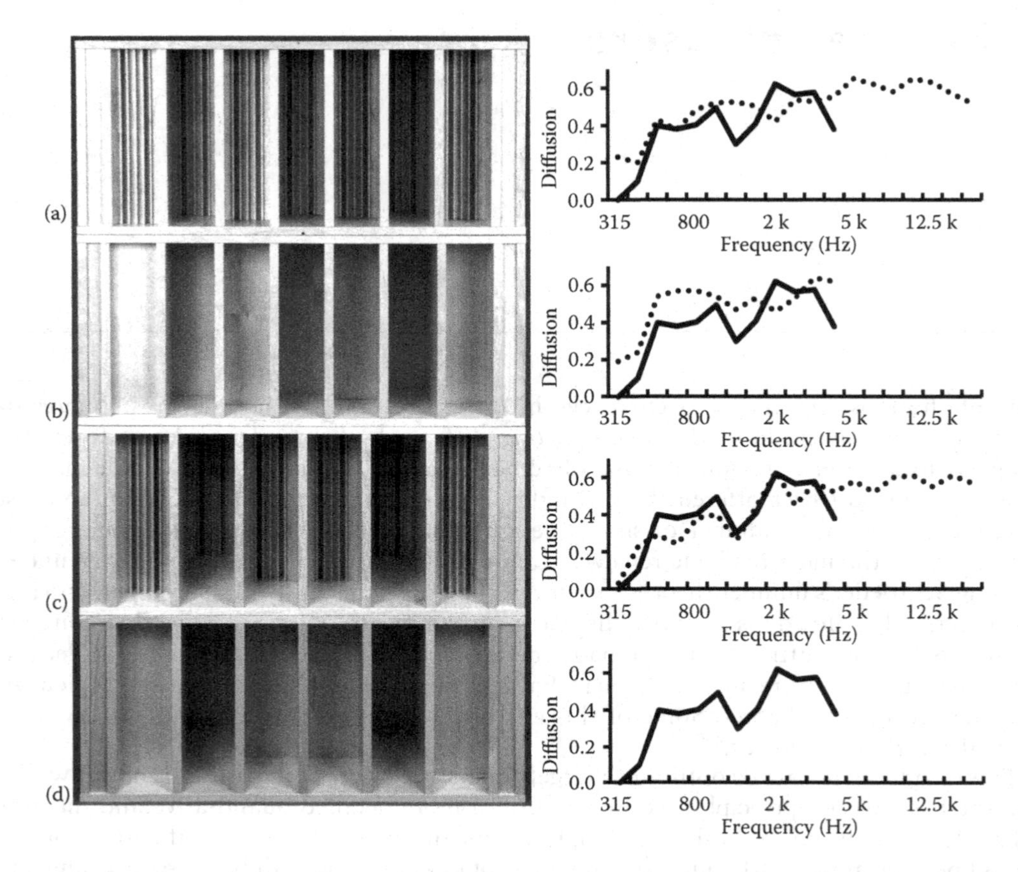

Figure 10.1 The evolution of the Schroeder diffuser. The images show the (a) Modffractal, (b) Modffusor, (c) Diffractal, and (d) QRD. The dotted lines show the normalized diffusion coefficient response for each of these, compared to the original QRD (solid line).

Single-plane diffusers cause scattering in one plane; in the other direction, the extruded nature of the surface makes it behave like a flat surface. Because of this, it is normal to just consider a cross-section through the diffuser (Figure 10.2) that contains the plane of maximum dispersion. Multiplane diffusers are possible, as shown in Figure 5.6, and are discussed later in Section 10.7.

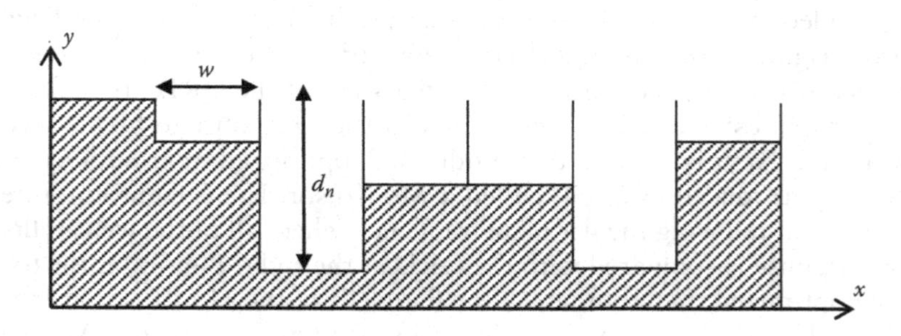

Figure 10.2 A cross-section through an $N = 7$ QRD.

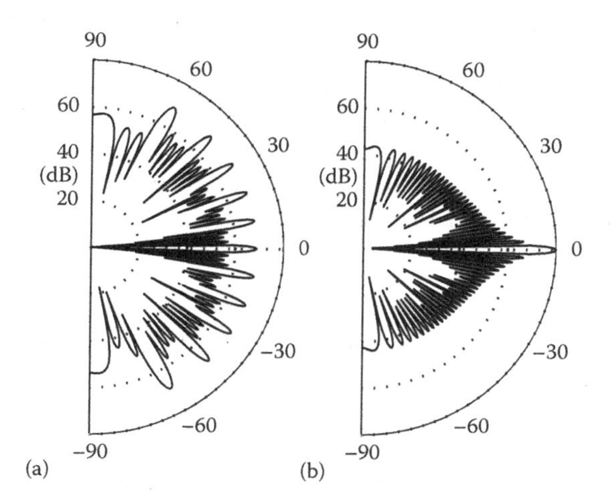

Figure 10.3 Scattered pressure level from (a) a Schroeder diffuser and (b) a plane surface of the same size.

There is a wide choice of construction materials for reflection phase gratings. Any smooth, reflective, non-diaphragmatic material is suitable. Over the years, diffusers have been manufactured from a wide variety of materials, such as wood in various species and finishes, light transmitting plastics to provide visibility, thermoformed plastics, glass-reinforced gypsum, and expanded polystyrene. Section 1.2 outlines some of the sustainability issues that need considering during manufacture and installation. To reduce absorption, it is important that any joints within the construction are well sealed. Figures 3.1, 3.12, 3.13, 3.17, 3.34, and 3.35 show some example applications.

Consider a mid-frequency plane wave incident onto the diffuser. Plane wave propagation within the wells occurs in the y direction. The plane waves are reflected from the bottom of the wells and eventually reradiate into the space. For now, it will be assumed that there is no loss of energy. The pressure at a point external to the diffuser is therefore determined by the interference between the radiating waves from each well. All these waves have the same magnitude but a different phase because of the phase change due to the time it takes the sound to go down and up each well. Consequently, the polar distribution of the reflected pressure from the whole surface is determined by the choice of well depths. Schroeder showed that by choosing a *quadratic residue sequence*, the energy reflected into each diffraction lobe direction is the same. In Figure 10.3, an example of the reflection from the surface is given, as calculated by the simplest and most approximate theory at a frequency where optimum diffusion is achieved. Eleven lobes of the same energy are found in this case. These *grating* (or *diffraction*) *lobes* are generated because the surface is periodic.

10.2 DESIGN EQUATIONS

For the design theory to be correct, plane wave propagation within the wells must dominate. Consequently, an upper frequency for the diffusion to follow the simple design principles can be found from

$$w = \lambda_{min}/2, \tag{10.1}$$

where λ_{min} is the minimum wavelength before cross-modes in the wells appear and w is the well width. For frequencies above this limit, dispersion will still occur because these are complicated structures. Consequently, this is just a limit on the applicability of a theory and not necessarily for the diffusion quality.

This requirement for plane wave propagation explains the need for fins to separate the different wells. The fins should be as narrow as possible, but not so narrow that they vibrate and cause significant losses.

A quadratic residue sequence is the most popular mathematical sequence used to form the well depths. The sequence number for the *n*th well, s_n, is given by

$$s_n = n^2 \text{ modulo } N, \tag{10.2}$$

where *modulo* indicates the least non-negative remainder and is often written as *mod* for short. N is the number generator, which in this case is also a prime and the number of wells per period. For example, one period of an $N = 7$ QRD has $s_n = \{0, 1, 4, 2, 2, 4, 1\}$.

Schroeder diffusers work at integer multiples of a *design frequency*, f_0. The design frequency is normally set as the lower frequency limit. It is more convenient to present formulations in terms of the corresponding *design wavelength*, λ_0, however. The depth, d_n, of the *n*th well is determined from the sequence via the following equation:

$$d_n = \frac{s_n \lambda_0}{2N}. \tag{10.3}$$

The well depths consequently vary between 0 and approximately $\lambda_0/2$. The design frequency is not the lowest frequency at which the diffuser produces more dispersion than a flat surface; it is just the first frequency at which energy diffraction lobes with the same energy can be achieved. It has been shown that Schroeder diffusers reflect differently from a plane surface an octave or two below the design frequency.[5,6]

10.3 SOME LIMITATIONS AND OTHER CONSIDERATIONS

Given the previous equations, it is possible to design a diffuser to a desired bandwidth. There are some important details in the design that must be heeded to achieve the best possible diffusion.

If the period width (Nw) is too narrow, then at the first design frequency, there is only one major lobe, and so this concept of grating lobes having the same energy is irrelevant. The period or repeat width is often significant in determining performance, especially when the repeat width is small. This is illustrated in Figure 10.4, where the scattering from diffusers of different period widths are shown. These are both $N = 7$ QRDs with a design frequency of 500 Hz. The well widths are 3 and 10 cm, which means that the period widths are 21 and 70 cm, respectively. The number of periods for each diffuser is set so that the overall widths of the devices are the same for a fair comparison. For the narrow wells and period width, shown right, the low-frequency limit of diffusion is determined by the period width and not by the maximum depth. This is illustrated in Figure 10.5, where the normalized diffusion coefficient versus frequency is shown. The narrow well width diffuser starts causing significant diffusion over and above the plane surface only at 1.5 kHz, which is three times the design frequency. This is roughly the frequency at which the first grating lobe appears and so is the lowest frequency where significant scattering in oblique directions is achieved. For the wide well width, the first grating lobe appears below the design frequency, and so significant diffusion is created at 500 Hz and above.

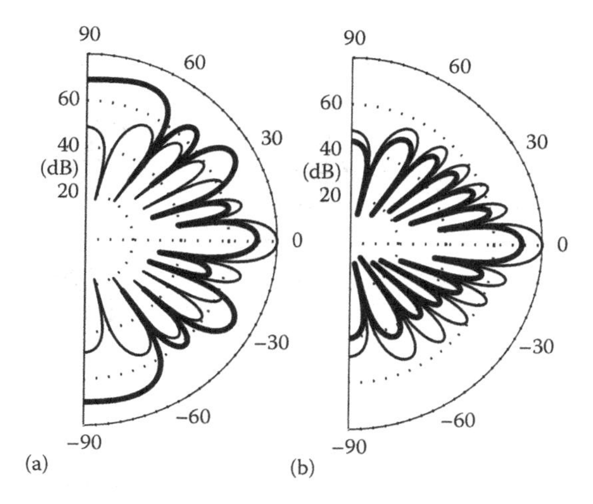

Figure 10.4 The pressure scattered from two QRDs at 1000 Hz for two well widths. (a) ▬▬ QRD w = 10 cm and (b) ▬▬ QRD w = 3 cm. On both figures, ——— is a plane surface. Overall width kept the same by changing the number of periods.

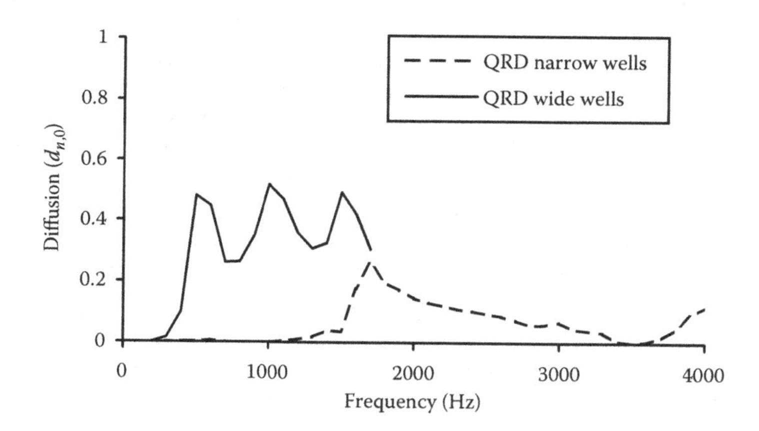

Figure 10.5 Normalized diffusion coefficient spectra for two QRDs showing that the lowest frequency at which significant diffusion occurs can be determined by period width rather than surface depth. The design frequency was 500 Hz.

For the diffuser to behave *optimally*, the device must be periodic. The lobes are generated by the periodicity of the surface. Without periodicity, all that the design equations portray is the fact that in certain directions, the scattering will have a similar level. This is illustrated in Figure 10.6 where the scattering from one and multiple periods of a diffuser is compared. The directions of similar level are marked. For the periodic cases, the directions of similar level align with the grating lobes. For the single-period case, they are just points of identical level in the polar response; the points do not align with the lobes. In this case, saying the levels are identical in some directions is meaningless because in most polar responses, there will be angles where the scattering is the same as other angles. Consequently, using one period of the device spoils the point of using the quadratic residue sequence. Using one period therefore causes problems with the mathematical make-up and definition of Schroeder diffuser. However, the scattering from a single-period diffuser

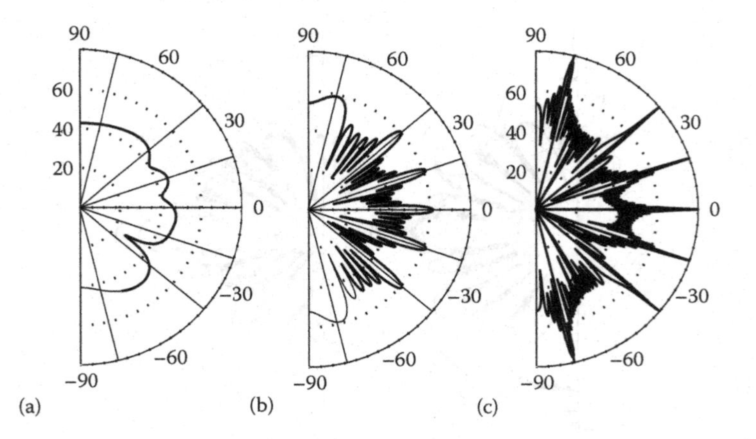

Figure 10.6 The scattering from $N = 7$ QRDs at 3000 Hz for a different number of periods. (a) 1 period; (b) 6 periods; (c) 50 periods. Locations of lobes and directions of similar level marked by radial lines at ±76°, ±40°, ±19°, and 0°.

is often more uniform than a periodic device, as Figure 10.6 shows. This issue will be returned to later in Section 10.5.

If too many periods are included, then the grating lobes become rather narrow; this leads to uneven scattering because there are large nulls present (see Figure 10.6). Manufacturing and installation constraints mean that a narrow base shape with a large number of repeats is usually cheapest, however. Furthermore, periodicity might also be preferred visually.

The points made in the last three paragraphs mean that the best design is one with a small number of periods, say five, to ensure periodicity, but with the grating lobes not too narrow. The period width must be kept large to ensure a large number of grating lobes, which then implies a large number of wells per period. Making the well width wide does not work as it risks specular reflections at high frequencies. Alternatively, modulation schemes can be used, as discussed later in Section 10.5.

From the maximum frequency calculated via Equation 10.1, it might appear as though a Schroeder diffuser should have very narrow wells to get the widest possible frequency range, but cost of manufacture and absorption need to be considered. As the diffuser wells become narrower, the viscous boundary layer becomes significant compared to the well width and the absorption increases (see Section 10.8). Consequently, practical well widths are at least 2.5 cm, and usually around 5 cm.

The choice of prime number is limited by manufacturing cost, low-frequency performance, and critical (flate plate) frequencies. For a given maximum depth, d_{max}, the design frequency achieved is

$$f_0 = \frac{s_{max}}{N} \frac{c}{2d_{max}}, \tag{10.4}$$

where s_{max} is the largest number in the quadratic residue sequence. The ratio of the largest sequence number to the prime number determines the low frequency efficiency of the device.[7] To take two examples: $N = 7$, $s_{max}/N = 4/7$; $N = 13$, $s_{max}/N = 12/13$. Consequently, an $N = 7$ diffuser will have a design frequency nearly an octave below that of an $N = 13$ diffuser (for a given maximum well depth). It is possible, however, to manipulate some sequences

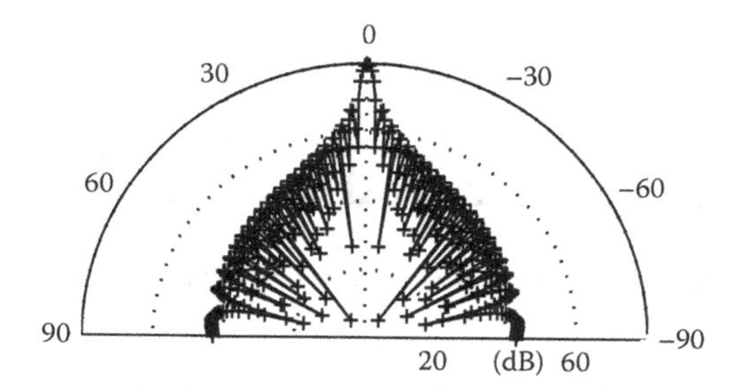

Figure 10.7 The scattering from a QRD at a critical frequency compared to a plane surface: ——— plane surface; + QRD. The data overlay each other.

and increase the bass response. A constant phase shift can be introduced to yield a better bass response:

$$s_n = (n^2 + m) \text{ modulo } N, \tag{10.5}$$

where m is an integer constant. Consider two $N = 13$ diffusers:

$$m = 0,\ s_n = \{0, 1, 4, 9, 3, 12, 10, 10, 12, 3, 9, 4, 1\},\ s_{max}/N = 12/13$$

and

$$m = 4,\ s_n = \{4, 5, 8, 0, 7, 3, 1, 1, 3, 7, 0, 8, 5\},\ s_{max}/N = 8/13.$$

Consequently, the design frequency has been lowered by two-thirds by this simple manipulation. This increased performance may not be realized, however, if the repeat width is too narrow.

For a quadratic residue diffuser, *critical frequencies* occur at mNf_0, where $m = 1, 2, 3....$ These are frequencies where the diffuser behaves like a flat surface, because all the wells reradiate in phase. This occurs when all the depths are integer multiples of half a wavelength. Figure 10.5 illustrates such a critical frequency at 3.5 kHz in the diffusion spectrum for the narrow diffuser. Figure 10.7 shows the scattering at this frequency. To avoid these critical frequencies, it is necessary to place the first critical frequency above the maximum operating frequency of the device defined by Equation 10.1, i.e.,

$$N > \frac{c}{2wf_0}. \tag{10.6}$$

10.4 SEQUENCES

10.4.1 Maximum length sequence diffuser

Schroeder began his work by investigating *maximum length sequences* (MLSs).[1] Figure 10.8 shows one period of such a surface based on the sequence {0, 0, 1, 0, 1, 1, 1}. Schroeder chose an MLS because it has a flat power spectrum at all frequencies (except DC). There is a close relationship between the power spectrum and the surface scattering; indeed, it is well

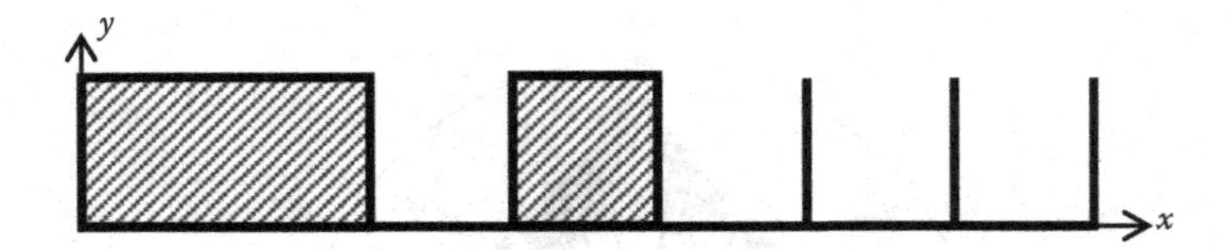

Figure 10.8 A cross-section through one period of an $N = 7$ MLS diffuser.

established in optics that the far field scattering can be found by applying a Fourier transform to the *surface*. Equation 9.32 gave the scattering in terms of the pressure magnitude $|p|$ from a surface when the Fraunhofer far field approximations are made:

$$|p(\theta,\psi)| = \left|A[\cos(\theta)+1]\int_s R(x)e^{jkx[\sin(\theta)+\sin(\psi)]}\,dx\right|, \tag{10.7}$$

where $R(x)$ is the pressure reflection coefficient, θ is the angle of reflection, A is a constant; ψ is the angle of incidence, and k is the wavenumber.

This theory is based around representing the complex diffuser shape by a simple box with a spatially varying reflection coefficient on the front face—the reflection coefficient being determined by considering plane wave propagation in the wells. This is essentially the same theory used originally by Schroeder, except for the term in $[\cos(\theta) + 1]$. This extra term is an extension that makes the optical Fraunhofer theory more applicable for oblique incidence and reflection. For convenience and compatibility with previous work, this term will be ignored:

$$|p(\theta,\psi)| \approx \left|A\int_s R(x)e^{jkx[\sin(\theta)+\sin(\psi)]}\,dx\right|. \tag{10.8}$$

Equation 10.8 can be interpreted as a Fourier transform, but the transform is in the variable x and transforms into $k[\sin(\theta) + \sin(\psi)]$ space (rather than the more familiar time to frequency transformation). If the reflection coefficients $R(x)$ are chosen to have a flat power spectrum with respect to x, then the amplitude is constant with respect to the transform variable $k[\sin(\theta) + \sin(\psi)]$. This does not relate to a constant scattering in all directions for a fixed wavenumber k, as the transform variable is not a simple function of θ and ψ; instead, even energy lobes are achieved.

If the surface is assumed to be periodic, then there will be scattering directions where *spatial aliasing* produces *grating (diffraction) lobes*. These are directions where the path length difference from the source to receiver via parts of the panel exactly one period apart is an exact multiple of a wavelength. This is illustrated in Figure 10.9, where periodicity lobes are generated when $|\mathbf{r}_1 + \mathbf{r}_2| - |\mathbf{r}_3 + \mathbf{r}_4| = m\lambda$, where m is an integer. By considering the geometry of Figure 10.9, it is possible to show that these grating lobes appear in the far field at the following angles:

$$\sin(\theta) = \frac{m\lambda}{Nw} - \sin(\psi);\ m = 0, \pm 1, \pm 2, \ldots, \tag{10.9}$$

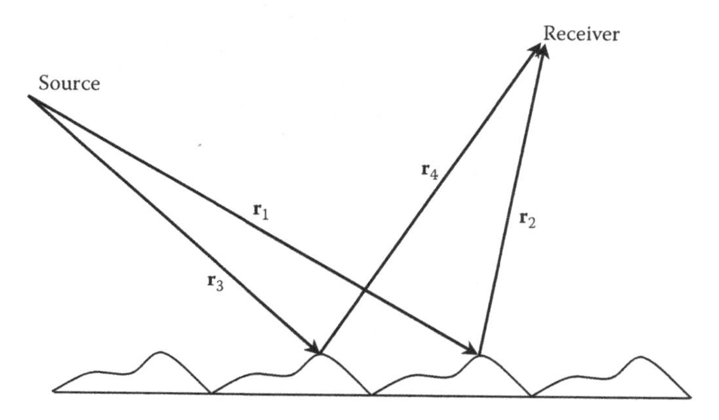

Figure 10.9 Geometry for generation of periodicity lobes.

where m is the order of the lobes. If Equation 10.9 is substituted into Equation 10.8, the following results at the design frequency:

$$\left|p_m\right| \approx \left|A\int_s R(x)e^{j2\pi xm/Nw}\,dx\right| = \left|A\sum_{n=1}^{N} R_n e^{j2\pi nm/N}\right|, \tag{10.10}$$

where it is assumed that each well radiates as a point source. Strictly speaking, a sinc function should be introduced to allow for the pistonic radiation, but for now, the wells will be considered to be relatively narrow compared to wavelength, $w \le \lambda/4$. Consider a seven-well design. A length 7 MLS is {1, 1, 0, 1, 0, 0, 0}, so R_n values are {1, 1, −1, 1, −1, −1, −1}. In this case, it can be shown that

$$\begin{aligned} \left|p_m\right| &= A \qquad m = 0,\ \pm N,\ \pm 2N \\ &= A\sqrt{N+1} \qquad \text{otherwise.} \end{aligned} \tag{10.11}$$

In other words, the grating lobes radiating into the far field ($|m| > 0$ and $|m| < N$) have the same level, whereas the main zeroth order lobe ($m = 0$) is lower by $10\log_{10}(N + 1)$. Figure 10.10 shows the scattering from the MLS diffuser where the depth is a quarter of

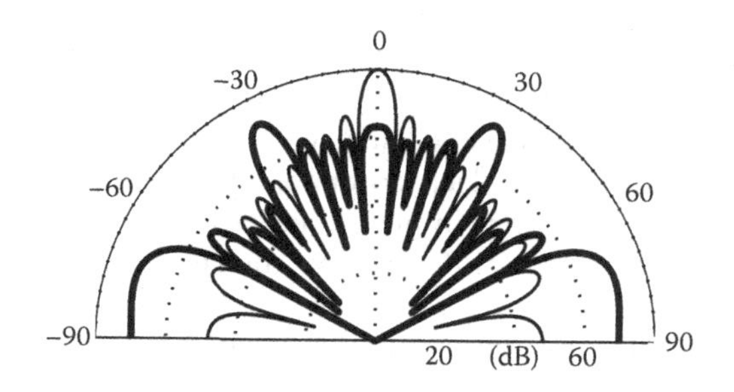

Figure 10.10 The scattering from five periods of an $N = 7$ MLS diffuser at its design frequency: ——— MLS diffuser and ——— plane surface.

the wavelength, compared to a hard plane surface of the same size. At this frequency, there are five lobes, with the central lobe being suppressed by $10\log_{10}(8) = 9$ dB as expected. At an octave higher, however, when the depth is half a wavelength, the surface behaves like a flat plate because all waves reradiate with the same phase—this is a critical frequency—the scattering will be more like that shown in Figure 10.7. Consequently, the MLS diffuser is useful only over an octave. This problem can be overcome, however, by using different number sequences.

10.4.2 Quadratic residue sequence

When a quadratic residue sequence is used, the grating lobe pressure amplitudes are given by

$$|p_m| \propto \sqrt{N};\ m = 0, \pm 1, \pm 2, \ldots \quad (10.12)$$

Consequently, all lobes will have the same energy, as has been shown in previous polar responses such as Figure 10.3.

10.4.3 Primitive root sequence

A primitive root sequence is generated using the function

$$s_n = r^n \bmod N;\ n = 1, 2, \ldots N - 1, \quad (10.13)$$

where N is an odd prime and r is a *primitive root* of N, and the diffuser will have $N - 1$ wells per period. A primitive root is one where s_n values for $n = 1, 2, \ldots N - 1$ are all unique.[8] For example, $N = 7$ has a primitive root of 3, so $s_n = \{3, 2, 6, 4, 5, 1\}$, which generates every integer from 1 to $N - 1$. Primitive roots can be found by a process of trial and error; alternatively, tables can be found in texts, such as Reference 8. Equation 10.13 can be rewritten as a recursive relationship:

$$s_n = (r \cdot s_{n-1}) \bmod N. \quad (10.14)$$

This form is useful because Equation 10.13 can cause overflow problems when being computed for large N.

The *primitive root diffuser* (PRD) is meant to reduce the energy reflected in the specular reflection direction and so produce a notch diffuser. In addition, it should have even energy for the other diffraction lobes. As with the QRD, the PRD achieves these performance criteria at integer multiples of the design frequency. At these frequencies, the specular direction amplitudes from the PRD are attenuated by $20\log_{10}(N)$ in comparison with a flat surface. It is noted, however, that virtually any welled surface will achieve a reduced specular energy provided that wells are significantly deep compared to wavelength. Equation 10.10 is a maximum when all radiating waves are in phase, as is the case for a flat surface. As soon as a depth sequence is introduced, partial destructive interference occurs between the waves, leading to a suppressed specular reflection.

The performance of the PRD in suppressing the specular reflection improves as the prime number, N, increases. This is shown in Figure 10.11, where the pressures scattered from two

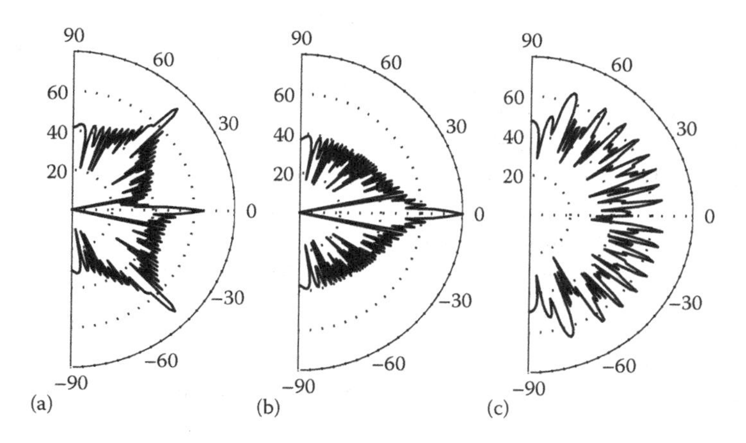

Figure 10.11 Scattering from two PRDs and a plane surface for normal incidence, showing that a large number (N) is required to get a significant notch in the specular reflection direction (0°). (a) N = 7 PRD; (b) plane surface; (c) N = 37 PRD.

PRDs are compared to a plane surface. A large number of wells, say greater than 20–30, are needed before a pressure minimum appears at the specular reflection angle.

The pressure amplitude of the PRD lobes can mathematically be expressed as

$$\begin{aligned} |p_m| &= A \qquad m=0,\ \pm N,\ \pm 2N \\ &= A\sqrt{N} \qquad \text{otherwise.} \end{aligned} \tag{10.15}$$

Note that although there is an implication of a series of suppressed lobes for $m = \pm N$, $\pm 2N$,... these are not seen in the far field. The frequencies at which the high-order suppressed modes occur will always be greater than the cut off frequency for plane wave propagation in the wells and so can be ignored.

The specular reflection is attenuated, but it is not a pressure null at integer multiples of the design frequency. Feldman[9] developed a modified primitive root sequence to overcome this issue. The Feldman modified PRD contains an extra zero-depth well so the sequence contains all integers from 0 to $N - 1$ (instead of from 1 to $N - 1$). This spaces the reflection coefficients evenly around the unit circle, on an argand diagram, for multiples of the design frequency, leading to an exact null in the specular reflection direction. This modification will, however, alter the evenness of the non-zero-order lobes.

A PRD does achieve nulls, but not at integer multiples of the design frequency. The nulls appear at frequencies given by $mNf_0/(N - 1)$, where $m = 1, 2, 3,...$ but it is at the integer multiples that the non-zero-order lobes have the same energy. This realization led to Cox and D'Antonio[10] devising a revised formulation for notch diffusers. The technique is to introduce an effective frequency shift to align the reflection coefficients appropriately around the unit circle in an argand diagram, at multiples of the design frequency to achieve nulls. This is done by rewriting Equation 10.3 as

$$d_n = \frac{s_n \lambda_0}{2(N-1)}\,. \tag{10.16}$$

This will be referred to as the Cox and D'Antonio modified PRD. Figure 10.12 illustrates the two modified PRDs compared to a PRD and a plane surface. This demonstrates the nulls that the modified schemes achieve. Also shown is a notch filter designed through optimization, a subject that will be returned to in Section 10.10.

It is important to reiterate that these notches are produced only at discrete frequencies. Figure 10.13 shows the scattering from a modified PRD, but not at an integer multiple of the design frequency. No notch is found. While not achieving optimum scattering from a QRD at all frequencies is disappointing, it can be expected that between the frequencies of optimum diffusion, the dispersion from a QRD will still be reasonable. The fact that PRDs work only at discrete frequencies, however, means the PRDs are impractical notch diffusers. This problem can be overcome to a certain extent by optimization,[10] to form a broader notch over a wider frequency range. Alternatively, triangles or pyramids may be used to get a more broadband notch, as discussed in Section 11.2, but then there are restrictions on the angle of incidence.

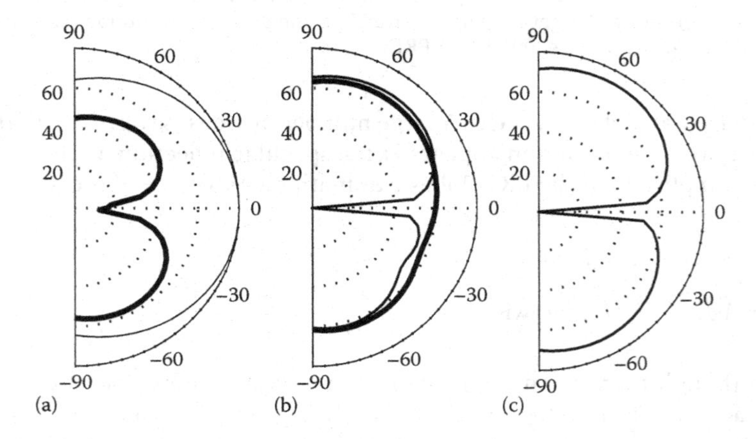

Figure 10.12 Scattering from a PRD; Feldman modified PRD (FMPRD); Cox–D'Antonio-modified PRD (CDMPRD); plane surface, and optimized surface. $N = 11$. One period, $w = 5$ cm, normal incidence source, at the design frequency of 500 Hz. (a) ▬▬ optimized and ——— plane; (b) ▬▬ PRD and ——— FMPRD; (c) ——— CDMPRD. (Data from Cox, T.J., D'Antonio, P., *Appl. Acoust.*, 60, 167–186, 2000.)

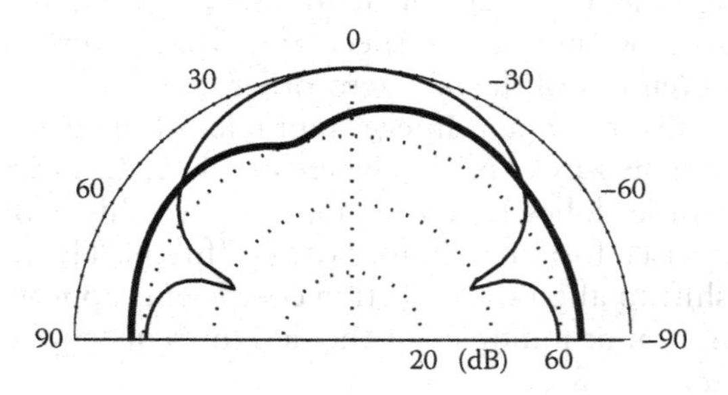

Figure 10.13 Scattering from a modified PRD and a plane surface. Design parameters given in Figure 10.12. At 750 Hz; not an integer multiple of the design frequency. ▬▬ modified PRD and ——— plane surface.

10.4.4 Index sequences

Schroeder[11] formed a complex Legendre sequence based on the index function. This has the following reflection coefficients:

$$R_n = \begin{cases} 0 \text{ for } n \equiv 0 \bmod N \\ e^{2\pi j s_n/(N-1)}, \end{cases} \tag{10.17}$$

where s_n is the number theoretic logarithm or index function defined by

$$r^{s_n} \bmod N = n;\ n = 1,2,\ldots,N-1, \tag{10.18}$$

where r is a primitive root of N. For example, the $N = 7$, $r = 3$ sequence is {6, 2, 1, 4, 5, 3} as $3^6 = 1 \bmod 7$, $3^2 = 2 \bmod 7$, etc. Finding a sequence for a given value of N requires a certain amount of trial and error. As the reflection coefficient for the $n = 0$ well is 0, this well should be filled with absorbent. Consequently, the diffuser absorbs a nominal $20\log_{10}(N - 1)$ amount of energy. The other wells are like other Schroeder diffusers. Apart from the absorption, the performance of the sequences should be very similar to the PRD.

10.4.5 Other sequences

According to the *Wiener–Khinchine theorem*, the Fourier transform of an autocorrelation function gives the auto power spectrum. This can be related to diffusers and enable the use of other sequences to be understood. The Fourier transform of the surface reflection coefficients approximates to the scattered pressure distribution, although strictly speaking this is in $k[\sin(\theta) + \sin(\psi)]$ space. Applying the Wiener–Khinchine theorem to this, if a Fourier transform is applied to the autocorrelation of the surface reflection coefficients, the scattered energy distribution should result. Consequently, a good diffuser is one whose autocorrelation function for the reflection coefficients is a delta function. This then leads to an even scattered energy distribution. (Although constant with $k[\sin(\theta) + \sin(\psi)]$, which is not the same as being constant with θ and ψ.)

To demonstrate this, a familiar diffuser sequence can be considered. In Figure 10.14, the autocorrelation function for an $N = 13$ quadratic residue sequence is compared to that for a plane surface. It can be seen that the quadratic residue sequence has good autocorrelation properties with small side lobe energy; in other words, the autocorrelation for index $\neq 0$ is small. This is one reason that a quadratic residue sequence makes a good diffuser.

Another way of viewing this is as follows. Peaks in an autocorrelation function away from zero indicate a sequence that has some similarity at some displacement. In terms of scattering, there will be angles at which this similarity will produce lobes due to constructive interference. If all similarities can be removed, then in all directions, no complete constructive interference can take place, and so the scattering in all directions will be the same.

Given this scenario, one approach to finding an appropriate sequence is to look for sequences with good autocorrelation properties. This is not difficult, as sequences with optimal autocorrelation properties are vital to digital communication systems, whether that is error checking systems for digital audio, code division multiple access systems used in mobile telecommunications, or modulating waveforms for radar and sonar.

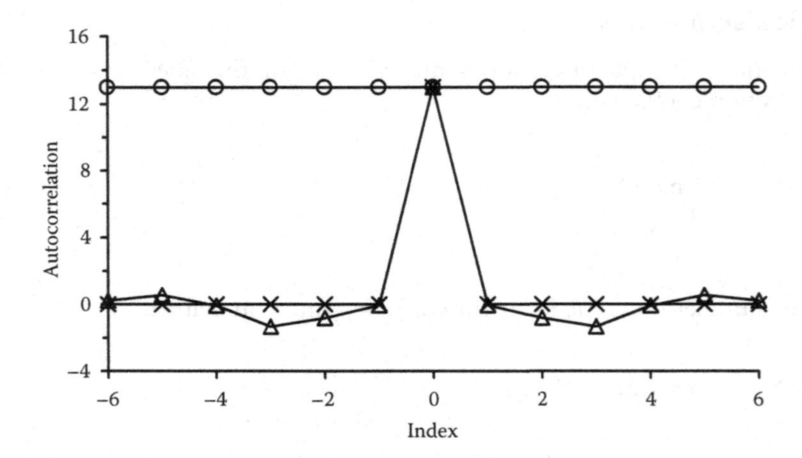

Figure 10.14 The autocorrelation for three different sequences: —△— quadratic residue, —×— Chu (perfect), and —○— constant (plane surface).

Dadiotis et al.[12] examined Lüke and power residue sequences, which both have good autocorrelation properties. In addition, these number sequences offer the opportunity to reduce problems due to critical frequencies where the diffuser radiates like a flat plate. Critical frequencies arise because there is a simple integer relationship between the different well depths. A way of mitigating this problem is to introduce an integer-based sequence, which, although having a small number of wells per period, is generated using a larger integer. For instance, short power residue sequences can be formed by undersampling longer primitive root sequences provided that certain rules are followed. Consider a primitive root sequence based on prime 73; this will be of length 72. By taking every ninth sample from this sequence, a shorter-length 8 sequence is formed. Although this power residue sequence displays slightly worse autocorrelation properties than the primitive root sequences do, the flat plate frequency will be nine times greater.

Another approach is to use an optimization algorithm to find sequences with good autocorrelation properties.[8] The principle of optimization will be discussed in more detail in Section 10.10, but the basic idea is to get a computer to search for a sequence with minimum side lobe energy. This works well for a small number of wells, but when the number of wells becomes large, the optimization is less effective.

The Chu sequence is a perfect polyphase sequence; in other words, the periodic autocorrelation function is a perfect delta function. Figure 10.14 shows the autocorrelation function for an $N = 13$ case showing that there is no side lobe energy. The elements of a Chu sequence can be generated by[8]

$$s_n = e^{j\phi_n}$$

$$\phi_n = \begin{cases} \dfrac{2\pi}{N}\left[\dfrac{1}{2}(n+1)n \mod N\right] & N \text{ odd} \\ \dfrac{2\pi}{N}\left[\dfrac{1}{2}n^2 \mod N\right] & N \text{ even} \end{cases} \tag{10.19}$$

$$0 \le n < N.$$

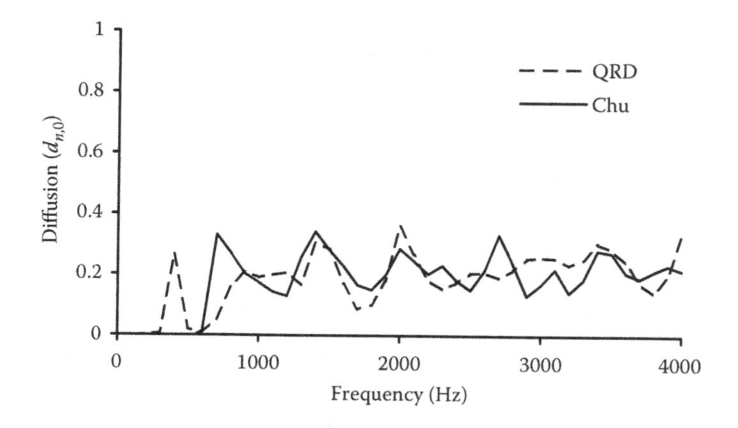

Figure 10.15 Normalized diffusion spectra for two diffusers.

The phase terms ϕ_n are converted to depths by equating the deepest depth to the design wavelength and maximum phase term, ϕ_{max}, i.e.,

$$d = \frac{\lambda_0 \phi_n}{2\phi_{max}}. \tag{10.20}$$

Figure 10.15 compares the normalized diffusion coefficient from an $N = 13$ Chu sequence and a QRD. Both diffusers have similar performance. Consequently, this presents an alternative design method, but not a better one.

10.5 PERIODICITY AND MODULATION

The scattered polar responses seen in Figure 10.6 are dominated by grating lobes generated because the diffusers are arranged in a periodic formation. The lobe energy may be constant, but there are broad minima between the lobes, except at high frequencies when the number of lobes becomes very large. The scattered energy is not even in all directions. For this reason, significantly better performance can be obtained if grating lobes can be removed by making the diffuser *aperiodic*. Alternatively, increasing the repeat distance will create more lobes at mid–high frequency. The small number of studies on the subjective effects of periodicity have all found that periodicity can cause audible coloration.[13] A phase grating diffuser that exploits number theory, such as a QRD, is in many ways cursed by periodicity. A QRD needs periodicity to form even energy grating lobes, yet these lobes cause uneven scattering.

One possibility is to use a single, long number sequence with good aperiodic autocorrelation properties to form a diffuser, giving just a single period without repetition. There are two problems with this solution: first, there are not many large aperiodic, polyphase sequences known, and second, it will usually be cheaper to manufacture a small number of base shapes and use each of these many times.

Angus[14–19] presented a series of papers outlining methods to deal with the problems of periodicity. She used two Schroeder diffuser types in a modulation scheme. Figure 10.16 shows such an arrangement for two QRDs, one based on $N = 7$, the other on $N = 5$. The

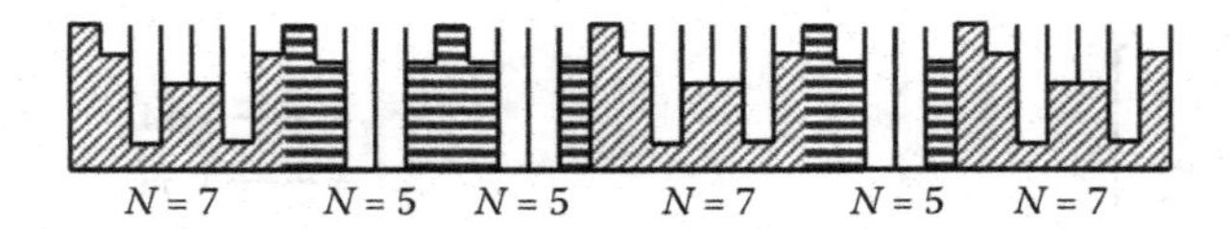

Figure 10.16 A cross-section through a modulation scheme using $N = 5$ and $N = 7$ quadratic residue diffusers and the modulation sequence {1, 0, 0, 1, 0, 1}.

idea is to use two or more diffuser types and arrange them according to a pseudorandom sequence so there is no repetition.

As discussed previously, the far field scattering is roughly given by the Fourier transform of the surface reflection coefficients. For a periodic device, the distribution of reflection coefficients can be expressed as the reflection coefficients over one period, convolved with a series of delta functions:

$$R(x) = R_1(x) * \sum_{n=-\infty}^{n=\infty} \delta(x - nW), \tag{10.21}$$

where $R_1(x)$ is the distribution of reflection coefficients over one period, n is an integer, * denotes convolution, $W = Nw$ is the width of one period of the device, and δ is the delta function.

Equation 10.21 and the following process are illustrated in Figure 10.17. When a Fourier transform is applied to Equation 10.21 to obtain the scattering in $k[\sin(\theta) + \sin(\psi)]$ space, the convolution becomes multiplication:

$$FT\{R(x)\} = FT\{R_1(x)\} \cdot FT\left\{\sum_{n=-\infty}^{n=\infty} \delta(x - nW)\right\}, \tag{10.22}$$

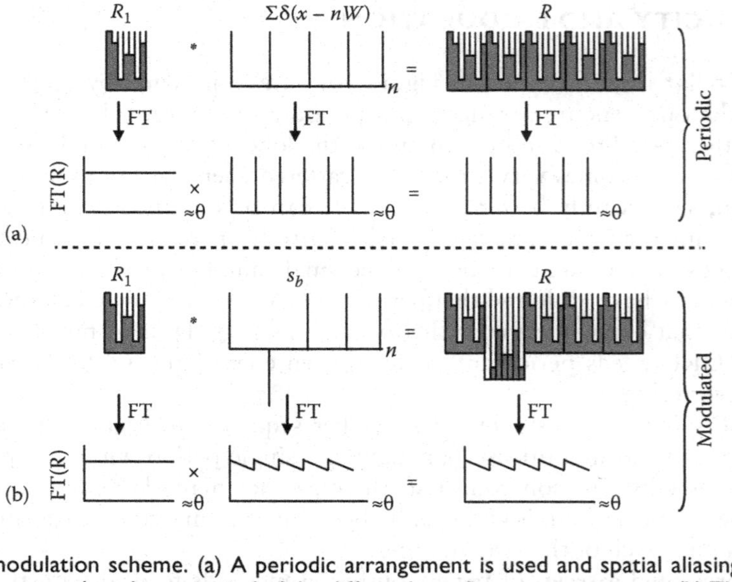

Figure 10.17 A modulation scheme. (a) A periodic arrangement is used and spatial aliasing causes grating lobes even though one period of the diffuser has a flat power spectrum. (b) The inverse of the diffuser is used in a modulation scheme to reduce periodicity effects and to get closer to the ideal flat power spectrum from an array of diffusers.

where $FT\{\}$ denotes a Fourier transform. The top two rows of Figure 10.17 illustrate Equation 10.22. The Fourier transform of a delta function series is another delta function series, and it is the spikes in this for $k[\sin(\theta) + \sin(\psi)] > 0$ that cause the grating lobes. Consequently, rather than use a delta function series to form a periodic device, another function should be used that has better Fourier transform properties. Again, what is needed is a sequence, s_b, with good autocorrelation properties. A good choice is a Barker sequence. This is a binary sequence with the flattest possible aperiodic Fourier Transform. How this could be used to form an array of diffusers is shown in the bottom two rows of Figure 10.17, where the –1 in the sequence is represented by a different Schroeder diffuser (that happens to be deeper). The response of the whole array of diffusers is now closer to the single diffuser alone than if a periodic arrangement is used. If a perfect binary sequence could be found, then the single diffuser response would be recovered, but there are no such 1D sequences.

Consider forming a QRD with five periods. The Barker sequence for $N = 5$ is {1, –1, 1, 1, 1}. Consequently, where a 1 appears in the Barker sequence, the normal $N = 7$ QRD should appear. Where a –1 appears, an $N = 7$ QRD is needed that produces the same reflection magnitude but with opposite phase. This can be done by using the rear of the normal $N = 7$ diffuser (provided the fins are extended far enough). Consequently, one QRD has a number sequence of {0, 1, 4, 2, 2, 4, 1} and the other QRD has a number sequence of {7, 6, 3, 5, 5, 3, 6}. This second sequence is found by subtracting the first sequence from N. This is equivalent to changing the phase change due to the well depths from ϕ to $2\pi - \phi$, i.e., obtaining a 180° out-of-phase surface.

Figures 10.18 and 10.19 show the scattering from the periodic arrangement of $N = 7$ QRDs compared to an arrangement according to the Barker sequence and a single diffuser for two frequencies. One of the frequencies where the diffusion improvement is most dramatic is 2 kHz; at other frequencies, the improvement is less marked. Figure 10.20 shows the normalized diffusion coefficient versus frequency. A clear improvement is seen in the diffusion, and grating lobes are much reduced. As Figure 10.20 shows, the diffusion from the periodic array only becomes significant compared to a plane surface at 1–1.5 kHz, an octave or so above the design frequency (500 Hz). This is a case of the diffuser width, rather than the diffuser depth, limiting the low-frequency response. At the design frequency, only

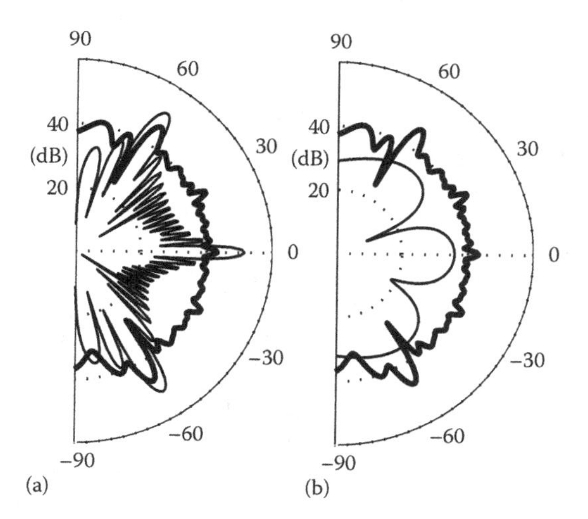

Figure 10.18 Scattered polar distribution from a single QRD, a periodic arrangement, and a Barker modulated array using the QRD and its inverse. 2000 Hz = $4f_0$. (a) ——— modulated and ——— periodic; (b) ——— modulated (same as [a]) and ——— one period.

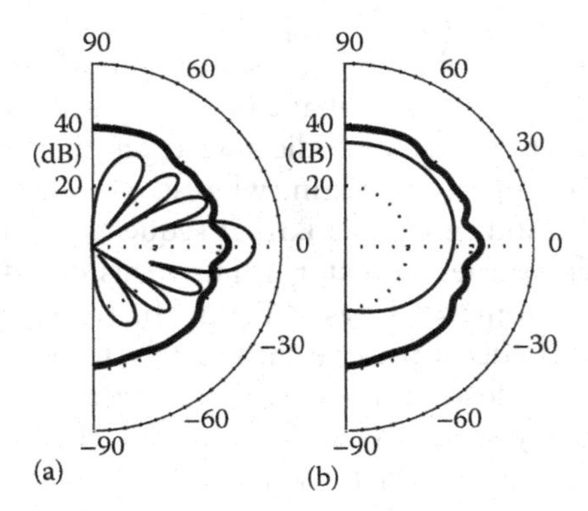

Figure 10.19 Same diffusers as Figure 10.18 but at 500 Hz, the design frequency. (a) ——— modulated and ——— periodic; (b) ——— modulated (same as [a]) and ——— one period.

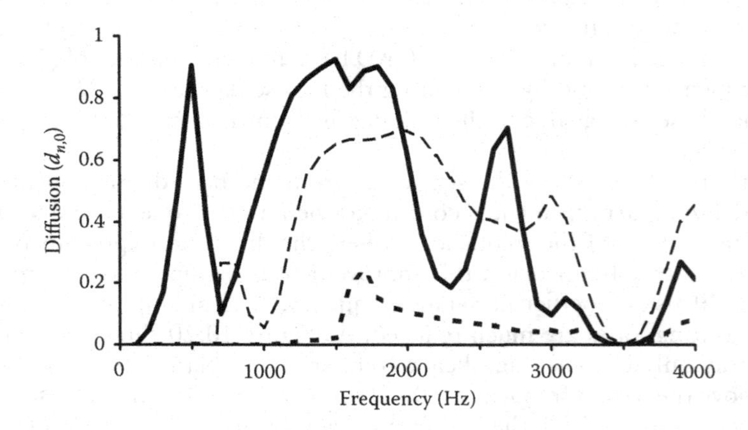

Figure 10.20 Normalized diffusion spectra for three diffuser arrangements: – – – – one narrow period of a QRD, — — — QRD periodic (a wide array), and ——— QRD Barker modulation (a wide array).

one lobe appears in the scattered polar distribution, as shown in Figure 10.19. The Barker sequence means that there is reduced periodicity in the arrangement, and so the low-frequency limit of the Barker modulated array is determined by the depth and not the periodicity. This is an important result, as it means that the low-frequency performance of some diffusers can be improved by modulation.

The polar responses show that the scattering from the Barker modulated array is more similar to the single diffuser response, which is as expected from the theory outlined previously. A single diffuser response is not recovered because the Barker sequence has good but not perfect autocorrelation properties. Figure 10.21 shows the autocorrelation properties for the diffuser arrays at the design frequency. The spikes for $n \neq 0$ arise because of the repeat distance of 7 for the $N = 7$ diffusers. The Barker modulated array has lower side lobe energy, which means the periodicity is reduced. Although the Barker sequence has reduced the side lobe spikes, they are not completely eliminated. This is why the single diffuser response is not completely recovered with this modulation.

There are a variety of number sequences that can be used for modulation. The Barker sequence is a good choice as it has good aperiodic autocorrelation properties. As the

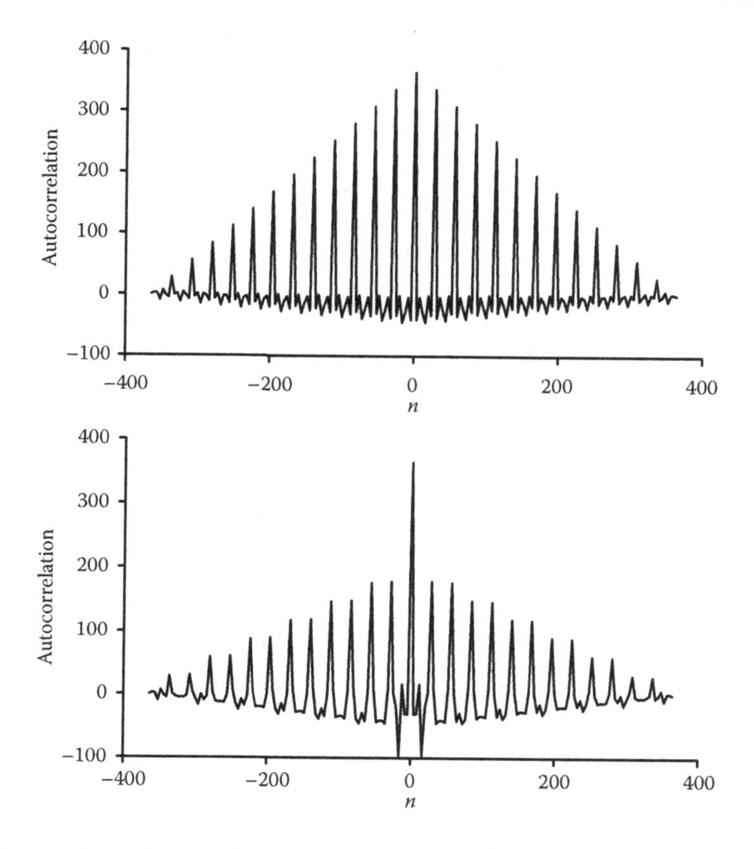

Figure 10.21 Autocorrelation function for two arrangements of QRDs. Top graph: periodic array; bottom graph: Barker modulated array.

modulation sequence does not repeat, a sequence with good periodic autocorrelation coefficient would not be optimal. Barker sequences analytically exist only for certain lengths: 2, 3, 4, 5, 7, 11, and 13, but computer-based search algorithms have been used to generate number sequences up to length 60.[8,20] For larger diffuser arrays, it may be necessary to use optimal periodic sequences such as MLS.

The modulation works best at multiples of the design frequency. Only at these frequencies do the diffuser and its inverse create exactly opposite pressures. At other frequencies, the modulation is likely to help with the scattering as it breaks up the grating lobes, but in a more uncontrolled manner.

At high frequencies, say greater than 5 kHz for many common diffusers, the dispersion by a modulated array summed over a one-third octave band is often worse than for a periodic arrangement. This happens because the number of grating lobes in the periodic case saturates, so when the polar responses are summed over a one-third octave band, the grating lobes average out. Having said this, the improvements generated by modulation in the more important low–mid frequency ranges far outweigh any slight decrease in performance at higher frequencies.

One other feature of note is that the critical frequency at Nf_0, where the diffusers behave as a flat plate, still remains even with the modulation. The flat plate frequency of 3500 Hz can be seen in the diffusion spectra of Figure 10.20. Both the QRD and its inverse have the same critical frequency, and consequently, this problem persists. One way to solve this is to use two diffusers with different critical frequencies in the modulation scheme. There are further advantages to using two different diffusers. The QRD works at discrete frequencies based on integer multiples of the design frequency. By using diffusers with two different design frequencies, it is possible to achieve more frequencies with better diffusion.[16]

Cox and D'Antonio[10] used a combination of $N = 11$ and $N = 7$ PRDs. Figure 10.22 shows the scattering from this arrangement. Not only does the modulated array still achieve a notch in the specular reflection direction, but also, the two dominant first-order lobes are broadened. The notch remains because each period of both diffuser types produces a null in the specular direction, and summing over all periods still leads to nothing scattered in the specular direction.

Angus[19] used a combination of $N = 5$ and $N = 7$ quadratic residue diffusers in an orthogonal modulation; Figure 10.23 shows the autocorrelation of the reflection coefficients at the

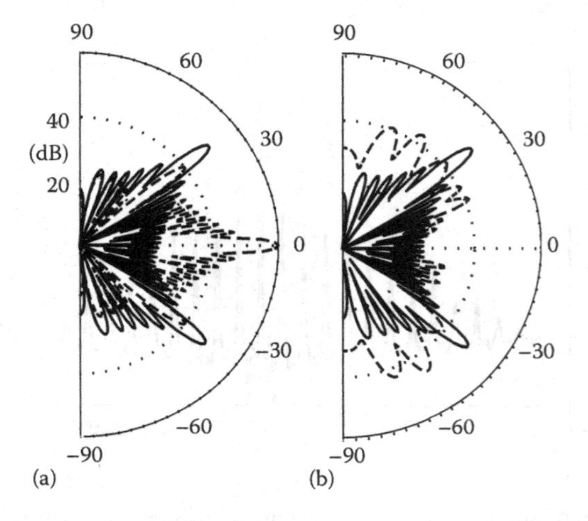

Figure 10.22 Comparison of a periodic arrangement of a modified PRD (CDMPRD); a modulated arrangement of two CDMPRDs and a plane surface. The periodic surface had 10 periods of $N = 11$ diffusers. $w = 0.05$ m, frequency is 1 kHz, and design frequency is 500 Hz. The modulated surface was formed from 12 periods of $N = 11$ and $N = 7$ placed in a random order. (a) ——— periodic and – – – – plane; (b) ——— periodic (same as [a]) and – – – – modulated. (Data from Cox, T.J., D'Antonio, P., *Appl. Acoust.*, 60, 167–186, 2000.)

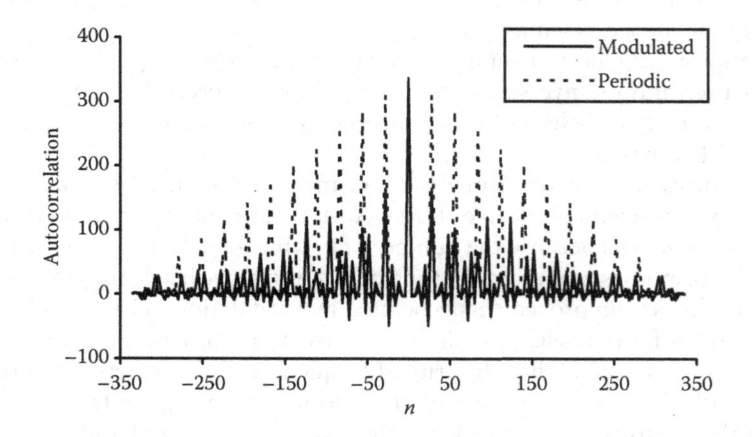

Figure 10.23 Autocorrelation function for a periodic arrangement of $N = 7$ QRDs and an orthogonal modulation using $N = 5$ and $N = 7$ QRDs and a Barker sequence.

design frequency for such an arrangement. It shows that the original periodicity lobes are reduced, but other smaller peaks are produced elsewhere.

The locations of the two diffuser types can be determined by flipping a coin, or better still by using a pseudorandom number sequence with good aperiodic autocorrelation properties, such as a Barker sequence. Figure 10.24 shows the normalized diffusion spectra. The effects of the flat plate frequency at 3.5 kHz are reduced by the modulation. Overall, the performance is not as good as the original Barker modulation using the QRD and its inverse described previously. Consequently, the best choice for modulation is to use a diffuser and its inverse, but choosing a diffuser with the critical frequency above the design bandwidth.

It is possible to achieve cheaper modulation using a single asymmetrical base shape. For example, an $N = 7$ PRD is {1, 3, 2, 6, 4, 5} can be rotated by 180° to form a new sequence {5, 4, 6, 2, 3, 1}. This modulation will not be successful with quadratic residue or Chu sequences, however, because these are symmetrical. Instead, asymmetrical sequences are needed such as primitive root or power residue sequences. Figure 10.25 shows the scattering from such a modulation using an $N = 7$ primitive root sequences. The modulation reduces the grating lobes at ±53°, while still preserving the null in the specular reflection direction.

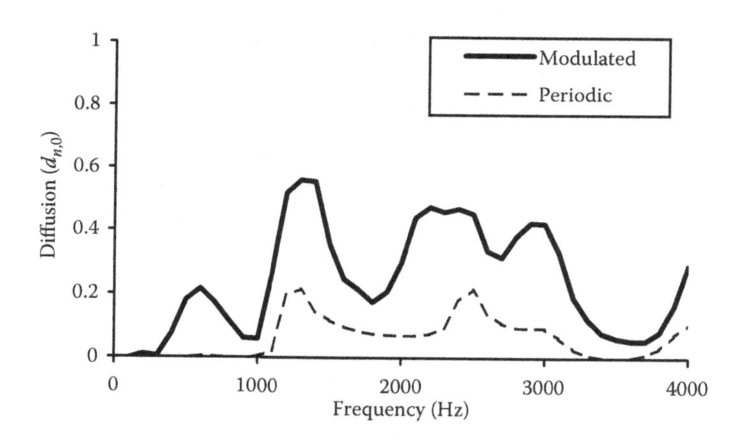

Figure 10.24 Normalized diffusion spectra for a periodic arrangement of $N = 7$ QRDs and an orthogonal modulation using $N = 5$ and $N = 7$ QRDs.

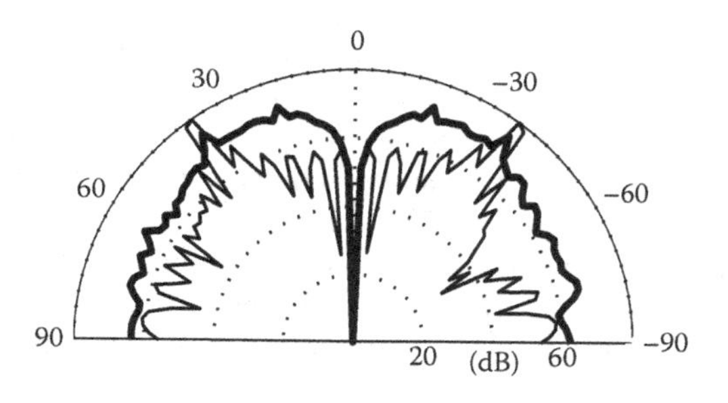

Figure 10.25 Scattering from a periodic and a modulated arrangement of PRDs at 2.5 kHz: ——— periodic and ━━━ modulated using asymmetry.

10.5.1 Fractal

There is another approach to modulation to reduce periodicity effects that also improves the low-frequency response. The bandwidth of a Schroeder diffuser is limited at high frequencies by the well width and at low frequencies by the maximum depth. To provide full spectrum sound diffusion in a single integrated diffuser, the self-similarity property of fractals can be combined with the uniform scattering property of Schroeder diffusers to produce a fractal diffuser. The surface consists of nested, scaled, self-similar diffusers, each of which covers a specific frequency range and offers wide area coverage (see Figure 10.26). Figure 3.12 showed an application of this device at the rear of a studio. Each diffuser provides uniform scattering over a specific range of frequencies so that the effective bandwidth is extended.

There are numerous natural phenomena exhibiting a macroscopic shape, which is repeated microscopically at progressively smaller and smaller scales. At each level of magnification, we find a scaled replica of the original is found. These scaled replications are *self-similar*; that is, they differ only in scale. The term *fractal* was first coined by Mandelbrot[21] to describe these structures, and hence, these diffusing fractals have been termed *Diffractals*. It is possible to carry out the scaling many times; typical commercial implementations use two magnifications of self-similarity. Another analogy to this is a two-way coaxial loudspeaker system.

The construction avoids using narrow deep wells to cover a wide bandwidth, and so decreases the absorption of the device. At low frequency, the small diffusers at the bottom of the wells have negligible depth compared to wavelength, and so only the bass diffuser needs to be considered. At high frequencies, the small diffusers act as a modulated array with phase modulation due to the depth of the deep diffuser. The phases introduced by the large diffuser on the small diffuser are conveniently also quadratic residues, and hence, when they sum with the small diffuser phases, the result is still a quadratic residue sequence. It is possible, therefore, to nest diffusers with overlapping frequency bandwidths.

Although the two layers of magnification appear to operate orthogonally, there is also an overlap region at mid frequencies where the situation is more complex. The cross-modes of the large wells will affect the scattering from the smaller diffusers. These difficulties make this surface impossible to model with a simple Fraunhofer approach; properly modelling requires a boundary element method (BEM) or finite difference time domain solution.

10.5.2 Diffusing covers

Placing a line of cylinders or spheres in front of a Schroeder diffuser can remove periodicity and so provide more scattering.[22] Figure 10.27 shows an example of where cylinders are placed in front of a 1D quadratic residue diffuser. To reduce periodicity effects, the cylinders should be arranged with spacing different from the period width of the Schroeder diffuser. Improvements to the bass response can also be achieved if the cylinders are large enough to scatter sufficient power at low frequencies. For maximum performance, the cylinders should have different sizes and their position determined by an optimisation algorithm (see Section 10.10). But even a simple line of evenly spaced cylinders, all the same size, can improve performance. This could also be more visually appealing, for example, with appropriate lighting, the array of cylinders could be used to hide the appearance of the Schroeder diffuser. Figure 10.28 shows how the normalised diffusion coefficient can be greatly improved by the application of a diffusing cover.

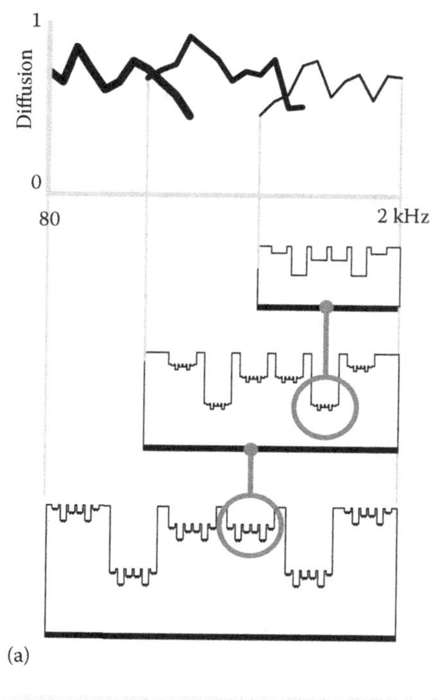

(a)

(b)

Figure 10.26 (a) How a fractal construction can be used to create diffusion over a wider bandwidth and (b) a Diffractal, which imbeds high-frequency diffusers within a low-frequency diffuser. ([a] After D'Antonio, P., Konnert, J., *J. Audio Eng. Soc.*, 40, 113–129, 1992.)

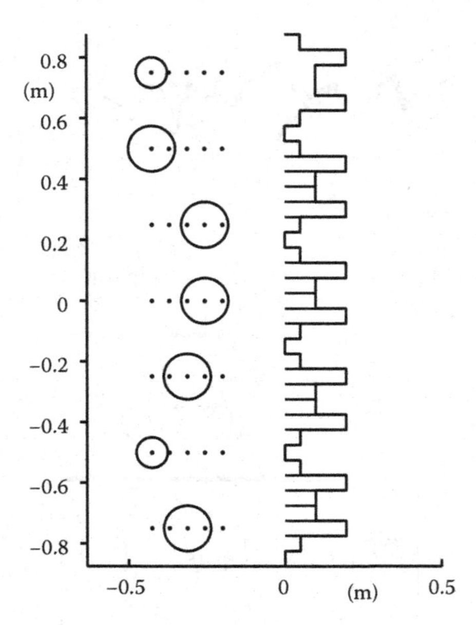

Figure 10.27 A set of cylinders in front of a periodic Schroeder diffuser. The position and size of the cylinders were determined using an optimisation algorithm. (After Pogson, M.A., Whittaker, D.M., Gehring, G.A., Cox, T.J., Hughes, R.J., Angus, J.A.S., *J. Acoust. Soc. Am.*, 128, 1149–1154, 2010.)

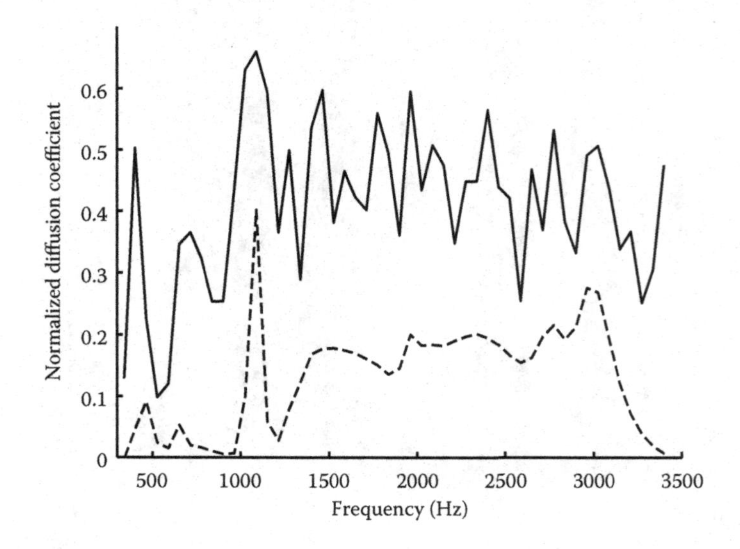

Figure 10.28 Normalised diffusion coefficient for a Schroeder diffuser - - - - - without and ———— with a set of cylinders in front. (Data from Pogson, M.A., Whittaker, D.M., Gehring, G.A., Cox, T.J., Hughes, R.J., Angus, J.A.S., *J. Acoust. Soc. Am.*, 128, 1149–1154, 2010.)

10.6 IMPROVING THE LOW-FREQUENCY RESPONSE

The depth of a diffuser is often limited by non-acoustic factors. Ultimately, the designer or architect will set the depth available for acoustic treatment. Furthermore, with the wavelength of sound extending to 17 m, it is impossible to construct a practical diffuser that can cover the full audible bandwidth. Consequently, various methods have been devised to extend the performance of diffusers to lower frequencies. Section 10.5.1 showed how this could be done using a fractal construction, and Section 10.5.2 examined using multilayer diffusers. Some other approaches are discussed here.

By using well folding to form L-shaped wells, it is possible to gain a greater bandwidth from a given overall depth. Jrvinen et al.,[23] Mechel,[24] Hargreaves and Cox,[25] and others have investigated such a modification that improves packing density. The standard Schroeder diffuser (Figure 10.2) has much wasted space at the rear, which can be better utilized.

Figure 10.29a shows the use of well folding to reduce the depth of an $N = 7$ quadratic residue diffuser. For low frequencies, the depth of the folded well should be calculated from the midline through the L-shaped well. So in the case shown, the number sequence determining the depth is a quadratic residue sequence of {2, 4, 1, 0, 1, 4, 2}. At high frequency, the sound does not pass around the bends, and therefore, the apparent depth is shallower,

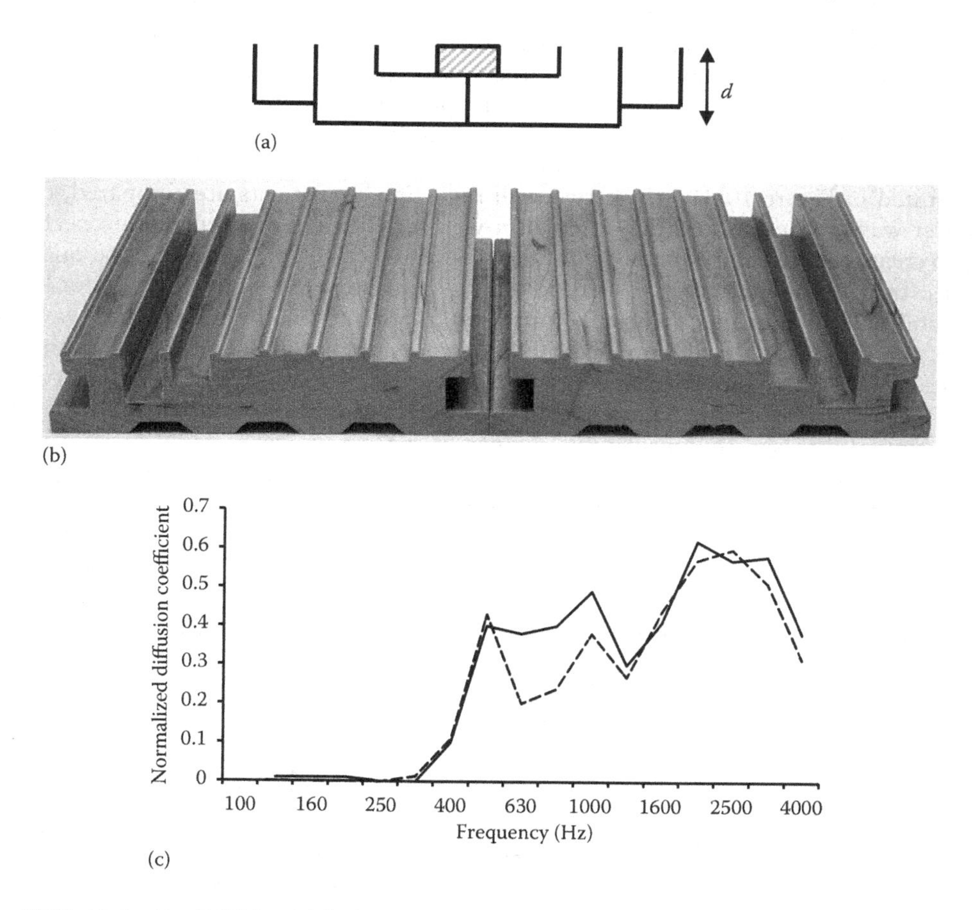

Figure 10.29 (a) An $N = 7$ QRD with L-shaped wells to reduce the depth to 63% of its original size. (b) An optimized diffuser where side-cuts are used to form a folded well. (c) The diffusion coefficient of two Schroeder diffusers, - - - - - folded and ——— not folded.

and the number sequence effectively becomes {2, 2½, 1, 0, 1, 2½, 2}. By bending the well such that the distance d shown in Figure 10.29 does not relate to the integers of the original quadratic residue sequence, it is possible to reduce the effect of critical frequencies where the surface reflects like a flat plate.

Under these assumptions, a simple Fraunhofer theory can be used. To properly model performance, especially at mid frequencies, a full BEM solution is needed. As the depth sequence varies between different frequency ranges, the application of number sequences is less straightforward. The problem with folded wells is that they are awkward to manufacture. A neat solution is to form the folded well by cutting into the sides of a diffuser,[26] as shown in Figure 10.29b. When the diffusers are stacked side by side, L-shaped wells are formed.

The plot at Figure 10.29c shows the measured normalized diffusion coefficient for two Schroeder diffusers, one utilizing a folded well, the other not. This shows that a folded well diffuser can have a similar performance to a more conventional design.

An alternative regime to gain more low-frequency performance is to use perforated sheets to add mass to the impedance of the wells, to lower the resonant frequencies and hence lower the design frequency. An example of such a device is shown in Figure 10.30, where the use of perforated sheets has enabled the longest wells to be shortened. Such an approach was tried for diffusion by Hunecke,[27] using microperforation, and for absorption by Fujiwara et al.[28] and Wu et al.[29] using larger-diameter perforations. For most diffusers, it is important that the perforation size is not too small; otherwise, significant losses may result.

The principle of the design is that for the first mode, the reflection coefficient of a Helmholtz resonator and of a quarter-wave tube can be made to be similar. Consequently, the quarter-wave resonator tubes of the Schroeder diffuser can be replaced by Helmholtz devices. This is illustrated in Figure 10.31, where two well reflection coefficients are compared, one for a quarter-wave resonator and the other for a well of half the depth but with a perforated sheet to create the correct resonant frequency. These reflection coefficients can be calculated using the transfer function matrix method outlined in Chapter 7. Unfortunately, for higher-order modes, the reflection coefficients for the Helmholtz resonator and quarter-wave tube diverge. Consequently, this design methodology works only around the design frequency.

Figure 10.32 shows the scattering at the design frequency, illustrating that the two diffusers behave similarly. The Schroeder diffuser with perforations is about half the depth of the original diffuser, so considerable savings of space have been made. A point to note when carrying out this calculation is that it is necessary to include the radiation impedance of the normal wells. This has been ignored for other Schroeder diffusers because it is a constant term for nearly all wells and therefore does not affect the diffusion. With a mixture of Helmholtz and quarter-wave devices, however, the correct radiation impedance must be included. Figure 10.33 shows the normalized diffusion coefficient, showing that for many frequencies, similar performance is obtained from the perforated device and the normal

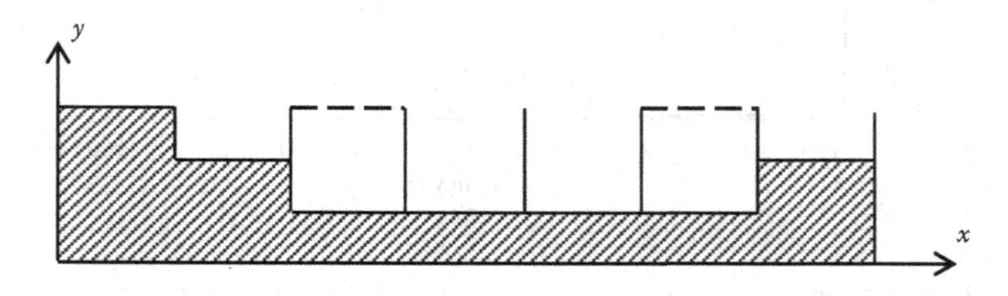

Figure 10.30 A cross-section through a Schroeder diffuser using perforated sheets to add mass to the surface impedance of the longest wells, therefore enabling the longest wells to be shortened.

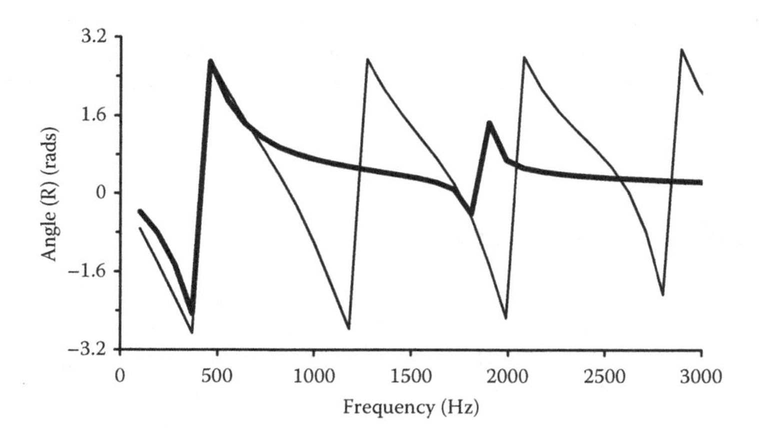

Figure 10.31 Reflection coefficient phase angle for two wells: ——— normal well and ▬▬▬ shorter well with perforated sheet to add mass.

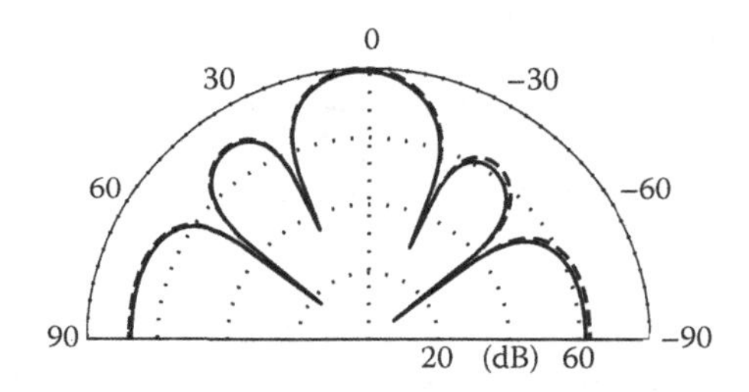

Figure 10.32 Scattering from an $N = 7$ QRD at the design frequency and a similar QRD where the longest wells have been shortened and a perforated sheet was used to add mass. ——— QRD and - - - - - QRD using perforated wells.

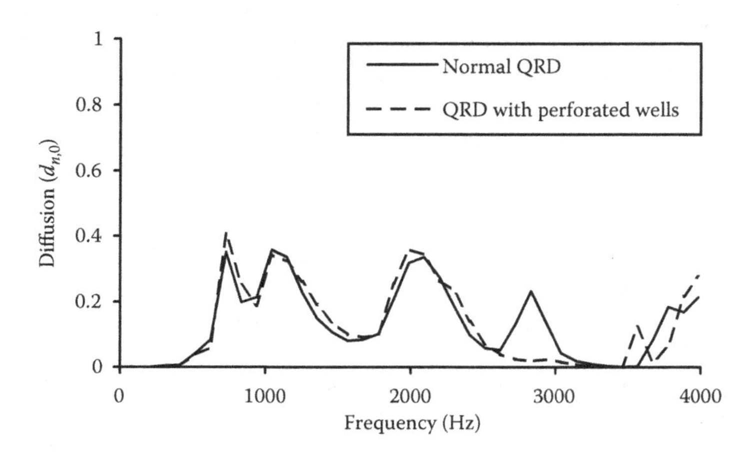

Figure 10.33 Normalized diffusion for a standard QRD and a QRD with perforated sheets to reduce the largest well depths.

Schroeder diffuser. At higher frequencies than shown, the scattering from the perforated sheet will become too specular and so care in design and application is needed.

To get a true broadband diffuser using this approach, optimization can be used (see Section 10.10). Note that an alternative to using Helmholtz resonators would be to add acoustic mass by putting limp membranes in the wells.

10.7 MULTIDIMENSIONAL DEVICES

The diffusers discussed thus far have been single-plane devices. They cause scattering into a hemidisc, acting as a plane surface in the other directions. While this is the preferred diffuser design for some applications, there is a need for a diffuser that scatters into a hemisphere. For a Schroeder diffuser, this can be achieved by forming a two-plane device, often referred to as a 2D diffuser. This device scatters optimally in the x and z directions and therefore gives even lobes on a hemisphere. Examples of such surfaces were shown at the top of the back wall in Figure 3.13 and in the ceiling of Figure 3.17.

As 2D diffusers scatter in multiple planes, a receiver in the bright zone will experience a scattered energy that is attenuated more than for a 1D diffuser (provided that multiple grating lobes are present). The number of grating lobes is squared if a 1D diffuser of width Nw and a 2D diffuser of size $Nw \times Nw$ are compared. Therefore, the energy in each lobe will reduce by $10\log_{10}(m)$, where m is the number of grating lobes present for the 1D diffuser.

There are two common processes for forming 2D diffusers. The first involves forming two sequences, one for the x direction one for the z direction, and amplitude modulating the x sequence with the z sequence. For a quadratic residue sequence, this can be expressed as[2]

$$s_{n,m} = (n^2 + m^2) \text{ modulo } N, \tag{10.23}$$

where n and m are integers and index the sequence for the nth and m^{th} wells in the x and z directions, respectively. A similar procedure can be used for PRDs:

$$s_{n,m} = (r^n + r^m) \text{ modulo } N. \tag{10.24}$$

It is even possible to have a quadratic residue sequence in one direction and a primitive root sequence in the other, provided that they are based on the same prime number, although it is hard to see why you would choose to do this.

A 2D quadratic residue sequence based on $N = 7$ is shown in Figure 10.34. In this case, the indexes n and m were started from 4 to place the zero-depth well in the middle of the diffuser. As the surface is periodic, it is possible to start the indexes n and m from any integer. Two-dimensional number theoretic diffusers will often have less bass diffusion efficiency than a 1D diffuser will, as the ratio $s_{\max}/N$ tends to be close to 1 for 2D devices.

Figure 10.34 also illustrates other sequences that can be used. On the diagonal of the diffuser, the following sequence appears: {4, 1, 2, 0, 2, 1, 4}. This is the original sequence {0, 1, 4, 2, 2, 4, 1}, but with every fourth element used. This new sequence has the same Fourier properties as the original sequence does due to the shift properties of quadratic residue sequences. This indicates that as well as producing good diffusion in the orthogonal directions x and z, good scattering in the directions along the diagonals should also be obtained.

The second method for making multidimensional diffusers is to use the Chinese remainder theorem.[30] This folds a 1D sequence into a 2D array while preserving the Fourier properties of the 1D sequence. The process is described in detail in Section 12.3.5, where it is applied to hybrid surfaces, but it can equally be applied to polyphase sequences.

4	6	3	2	3	6	4	4	6	3	2	3	6	4
6	1	5	4	5	1	6	6	1	5	4	5	1	6
3	5	2	1	2	5	3	3	5	2	1	2	5	3
2	4	1	0	1	4	2	2	4	1	0	1	4	2
3	5	2	1	2	5	3	3	5	2	1	2	5	3
6	1	5	4	5	1	6	6	6	1	5	4	5	1
4	6	3	2	3	6	4	4	4	6	3	2	3	6
4	6	3	2	3	6	4	4	6	3	2	3	6	4
6	1	5	4	5	1	6	6	1	5	4	5	1	6
3	5	2	1	2	5	3	3	5	2	1	2	5	3
2	4	1	0	1	4	2	2	4	1	0	1	4	2
3	5	2	1	2	5	3	3	5	2	1	2	5	3
6	1	5	4	5	1	6	6	6	1	5	4	5	1
4	6	3	2	3	6	4	4	4	6	3	2	3	6

Figure 10.34 A sequence array for a 7 × 7 quadratic residue diffuser; one period is shaded.

The requirement for coprime factors means that this folding technique cannot be applied to single periods of QRDs because there is a prime number of wells. This can be overcome by using an odd-number generator N for the quadratic residue sequence that is not prime. For example, a quadratic residue sequence based on $N = 15$ will work perfectly well at the design frequency and can be wrapped into a 3 × 5 array. The problem is that the surface will have flat plate frequencies at 3 and 5 times the design frequency (as well as 6, 10, 9, 15… times). Consequently, to use a non-prime N, it is necessary to make sure the factors of N are sufficiently large that the flat plate frequency is above the frequency of interest. For example, $N = 143$ has factors of 11 and 13 and so would be a good choice as the flat plate frequencies will be beyond the upper frequency limit of most diffusers. It is also possible to apply the Chinese remainder theorem to some primitive root and other mathematical sequences, such as the Chu sequence outlined previously.

It has also been suggested by Pollack and Dodds[31] that the wrapping can be carried out in a hexagonal configuration:

$$s_{n,m} = (n^2 + m^2 + nm) \text{ modulo } N. \tag{10.25}$$

Figure 10.35 illustrates a hexagonal QRD based on $N = 7$ generated using Equation 10.25.

Figure 10.36 illustrates the scattering from a 2D $N = 7 \times 7$ QRD and a plane surface. There are a regular set of grating lobes, but these are difficult to see unless the polar response can be animated and rotated. Figure 10.37 shows the data as a contour plot, where the grating lobes become more obvious. These grating lobes form a regular grid; the middle nine in a 3 × 3 grid are most obvious in the case shown. These contour plots are effectively the contour on the surface of the hemisphere, looking down onto the hemisphere. Consequently, the x and z axes shown are non-linear.

Figure 10.38 illustrates the scattering from a diffuser formed using the Chinese remainder theorem. A modified primitive root sequence based on the prime number $N = 43$ was generated, and so the sequence is 42 elements long. It was folded into a 6 × 7 array using the Chinese remainder theorem. Figure 10.38 shows the response at four times the design

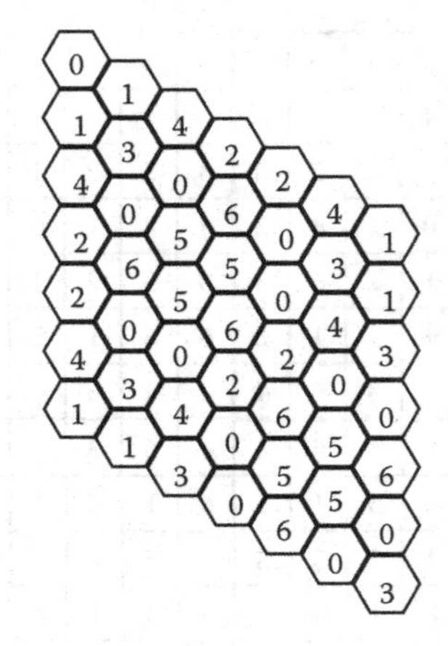

Figure 10.35 One period of a hexagonal quadratic residue diffuser based on $N = 7$.

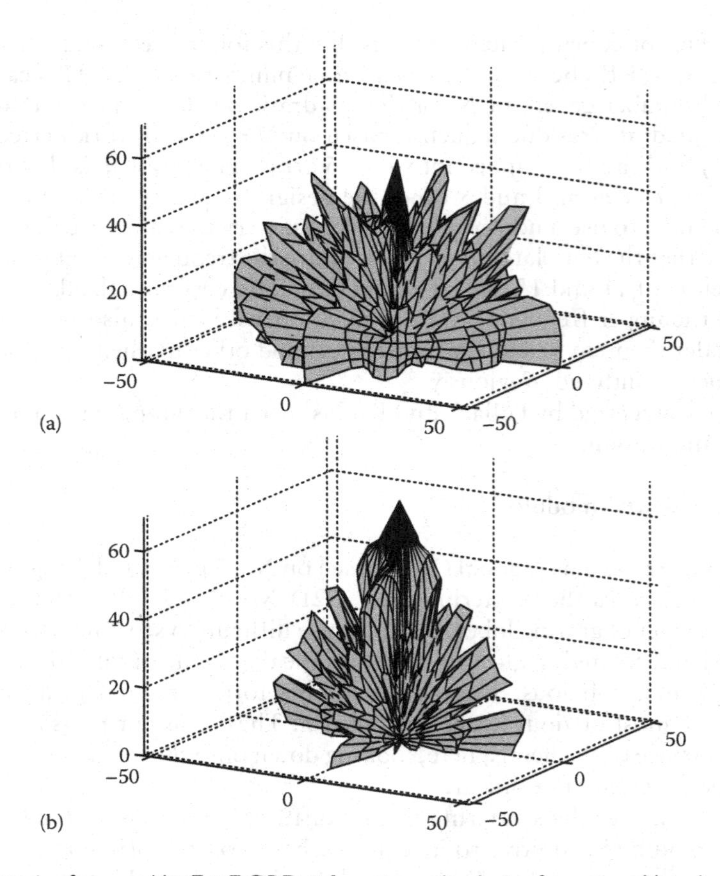

Figure 10.36 Scattering from an $N = 7 \times 7$ QRD at four times the design frequency (a) and a plane surface (b).

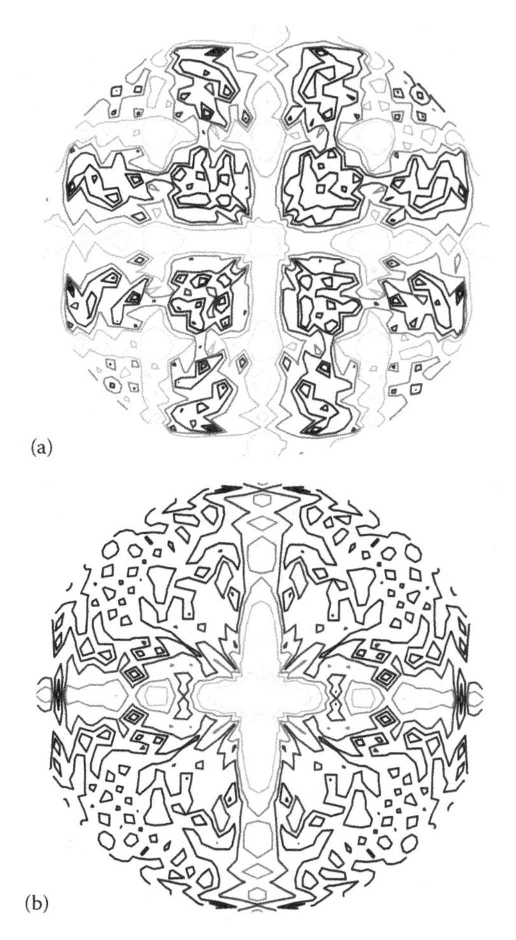

Figure 10.37 Contour plot of polar response shown in Figure 10.36 seen from above. The QRD (a) shows 13 grating lobes, where the 3 × 3 grid of lobes shown in the centre is clearest. The plane surface (b) just has a lobe in the specular reflection direction.

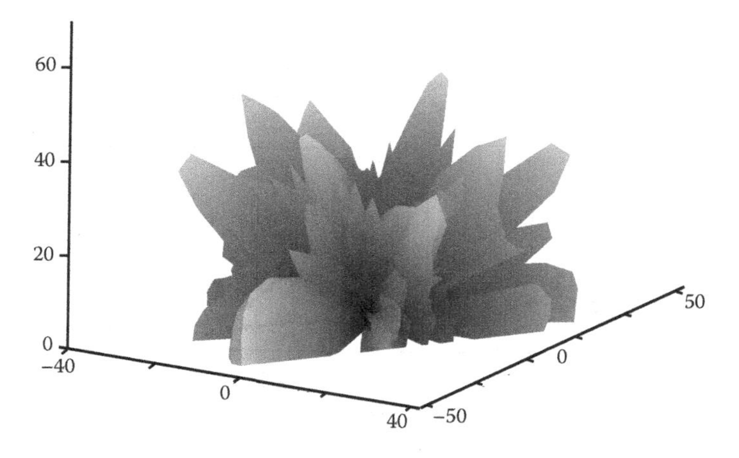

Figure 10.38 Scattering from a modified PRD based on prime 43 and wrapped into a 6 × 7 array using the Chinese remainder theorem.

frequency, and the specular lobe, which would normally be pointing straight up the page, is missing. This demonstrates that the folding technique succeeds in preserving the primitive root properties. When amplitude modulation is used to form primitive root sequence arrays (Equation 10.24), then planes of reduced scattering are produced. All the scattering in the directions given by $\varphi = 0°, 90°, 180°$, and $270°$ will be suppressed.

Little measurement or prediction work on multidimensional devices has been carried out. It appears, however, that the understanding developed from a 1D analysis can be extended to multiple dimensions. Issues such as lobes, periodicity, critical frequencies, and frequency limits are all similar.

10.8 ABSORPTION

Section 8.2 discussed in some detail how and why Schroeder diffusers absorb sound and how to make a phase grating device into an efficient absorber. Briefly, Schroeder diffusers primarily absorb because of (1) high-energy flows from wells in resonance to wells out of resonance and (2) quarter-wave resonant absorption in the wells, especially if the wells are narrow. Figure 10.39 shows the random incidence absorption coefficient for 1D and 2D commercial wooden and plaster Schroeder diffusers.

Figure 10.39 shows how important it is not to cover Schroeder diffusers with cloth, as this greatly increases the absorption. There is high particle velocity around the front face of a Schroeder diffuser and any cloth covering will cause excess absorption, as might be expected if resistive material is placed in a region of high particle velocity. Any cloth covering should be placed at least a well width away from the front face and should have the lowest possible flow resistivity.

Figure 10.39 also shows that 2D Schroeder diffusers absorb more sound than 1D devices do. It is assumed that this is due to the greater number of different well depths in the 2D device, leading to more energy flow between the wells, in addition to a greater density of quarter-wave resonances. Furthermore, because there are more well walls present than in a 1D device, more losses due to viscous boundary layer effects can occur.

It is important that the Schroeder diffuser is constructed to a high precision. Some data have been published showing very high absorption coefficients for these devices, but this is

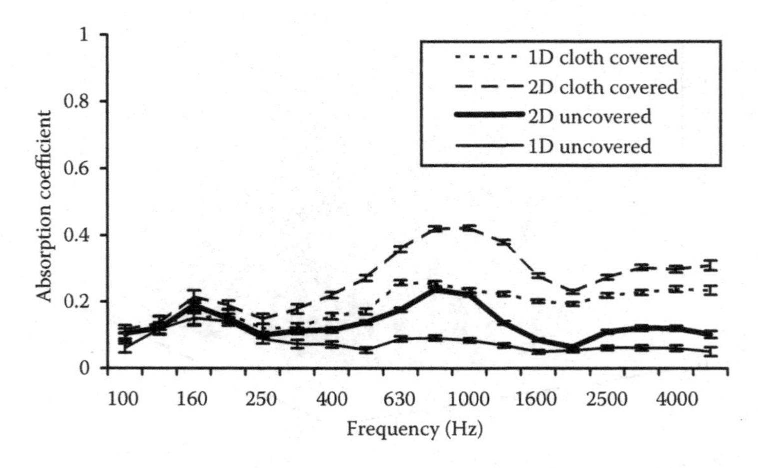

Figure 10.39 Random incidence absorption coefficients measured for 1D and 2D Schroeder diffusers based on $N = 7$ with and without cloth covering.

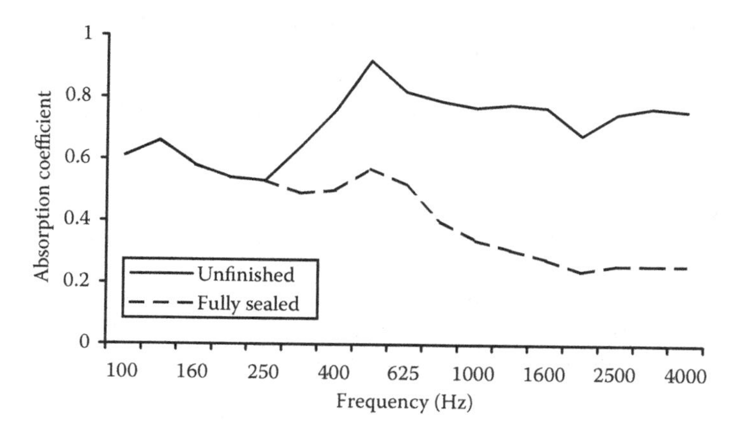

Figure 10.40 Random incidence absorption coefficients measured for a concrete masonry Schroeder diffuser before and after rough surfaces are sealed.

generally due to poor construction. Small cracks in the bottom of the wells—between the well sides and bottoms—are difficult to avoid unless care is taken. If any cracks open up to cavities behind, these can cause excess absorption as a Helmholtz absorber/resonator has been formed. Proper sealing with varnish or paint is vital. Construction materials are generally not that important unless rough surfaces are used. Figure 10.40 shows the absorption of a concrete masonry Schroeder diffuser before and after it is sealed with paint. Before being fully sealed, the rough and porous surfaces cause excess losses at the boundary layers and due to the energy flow between wells around the rough edges of the fins. Absorption is greatly reduced by sealing properly.

Commins et al.[32] experimentally investigated the absorption characteristics of a Schroeder diffuser. They showed that by sloping the bottom of the diffuser wells, the absorption could be reduced. The slope broadens the resonances of the wells and so decreases the energy flow within and between wells at resonance. Some early concert hall designs used very wide wells (30 cm), presumably to avoid the problems of absorption. The problem with such wide well widths is that at high frequencies, specular reflections from the bottom of the wells can cause problems.[33] Furthermore, the bandwidth of the device will be greatly reduced. Consequently, if the fear of absorption cannot be overcome, a fractal construction or a surface topology that does not generate strong resonances such as a curved surface are better solutions.

10.9 DO THE SIMPLEST THEORIES WORK?

All the previous analysis has relied on a simplified model of acoustic reflection—the Fraunhofer theory. The performance of this theory can be divided into three frequency ranges, as discussed in Chapter 9. At low frequencies, the theory is inaccurate, as the mutual interactions across the surface are not correctly modelled by the Kirchhoff boundary conditions. At mid frequencies, the theory is most accurate. At high frequencies, the theory becomes inaccurate again because application realistic sources and receivers are not in the far field. Berkhout et al.[34] pointed out that the theory used by Schroeder was approximate and concluded that the development of the diffuser should be based on a more complex model. While it is possible to use a more complex theory, it is only with the simplified Fraunhofer theory that the design can be carried out via simple formulations and using basic Fourier concepts. A more complex prediction model can be used in an optimization process, as will be discussed in Section 10.10, but this is a brute force technique, which often means that the designer learns little about the basic principles of good diffuser design.

One advantage of working with dispersing surfaces is that the laws of physics and the tendency to disorder are going to aid diffusion. This helps explain why Schroeder et al.[35] could reply to Berkhout et al. by pointing out that measurements and accurate theory were not too different from the approximate theory. The difference between approximate and accurate theories is probably most critical when exacting results such as notches formed from PRDs are required, or when there is a small number of wells. Even if the optimum diffusion as defined by Schroeder is not achieved, phase gratings are pretty complex surfaces, which will create some dispersion.

There are other limitations that apply to Schroeder's design, some of which have been touched on before:

1. The design methodology is based on an approximate model.
2. Losses are ignored.
3. The design is carried out for the far field, whereas most listeners are in the near field.
4. The wells are assumed to be local reacting.

Assumption 3 may not be that limiting. There is evidence that a diffuser that creates good dispersion in the far field also works well in the near field. In the near field, the path length differences from different points on the surface dominate and cause the scattering to have a large number of minima and maxima. Indeed, the polar plots for different diffusers have similar statistical features in the near field. Some studies have compared periodic and aperiodic arrangements of diffusers either subjectively or objectively[36] using application realistic near field listener positions. In these cases, the aperiodic arrangements, which will create more dispersion in the far field than the periodic cases do, are found to be more efficient diffusers in the near field. Consequently, it is assumed that Schroeder diffusers, which create good far field dispersion, will also be effective in the near field.

Schroeder gave an alternative take on the near and far field problem. He suggested that by bending the diffuser to follow a parabolic concave mirror, the far field scattering pattern could be focussed at near field receivers. A similar effect can be achieved more easily by modulating the well depths by the locus of a parabolic mirror. In this case, changing the phases of the waves reradiated from the wells causes the far field beam to be focussed into the near field. This process has been tested with a boundary element model and shown to work.[5]

Assumption 4 concerns whether the well admittances change due to the presence of the neighbouring wells. This assumption has undergone some limited tests. Cox and Lam[5] compared the admittance predicted by a BEM, which models the surface shape precisely, against the simple phase change admittance values derived from a reflection coefficient of $\exp(-2jkd_n)$. Figure 10.41 shows that reasonable agreement is found, indicating that the surface admittances are indeed local reacting to a degree. Some real parts are seen indicating losses or maybe mathematical inaccuracies in the BEM model. If these are true losses, they are due to evanescent waves, as the BEM model did not include any absorption. Cox and Lam also looked at the admittance variation along the elongated dimension of a 1D Schroeder diffuser. They again showed that the admittance from a BEM model

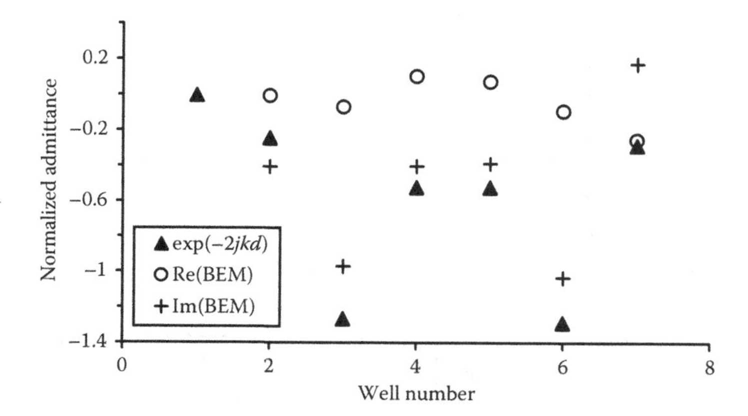

Figure 10.41 Admittance for an $N = 7$ QRDs at a single mid-frequency. A BEM prediction was used to generate accurate admittances for comparison with the simple phase change calculation $\exp(-2jkd)$. The first well is zero depth, and so no comparison is possible. (After Cox, T.J., Lam, Y.W., *J. Acoust. Soc. Am.*, 95, 297–305, 1994.)

approximately matched the simple phase change admittance, except for positions close to the ends of the wells.

The most significant limitations of these number theoretic designs are, however, as follows:

1. They work only at discrete frequencies.
2. Optimum diffusion means the same energy in the diffraction lobes; this is not the same energy in all directions.

As the number theoretic Schroeder diffusers are not truly broadband and do not completely disperse to all directions, it is possible to improve on the design. To do this, optimization algorithms can be used.

10.10 OPTIMIZATION

10.10.1 Process

de Jong and van den Berg[37] developed the idea of using an iterative method to produce both Schroeder-style diffusers. It was not until Cox[38] rediscovered this idea in the early 1990s, however, and D'Antonio[39] provided experimental evidence for the improved performance over traditional number theoretic Schroeder diffusers that this concept was widely exploited. de Jong and van den Berg used an approximate prediction model and narrow deep wells that were unrealistic for practical diffusers. Cox was able to apply greater computing power, more accurate BEMs, and also more realistic geometries.

The concept of optimization is illustrated in Figure 10.42. The idea is to get a computer to go through a trial-and-error process searching for the best well depth sequence. First, a starting well depth sequence is randomly chosen. Then the computer predicts the scattering from the surface and evaluates the quality of the scattering using a diffusion coefficient. The computer then adjusts the well depth sequence in an effort to improve the diffusion coefficient. When a maximum in the diffusion coefficient is achieved, the iteration process ends because a good diffuser has been found.

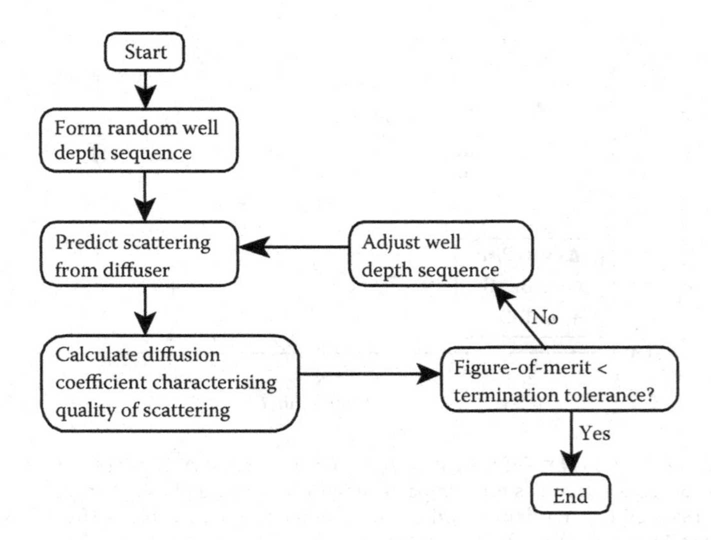

Figure 10.42 Flow diagram for optimizing the well depth sequence of a Schroeder diffuser.

Doing this requires the following:

1. A validated prediction model.
2. A diffusion coefficient to measure quality of scattering.
3. An optimization algorithm to change the well depth sequences.

A validated prediction model is needed, and for this, a BEM can be used. The disadvantage with a BEM is that it can be very slow to compute, but as computing power is constantly increasing, this is becoming less of a limitation. For computing power reasons, the diffusers that Cox originally optimized were rather narrow; now, it is possible to carry this out with wider diffusers in large arrays over a wider bandwidth. It is also possible to use the simpler Fraunhofer models, which means that the optimization is very fast but then the accuracy may be compromised. One possibility is to use simple models to carry out a course optimization and then use the more accurate models to focus on solutions.

The diffusion coefficient can be used to evaluate the quality of the scattering produced by a surface. The diffusion coefficient is evaluated at each frequency band of interest, say each one-third octave band. The reciprocal of the diffusion coefficients are then averaged across frequency to obtain a single figure-of-merit. (The reciprocal of the diffusion coefficient is used because numerical optimization algorithms minimize functions.) The risk with this simple averaging process is that the diffusion may be very uneven versus frequency. Frequencies with very good diffusion may compensate for frequencies with very poor diffusion, where a better solution might be moderate diffusion consistent across the whole frequency range. This problem is most easily solved by subtracting a standard deviation of the diffusion coefficients from the mean value across frequencies. This then penalizes surfaces with very uneven diffusion coefficient spectra.

An optimization algorithm is used to adjust the well depth sequence during the search. It is needed so that the different well depth sequences can be tried and tested in a logical manner rather than by a completely random trial-and-error basis. A usual analogy for a 2D optimization is finding the lowest point in a hilly landscape (while blindfolded). If a human was to carry out such a search, he/she would start by going downhill on the presumption that this will lead him/her to a lower point. The optimization algorithm must make similar decisions. It is vital that the solution is found in the fewest steps, as otherwise the optimization process becomes tedious. The landscape is an optimization problem with two degrees of freedom; the analogous diffuser would have only two well depths. Practical optimization problems involve many more degrees of freedom, which makes finding the minima more difficult and time consuming.

There are a variety of algorithms available for optimization.[40] The key decision is whether the optimization is to take place with only the figure-of-merit or with the figure-of-merit and its derivative. Knowledge of the derivative vastly speeds up the optimization processes and so should always be used if available. To continue with the landscape analogy, it is much quicker if the person is told the downhill direction; otherwise, he/she has to stagger around for a while trying to decide which way to go. The problem with diffuser optimization is that the derivative is not often known. With the Fraunhofer theory and a simple figure-of-merit, such as minimizing the specular zone energy, it is possible to formulate the derivative,[10] but that is an unusual case.

For most work on diffuser optimization, only function values are known. A downhill simplex algorithm has been often used, which is rather slow. It is, however, extremely robust to non-linear constraints, something that will become important for the geometric diffusers discussed in Chapter 11. There are other techniques like genetic algorithms or quasi-Newton gradient descent methods. The disadvantage of a genetic algorithm is that it requires tuning by appropriately choosing variables such as population size and mutation rate. Methods that calculate gradients with finite differences such as quasi-Newton methods can cause problems with solution techniques such as BEMs, which often have small numerical inaccuracies that can greatly affect the estimated gradient. A downhill simplex method may not be trendy, but it just needs plugging in, and it works.

When carrying out the optimization for a Schroeder diffuser, it is most efficient to use a BEM where the diffuser is modelled as a box with a spatially varying admittance on the surface. Then all that changes when the optimization is the front face admittance and not the surface profile. This means that the time-consuming processes of carrying out the interaction-matrix integrations need only to be done once at the start of the optimization; this greatly reduces computation time. In fact, it should be possible to get the derivatives of the figure-of-merit in this case.[41] Effort spent speeding up the prediction algorithm is time well spent; in a typical optimization process, the scattering is typically evaluated a thousand times, so unless each individual case takes a matter of seconds, the optimization process will become very slow.

In any optimization problem, there will be a large number of *local minima*, and somewhere, there will be the numerically lowest point—called the *global minimum*. To return to the landscape analogy, the blindfolded person might find a valley bottom and think the best point has been found, not realizing that over the next mountain ridge, there is a lower valley. The key to a good optimization algorithm is not to be trapped in poor local minima, but to continue to find deep local minima. Provided a good optimization algorithm is chosen, this should not be a problem, especially if the optimization is tried many times from different starting points, as is customary good practice.

When there are a large number of well depths to be optimized, the surface describing the variation of the figure-of-merit with the well depths becomes very complex. There will be a very large number of minima. It is virtually impossible to find the global minimum unless the optimization algorithm is run a large number of times starting with many different random well depth sequences. Fortunately, as the number of degrees of freedom increases, the need to find the global minimum becomes less important. Experience has shown that there are a large number of good local minima solutions available, and although the scattering will be different in each case, there is often negligible difference in performance between the good local minima. There is usually no magical global minimum where the quality of scattering produced is significantly better than good local minima.

10.10.2 Results

When testing the results of an optimization, it is important to look at frequencies, source, and receiver positions that are different from those used during the optimization. This checks to see whether the design process has found a robust solution. Also, if an approximate theory was used during the optimization, results should be checked with an accurate model or measurement. There is always a risk that the optimizer will overfit a poor solution; this is where the solution is good only for the specific geometries and frequencies used during the optimization.

Figure 10.43 shows the scattering from two optimized diffusers compared to an $N = 7$ QRD. One of the optimized diffusers had fins; the other had a stepped profile—essentially a Schroeder diffuser with the fins removed. Both the optimized diffusers produce more even scattering than the QRD does at this and other frequencies.

Cox[38] found that the optimized $N = 7$ diffuser with fins outperforms the QRD over a wide variety of frequencies. When the number of wells was increased to about 36 and compared to two periods of an $N = 17$ QRD, however, the gains were less marked. The scattering from the QRD was already fairly uniform at low frequencies, and so the room for improvement was relatively small. This was not, however, particularly due to the use of a quadratic residue sequence—even a diffuser with randomly determined well depths gave reasonable diffusion.

Removing the constraints on geometry imposed by a Schroeder-style diffuser and forming a stepped surface produced better diffusion. The magic of Schroeder's diffuser geometry is not that it produces diffusion, but that it enables simple design methods to be used. Removing the fins enables a simpler geometry that is cheaper to make. It also removes the resonant wells and so has lower absorption. The improved performance was seen for both the $N = 7$ and $N = 36$ cases. Interestingly, the $N = 7$ optimized stepped diffuser looked rather like a faceted semicylinder.

D'Antonio[39] carried out a thorough experimental evaluation of the work by Cox. The measurements confirmed that optimization produced better diffusers than number theory sequences. D'Antonio also looked at the performance of the diffusers outside the domain of optimization: at higher frequencies, at oblique angles of incidence, for different receiver radii, and for a periodic arrangement. Outside the domain of optimization, the optimized diffusers

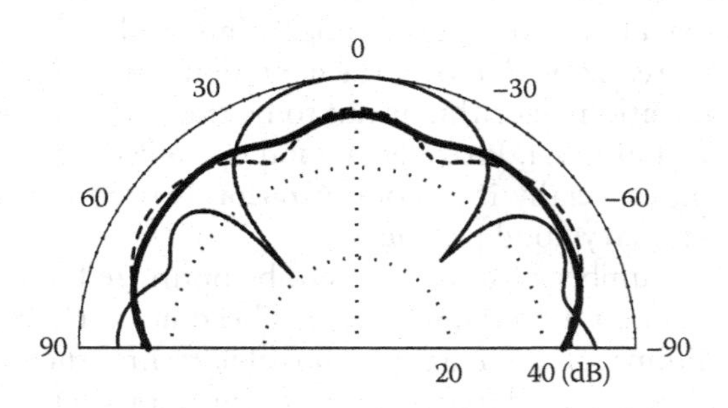

Figure 10.43 The scattering from an ——— $N = 7$ Schroeder diffuser, - - - - - an optimized Schroeder diffuser, and ▬▬▬ an optimized stepped diffuser at 1050 Hz. (Data from Cox, T.J., *J. Acoust. Soc. Am.*, 97, 2928–2941, 1995.)

were found to give roughly the same diffusion as the Schroeder diffusers—sometimes worse, sometimes better. The solution to this problem is to carry out the optimization including all frequencies, angles of incidence, and source and receiver radii of interest.

The original optimization work was limited because of computing power; now it can be applied to diffusers spread over larger areas. Figure 10.44 compares the normalized diffusion coefficient from an $N = 7$ QRD to an optimized modulated arrangement of phase grating diffusers. The improvement in diffusion is quite marked.

It is also possible to optimize for a non-even scattering distribution. For example, Cox and D'Antonio[10] tried to minimize the energy in a particular direction, to produce a notch in the specular reflection direction to create an improved PRD. In Figure 10.45, the results from trying to optimize a diffuser to work from 500 to 3000 Hz for an angular range of ±5° about the specular reflection direction are shown. The diffuser is labelled *optimized* and is compared to a plane surface and a modified PRD. Across the optimization range, the specular reflection is reduced by about 25 dB compared to a plane surface and by 10 dB when compared to the PRD. This was for a single period of the diffuser. When multiple periods

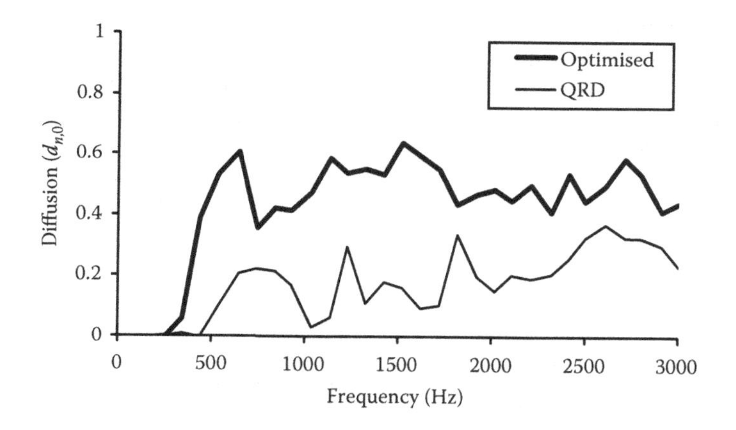

Figure 10.44 Normalized diffusion spectra for multiple periods of a Schroeder diffuser and an optimized phase grating diffuser.

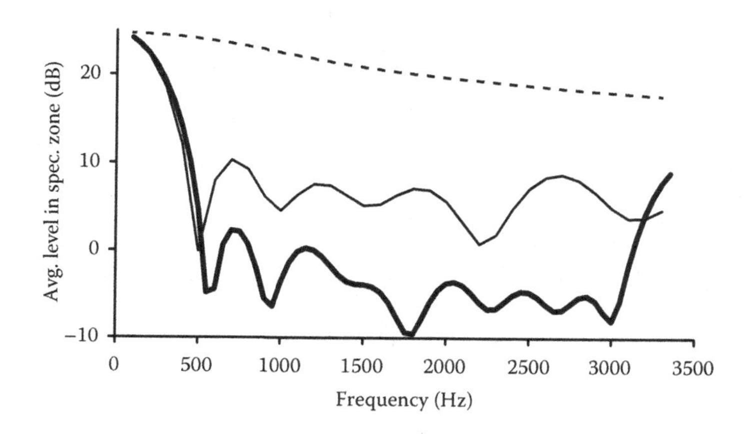

Figure 10.45 The average energy in the specular zone for three surfaces: - - - - - plane, ▬▬▬ optimized, and ——— modified PRD. (After Cox, T.J., D'Antonio, P., *Appl. Acoust.*, 60, 167–186, 2000.)

were attempted, the results were less impressive. Over the bandwidth 500–3000 Hz, the best optimized diffusers that could be achieved was an extra 4 dB attenuation of the specular and near-specular energy when compared to the modified primitive root. Experience has shown that using optimization to shape polar responses or create notches is fraught with difficulty. Optimization is most successful when trying to create uniform dispersion.

10.11 SUMMARY

Schroeder diffusers have been very successful, thanks to their simple, ingenious concept, straightforward design equations, and their commercial exploitation. There has always been a certain reticence among some designers to use this type of diffuser, however. The rumours of absorption have continued since their inception, with high absorption coefficients being published for poorly constructed surfaces. Now that there is a proper understanding of the absorption mechanisms, this should no longer be a problem. A few people have claimed to hear strange artefacts from Schroeder diffusers, but the designs they have been listening to have not followed some of the important design principles discussed in this chapter. Following proper design principles and applying all the current knowledge result in high-quality sound from Schroeder diffusers. One of the main stumbling blocks to their use is, however, their visual appearance. Consequently, other diffusers are needed that have defined acoustical properties but with different visual aesthetics. This will be addressed in the next chapter.

REFERENCES

1. M. R. Schroeder, "Diffuse sound reflection by maximum-length sequences", *J. Acoust. Soc. Am.*, **57**(1), 149–50 (1975).
2. M. R. Schroeder, "Binaural dissimilarity and optimum ceilings for concert halls: More lateral sound diffusion", *J. Acoust. Soc. Am.*, **65**, 958–63 (1979).
3. P. D'Antonio and J. Konnert, "The reflection phase grating diffusor: Design theory and application", *J. Audio Eng. Soc.*, **32**(4) (1984).
4. T. J. Cox and P. D'Antonio, "Schroeder diffusers: A review", *Bldg. Acoust.*, **10**(1), 1–32 (2003).
5. T. J. Cox and Y. W. Lam, "Prediction and evaluation of the scattering from quadratic residue diffusors", *J. Acoust. Soc. Am.*, 95(1), 297–305 (1994).
6. W. R. T. Tenkate, "On the bandwidth of diffusors based upon the quadratic residue sequence", *J. Acoust. Soc. Am.*, **98**(5), 2575–9 (1995).
7. P. D'Antonio and J. Konnert, "The QRD diffractal: A new one- or two-dimensional fractal sound diffusor", *J. Audio Eng. Soc.*, **40**(3), 113–29 (1992).
8. P. Fan and M. Darnell, *Sequence Design for Communications Applications*, John Wiley & Sons, New York, 49 (1996).
9. E. Feldman, "A reflection grating that nullifies the specular reflection: A cone of silence", *J. Acoust. Soc. Am.*, **98**(1), 623–34 (1995).
10. T. J. Cox and P. D'Antonio, "Acoustic phase gratings for reduced specular reflection", *Appl. Acoust.*, **60**(2), 167–86 (2000).
11. M. R. Schroeder, "Phase gratings with suppressed specular reflections", *Acustica*, **81**, 364–9 (1995).
12. K. Dadiotis, T. J. Cox, and J. A. S. Angus, "Lüke and power residue sequence diffusers", *Proc. 19th ICA*, Madrid, RBA-11-005 (2007).
13. D. Takahashi and R. Takahashi, "Sound fields and subjective effects of scattering by periodic-type diffusers", *J. Sound Vib.*, **258**(3), 487–97 (2002).

14. J. A. S. Angus, "Large area diffusors using modulated phase reflection gratings", *Proc. 98th Convention Audio Eng. Soc.*, preprint 3954 (D4) (1995).
15. J. A. S. Angus, "Using modulated phase reflection gratings to achieve specific diffusion characteristics", *Proc. 99th Convention Audio Eng. Soc.*, preprint 4117 (1995).
16. J. A. S. Angus and C. I. McManmon, "Orthogonal sequence modulated phase reflection gratings for wideband diffusion", *Proc. 100th Convention Audio Eng. Soc.*, preprint 4249 (1996).
17. J. A. S. Angus and A. Simpson, "Wideband two dimensional diffusers using orthogonal modulated sequences", *Proc. 103rd Convention Audio Eng. Soc.*, preprint 4640 (1997).
18. J. A. S. Angus and C. I. McManmon, "Orthogonal sequence modulated phase reflection gratings for wide-band diffusion", *J. Audio Eng. Soc.*, **46**(12), 1109–18 (1998).
19. J. A. S. Angus, "Using grating modulation to achieve wideband large area diffusers", *Appl. Acoust.*, **60**(2), 143–65 (2000).
20. Stephan Martens—Research, http://www-e.uni-magdeburg.de/mertens/research/, accessed 7 January 2016.
21. B. B. Mandelbrot, *The Fractal Geometry of Nature*, Freeman, San Francisco (1983).
22. M. A. Pogson, D. M. Whittaker, G. A. Gehring, T. J. Cox, R. J. Hughes, and J. A. S. Angus, "Diffusive benefits of cylinders in front of a Schroeder diffuser", *J. Acoust. Soc. Am.*, **128**(3), 1149–54 (2010).
23. A. Jrvinen, L. Savioja, and K. Melkas, "Numerical simulations of the modified Schroeder diffuser structure", *J. Acoust. Soc. Am.*, **103**(5), 3065 (1998).
24. F. P. Mechel, "The wide-angle diffuser—A wide-angle absorber?" *Acustica*, **81**, 379–401 (1995).
25. J. A. Hargreaves and T. J. Cox, "Improving the bass response of Schroeder diffusers", *Proc. IoA(UK)*, **25**(7), 199–208 (2003).
26. P. D'Antonio and T. J. Cox, *Extended Bandwidth Folded Well Diffusor*, US Patent 7,322,441 (2006).
27. J. Hunecke, "Schallstreuung und Schallabsorption von Oberflächen aus Mikroperforierten Streifen", PhD thesis, University of Stuttgart (1997).
28. K. Fujiwara, K. Nakai, and H. Torihara, "Visualisation of the sound field around a Schroeder diffuser", *Appl. Acoust.*, **60**(2), 225–36 (2000).
29. T. Wu, T. J. Cox, and Y. W. Lam, "A profiled structure with improved low frequency absorption", *J. Acoust. Soc. Am.*, **110**, 3064–70 (2001).
30. M. R. Schroeder, *Chaos, Power Laws: Minutes from an Infinite Paradise*, W. H. Freeman & Co, New York (1991).
31. J.-D. Pollack and G. Dodd, personal communication.
32. D. E. Commins, N. Auletta, and B. Suner, "Diffusion and absorption of quadratic residue diffusers", *Proc. IoA(UK)*, **10**(2), 223–32 (1988).
33. T. J. Cox and Y. W. Lam, "The performance of realisable quadratic residue diffusers (QRDs)", *Appl. Acoust.*, **41**(3), 237–46 (1994).
34. A. J. Berkhout, D. W. van Wulfften Palthe, and D. de Vries, "Theory of optimal plane diffusers", *J. Acoust. Soc. Am.*, **65**(5), 1334–6 (1979).
35. M. R. Schroeder, R. E. Gerlach, A. Steingrube, and H. W. Strube, "Response to 'theory of optimal planar diffusers'", *J. Acoust. Soc. Am.*, **65**(5), 1336–7 (1979).
36. T. J. Hargreaves, "Acoustic Diffusion and Scattering Coefficients for Room Surfaces", PhD thesis, University of Salford (2000).
37. B. A. de Jong and P. M. van den Berg, "Theoretical design of optimum planar sound diffusers", *J. Acoust. Soc. Am.*, **68**(4), 1154–9 (1980).
38. T. J. Cox, "Optimization of profiled diffusers", *J. Acoust. Soc. Am.*, **97**(5), 2928–41 (1995).
39. P. D'Antonio, "Performance evaluation of optimized diffusors", *J. Acoust. Soc. Am.*, **97**(5), 2937–41 (1995).
40. W. H. Press et al., *Numerical Recipes, The Art of Scientific Computing*, Cambridge University Press, Cambridge, 289–92 (1989).
41. R. D. Ciskowski and C. A. Brebbia (Eds), *Boundary Element Methods in Acoustics*, Kluwer Academic Publishers, Netherlands (1991).

Chapter 11

Geometric reflectors and diffusers

Curved surfaces such as cylinders, pyramids, wedges, and flat surfaces are all common forms in modern architecture. Example treatments using some of these geometric shapes are shown in Figure 11.1. As Figures 3.2, 3.26 through 3.29, 3.39, and 3.40 illustrated, curved geometric shapes find applications in many different places in production and reproduction rooms. In this chapter, the performance and design of geometric surfaces including those shown in Figure 11.1 will be considered.

Triangles (wedges) or pyramids can produce dispersion, redirection, and specular reflection depending on their geometry. Applied correctly, triangles and pyramids can form notch diffusers, where for some source angles, the energy in certain receiver directions is much reduced. Curved surfaces are more obviously diffusers and more universally used; indeed, a simple sphere or cylinder is very effective at spatially spreading reflections, but this is not the only ingredient needed for a good diffuser. Furthermore, in practice, many spheres or cylinders are needed to cover a reasonable area, then the scattering is as much about how the objects are arranged, periodically or randomly, as about the scattering characteristic of the individual sphere or cylinder. A well-designed curved surface has the advantage of blending with modern architectural designs.

While most diffuser design is about breaking up wavefronts by surface roughness or impedance changes, it should be remembered that even a plane surface can cause significant diffraction from its edges, provided that the surface has a similar size to the acoustic wavelength. Furthermore, an understanding of scattering from finite-sized surfaces is fundamental to diffraction and diffuse reflection; consequently, this chapter starts by looking at this.

11.1 PLANE SURFACES

Whether by accident or design, plane surfaces are probably the most common architectural form. Consequently, understanding the reflection effects of finite-sized plane surfaces is important. Without surface roughness, any dispersion is generated by edge diffraction. The effects of edge diffraction from more complex surface topologies can also be partly or fully explained using the concepts given here.

11.1.1 Single-panel response

Consider the geometry shown in Figure 11.2, where a source and a receiver are near a finite-sized plane surface. The surface is assumed rigid, hard, and non-absorbing. If the source and receiver are chosen so that the geometric reflection point, the point at which the angle of incidence equals the angle of reflection, lies on the panel, then the scattered pressure as a function of frequency, as shown in the top line in Figure 11.3, resembles an approximate

Figure 11.1 Top left to right: Golden GRG pyramid, Thermoformed Kydex curvilinear optimized bicubic diffusor (Harmonix); Bottom left to right: Optimized GRG spline (Waveform Spline), GRG randomized wrinkle (Planarform Wrinkle). (Courtesy of RPG Diffusor Systems, Inc.)

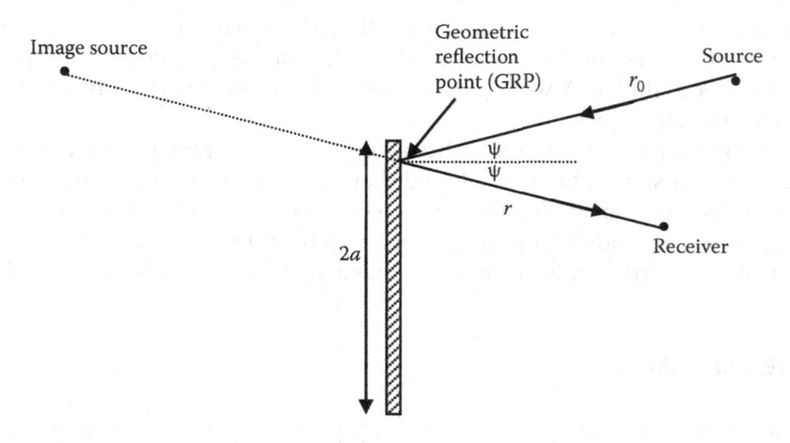

Figure 11.2 Geometry for sound reflecting from a plane surface.

high pass filter response. At very low frequencies, when the wavelength is very large compared to panel size, little or no sound is reflected from the surface. At very high frequencies, when the wavelength is small compared to the surface, specular reflection results.

It is useful to define the cutoff frequency for the plane reflector, which gives the transition frequency between specular reflection and significant diffraction. To continue with the filter analogy, the transition frequency can be taken as the –3 dB point of the high pass filter.[1–4] This gives an approximate frequency below which the panel most effectively scatters sound in all directions and above which the panel produces more specular-like reflections. Rindel[4] has derived a simple and useful formulation for this. He used a simplified Fresnel solution for the scattering from a plane surface, with the Fresnel integrals approximated by simple

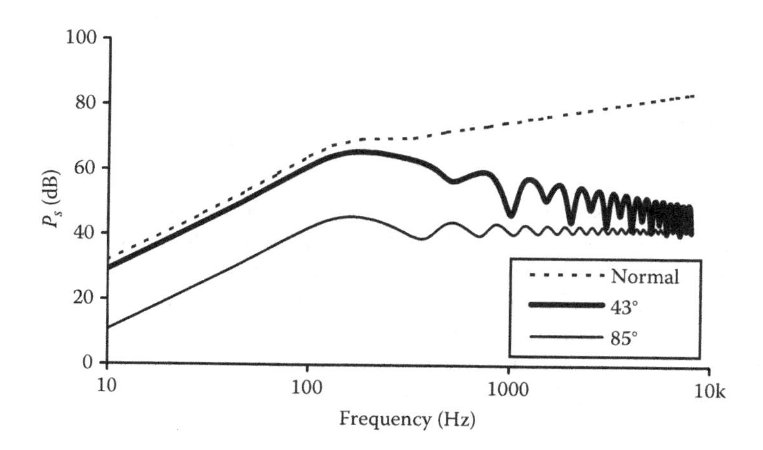

Figure 11.3 Scattered pressure level from a plane surface for three different receiver angles as shown in the legend. Normal incidence sound.

mathematical functions. This enabled Rindel to define a transition frequency above which the Fresnel integrals remain roughly constant. He defined this point as the cutoff frequency. For a plane panel, it is given as

$$f_{-3\mathrm{dB}} = \frac{cr^*}{8a^2\cos^2(\psi)}, \tag{11.1}$$

where r^* is given by

$$r^* = \frac{2rr_0}{r + r_0}, \tag{11.2}$$

where r_0 is the distance from the source to the panel centre, r is the distance from the receiver to the panel centre, $2a$ is the panel width, c is the speed of sound, and ψ is the angle of incidence.

The use of a cutoff frequency is most valid for receivers close to the specular reflection direction. Figure 11.3 shows predictions for reflection from a plane surface for two oblique reflection angles. This shows that representing the reflected pressure by a simple high pass filter does not work for every direction. In this case, there is often a complicated pattern of minima and maxima. When the geometric point of reflection lies on the surface of the panel, it is reasonable to assume that the reflected pressure at high frequencies is going to be dominated by specular behaviour. When the geometric reflection point does not lie on the surface, however, the reflected pressure is entirely due to diffraction. In this case, the diffracted energy reaching the receiver will decrease as the frequency increases. Consequently, the frequency response for these receivers is more likely to follow something closer to a band pass filter response. This is illustrated by the 43° receiver in Figure 11.3. This is not always true, however, as the grazing reflection case illustrates. A rough guide to the region over which the cutoff frequency representation works is therefore the region over which the geometric reflection point lies on the panel. Incidentally, to simplify the calculation of these angles, an image source construction is a good idea as it greatly reduces the complexity of the trigonometry; this is shown in Figure 11.2.

For a plane panel, receivers close to the specular reflection direction are usually of most interest, as they will have the largest amount of the reflected energy at high frequencies. Nevertheless, with significant energy scattered into other angles at low frequencies, the use of a cutoff frequency should be carried out with caution. Equation 11.1 has assumed either a square panel, where the azimuth and elevation incident angles are the same, or a 2D world. For rectangular panels, and arbitrary incidence angles with square panels, there will be two different cutoff frequencies to consider. For circular or odd-shaped panels, the transition will be more complicated, but similar general principles to that shown in Figure 11.3 apply.

Figure 11.4 shows the total sound field impulse response—incident (direct) plus reflection sound—for a plane surface. The direct and reflected sounds are clearly distinguishable, as is the edge diffraction wave, which has a negative magnitude. Figure 11.5 shows the frequency response for the total sound field. Unless the panel is small, the reflected sound from a plane panel is very similar to the incident sound, and so the frequency response shows *comb filtering*. Comb filtering is characterized by minima and maxima at a regular spacing in frequency. The ear is particularly sensitive to this emphasis and deemphasis of frequency components, and when audible, listeners will complain of harshness or glare from these reflections (see Section 5.4.2).

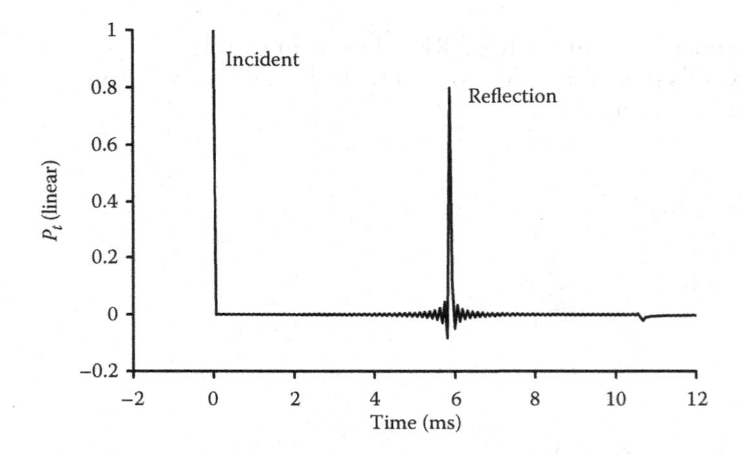

Figure 11.4 Time response for incident and reflected sound from a plane surface (N.B., not the same case as Figure 11.3).

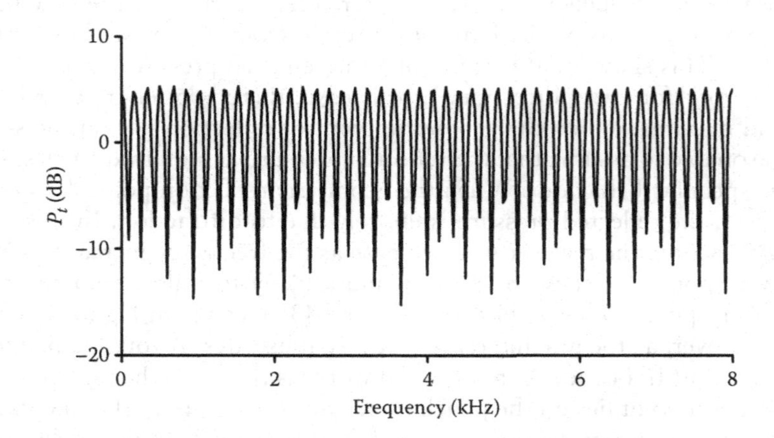

Figure 11.5 Total field frequency response for a plane surface.

In diffuser design, the ability of a surface to disperse the sound spatially is often monitored. Figure 11.6 shows the reflection from a plane, thin, rigid surface for several frequencies in the far field. For the largest wavelengths (lowest frequencies), the scattered response is exactly the same as that produced by a dipole, following a $|\cos(\psi)|$ function. At grazing angles, there is zero pressure. This happens only for the infinitesimally thin surface, since surfaces with finite thickness produce finite pressure. The pressure for all angles is relatively low for large wavelengths, as destructive interference is the dominant phenomenon, as is true of dipoles. To put this in less technical language, when the wavelength is much larger than the panel size, the wave does not *see* the panel and propagates largely undisturbed.

As the frequency increases and the wavelength becomes comparable and then smaller than the panel size, eventually, a specular reflection becomes apparent. Energy is concentrated in the specular reflection direction obeying the reflection law, where the angle of incidence equals the angle of reflection. This is a special case of Fermant's principle, where the specular reflection direction is the shortest possible path length and so is preferred.

Figure 11.6 presents the far field response. In real spaces, however, listeners and sources can be quite close to surfaces. Figure 11.7 shows that the reflected pressure distribution varies for a high frequency, as the receiver distance varies. At 0.8 m from the panel, the receiver arc diameter is actually smaller than the panel width. For all receivers on this arc, the reflected pressure is high because for every receiver, there is a geometric reflection point on the panel, giving a strong specular reflection. As the receiver arc moves further from the panel, fewer receivers get a strong reflection; eventually, the far field response is achieved.

Figure 11.7 implies that, close to the panel, the flat surface is good at dispersing sound. In particular, good coverage is achieved because all receivers get similar energy in the reflection. This does not mean, however, that the plane surface is a good diffuser. In reality, the

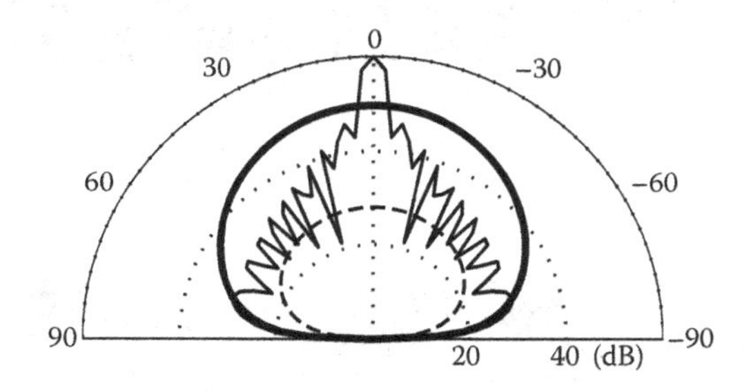

Figure 11.6 Scattered pressure level from a plane panel for three different frequencies: - - - - $\lambda = 20a$; ▬▬ $\lambda = 2a$, and —— $\lambda = 0.2a$. Thin panel BEM prediction.

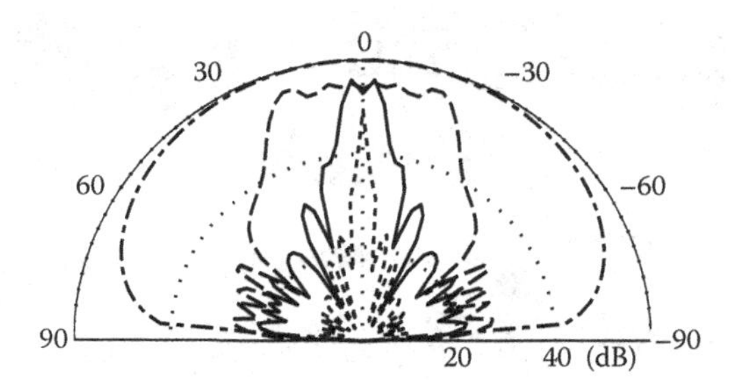

Figure 11.7 High-frequency scattering from a plane surface 1 m wide (= 30λ) for four different receiver radii: -·-·-·- 0.8 m, - - - - 1.2 m, —— 5.0 m, and ······· 1000 m.

plots in Figure 11.7 are telling only part of the story. The polar plots of scattered energy do not show how the direct and reflected sounds interfere and the effect that this has on the sound heard by the listener. In fact, a comb filter response would result, and this is likely to colour the sound.

Figures 11.8 and 11.9 show two applications of flat reflective surfaces. Figure 11.8 shows the surface treatment at Centro de Bellas Artes Humacao in Puerto Rico. The side walls and balcony front are treated with large, constrained layer damped, angled wooden panels. The

Figure 11.8 Angled side wall panels at Centro de Bellas Artes Humacao, Puerto Rico, Acoustician: Jorge Rocaforte Consulting. (Courtesy of Jorge Rocaforte Consulting.)

Figure 11.9 Angled reflective side wall panels at Baldwin High School Auditorium, Pittsburgh, PA. Acoustician: vizzAcoustics. (Courtesy of vizzAcoustics.)

acoustic consultant wanted a rigid reflective surface that would not be diaphragmatic and so a constrained layer damped panel consisting of viscoelastic material sandwiched between 19 mm plywood was used. The consultant determined the angled design to provide uniform reflection coverage from the large panel sections. In the second example, Figure 11.9, the consultant meticulously developed a set of angled panels to provide a uniform temporal distribution of high-level specular reflections to the audience in this school auditorium. The rigid panels were provided with angled mounts for simple installation in the field to ensure that the desired angles were insured.

11.1.2 Panel array response: far field arc

When multiple plane panels are used in an array, the reflection is a combination of both the response of a single panel and the effect of the periodic arrangement. Figure 11.10 shows a sketch of an array that will be used to demonstrate this. For simplicity, reflection in one plane predicted using a 2D model will be used. The findings can be generalized to a 3D array, as the principles are the same. Using the simple Fourier theory detailed in Section 9.4, it is possible to represent the array response, p_a, as a multiplication of the single-panel response and a set of delta functions:

$$p_a(\beta) = p_1(\beta) \sum_{m=-\infty}^{\infty} \delta\left(\beta - \frac{m\lambda}{W}\right), \tag{11.3}$$

and

$$\beta = \sin(\psi) + \sin(\theta), \tag{11.4}$$

where β is the transform variable, as discussed in Chapter 10; ψ and θ are the incidence and reflection angles, respectively; p_a is the pressure from the array; p_1 is the pressure from a single panel; m is an integer; λ is the wavelength; $2a$ is the single-panel width; W is the repeat distance; and δ is the delta function.

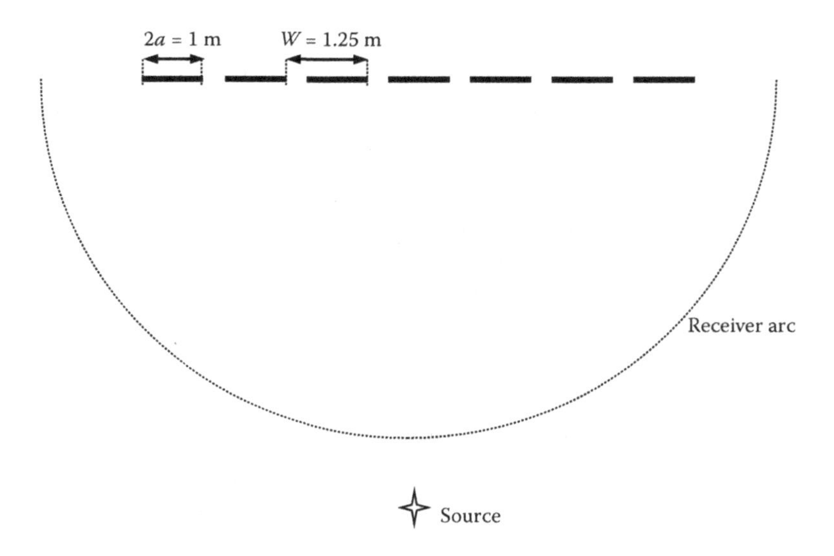

Figure 11.10 Sketch of an array of plane panels tested (source and receiver positions not to scale).

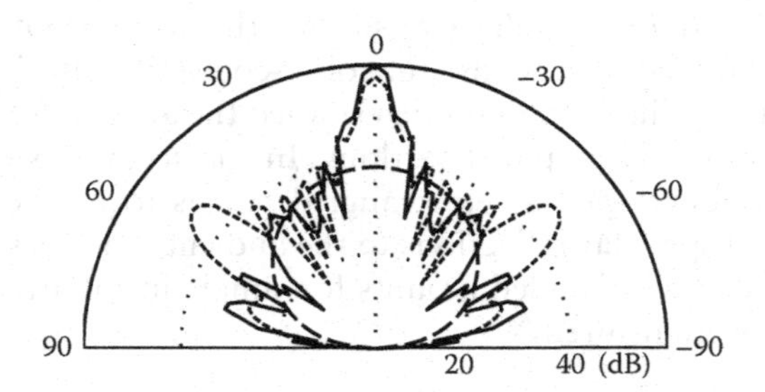

Figure 11.11 Scattering from an array of plane panels for three different frequencies: – – – 34 Hz (λ = 10·2a), ------ 340 Hz (λ = 2a), and —— 10 kHz (30λ = 2a). The $1/\sqrt{f}$ variation that naturally occurs with line sources has been removed to make comparison easier.

This formulation is for the far field. It is an approximate representation, and so the graphs that are being shown are actually generated by an accurate boundary element method (BEM) model (see Section 9.1). Equation 11.3 is being used purely to aid in understanding of the physical processes. This formulation uses similar arguments to those used for modulated forms of Schroeder diffusers in Section 10.5, where the concept is given in a little more detail.

Figure 11.11 shows the reflections for three contrasting frequencies. The last term in Equation 11.3 means that it would be expected that whenever

$$\beta = \sin(\psi) + \sin(\theta) = \frac{m\lambda}{W}, \tag{11.5}$$

there should be a reflection similar to the single panel alone. For the middle frequency in Figure 11.11, Equation 11.5 predicts lobes at 0°, ±53°, and this is borne out by the prediction of three grating lobes. Consequently, at mid frequencies, periodicity effects will often dominate, and Equation 11.5 will predict the location of major lobes.

At low frequencies, the reflection from a single panel is relatively small (20 dB less in the specular reflection direction) and follows a dipole response as the wavelength is large compared to panel size. In this case, the single-panel response, $p_1(\beta)$, dominates the reflected level. The array produces a polar response that is very similar to a single panel, albeit with an increased power due to the greater surface area of the array of panels compared to a single panel. There are no periodicity lobes because the wavelength is so large that only the zeroth-order mode ($m = 0$) can exist in the far field.

At the highest frequency, a strong specular reflection dominates. Equation 11.5 predicts a large number of side lobes (70), but these are not seen. The reason for this is that the response of the single panel, p_1, is highly directional, as was shown in Figure 11.6 ($\lambda = 0.2a$ line). Consequently, most of the side lobes have very low levels. In fact, the reflection from the array of panels is not too dissimilar to that of a single panel, except for a change in radiated power due to the greater surface area of the array.

11.1.3 Panel array response: near field

Simple reflector arrays are often used above stages and audiences in auditoria. In this case, it is not just the response in a far field arc that should be considered, but also the response at application-realistic source and receiver positions. In many cases, this will be along a straight line 5–12 m below the reflector array. This produces a response that is quite different in character to the far field arc. Figure 11.12 shows the reflection from the same array as shown in Figure 11.10 with a far field source and a line of receivers 8 m below the array running parallel to the array. (A far field source is used to simplify matters, but in reality, the source would also be in the near field.) Figure 11.12 also indicates the panel locations.

At high frequency, the specular reflection from each panel is apparent. The scattered pressure is uneven, with a minimum where the geometric reflection point for a receiver is between panels, and a maximum where the geometric reflection point lies on a panel. For most designs of overhead canopies, this uneven response is undesirable. Due to these absences between reflectors at high frequency, and the strong specular reflections between, shaped elements are usually used, such as arcs, to disperse energy more evenly to all receivers. This will be discussed in more detail in Section 11.5.5.

For the middle two frequencies (340 and 3.4 kHz), the response is a complex pattern of minima and maxima. These are near field effects (the 10 kHz case was also in the near field, but the high directivity of the individual panel response made the near field effects less important). The rapidly changing path length differences from the array to the receiver, as the receiver location is moved along the x axis, cause a multitude of minima and maxima. The lowest frequency is in the far field, so something like the dipole response seen previously for the far field arc is obtained.

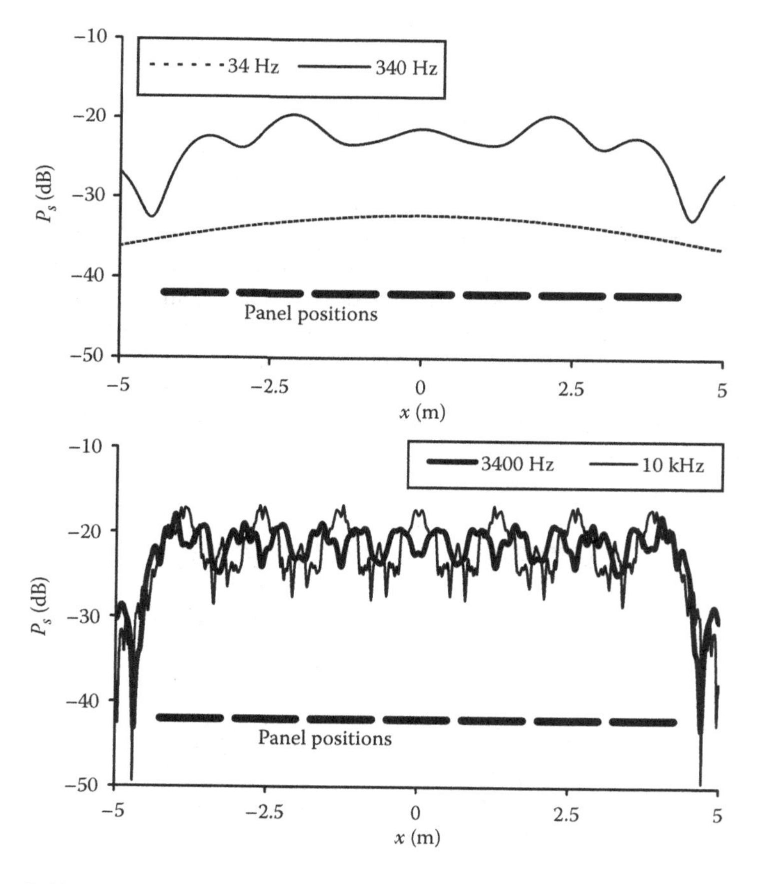

Figure 11.12 Near field scattered level from an array of plane panels along a straight line for various frequencies. The panel positions in the x direction are shown at the bottom of both pictures.

Rindel[5] used Fresnel theory to investigate arrays of ceiling reflectors. He used square reflectors and investigated the effect of reflector density on the frequency response. He found that if the geometric reflection point lay on a panel, a high pass characteristic with some similarity to the single-panel response was obtained. Due to the fact that the reflections come from multiple panels, the actual frequency response had many more local minima and maxima than was the case for the single panel alone.

If the geometric reflection point was between panels, however, the scattering had a low pass filter response. In the latter case, the energy is greatest when the diffraction is greatest, and this occurs at the low frequencies. The reflected energy for these receivers is small at high frequencies because the energy is concentrated in specular directions. Rindel showed that using smaller panels was advantageous, as it reduced the roll-off at high frequencies for receivers away from the geometric reflection points.

Either the size of the reflectors or the panel density (1 – the fractional open area of the array) determines the low-frequency performance. It is also possible to imagine cases where it is a combination of these effects that is important. The mid- and high-frequency performance is dominated by strong local variations due to the size of the reflectors and the repeat distance between them. The solution to this is to use corrugated surfaces, as shall be discussed in Section 11.5.5. Alternatively, a pseudorandom arrangement of different panel sizes and spacings could be used, but only if the tilt angle is also randomized. To reduce the effects of edge diffraction on the reflection from canopies, it has been suggested that the edges of the reflectors can be treated with diffusing elements.[6]

11.2 TRIANGLES AND PYRAMIDS

Triangles, wedges, and pyramids can display a wide variety of scattering behaviour, ranging from a good diffuser to a surface that generates specular reflections; it depends on the geometry. The reflection from an array of triangles or pyramids is very much determined by the steepness of the side slopes. For simplicity, the analysis here considers only a 2D case with triangles, but the arguments can easily extend to 3D surfaces, such as pyramids.

A simple ray tracing yields much about how a triangle reflects sound; Figure 11.13 shows some examples. As the angle (χ) of the triangle varies, the reflection characteristic shifts between a notch response, diffuse reflection, and a specular response. To understand a triangular arrangement, very simple prediction theories based on the Kirchhoff boundary conditions will not always work because they do not model second- and higher-order reflections (see Figure 9.10). Consequently, it is best to use a BEM model or finite difference time

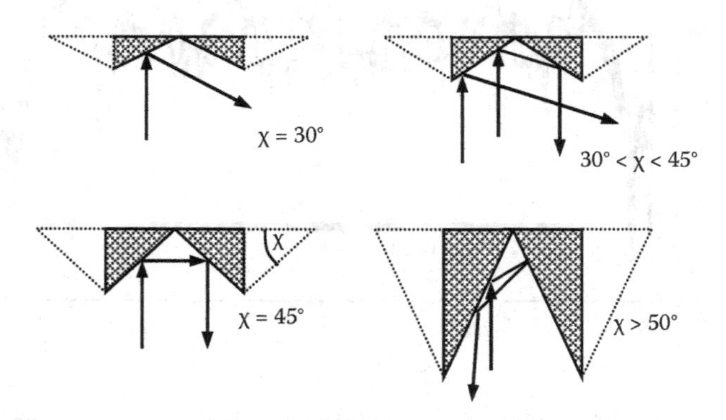

Figure 11.13 Ray tracing of sound reflecting from the centre of pairs of triangles.

domain. Initially, only high frequencies will be considered at normal incidence. A further simplification is that the response for only the centre portion of two triangles will be considered, as this is more representative of what happens when an array of triangles is used.

For shallow angles ($\chi \le 30°$), the ray tracing shows only a single reflection from each side as shown in Figure 11.13. This results in two distinct lobes being generated at angles of $\pm 2\chi$. An example of the far field scattering is shown in Figure 11.14. This then forms a notch response, with the energy in the specular direction being reduced. This is effectively a redirecting surface that generates two strong reflections in different distinct directions. Unlike primitive root diffusers discussed in Chapter 10, this notch extends over a relatively wide frequency range, although the performance will be compromised at low to mid frequencies, when finite-sized panel effects become important. Equation 11.1 could be used as a guide as to when each side of the triangle will produce specular-like reflections and so produce a notch response and when the reflection will be more dominated by edge diffraction. This finite panel effect is true for all the discussion in this section, but these comments will not be repeated again. With a $\chi \le 30°$ triangle, there is only a notch for certain incident angles, whereas a primitive root diffuser worked for any angle of incidence, but only at a few distinct frequencies.

For $30° < \chi < 45°$, a mixture of single and double reflections are apparent in Figure 11.13. The single reflections will again form lobes in the directions of $\pm 2\chi$; the double reflection directions will be in the directions of $\pm(180 - 4\chi)$. Figure 11.14 shows four distinct lobes in the polar response. By choosing an appropriate angle for the triangle, it is possible to have a notch in the specular reflection direction, but now, the reflected energy is spread over four lobes, which is often more desirable. However, the range of incidence angles over which this notch is achieved will be reduced.

$\chi = 45°$ is a special case because the energy is returned back towards the source, as shown in Figures 11.13 and 11.14. This is sometimes termed a *corner reflector.* The ability of a corner reflector to return energy back to sources is relatively well known and has been exploited in some auditoria as a way of reflecting sound back onto the stage to help musicians and actors hear themselves and others. An example of this can be seen in Figure 11.15. Figure 3.36 shows an outdoor stage shell that exploits the same idea, which was designed via optimization using an image source model.

For $45° < \chi < 54°$, double reflections always occur, but these generate only two lobes. An example is shown in Figure 11.16, for $\chi = 50°$. As χ increases beyond 54°, the number of reflections that a ray undertakes before escaping the surface rises. A varying number of

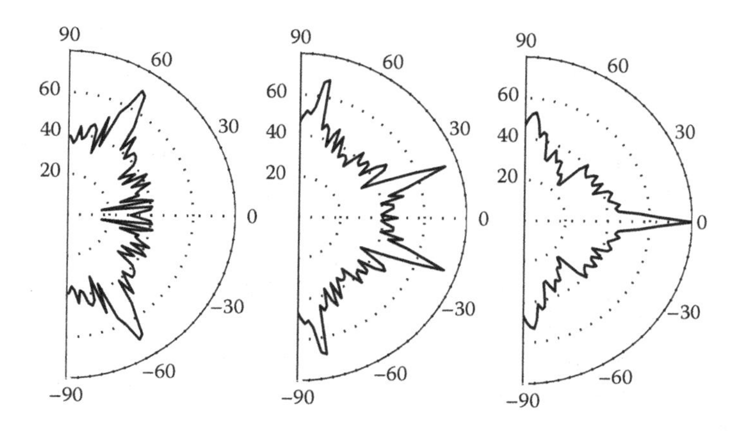

Figure 11.14 Scattered level from the centre of two triangles for three side angles, from left to right: χ = 30°, 40°, and 45°.

Figure 11.15 The upstage wall at Nicholas Music Center, Rutgers University, showing an array of protruding shelves to provide corner cube reflections back to the musicians for support. (Courtesy of RPG Diffusor Systems, Inc.)

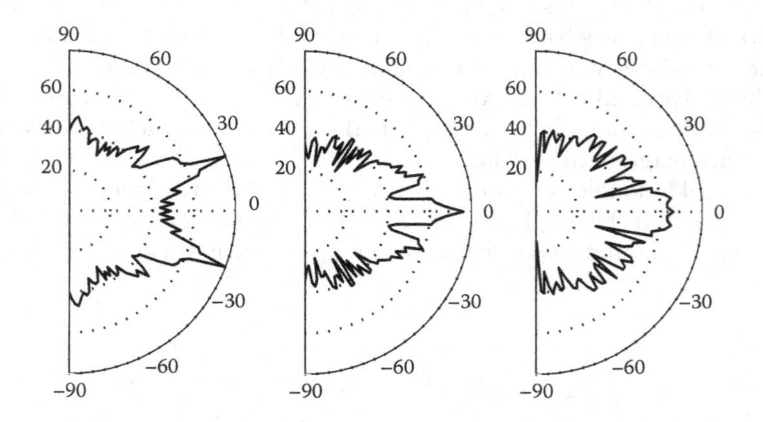

Figure 11.16 Scattered level from the centre of two triangles for three side angles, from left to right: χ = 50°, 80°, and 85°.

clear distinct lobes are still generated and simple ray tracing techniques can still be used to locate the directions of the most significant lobes. The relative level of the lobes varies, however, depending on the reflection paths. When the angle becomes very large ($\chi > 85°$), a single fairly broad lobe appears. The surface is acting like a horn loudspeaker in that a highly directional response is obtained. This occurs because the escape angles for the rays are limited to $\pm(90 - \chi)$. This then returns the energy back to the source, but in a more diffuse manner than occurs with a $\chi = 45°$ surface. The simplistic analysis used here needs to be read with a little caution. Once these devices become very narrow and a great number of reflections occur, then a resonant structure has been formed. Consequently, there is a risk of resonant absorption.

11.2.1 Arrays of triangles

Single triangles are not that useful because, usually, large areas are needed to be covered and then the surface will become too deep, unless this is incorporated into the overall room shape in some way. Consequently, arrays of triangles need to be considered. Figure 11.17 shows the scattering from an array of $\chi = 15°$ triangles compared to a single triangle of the same size. Also shown is the response of two plane panels. An impressively large notch of almost 30 dB is generated for the triangle array case to the plane surface. Additional lobes arise because of periodicity. The location of these can be predicted from Equation 11.5; however, not all the lobes appear. The $m = 0$ and $m = 3$ lobes are attenuated because the single triangle response is weak in those directions. These periodicity lobes can be reduced by using modulation similar to that used for Schroeder diffusers and discussed in Section 10.5. For example, two different triangle sizes could be chosen and arranged according to a pseudorandom number sequence.

Figure 11.17 actually represents a frequency where the notch diffuser is working well. Figure 11.18 plots the drop in the specular direction level as a function of frequency for the single triangle and the array. This shows that at other frequencies, the attenuation is not as good; for example, at 1.5 kHz, the attenuation is only 13 dB. While this attenuation is likely to be audible, it would be better if the performance could be improved. For example, the reflection free zone concept for small room design[7] would typically be trying to achieve a 18–20 dB drop in the first-order specular reflection level.

A brief parametric study looking at different triangle sizes shows that the peaks and dips in Figure 11.18 relate to the triangle depth. The maximum attenuation occurs when the depth is a multiple of $\lambda/2$. Unfortunately, if χ is increased further, say to 30°, this neat story relating the depth of the triangle to the wavelength of the minima and maxima is no longer true. Nevertheless, these results again lead to the thought that orthogonal modulation could be useful, following the ideas developed by Angus for Schroeder diffusers outlined in Section 10.5. By using two or more different depth triangles, so that their frequency bands of higher specular reflection energy are different, it should be possible to deepen the notch

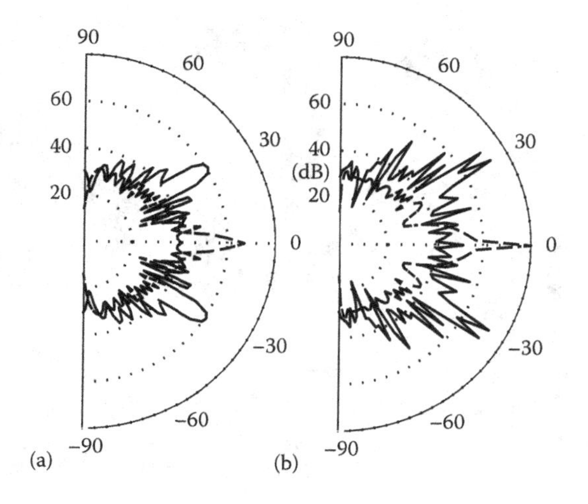

Figure 11.17 Scattered level from triangles and plane surfaces. - - - - plane surface, —— triangles. (a) One triangle; and (b) five triangles.

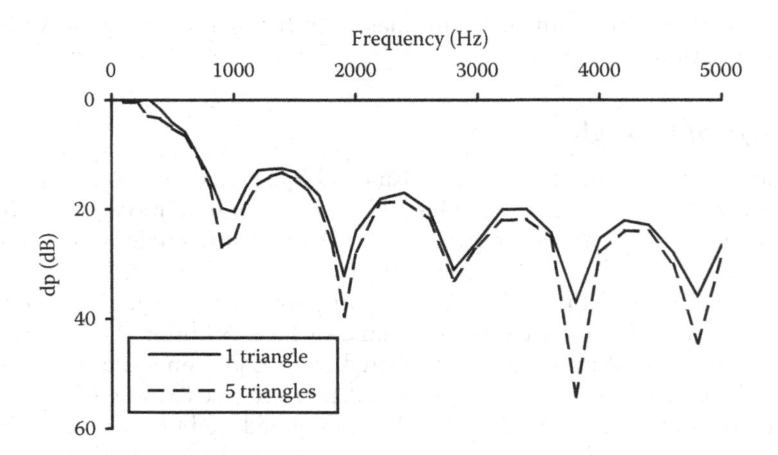

Figure 11.18 Decrease in specular zone pressure level when a triangle diffuser replaces a plane surface for different number of triangles. (Note: *y* axis plotted in reverse order.)

generated. By using two different depth triangles it might be expected that the improvement would be of the order of 3 dB, so a large number of different triangles would need to be used. Another solution to the problem is just to make the triangles much deeper, although overly deep surfaces are often not possible for non-acoustic reasons such as cost, weight, and visual appearance.

An example of a pyramid ceiling treatment can be seen in Figure 11.19. The acoustician wanted to use a pyramidal ceiling design for economy, aesthetics, and acoustics. The 1.2 × 1.2 m asymmetric pyramids are fabricated from glass reinforced gypsum to minimize diaphragmatic absorption and the sides are designed according to the Golden Ratio. The offset

Figure 11.19 Pyramid ceiling treatment at Southland Christian Church, Nicholasville, KY, Acoustician: Michael Garrison Associates. (Courtesy of Michael Garrison Associates.)

vertex allows the pyramids to be aperiodically mounted in the ceiling, by rotating adjacent pyramids by 90°. This spreads reflections as uniformly as possible to the congregation and reduces grating lobe effects. An offset vertex also avoids all the pyramid sides having the same angle, which can lead to sound being reflected into only a few directions (see examples in Figures 11.14 and 11.16).

11.3 CONCAVE SURFACES

Concave surfaces such as domes are often an acoustician's nightmare. Used wrongly, they lead to focussing effects that generate strong reflected energy in certain places. This can lead to uneven energy distribution across the room, as well as echoes and coloration of timbre.

Whether an arc causes problems depends on the positions of the sources and receivers and the radius of the arc. Figure 11.20 shows the reflection from a concave arc at a mid–high frequency for different receiver radii. Figure 11.21a, schematically shows a ray tracing of the surface scattering. Figures 11.20 and 11.21 show that the focussing effect of the curved

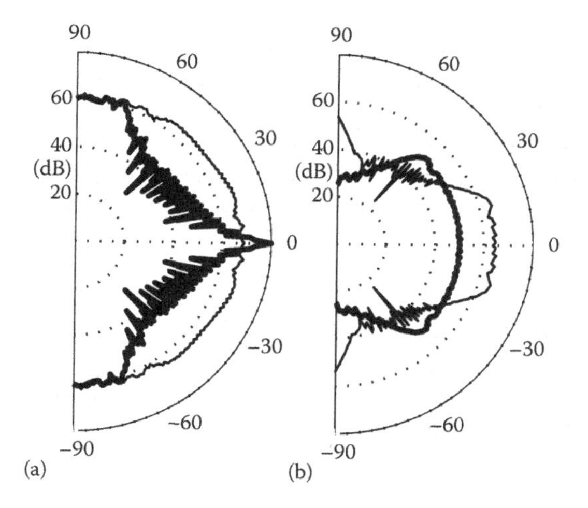

Figure 11.20 Reflected level from a concave arc for various receiver radii, r. r_f is the focal length of the concave surface. (a) —— $r < r_f$, ▬▬ $r = r_f$; (b) —— $r > r_f$, ▬▬ $r >> r_f$.

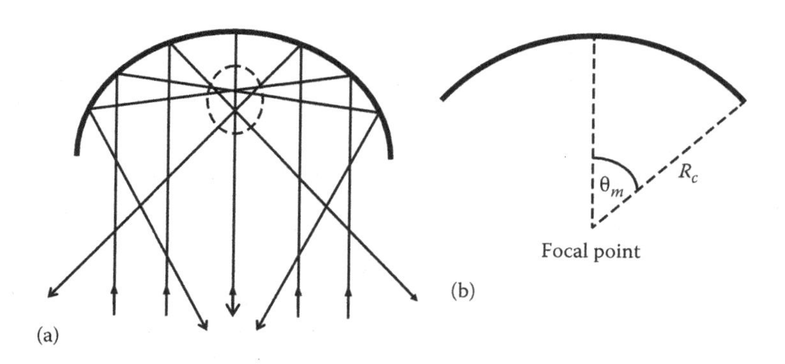

Figure 11.21 (a) Ray tracing for reflection from a concave arc. The region marked with a dashed line will receive higher reflection levels than elsewhere. (b) Definition of geometry for formulations.

surface is a problem only for some receiver distances, close to the focal length of the arc. Consequently, it is possible to use a curved surface provided the focus of the surface is away from the listeners. For example, any concave ceiling in an auditorium should focus the sound well above or below the audience. If the focus is well above the audience, then the concave surface can paradoxically cause dispersion, but not as well as many other surface topologies. On the other hand, if the focus is below the audience, although the focussing may not be heard, the concentration of median plane sound from above may not be desirable for other acoustic reasons.

Vercammen[8,9] has carried out an extensive study into sound focussing from spherically concave surfaces, such as hemispheres or segments of spheres, including developing formulations to predict the reflected sound pressure level. For a deep concave surface made from a section of a sphere, the width of the focal *point* and the peak pressure are functions of wavelength. For example, for a hemisphere, the area of the focus extends over about half-a-wavelength for sources and receivers not far from the centre of the hemisphere. This means that at high frequency, the extent of the focus is small, but the amplification is high. But because the extent is small, this amplification is not as much of an audible problem as it is at low and mid frequencies.

Away from the focal point or plane, the average sound pressure level from segments of cylinders and spheres can be estimated via geometric techniques. Common geometric methods such as image source, beam tracing, and ray tracing struggle to predict the pressure and extent of the focal area. Wave-based models, in contrast, can accurately predict the pressure at the focal point. For a segment of a sphere, the focal energy is a function of the wavelength and opening angle of the sphere section; it is independent of the sphere radius. Vercammen found that the increase in sound pressure level, ΔL, relative to the direct sound is

$$\Delta L = 20 \log_{10} (kr_d[1 - \cos(\theta_m)]), \tag{11.6}$$

where k is the wavenumber, θ_m is the opening angle (see Figure 11.21, right), and r_d is the distance from the source to receiver.

While a sphere returns all sound to its focus, a cylinder has dissipation along its axis. This means that for a cylinder, the formulation includes the radius of the cylinder, R_c. For a half circular cylinder, the increase in energy is

$$\Delta L = 10 \log_{10}\left(\frac{\pi k r_d^2}{4R_c}\right). \tag{11.7}$$

The focal energy is much higher for a hemisphere than a cylinder. For example, if the source and receiver are 1 m apart near a hemisphere, the focussed reflected energy exceeds the direct sound by 19 dB at 500 Hz. In contrast, for a 7 m hemicylinder, the direct and focussed reflections have a similar sound pressure level.

If a concave arc with a focus on listeners is inevitable, there are a few solutions: either treat the surface with an absorber or a diffuser or use reflectors to prevent sound from reaching the curved surface. Absorbers can be placed on the surface to remove the reflection, although Vercammen[9] shows that while absorption can be used to treat cylindrical surfaces, the focusses from large spherically concave surfaces such as hemispheres are too powerful for surface absorbers to work alone. The reverberation time requirements of the room should also be considered when applying absorption. Also, there might be a desire to produce some reflected energy from the concave surface, perhaps to provide ensemble reflections to musicians or early reflections to the audience to improve spaciousness or clarity.

Diffusers can be placed on the concave surface to break up the reflected wavefront and so disperse the focus, without losing acoustic energy. Figure 3.38 showed a scattered polar distribution for a concave arc before and after treatment with an optimized curved diffuser. Figure 3.39 showed the curved surface used. The reduction in focussing is dramatic. For large spherically concave surfaces, these diffusers need to be designed to scatter energy away from the focal point. They cannot be general diffusers designed to scatter in all directions because then they do not provide enough attenuation at the focal point.

A final treatment is to use volumetric diffusers as described in Section 8.3. An example would be the *mushrooms* that are hung in the Royal Albert Hall, London (Figure 3.37). These are some distance from the grand dome and prevent much of the sound reaching the concave surface, consequently reducing the focussed echoes.

11.4 CONVEX SURFACES

A single cylinder is an efficient disperser of sound in one plane and a single sphere is efficient at dispersing in all directions. They generate responses that mimic the behaviour of radiating line and point sources, respectively. Figure 11.22 shows the scattering from a semicylinder for three frequencies. The radial axis range of this graph is only 20 dB so at all frequencies, the response from the semicylinder is fairly omnidirectional. The 400 Hz response is where the wavelength is roughly the width of the semicylinder and so some edge diffraction effects are seen. The lowest frequency, 40 Hz, is not omnidirectional because the rear of the semicylinder becomes important—it would be more omnidirectional if a cylinder had been modelled. At the two highest frequencies, the reflected level varies by only 2–3 dB over the receiver arc.

It might appear that the cylinder is the ideal diffuser—the Holy Grail of diffuser designs—but this unfortunately is not the case. A single cylinder on its own is rarely of much use. The example given in Figure 11.22 was 0.5 m deep—already deeper than many architects allow—and it was only 1 m wide, which is not wide enough for most applications. One solution is to use multiple cylinders in an array. Then the reflected response is dominated by how the cylinders are arranged, and the perfect response from a single cylinder becomes a secondary and less important issue. Another solution is to flatten the cylinder into an ellipse, but then, the perfect angular dispersion will be lost for oblique sources.

There are also issues with cylinders and the total sound field response. Figure 11.23 shows the impulse response for a direct sound and a reflection from a large semicylinder. The

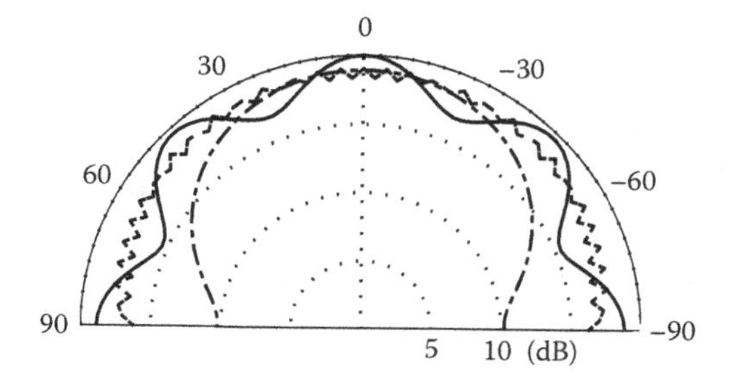

Figure 11.22 Scattered pressure from a 1 m-diameter semicylinder for various frequencies (the $1/\sqrt{f}$ change in overall sound pressure level due to the Green's function has been removed by normalization): -·-·-·- 40 Hz ($\lambda = 8.5d$, d = cylinder diameter), —— 400 Hz ($\lambda = 0.85d$), and - - - - 4000 Hz ($\lambda = 0.085d$).

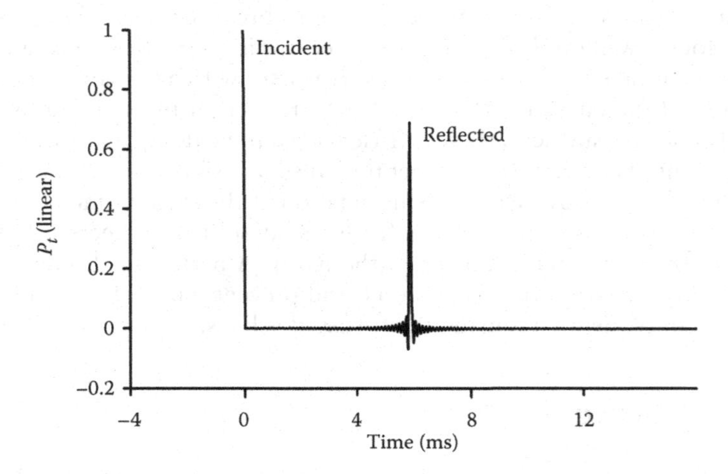

Figure 11.23 Incident and reflected time response for a semicylinder the same width as the plane surface used for Figure 11.4.

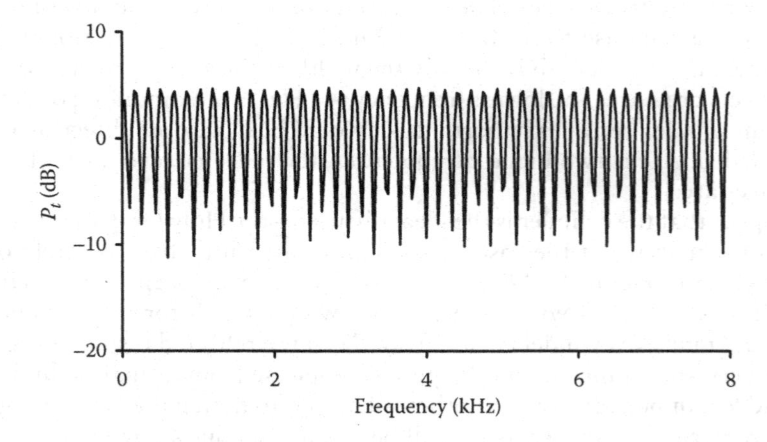

Figure 11.24 Frequency response for the direct sound plus a reflection from a large semicylinder.

semicylinder has the same width as the plane panel used for Figure 11.4 to allow direct comparison between the surface types. The reflected sound is attenuated because of spatial dispersion, but the time signature is still very similar to the reflection from the plane surface. Consequently, large semicylinders produce comb filtering similar to that generated by plane surfaces, as shown in Figure 11.24; however, the minima and maxima variation is over a smaller magnitude range for the semicylinder. The comb filtering is thought to give rise to the harsh sound that some large semicylinders generate, although a more detailed set of perceptual tests would be needed to confirm this. Certainly, semicylinders are an enigma; they appear to be a near perfect diffuser from dispersion graphs, but they do not sound like a perfect diffuser.

11.4.1 Geometric theory and cutoff frequencies

One method that has been proposed to predict the reflection from a curved surface is geometric theory.[3,10] The reflection is split into two processes. First, the diffraction from the finite-sized panel is considered and then the effect of curvature is added. The finite-sized surface effect is predicted using Fraunhofer or Fresnel theory. The effects of curvature are accounted for by beam tracing. If a curved surface is illuminated by a beam with parallel sides, the reflection will diverge due to the curvature of the surface as shown in Figure 11.25.

If the wavelength is assumed to be small compared to surface size, then simple geometric constructions can be used to calculate the attenuation due to curvature. For plane waves, this is given by[3]

$$\text{attenuation} = \left| 1 + \frac{r^*}{R_c \cos(\psi)} \right|, \tag{11.8}$$

where the composite radius r^* is defined in Equation 11.2, R_c is the radius of curvature of the panel, and ψ is the angle of incidence and reflection. It is also possible to produce a formulation for spherical waves.[10] Figure 11.26[11] shows the reflected pressure versus frequency for the specular reflection direction for the two theories and also a prediction by a 3D BEM. It is assumed that the BEM gives accurate results, and so this indicates that the geometric

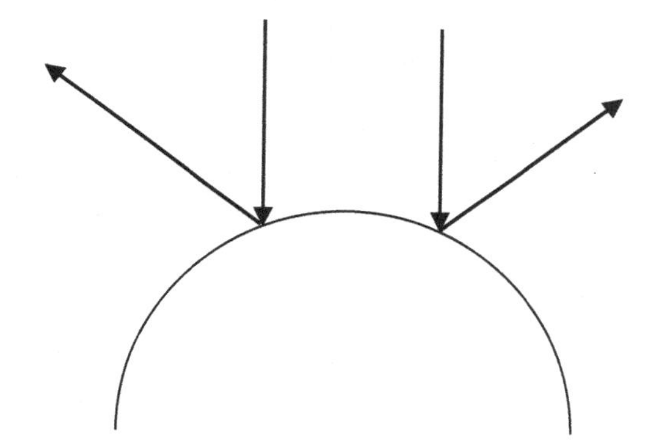

Figure 11.25 Effect of curvature on a sound beam.

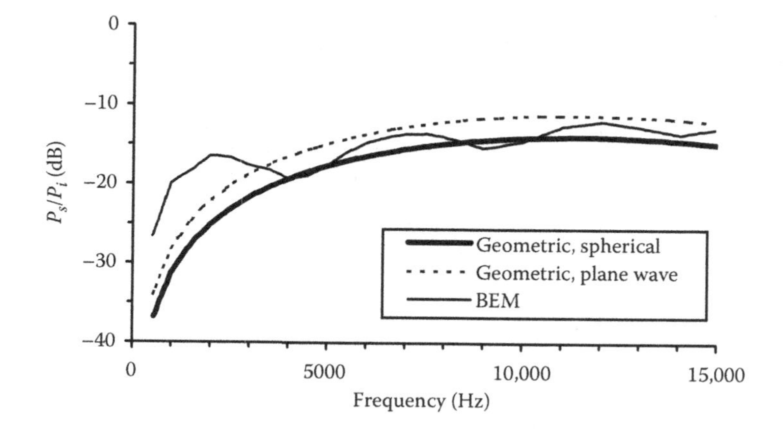

Figure 11.26 Prediction of scattered levels for a curved surface using three theories. (After Cox, T.J., "Objective and Subjective Evaluation of Reflection and Diffusing Surfaces in Auditoria", PhD thesis, University of Salford, 1992.)

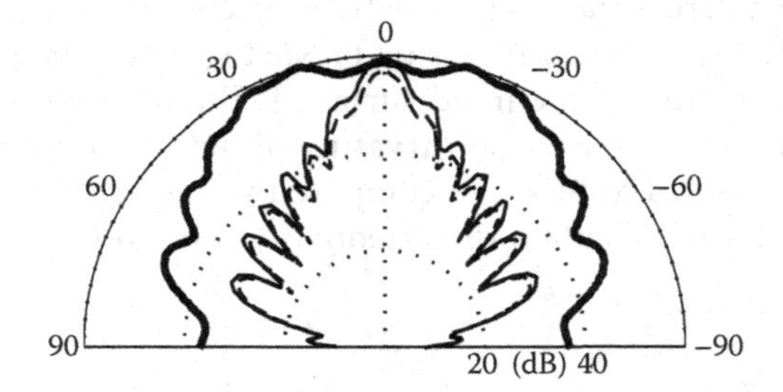

Figure 11.27 Prediction of scattered levels from a curved surface using three theories: —— BEM; ▬▬ geometric, plane wave and geometric; and - - - - spherical wave. Normal incidence. (Data from Cox, T.J., "Objective and Subjective Evaluation of Reflection and Diffusing Surfaces in Auditoria", PhD thesis, University of Salford, 1992.)

model works to a certain degree—the magnitude is approximately right, but the ripples are not predicted. Figure 11.27 shows a scattered polar response, demonstrating that geometric theories are only useful for receivers close to the specular reflection direction. Geometric theories do not work for receivers where the geometric reflection point is not on the surface, because the formulation incorrectly applies an attenuation. For these receivers, the effect of adding curvature should be to increase the scattered pressure, as energy is moved away from specular reflection angles to these receivers.

The cutoff frequency for plane panels was a simple concept that readily allows some simple design principles to be applied. Investigations have shown[11,12] that Equation 11.1 also works for curved surfaces, provided the receivers are close to the specular reflection direction.

11.4.2 Performance of simple curved reflectors

For normal incidence, reflectors based on a segment of a circle have good dispersion performance. Figure 11.28 compares the predicted scattering from a semicylinder and a flattened semicylinder (an ellipse) for normal and oblique incidence. For normal incidence, the semicylinder disperses the sound well, as it generates a virtual line source. For oblique

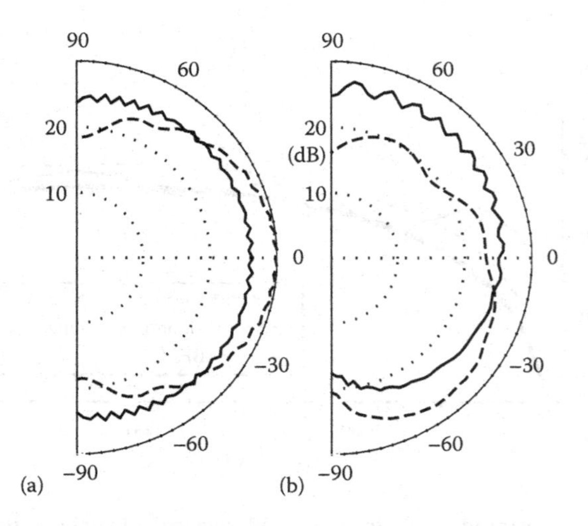

Figure 11.28 Scattered level for two surfaces for normal (a) and oblique (b), 60° incidence: —— semicylinder and - - - - flattened semicylinder.

incidence, however, the performance is poorer for the flattened semicylinder. The flattened semicylinder also has worse performance for normal incidence. Incidentally, the trends are similar if a section of a circle is used instead of a flattened circle—only a complete semicylinder at normal incidence produces near-perfect spatial dispersion. Consequently, if a whole semicylinder cannot be used in an application, a better solution than flattening the semicylinder or taking part of a circle is needed. There are two possibilities: either use an array of semicylinders or use optimization to construct a more complicated curved shape. Arrays are considered in the following section, and complicated curved shapes, in Section 11.5.

11.4.2.1 Arrays of semicylinders

Once cylinders are arranged in an array, the performance is dominated by a combination of the single cylinder response and the effects of periodicity. If the simple analysis surrounding Equation 11.3 is considered, then getting good diffusion is mostly about how the cylinders are arranged. The scattered pressure distribution from one cylinder, $p_1(\beta)$, is constant if second-order reflections are not considered. Consequently, the sum of the delta functions in Equation 11.3 dominates the scattered pressure distribution. Once again, a modulation technique, where the cylinders are not arranged periodically, is needed to change the functional form of the last term in the Equation 11.3 to give more even scattering.

The simplest method to follow is the modulation techniques described for Schroeder diffusers in Section 10.5. This means using two or more semicylinders and arranging them randomly or pseudorandomly on the wall. This will reduce periodicity and so improve dispersion. Figure 11.29 compares a periodic and random array of semicylinders. At mid frequencies, where there are some grating lobes, but not too many, the aperiodic arrangement helps to create extra lobes, thereby improving dispersion.

In this case, a simple diffraction grating with one point source in the middle of each diffuser can produce reasonably accurate predictions of the scattering. At mid frequencies, many simple geometric base shapes mimic point sources.

The overall envelope shows some reduction for reflection angles far from the specular reflection direction. This tailing off is probably due to second- and higher-order reflections from the array—neighbouring semicylinders get in the way of reflected sound. Similar results are seen for oblique incidence.

At high frequency, however, the modulation of the cylinders does not create additional dispersion. An example is shown in Figure 11.30. At these frequencies, the grating lobes are close to one another, and the variation in the minima and maxima is similar for periodic, modulated, and random arrangements. Again, as in the mid-frequency range, there is a gradual tailing off at the edges of the overall envelopes of the polar responses. In the example

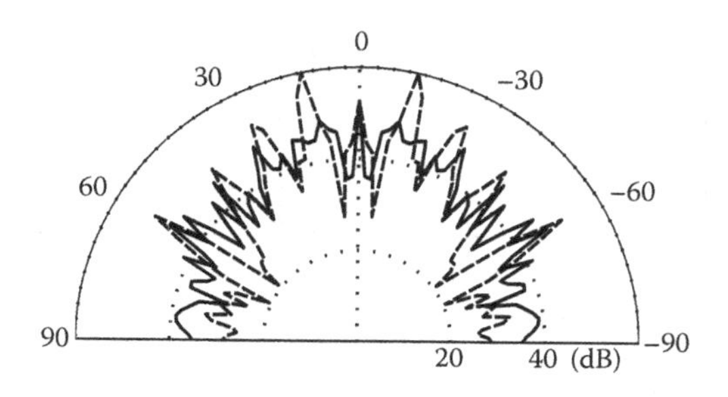

Figure 11.29 Scattered pressure distribution for a periodic and aperiodic arrangement of semicylinders. For the periodic set, $4\lambda \approx d$, where d is the diameter (chord) of the semicylinder: ---- periodic and —— aperiodic.

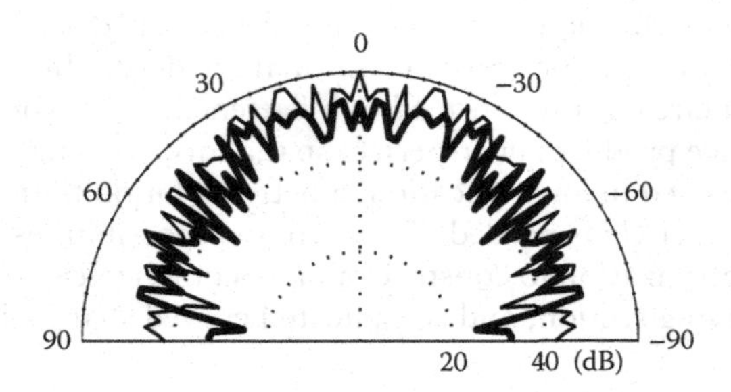

Figure 11.30 Scattered pressure distribution for a periodic and aperiodic (modulated) arrangement of semicylinders. Normal incidence. For the periodic set, $15\lambda \approx d$, where d is the diameter (chord) of the semicylinder: —— periodic and ▬▬ aperiodic.

shown in Figure 11.30, there is roughly a 15 dB drop from normal to grazing receivers. At oblique incidence (result not shown), the modulated arrangement is better at controlling the overall envelope, and a clear improvement on a periodic arrangement is achieved.

As previously discussed, dispersion is only one aspect of diffuser performance, although possibly the most important one. The total field response, incidence plus reflection, should also be considered. Figure 11.31 shows the total field for a periodic set of four cylinders. Only two arrivals are shown because the array was set up symmetrically and so there are only two unique arrival times. Figure 11.32 shows the frequency response for the total sound field. This can be compared to previous graphs, Figures 11.4 and 11.24, for plane and single semicylinder scattering. Using an array of cylinders has not destroyed the comb filtering, but it has been reduced it. It might be expected that coloration will be less audible but it may be still present. Figures 11.33 and 11.34 show the total sound field time and frequency responses for a complicated arrangement of many different-sized semicylinders. The use of a random arrangement of cylinder sizes and shapes has further broken up the regularity of the frequency response, making it less likely that coloration will be heard.

In conclusion, several key features will determine the performance of semicylinder arrays. The low-frequency limit of the diffuser will be determined either by the repeat distance or the diffuser depth. Multiple grating lobes need to be present for dispersion, and hence, repeat distance is important. The issue of depth has not been discussed before for semicylinders and unfortunately

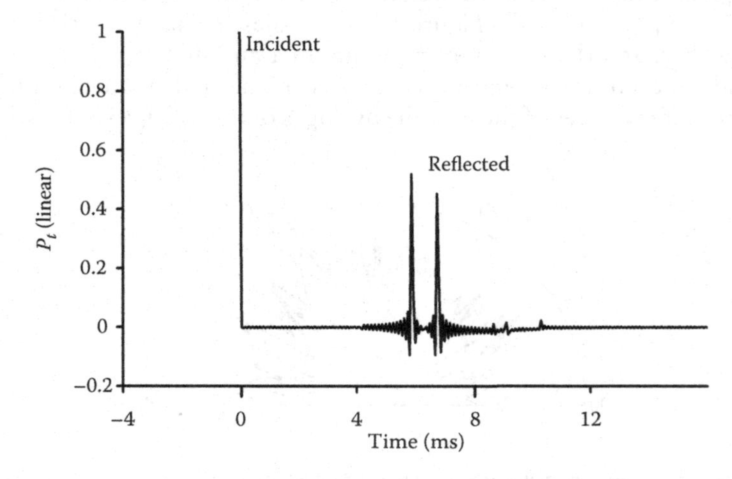

Figure 11.31 Scattered pressure from an array of four cylinders symmetrically arranged about the source (and hence only two unique reflection arrival times).

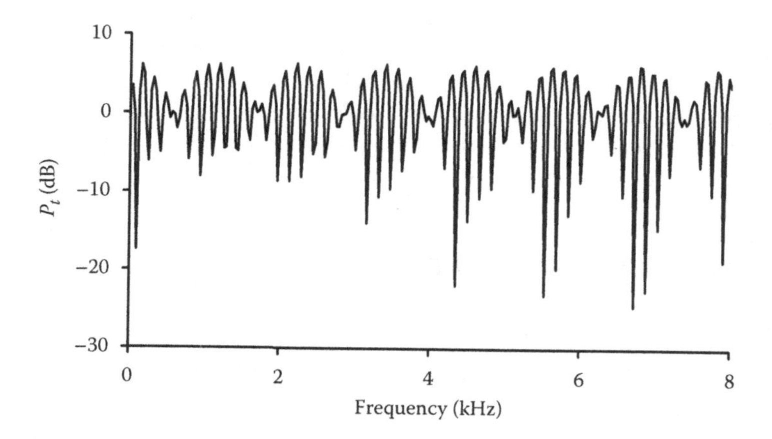

Figure 11.32 Frequency response for four periodic cylinders.

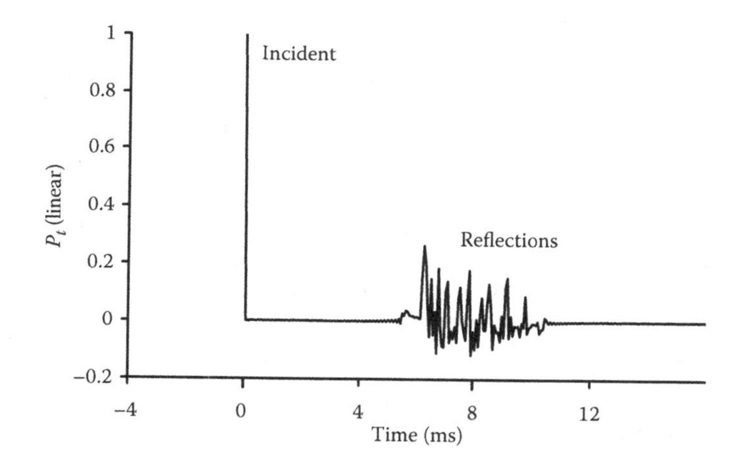

Figure 11.33 Total sound field for a complicated random array of many sized semicylinders.

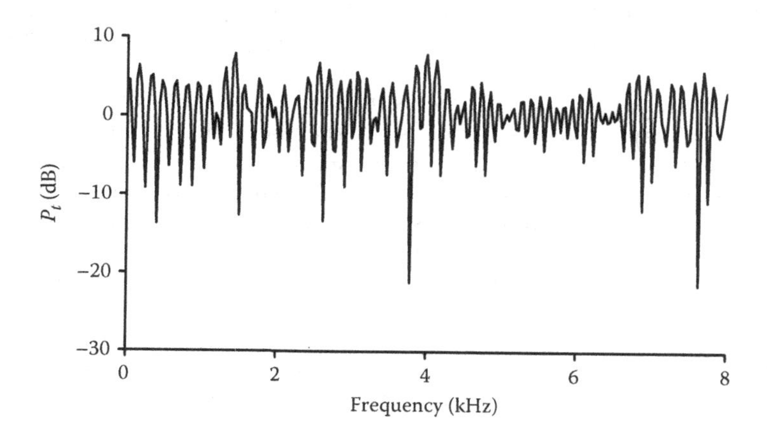

Figure 11.34 Frequency response of total sound field for a complicated random array of many sized semicylinders.

is as likely as not to be set by non-acoustic requirements. If the depth is a determining feature, then empirical results have shown that curvature produces significantly more scattering than a plane panel when the depth is greater than ≈λ/10. The mid-frequency performance is dominated by the arrangement of the semicylinders. Least coloration is achieved by avoiding periodicity or by ensuring that the repeat distance is as large as possible. The high-frequency performance when the number of grating lobes is very large seems difficult if not impossible to control.

11.5 OPTIMIZED CURVED SURFACES

11.5.1 Example application

When designing a diffuser, the requirements of visual aesthetics and acoustics must be considered, and these are often in conflict. Schroeder diffusers may have well-defined acoustic performance, but they do that with a visual appearance that is not always liked. Unless diffusers are visually acceptable to the architect, they are unlikely to be used. Curved surfaces are common in modern architecture. Spurred on by the availability of new materials and computer-aided design, architects are increasingly designing prestigious buildings where large flat surfaces appear to be outlawed in principle.

Curved diffusers are therefore appealing because they can complement modern architectural trends. They can have a form that blends with other structures in a building and they do not have to look like an obvious add-on. The Hummingbird Centre in Toronto is an interesting example[13] (shown in Figure 3.2). Diffusers were required on the side walls of the auditorium because a sound enhancement system was being installed, which would generate echoes across the room unless some surface treatment was applied. The original design specification was for Schroeder diffusers, which were to be cloth wrapped to hide them. After much discussion, a curved surface designed by optimization was accepted instead, and because this fitted the visual appearance of the room, there was no need to hide the diffusers. Other examples of optimized curved surface were shown in Figures 3.26 through 3.29, 3.39 and 3.40.

11.5.2 Design process

Section 10.10 described how it is possible to get a computer to search for the best possible depth sequence for a Schroeder diffuser. It is also possible to get a computer to search for the best curved shape to generate dispersion.[14] For those unfamiliar with optimization, it may be necessary to read Section 10.10 first because the background details of how optimization works are not repeated here.

As with any diffuser optimization process, it is necessary to have a set of numbers that describe the surface shape. These parameters can then be changed by the optimizer to allow the computer to search through possible surface shapes. In the case of Schroeder diffusers, this was straightforward: the parameters were the well depths. For a curved diffuser, a different regime is needed. Any topology can be represented by a Fourier series and so the surface displacement, y, can be represented by

$$y(x) = \sum_{n=1}^{N} a_n \cos(k_x x) + b_n \sin(k_x x), \tag{11.9}$$

where a_n and b_n are the shape parameters that are altered to change the surface shape. k_x is usually set so that the harmonic for $n = 1$ corresponds to half a wavelength across the panel in the x direction. N is the number of harmonics used. This gives a single plane diffuser that has modulation in the y direction only. It is also possible to use a 2D Fourier series to form the shape in two dimensions—the only cost is computation time as the number of shape parameters to be optimized increases.

From Fourier theory, if an infinite series is used, any diffuser shape can be produced. In reality, it is necessary to truncate the series at some point as every extra element in the series gives a new dimension to the optimization process. Too many dimensions and the minimization becomes too slow. Furthermore, one advantage of curved diffusers over more complex surfaces is their simpler construction leading to potentially lower costs and lower absorption. An increase in the number of harmonics in the series increases the complexity of the diffuser shape, which may cause excess absorption and increased cost. For these reasons, four to six harmonics are typically used. Once $y(x)$ from Equation 11.9 has been calculated, it is necessary then to scale the shape to fit the maximum displacement in the y direction required (i.e., fixing the maximum diffuser depth).

There are other mathematical representations of curved surfaces that can be used. For example, it is possible to define a number of movable points on the surface and use a cubic spline algorithm in one plane or a bicubic spline algorithm in two planes[15] to form a smooth curved surface between the points. It is possible to construct a harmonic series not based on sinusoidal basis functions. Frequency and amplitude modulation processes can also be used to generate many different shapes. Although there are many possibilities, the essential principle of needing shape parameters remains.

Problems sometimes arise when the best surface found by the computer does not meet the visual requirements of the architect. A curve is wanted, but the solutions produced are not quite what the designer originally envisaged. In addition, it is often necessary to ensure that the surface avoids other objects in the room or has appropriate breaks to allow for lighting. In this case, non-acoustic constraints must be used in the optimization process to force the shape to meet visual and physical constraints. This can be done via a set of fuzzy coordinates through which the surface must pass. Figure 11.35 illustrates how such a system can be used to force a surface to pass through particular points. The error parameter in the optimization becomes a combination of the diffusion coefficient that measures the scattering quality and a penalty value that measures how close the surface is to the constraint points (this can be an additive or multiplicative penalty). This is often used to ensure that edges of diffusers meet walls as illustrated in Figure 11.35c. In addition, this technique can be used to ensure that

- Cusps are not formed between adjacent periods of periodic diffusers;
- The left and right edges of diffusers are at the same displacement so that periodic diffusers' edges will meet without a discontinuity, and
- Obstructions, such as pillars, are avoided.

While using such a constraint system is straightforward for physical problems, such as avoiding cusps, it is more problematic when trying to force the shape of the curve into the visual aesthetic demanded by the designer. Often, during room design, the interior designer has a definite idea about the general shape required for the diffuser, e.g., 'We would like an S-shaped diffuser'. Trying to come up with a suitable set of constraint points for this is possible but involves some trial and error. Furthermore, the constraint point system lacks elegance and will slow down the optimization process by increasing the complexity of the error function to be searched. One solution is to use a spline construction, as then linear constraints on the shape parameters can be used.

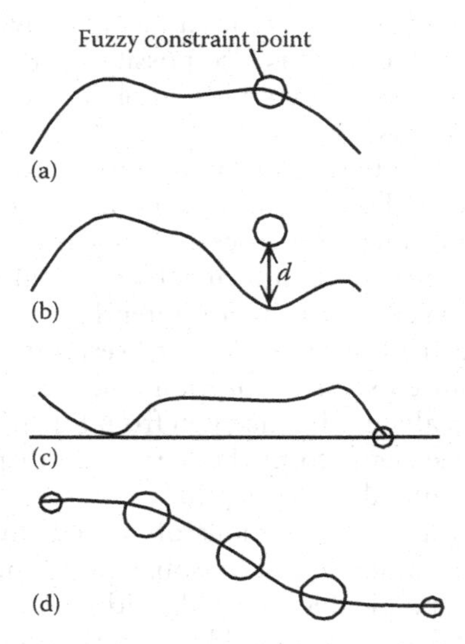

Figure 11.35 Use of fuzzy constraints to ensure optimized curved diffuser meets visual and physical constraints. (a) Surface passes through fuzzy constraint point, no penalty applied, (b) surface misses fuzzy constraint point, penalty applied proportional to *d*, (c) use of fuzzy constraints to ensure diffuser meets a wall at edges, and (d) use of fuzzy constraints to ensure an s-shape. (After Cox, T.J., D'Antonio, P., *Proc. IoA(UK)*, 21, 201–206, 1999.)

A superior system is one where the surface is designed from shape variables in such a way that the only surfaces generated are ones that satisfy the visual constraints. One way to do this is distortion.[16] The architect supplies a base shape and distortion is used to change the acoustical performance of the diffuser while retaining the visual integrity. Such processes are familiar in image processing as techniques for adding effects to photographs. This is illustrated in Figure 11.36. In the distorted pictures, it is still possible to recognize the picture as being a person, the rough visual appearance is maintained, yet radically different pictures are obtained. For diffusers, compression, modulation, and warping techniques taken from image processing are used to keep the general visual appearance of the device while allowing the acoustic performance to be optimized.

Figure 11.36 Distortion in image processing. (Courtesy of Cox, T.J., D'Antonio, P., *Proc. IoA(UK)*, 21, 201–206, 1999.)

11.5.3 Performance for unbaffled single optimized diffusers

Initial work on curved diffusers examined whether they could perform better than arcs of a circle.[14] It was found that for all depths and widths tested, the optimized diffusers were as good or better than arcs of circles. When the maximum allowable depth and width of the surface were similar, allowing a rough semicircular diffuser to be formed, this was the shape found via optimization. This happened because no better solution was possible. When the geometric constraints meant that a semicircle was not a possible solution, then the optimized surface found different, more complex shapes that were better at generating spatial dispersion than a segment of a circle.

The failure to improve on the simple arc does not show a fundamental weakness in the optimization process—it in fact illustrates how well it works! What has been shown is that, for the geometries tested, approximately semicircular diffusers have near optimal spatial dispersion and there is little room for improvement. (Remembering that spatial dispersion is not the only consideration for diffusers, time dispersion should also be considered.) This is illustrated in Figure 11.37, where the scattered pressures from two arcs and one optimized surface are shown. For the wide arc (Figure 11.37b), the scattered pressure shows a noticeable fall off at large angles of incidence. The optimized diffuser provides more uniform scattering. For the narrower surface (Figure 11.37a), the arc is nearly semicircular; the diffusion is fairly uniform, and it would be difficult for any surface to improve on the scattering produced in terms of spatial redistribution. Note, a single semicylinder is unlikely to give the necessary temporal dispersion and will not cover a large enough area, so the case shown in the left graph is not a practical solution.

Typical examples of a standard deviation (STD) diffusion parameter, as a function of incident angle for an arc and optimized surfaces, are shown in Figure 11.38.[17] The figure is using an old diffusion evaluation technique, and the lower the value of the *STD diffusion*, the greater the spatial dispersion. In this case, the optimized diffuser has sacrificed a little

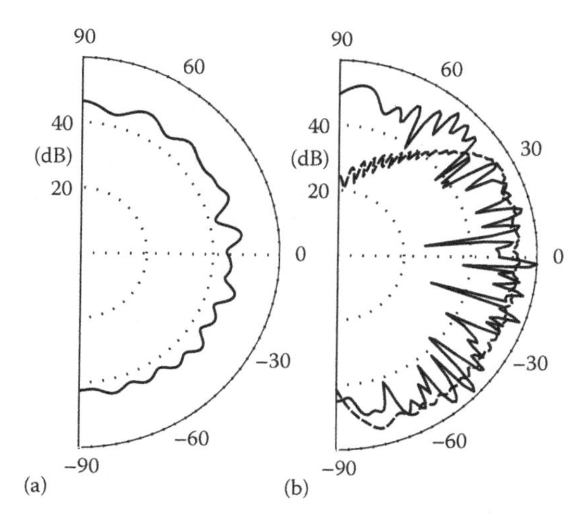

Figure 11.37 Scattered level from three surfaces at 2.8 kHz, 30° incident source. (a) —— standard curved diffuser (same as optimized diffuser) 1 × 0.4 m. (b) - - - - standard curved diffuser, dimensions 4 × 0.4 m; ▬▬ optimized diffuser 4 × 0.4 m. (Data from Cox, T.J., *J. Audio Eng. Soc.*, 44, 354–364, 1996.)

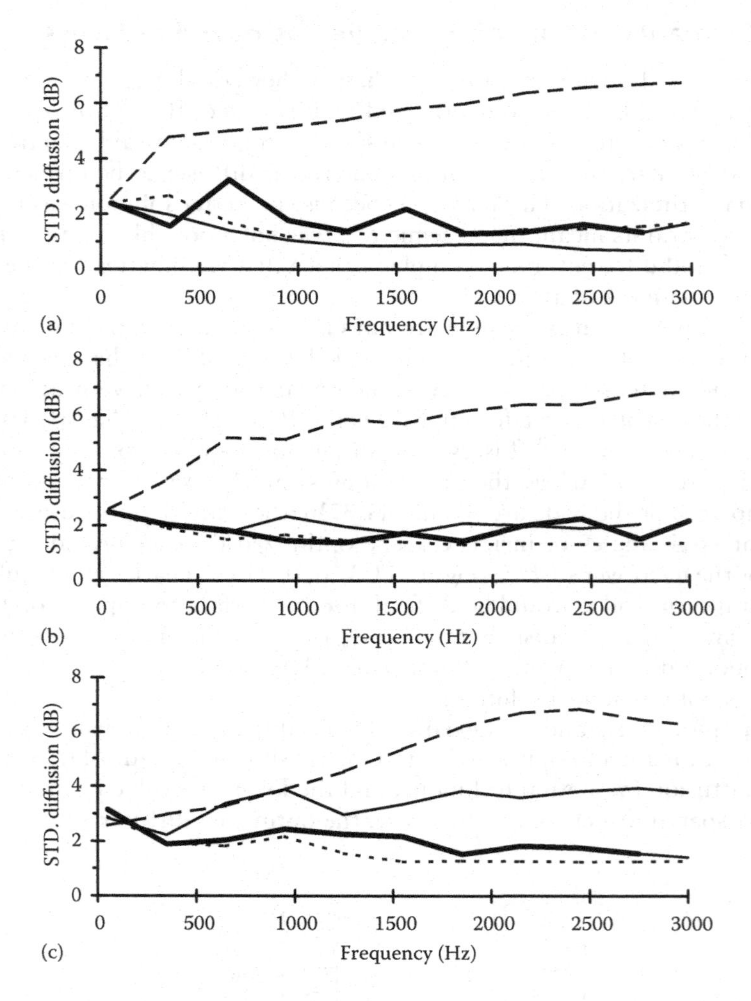

Figure 11.38 Diffusion from four surfaces; incidence sound angles are different for each graph. (a) Normal; (b) 30°; and (c) 60° incidence sound. This uses an old STD measure, where complete diffusion is when the standard deviation is 0. – – – plane surface, —— arc of a circle, ▬▬ optimized fractal, and - - - - optimized curved surface. (After Cox, T.J., D'Antonio, P., *Proc. 103rd Convention Audio Eng. Soc.*, preprint 4578, 1997.)

performance for normal incidence sound, to improve the scattering for oblique sources. This shows the ability of the new surface to produce more uniform scattering for random incident sound.

11.5.4 Periodicity and modulation

When placed in a periodic arrangement, the quality of scattering at low to mid frequencies is dominated by grating lobes generated by surface repetition. One way of achieving more even scattering is to remove the periodicity completely by using a very large wavy surface. Unfortunately, this is likely to be an expensive solution, and consequently, a modulation scheme similar to that for Schroeder diffusers (Section 10.5) can be used.

Consider a single asymmetrical diffuser base shape shown in Figure 11.39a in bold. If this base shape were arranged in a periodic fashion, grating lobes will arise. If, however, some of the periods are rotated, then the periodicity can be reduced. In the case shown in Figure 11.39, the repeat distance has been doubled by this modulation.[18] Figure 11.39 also shows a 3D rendering of the idea. For this type of modulation to improve diffusion, the base shape needs to be sufficiently asymmetrical, so that flipping the shape produces completely different scattering. A further complication is the need to ensure that neighbouring diffusers tile together without discontinuity in surface displacement or gradient. By forming surfaces with zero gradient and the same surface displacement at both ends, it is possible to form a surface that will tile in any orientation. Then the architect can decide what pattern to form. More importantly, pseudorandom arrays enable diffusers of considerable extent to be created from small base shapes. If there is a discontinuity in gradient between panels, then a cusp results. An inward facing cusp can be useful in generating additional scattering, but the appearance may not be desirable.

In its extreme, this modulation can result in a surface where the individual base shape is not clearly distinguishable.[19] Figure 11.40a shows a single period of a 3D curved surface in various orientations and the bottom shows the same shape in a modulated array. The single period device has the same symmetrical shape on each edge, and the gradient around the perimeter is 0. This allows the surface to be tiled in any orientation. If the surface depth is

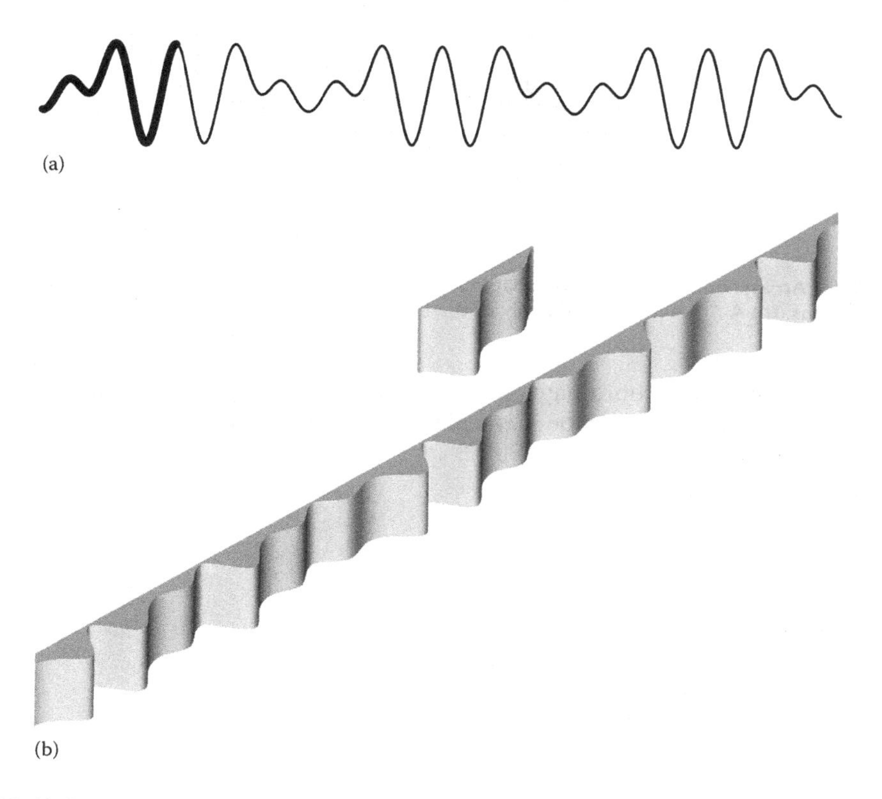

Figure 11.39 (a) Curved diffuser (not optimized for acoustics). The base shape is shown in bold. By changing the orientation between periods, it is possible to increase the repeat length of the diffuser. (b) The same principle rendered in 3D showing one base shape and a modulated array.

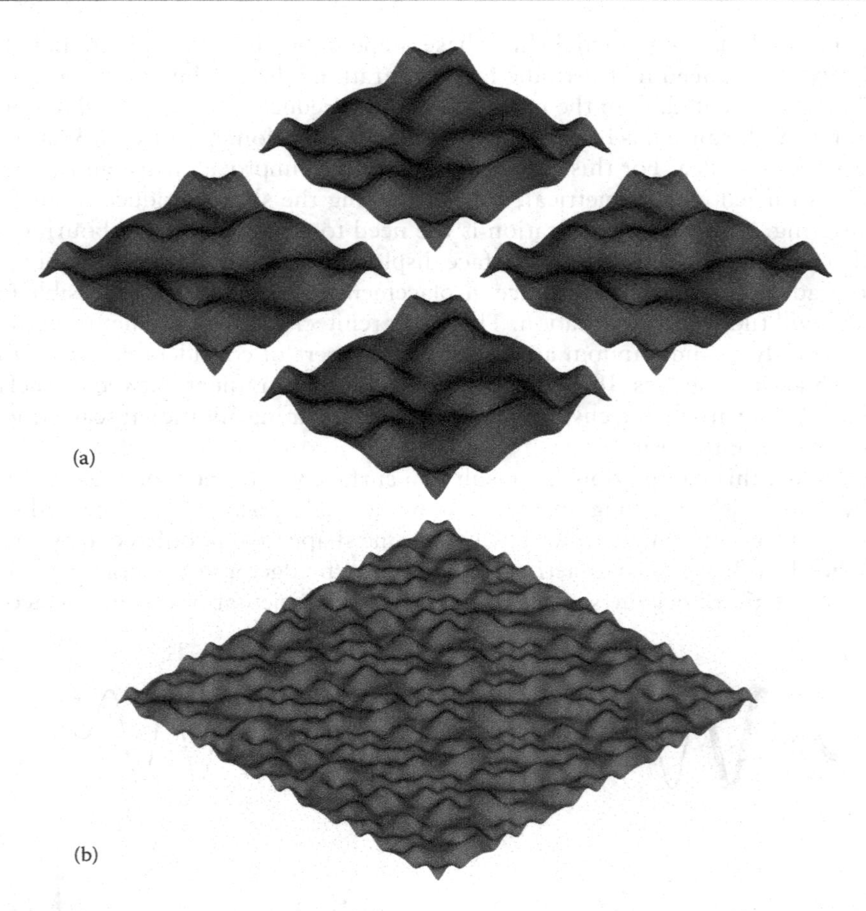

Figure 11.40 (a) An asymmetric single base shape used in modulation shown in different possible orientations. (b) A 4 × 4 modulated array of the base shape.

z and is a function of the coordinates across the width x and length y of the diffuser, $z = F(x,y)$, then the requirements can be mathematically expressed as

$$\begin{aligned} &z(0,y)=z(L,y)\\ &z(x,0)=z(x,L)\\ &z(x,0)=z(L-x,0)\\ &z(x,0)=z(0,x)\\ &\frac{\partial z}{\partial x}=0 \quad x=0\vee L\\ &\frac{\partial z}{\partial y}=0 \quad y=0\vee L, \end{aligned} \tag{11.10}$$

where L is the width (or length) of the diffuser.

When placed in a modulated array, the base shape disappears in a complex pattern of minima and maxima. This allows the use of one base shape and reduces manufacturing

Figure 11.41 Modulated curved diffusers used in the ceiling of a radio production studio at KTSU, Texas Southern University, Houston, TX. Acoustician: HFP Acoustical Consultants. Insert shows close-up of ceiling.

costs. Figure 11.41 shows an example application of modulated curved diffusers on the ceiling of a radio studio.

A periodic look is often favoured, however. It seems that a periodic object enables the brain to more easily decode the design. A completely random surface can be too difficult to interpret and hence not pleasing to the eye. This is, of course, a generalization, but it is more common for a periodic entity to be specified. Using this asymmetrical base shape gives designers control over the appearance. It can be made to look random or periodic, but the designers have to remember that short repeat distances will result in worse dispersion. Figure 11.42 shows the scattering from three surfaces showing how modulation can improve the performance over a periodic array.

11.5.5 Stage canopies

Overhead stage canopies are often designed with arrays of gently curved panels. Often, the primary role of stage canopies is to provide reflections back to the musicians or actors to allow them to hear others. Canopies may also be used to distribute some energy to the audience. There might be only a short delay between the direct sound and the overhead canopy reflection in the audience area, however, and so care should be taken to avoid coloration due to comb filtering. If a completely flat large surface is used above the stage, plenty of energy will return

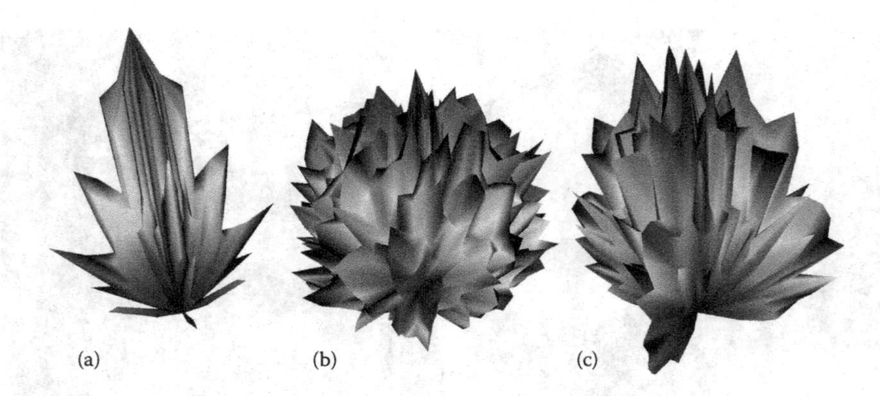

Figure 11.42 Three polar responses for scattering at 2 kHz. (a) Plane surface; (b) optimized modulated array; and (c) periodic arcs.

to the musicians. Unfortunately, the energy will be too strong and likely to cause coloration. Consequently, a canopy often needs to be shaped to create temporal diffusion to reduce coloration. Figures 3.27 through 3.29 showed examples of curved canopies designed using optimization.

Canopies appear to fall into two rough categories, which are determined by the amount of open area in the canopy. Sometimes, canopies completely cover the stage (virtually no open area). This is most common where the canopy is being used to block sound entering a fly tower as was shown in Figure 3.34. Other canopies use elements more sparsely, with more open area between the canopy diffusers or reflectors, as is the case for the reflectors shown in Figure 3.29.

An optimization study[20] was carried out to investigate whether this design method could be used for stage shells with a reasonable amount of open area. Figure 3.27 showed a canopy designed using this principle. The canopy is such that there is a little open space between each of the reflectors to allow some energy to pass up to the void above. The canopy is about 7 m above the stage so that the reflections are delayed by an amount known to give a good chance of them aiding ensemble and support.[21] For the optimization study, five diffusers were arranged on the arc of a large circle from the front to the back of the stage. The diffusers were assumed to extend across the full width of the stage. The design criterion in the study was that for any source position on the stage, as even as possible energy distribution would be created to all positions on the stage. In other words, each musician has an equally good chance of hearing others, as far as the stage canopy reflections are concerned. This is a criterion that can be used in an optimization process described previously.

The optimized design was compared to several other reflector shapes, such as a plane surface and arcs of circles. The optimized surface outperformed the other shapes in producing the most even energy distribution across the stage area. Figure 11.43 shows the reflected sound pressure level from the plane reflector canopy at 2 kHz for a source in the middle of the stage. The variation in pressures across the width is small because this was modelled as a large flat surface (any variation is due to spherical spreading and path length differences). The reflected sound pressure level directly below the source is large, as much of the energy is being reflected straight back to the musician who is playing the instrument. Dips in the reflected pressure level occur where there is no geometric reflection point on one of the reflectors due to the spaces in the canopy design. As discussed previously, shaping the reflectors can reduce the effects of pressure minima, and this is often done by forming convex arcs.[5]

The effect of using a more complex optimized curved surface is shown in Figure 11.44. The curved diffuser reflects energy to receivers where specular reflections are missing. It also

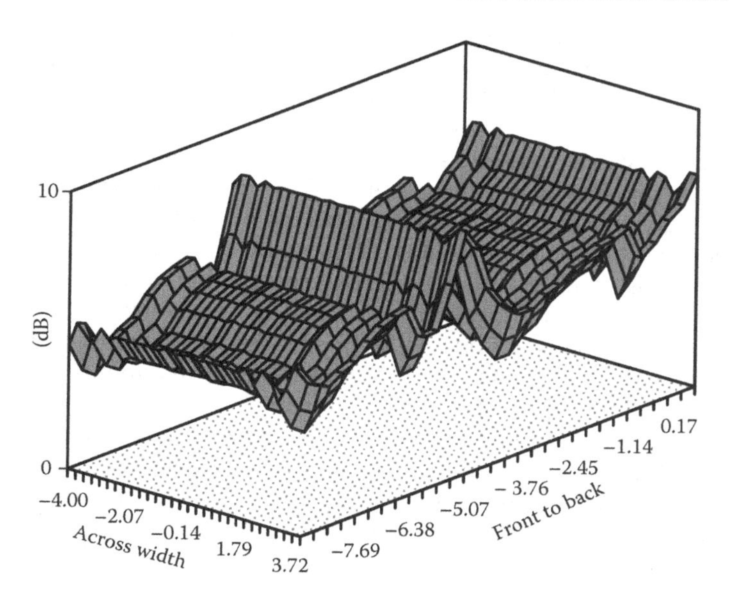

Figure 11.43 Scattering from a stage canopy with plane reflectors. (After Cox, T.J., D'Antonio, P., *Proc. IoA(UK)*, 19, 153–160, 1997.)

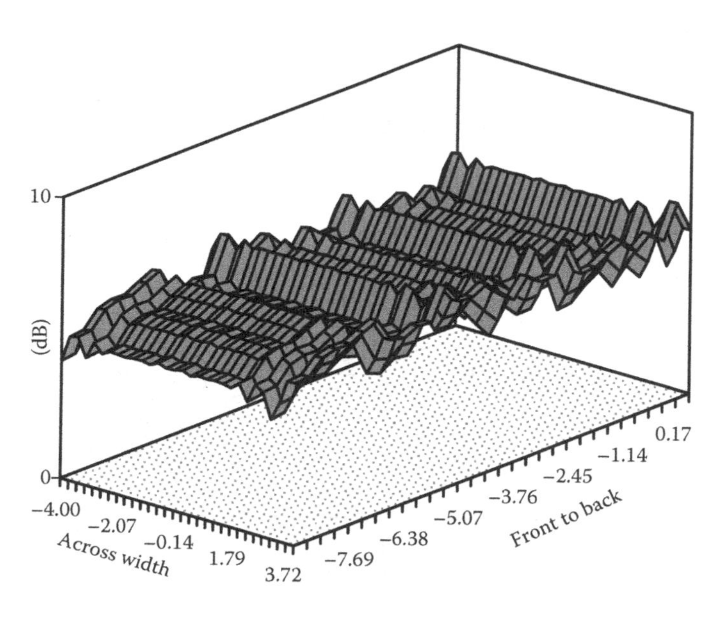

Figure 11.44 Reflection from a stage canopy made from optimized curved diffusers. (After Cox, T.J., D'Antonio, P., *Proc. IoA(UK)*, 19, 153–160, 1997.)

greatly reduces the strong specular energy being reflected straight back to the musician. The reflected energy is more evenly distributed across the stage—all pressure levels along the front to back axis lie within 3 dB of each other.

The criterion of even energy across the stage needs to be used with caution, however, and is not appropriate for all cases. Consider a canopy with a very small open area, in other words a large surface with the same width and depth as the stage. A flat surface will give very good coverage on the stage, as noted before, but this does not make it a good stage canopy because the overhead reflections will be too strong. Diffusers are used in this case to reduce the coloration produced by these reflections. It is therefore usually better to design this sort of stage canopy to promote maximum dispersion across a complete arc from -90 to +90°, in other words asking for maximum dispersion not just on the stage. For complex surfaces, this maximizes the temporal dispersion from the overhead canopy and so minimizes coloration effects.

It is also possible to get the optimizer to look at the best locations, angles, number, and sizes of canopy elements, alongside examining the detailed surface shape of each element. While this can be carried out on a case-by-case basis, a study was also undertaken to explore what the underlying principles in canopy design might be and so develop some rules of thumb.[22] The optimizer was run many times with different target values for the reflection level on stage—this was done using the support measure[21]—resulting in many different canopy designs with different open areas.

Figure 11.45 shows how the width and depth of the best canopy designs varied with the open area of the canopy. Figure 11.46 shows the variation in the number of canopy elements

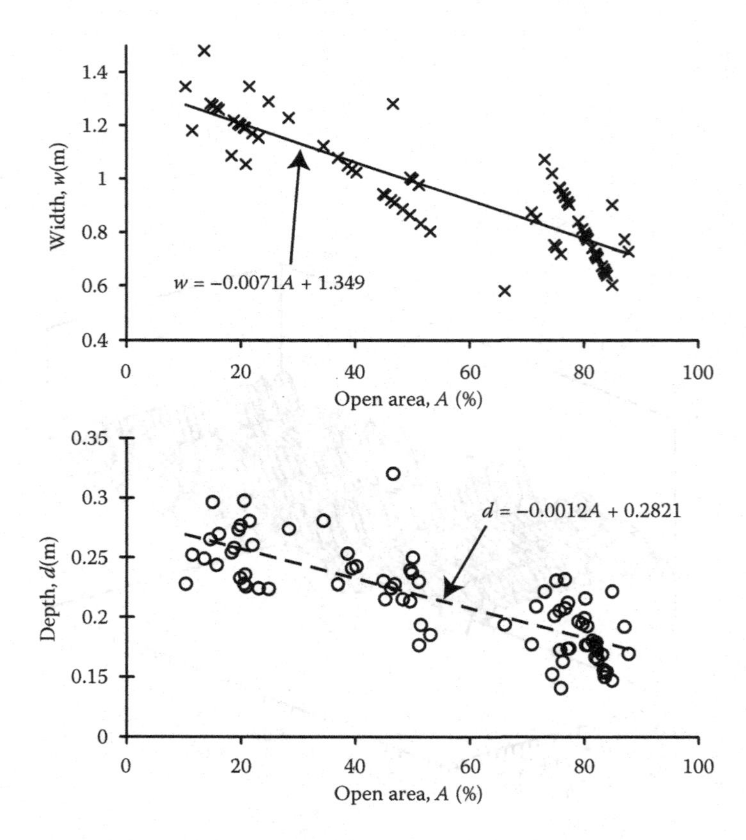

Figure 11.45 Changes in optimized reflector width and depth as a function of percentage open area, A. The best fit lines show the underlying relationship between optimum reflector size and canopy open area.

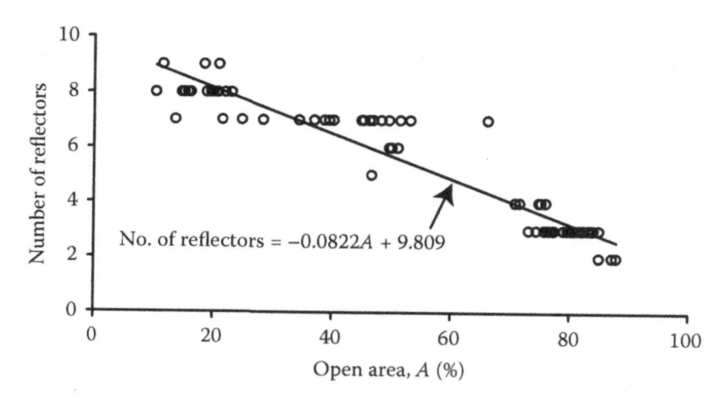

Figure 11.46 Variation in number of stage canopy reflectors with percentage open area, *A*. A least squares best fit straight line is also shown.

with open area. Fuller canopies with smaller open areas are shown towards the left of Figures 11.45 and 11.46. These canopies provide more reflections back onto the stage and have higher support. To the right of the figures, the canopies provide lower level of reflections because the open area of the canopies is quite large. For the fuller canopies with small open area, the best designs tend to have more canopy elements that are wider and deeper in comparison to more open canopies. Furthermore, fuller canopies have simpler, less wiggly shapes than do canopies with larger open areas. Because fuller canopies have only a small open area, there is less need for dispersion from the canopies to provide reflections to receivers that lack specular reflections.

It seems that relatively small panels are preferred. This supports the findings of Rindel,[5] who studied the effect of density and panel size on array performance for flat elements. He concluded that panels needed to be relatively small so that energy is diffracted to receivers who do not receive a specular reflection from the canopy array. For this reason, high-frequency performance was determined by panel size. The low-frequency performance of the array was determined by the canopy density. The optimization study has also produced a new finding: that the desirable reflector width and depth are dependent on the open area of the canopy.

When a canopy is completely full and there are virtually no gaps between the panels, then the design issue is how to reduce coloration caused by the inevitable strong specular reflections. In this case, covering the surface with diffusing surfaces may be a good option. The situation is very similar to rear and side stage enclosure design, which was discussed in Section 3.8.2.

11.6 FRACTALS

Fractal mathematics is used to create natural objects in computer-generated graphics, for example to make landscapes for animated films. Fractals are surfaces with a different visual aesthetic compared to common sound diffusers and so offer the possibility of expanding the pallet of surfaces available to designers.[23]

There is reason to believe that fractal surfaces may have good acoustic properties. Fractals are self-similar or self-affine; as a surface is magnified, a similar looking surface is found. Consequently, the rough surfaces at different magnifications can be used to scatter different frequency ranges in an analogous way to the use of drivers of various sizes to radiate different bandwidths from multiway loudspeakers. This is the principle of the Diffractal,[23] as shown in Figure 10.26, which imbeds small-scaled copies of an $N = 7$ QRD at the bottom of a larger $N = 7$ QRD. The small QRDs scatter mid–high frequencies, and the large QRDs, the bass frequencies. The construction is precisely self-similar; the exact shape is found upon magnification. This will not be true for surfaces considered in the following sections, where the surfaces are simply statistically self-similar or self-affine.

There are many established techniques for generating finite sample approximations to mathematically pure fractal shapes.[24,25] Single-plane diffusers are described and are made from extruded 1D fractals. Construction methods for higher-dimensioned surfaces are also available but are ignored here for conciseness.

11.6.1 Fourier synthesis

Fractal surfaces can be constructed from spectral shaping of a Gaussian white noise source. Figure 11.47 is a schematic showing how the surfaces can be generated; such a scheme is more familiar in digital signal processing as a time signal filtering process.[17] The Gaussian white noise is passed through a filter, which is implemented using simple Fourier techniques. The shaping of the spectrum is done using a roll-off with a defined number of decibels per octave. The decrease in spectral content per octave is characterized by the gain of the filter at each spatial frequency. The filter gain $A(X)$ is given by

$$A(X) = \frac{1}{X^{\beta/2}}, \tag{11.11}$$

where X is the spatial frequency and β is the spectral density exponent, which takes values between 1 and 3. This then ensures that the dimension lies between 1 and 2, as required for a 1D fractal shape. Figure 11.48 illustrates some typical surface shapes that were generated by such a scheme.[17] The bottom line is the input Gaussian white noise. The middle line was generated with a roll-off of 3 dB/octave ($\beta = 1$), which gives $1/f$ noise, which in acoustics is termed *pink noise*. It is also a commonly occurring spectrum in other natural phenomena. The top line is formed by a steeper roll-off of 6 dB/octave ($\beta = 2$) and is characteristic of Brownian motion, random walk, or *brown noise*.

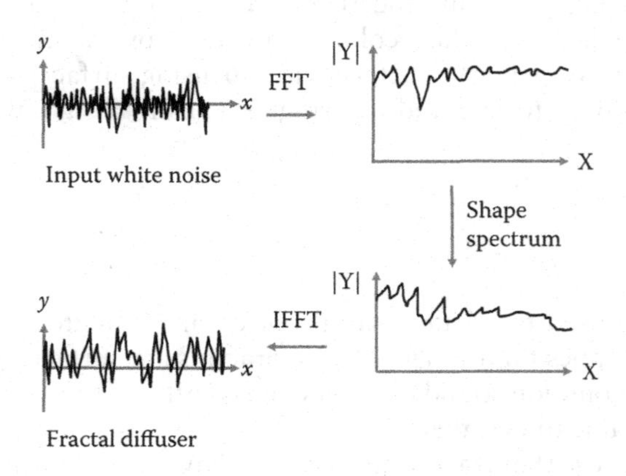

Figure 11.47 Schematic diagram showing Fourier synthesis construction technique where FFT and IFFT indicate a Fourier transform and its inverse, respectively. (After Cox, T.J., D'Antonio, P., *Proc. 103rd Convention Audio Eng. Soc.*, preprint 4578, 1997.)

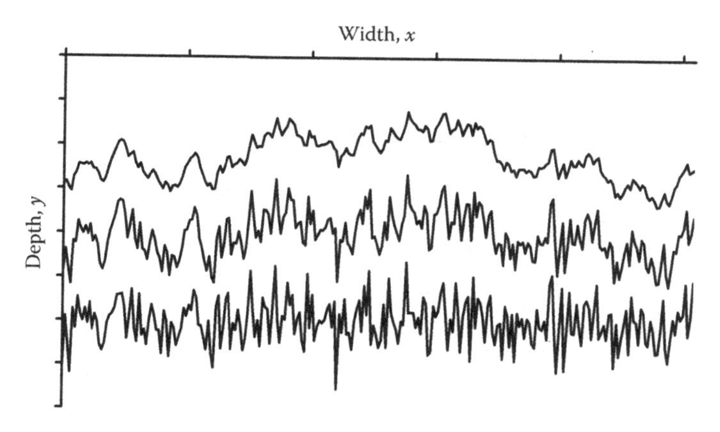

Figure 11.48 Different fractal surfaces generated by Fourier synthesis. Bottom line: input white noise; middle line: pink noise; top line: brown noise. (After Cox, T.J., D'Antonio, P., *Proc. 103rd Convention Audio Eng. Soc.*, preprint 4578, 1997.)

The shapes given in Figure 11.48 can be made into diffusers by extrusion. By varying the spectral density exponent, the spikiness of the surface shape can be altered. In the most general terms, to get the best low-frequency scattering performance, the spectral density exponent should be large, leading to a smooth shape. To get the best high-frequency performance, however, a low spectral density exponent is needed as this makes a spiky shape. The story is, however, more complicated because while the spectral density exponent does determine the scattering quality at low and high frequencies to a certain degree, the correct choice of the white noise sequence is dominant. Unfortunately, it is not possible to optimize these surfaces, as there are too many governing shape parameters.

11.6.2 Step function addition

Brownian motion can also be simulated by random midpoint displacement[26] or by adding a set of randomly displaced step functions. Although this is not always as mathematically pure as a Fourier synthesis technique, it facilitates a reduction in the number of parameters required to represent the surface shape.

Getting proper Brownian motion requires the addition of an infinite number of step functions. Each step function has a random amplitude and random step position along the width of the diffuser. The displacement of the diffuser from a flat surface y at a distance x along the diffuser is given by

$$y(x) = \sum_{i=1}^{\infty} A_i f(x - x_i), \quad \text{where } f(\alpha) = \begin{cases} 0, \alpha < 0 \\ 1, \alpha \geq 0, \end{cases} \tag{11.12}$$

where A_i is a set of Gaussian distributed random amplitudes and x_i is a uniformly distributed random set of positions on the diffuser with $-a \leq x_i \leq a$, where $2a$ is the diffuser width. Figure 11.49 illustrates the random addition of step functions for a few terms.

To utilize this generation method efficiently, the infinite sum in Equation 11.12 needs to be truncated. In acoustics, the difference in wavelength from the highest to the lowest frequency of interest is not so great that a sound diffuser requires self-similarity over a large

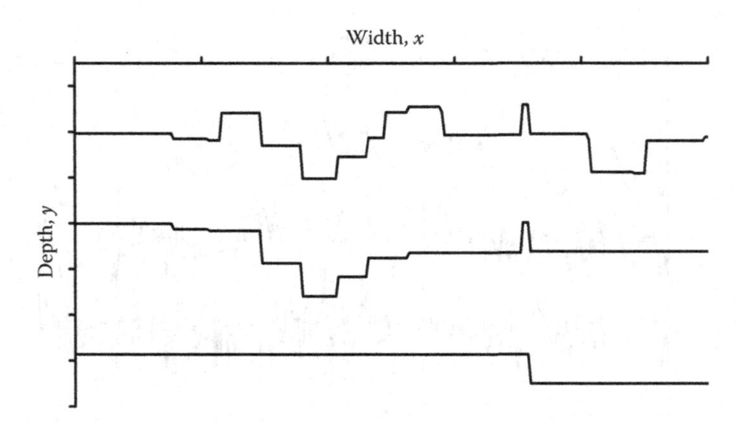

Figure 11.49 Fractal generation by step function addition. Bottom line: 1 step function; middle line: 10 step functions; top line: 20 step functions. (After Cox, T.J., D'Antonio, P., *Proc. 103rd Convention Audio Eng. Soc.*, preprint 4578, 1997.)

range of magnifications. Instead of an infinite sum of terms, a finite number, N, has been used, with each successive term having a decaying amplitude.

$$y(x) = \sum_{i=1}^{N} B_i f(x - x_i), \quad \text{where } B_i = \frac{1}{i^{\alpha}}; 0 \le \alpha \le \infty. \tag{11.13}$$

The decaying amplitude function produces a similar probability distribution of values as the original Gaussian random values A_i, except at the most extreme values of A_i. The other difference is that the amplitudes can now only be regularly spaced values, rather than being truly random. This reduces the number of independent parameters to $N + 1$, the location of the steps x_i, and the amplitude decay rate α and enables optimization to be used.

Using only a few step functions can lead to flat areas, which might be prone to producing specular reflections at high frequencies, or sharp spikes, which are undesirable. This can be seen in the 20 step function surface shown in Figure 11.49. Consequently, rather than step functions, $f(x)$ can be replaced with a function with a more graceful transition. For example, a hyperbolic tangent can be used for this

$$f(x) = \tanh(\gamma_i x). \tag{11.14}$$

The effect of using a hyperbolic tangent function is to round the top and the bottom of the step functions. The γ_i value changes the rate of transition from the top to the bottom of the step and the amount of rounding. γ_i may either be a constant for all terms in Equation 11.13 or alternatively may be allowed to decay or increase

$$\gamma_i = \frac{1}{i^{\xi}}; \ 0 \le \xi \le \infty. \tag{11.15}$$

The *fractal* shape is now determined by $N + 2$ parameters. These are known as random addition diffusers.

When optimizing and testing these surfaces, it was surprising to find that this generation technique can make semicylinders when the depth was roughly half the width. As discussed previously in connection with curved surface optimization, the semicylinder is very good at spatially dispersing sound provided it is on its own. This demonstrates that the optimization is working but does raise the question as to whether the optimized shape can really be called a fractal!

When a more application realistic wide diffuser is optimized, the arc of a circle is no longer optimal, and better fractal diffusers are found. Figure 11.38 compared different diffusers using a STD diffusion parameter (small is best). The fractal is better than the arc of a circle for random incidence sound. It seems that optimized fractal surfaces do produce reasonable diffusers; however, optimized curved surfaces often have better dispersion. Consequently, fractal construction techniques may produce different visual aesthetics to optimized curved surfaces, but not necessarily better diffusers.

11.7 MATERIALS

Geometric reflectors and diffusers can be made from a variety of materials. Examples include wood in various species and finishes, light transmitting or thermoformed plastics, fibre reinforced gypsum, concrete, high-density polystyrene, and metal. The choice of material depends on factors such as visual appearance, weight, and acoustic absorption. Nowadays, environmental effects are an additional consideration in choosing a suitable material. When using wood and plastics, fire resistance is an important factor, as are indoor air quality standards for low-emitting products, such as formaldehyde, volatile organic compounds, and other carcinogenic materials. In addition, recycled content, locally supplied materials, and other characteristics are also important in sustainable or green design. The discussions in Section 1.2 about sustainability are pertinent here.

Geometric reflectors have potential to resonate and cause absorption. The resonant frequency of a simply supported rectangular isotropic panel of width a and length b is given by[27]

$$f_0 = \frac{\pi}{2}\sqrt{\frac{B}{m}}\left(\frac{i^2}{a^2}+\frac{n^2}{b^2}\right); \quad i,\ n = 1,2,3,\ldots, \tag{11.16}$$

where m is the surface density (kgm^{-2}) and B is the bending stiffness, given by

$$B = \frac{Eh^3}{12(1-v^2)}, \tag{11.17}$$

where E is Young's modulus, h is the panel thickness, and v is Poisson's ratio. The fundamental resonance, $i = n = 1$, is of primary concern here because this is usually the most easily excited and so has the greatest potential for absorption.

For large panels, the coincidence phenomenon may also be important at higher frequencies. When the wavelength of the (bending) waves propagating in the reflector is the same as the wavelength of sound in air, then the panel is easily vibrated by the airborne sound, and absorption will result. This happens at the coincidence frequency, which is given by

$$f_c = \frac{c^2}{2\pi}\sqrt{\frac{m}{B}}, \tag{11.18}$$

where c is the speed of sound in air. If surfaces are orthotropic, for instance, if a corrugated surface is used, then the coincidence effect will extend over a significantly wider bandwidth.

Some favour having stage enclosures with low-frequency absorption, in which case the resonances in the above formulations need to be placed at appropriate bass frequencies and the panel should not be overly damped. Proponents of these *diaphragmatic* enclosures, which are typically not sealed, argue that the surfaces surrounding an orchestra can be used to provide low-frequency absorption to create more clarity on the stage. Furthermore, in a small hall, a sealed shell may overpower a small audience, and typically, these shells contain openings and/or diaphragmatic surfaces to dissipate energy. There are also occasions in small halls where the low-frequency absorption from surfaces surrounding the audience can be used to control bass reverberance.

In a large hall, however, it is usual to preserve as much sound energy as possible. So surfaces around the stage and the audience should have low absorption. One way of achieving this is to make the surface heavy, placing the first resonance of Equation 11.16 at such a low frequency that resonant absorption is no longer important. However, the coincidence frequency of Equation 11.18 needs to be checked for large surfaces unless the bending waves are highly damped.

However, heavy elements are often undesirable because they increase costs as the weight must be supported by the building somehow. This is especially true for suspended reflectors. An approach often used to lighten the load is using a honeycomb core with two lightweight faces. This laminated structure is very stiff and has low mass. However, Equation 11.16 shows that a high stiffness and low mass result in resonance at an elevated frequency. Furthermore, Equation 11.18 shows that this also lowers the frequency of coincidence. Consequently, increasing stiffness on its own risks increased absorption from both resonance and coincidence.

A better solution for lightweight surfaces is to heavily damp the resonance, which will prevent sound from significantly vibrating the material and consequently minimize absorption. Adding damping to increase reflectivity might seem counterintuitive. But by damping the resonances, the impedance mismatch between the air and the surface increases, leading to less absorption. Damping might be achieved by firmly bonding two materials together to form a constrained-layer damping system. For instance, rubber, cork-rubber, or a viscoelastic material might be sandwiched between two layers of fibre reinforced gypsum or wood. Alternatively, some materials are inherently highly damped.

11.8 SUMMARY

This chapter has covered a wide range of different diffuser types. Starting with plane surfaces, it looked at the effects of edge diffraction. The performance of arcs, triangles, pyramids, fractals, and curved surfaces was then considered. Current state-of-the-art is to use numerical optimization to allow the surface shape to meet both the acoustic and visual requirements. With good design, there are many shapes that can make good diffusers, but only a few of these will ever be visually acceptable in a particular project. The next chapter looks at combining absorption and diffusion into a single device.

REFERENCES

1. L. Cremer, "Early reflections in some modern concert halls", *J. Acoust. Soc. Am.*, **85**, 1213–25 (1989).
2. L. Cremer, "Fresnel's methods of calculating diffraction fields", *Acustica*, **72**(1), 1–6 (1990).

3. J. H. Rindel, "Attenuation of sound reflection from curved surfaces", *Proc. 25th Conference on Acoustics* (1985).
4. J. H. Rindel, "Attenuation of sound reflections due to diffraction", *Proc. Nordic Acoustical Meeting*, 257–60 (1986).
5. J. H. Rindel, "Design of new ceiling reflectors for improved ensemble in a concert hall", *Appl. Acoust.*, **34**, 7–17 (1991).
6. A. Szeląg, A. Pilch, T. Kamisiński, and J. Rubacha, "Overhead stage canopy effective over a wide frequency range", *Proc. Forum Acusticum* (2014).
7. P. D'Antonio and J. Konnert, "The RFZ/RPG approach to control room monitoring", *Proc. Convention Audio Eng. Soc.*, preprint 2157 (I-6) (1984).
8. M. Vercammen, "Sound reflections from concave spherical surfaces. Part I: Wave field approximation", *Acta Acust. Acust.*, **96**(1), 82–91 (2010).
9. M. Vercammen, "Sound reflections from concave spherical surfaces. Part II: Geometrical acoustics and engineering approach", *Acta Acust. Acust.*, **96**(1), 92–101 (2010).
10. A. D. Pierce, *Acoustics: An Introduction to Its Physical Principles and Applications*, McGraw-Hill (1981).
11. T. J. Cox, "Objective and Subjective Evaluation of Reflection and Diffusing Surfaces in Auditoria", PhD thesis, University of Salford, Salford (1992).
12. T. J. Cox and Y. W. Lam, "Evaluation of methods for predicting the scattering from simple rigid panels", *Appl. Acoust.*, **40**, 123–40 (1993).
13. J. P. O'Keefe, T. J. Cox, N. Muncy, and S. Barbar, "Modern measurements, optimized diffusion, and electronic enhancement in a large fan-shaped auditorium", *J. Acoust. Soc. Am.*, **103**(5), 3032–3, also *Proc. 16th ICA* (1998).
14. T. J. Cox, "Designing curved diffusers for performance spaces", *J. Audio Eng. Soc.*, **44**(5), 354–64 (1996).
15. W. H. Press et al., *Numerical Recipes, the Art of Scientific Computing*, Cambridge University Press, Cambridge, 289–92 (1989).
16. T. J. Cox and P. D'Antonio, "Holistic diffusers", *Proc. IoA(UK)*, **21**(6), 201–6 (1999).
17. T. J. Cox and P. D'Antonio, "Fractal sound diffusers", *Proc. 103rd Convention Audio Eng. Soc.*, preprint 4578 (1997).
18. P. D'Antonio and T. J. Cox, "Aperiodic tiling of diffusers using a single asymmetric base shape", *Proc. 18th ICA*, Mo2.B2.3 (2004).
19. P. D'Antonio and T. J. Cox, *Embodiments of Aperiodic Tiling of a Single Asymmetric Diffusive Base Shape*, US patent 6,772,859 (2004).
20. T. J. Cox and P. D'Antonio, "Designing stage canopies for improved acoustics", *Proc. IoA(UK)*, **19**(3), 153–60 (1997).
21. M. Barron, *Auditorium Acoustics and Architectural Design*, 4th edn, Taylor & Francis, Oxon, UK (2009).
22. P. D'Antonio and T. J. Cox, "Canopy arrays: Density, size, shape and position", *Proc. IoA(UK)*, **28**(2) (2006).
23. P. D'Antonio and J. Konnert, "The QRD Diffractal: A new 1 or 2-dimensional fractal sound diffusor", *J. Audio Eng. Soc.*, **40**(3), 117–29 (1992).
24. H.-O. Peitgen and D. Saupe (Eds), *The Science of Fractal Images*, Springer-Verlag, New York (1988).
25. J. Feder, *Fractals*, Plenum Publishing Corporation, New York (1988).
26. D. Bradley, E. O. Snow, K. A. Riegel, Z. D. Nasipak, and A. S. Terenzi, "Numerical prediction of sound scattering from surfaces with fractal geometry: A preliminary investigation", *POMA* **12**, 015010 (2011).
27. D. A. Bies and C. H. Hansen, *Engineering Noise Control*, 4th edn, CRC Press, Abingdon, UK (2009).

Chapter 12

Hybrid surfaces

A diffuser needs to break up the reflected wavefront. This can be achieved by both shaping a surface and making the surface impedance vary across the front face of the device. Indeed, Schroeder diffusers are often interpreted as a set of wells that create a spatially varying surface impedance. In this chapter, varying impedance is achieved using patches of absorbent alongside reflecting materials. Unlike the Schroeder diffuser, these hybrid devices cannot be designed for minimum absorption. These surfaces are somewhere between pure absorbers and non-absorbing diffusers. Partial absorption is inherent in the design while any reflected sound is dispersed.

12.1 EXAMPLE DEVICES

Figure 12.1 shows various different designs to scatter in one plane. The front view of the binary device shown at the top builds on a tradition in studios to set out absorbent in patches to generate dispersion. In the 1990s, a new breed of surface was produced, with smaller absorbent patches arranged according to a pseudorandom sequence.[1]

Flat hybrid surfaces generate a coherent specular reflection, albeit attenuated because it is partially absorbed. The hard parts still generate reflected waves that arrive in phase in the specular direction. One solution to this is to shape the surface, as then the specular reflection can be significantly dispersed to other angles, to achieve more uniform diffuse reflections. Figure 12.1b is a side view of a single-plane device with a simple curve. Figure 12.2 shows a hemispherical device made from a metal mask on top of curved porous absorbent, using a 2D pseudorandom sequence to determine the location of the holes. An application of the device is also shown. The absorption coefficient of curved hybrid surfaces will be similar to planar equivalents, but the dispersion will be greatly increased. This curved construction has found favour in recording studios as it allows treatment away from the extremes of complete absorption, specular reflection, or diffuse reflection. This enables the sweet spot, the place where the room acoustics are at their best, to be expanded. Figure 12.3 shows an application of a curved hybrid surface[2] on the ceiling and walls of a postproduction studio.

Chapter 11 explained how optimization can be used to design rigid curved surfaces. This process can also be used to make hybrid curved surfaces. A binary sequence with good autocorrelation properties is chosen and then the computer is tasked to find the best shape.

Two other ways of achieving more dispersion than flat hybrid surfaces are shown in Figures 12.1c and d. As noted previously, a problem with planar hybrid surfaces is that energy can only be removed from the specular reflection by absorption. So, the specular reflection is only attenuated by about 6 dB for a surface with a 50% absorptive area. To improve performance, it is necessary to exploit interference and reflect waves out-of-phase

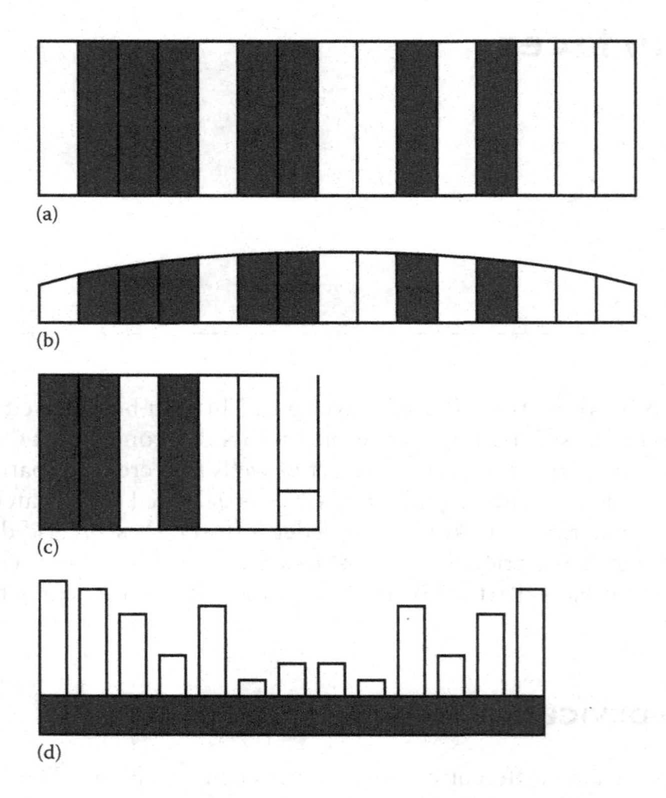

Figure 12.1 Various single-plane hybrid devices where the grey-shaded material is porous absorption and the unshaded parts are hard and reflecting: (a) front view of a flat binary device, (b) side view of a curved binary device, (c) side view of a ternary device, and (d) side view of a slatted hybrid device.

with the specular reflection. This can be achieved by adding wells to the surface.[3] An example of this is shown in Figure 12.1c. The reflection coefficients of this surface are then made up of three values, which at the design frequency are –1, 0, and +1 representing inverted reflection, absorption, and normal reflection, respectively. As a result, optimal ternary number sequences are required. By adding wells to hybrid surfaces, a very useful improvement in performance is achieved for a modest depth penalty.[4] Performance is even better if two different well depths are used, in which case a quadriphase (four-level) sequence is used.

The device in Figure 12.1d has different sized slats in front of a piece of porous absorbent to alter the reflected path lengths from the rigid parts of the diffuser. The varying depths of the slats break up any reflections, while the slots opening up to the porous material provides absorption. The need to vary the depths of the slats to improve dispersion is illustrated in Figure 12.4. The graph shows the normalised diffusion coefficient for the devices shown as inserts. When the device is just a regularly spaced set of slats, then at best, a coefficient of 0.2 is achieved. A common motif in architecture is to use slats that are spaced unevenly. Two examples are shown. These both have improved diffusion, but at most, the normalised diffusion coefficient is 0.5. By varying the depth of the slats, the maximum diffusion coefficient is almost 0.7. The example shown used a quadratic residue sequence to determine the slat depths. Figure 12.5 gives an example application of similar devices in a recording room.[5]

Figure 12.2 A curved hybrid surface and its application in a control room at the Universal Music Mastering Studios in New York City. Room design: Francis Manzella—FM Design Ltd. (Courtesy of George Roos.)

As porous absorbents have a lower speed of sound than air, hybrid surfaces have the ability to perturb the sound field at lower frequencies when compared to hard diffusers of the same depth. But at low frequency, the effect of these surfaces can be dominated by absorption. Nevertheless, it appears that hybrid surfaces can make efficient use of a limited depth. As absorption is inevitable in these devices, they are not useful in spaces where absorption must be minimized, such as large auditoria for symphonic music.

Figure 12.3 A curved hybrid diffuser applied to the ceiling and side walls of SonyM1, New York. Acoustician: Harris Grant Associates. (Courtesy of Paul Ellis of The M Network Ltd.)

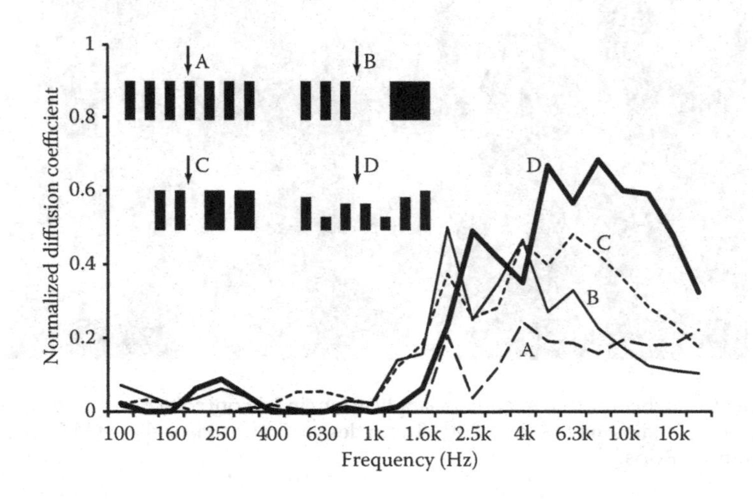

Figure 12.4 Normalised diffusion coefficient for a variety of slat patterns shown as inserts, with the arrow indicating the angle of incidence. Each slat is 1.9 cm wide and 7.6 cm deep. The rear of the slats open up to free space, mimicking the effects of a backing layer of an optimal porous absorber. Slat patterns: – – – (A) regular array of slats, ——— (B) an irregular arrangement according to an optimal aperiodic sequence (Mertens, S., *J. Phys.* A, 29, L473, 1996.), - - - - (C) an irregular arrangement using a truncated MLS, and ━━━ (D) an arrangement with different heights following a quadratic residue sequence.

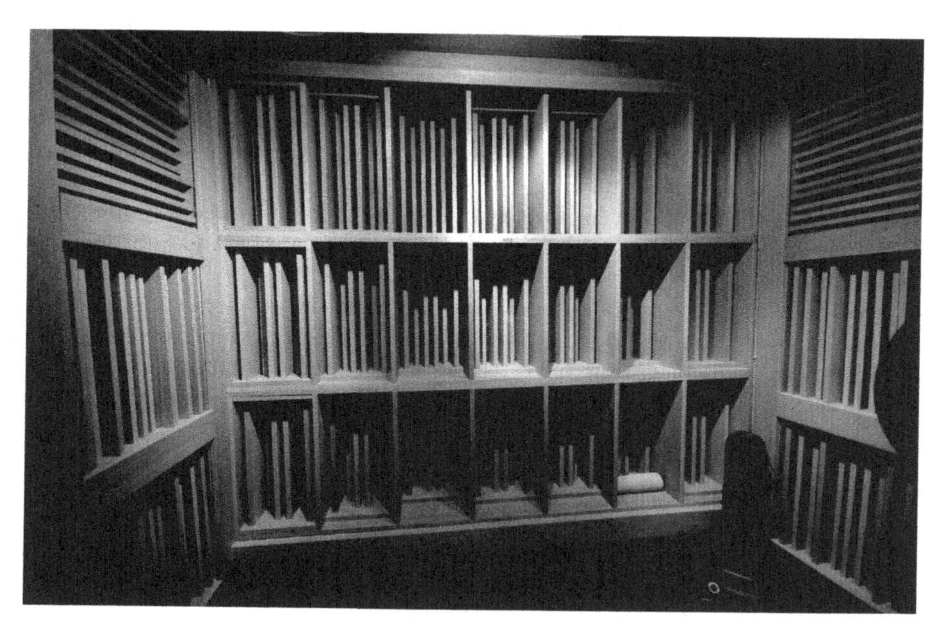

Figure 12.5 A hybrid diffuser made from slats in front of porous absorbent as applied in a recording room. (Courtesy of Bogić Petrović and Zorica Davidović.)

12.2 CONCEPTS

First, a discussion of the hybrid surfaces using the simplest Fourier theory discussed in Chapter 9 will be given. Some useful concepts can be developed using the simplest equations, but ultimately, more exact theories will be necessary. For simplicity, consider a 2D world so that scattering is in one plane only. The pressure amplitude, $|p_s|$, reflected from a planar surface with a spatially varying impedance, represented in the theory by the reflection coefficient, is given by

$$\left|p_s(\theta,\psi)\right| \approx \left|A\int_s R(x)e^{jkx[\sin(\theta)+\sin(\psi)]}\,dx\right|, \tag{12.1}$$

where ψ and θ are the angles of incidence and reflection, respectively, $R(x)$ is the reflection coefficient at a distance x along the surface, k is the wavenumber, A is a constant, and s is the diffuser surface.

This is an approximate far field theory, which forms the basis of the hybrid surface design. This simple prediction theory and the subsequent design process are applicable only at mid and high frequency. At low frequency, the mutual interactions across the surface make the prediction model inaccurate, as shall be discussed later. Equation 12.1 is a Fourier transform in x and transforms the reflection coefficients into $k[\sin(\theta) + \sin(\psi)]$ space. To a first approximation, the absorptive parts of the hybrid surface will have a reflection coefficient of $R(x) = 0$, and the reflective parts have $R(x) = 1$. A pseudorandom number sequence with good autocorrelation properties is used to determine the spatial distribution of the hard and soft patches. For example, the number sequence might be {0, 1, 0, 1, 0, 1, 0, 0, 0, 0, 1, 0, 1, 0, 0, 0, 0, 0, 0, 0}, in which where there is a 0 in the sequence, a patch of absorption is used, and where there is a 1 in the sequence, the surface is reflective. Figure 12.1a illustrates a surface where the impedance varies in one plane only and the strips of absorption or reflection are extruded in one direction.

A number sequence with good autocorrelation properties will have a flat power spectrum with respect to x. This means that the pressure amplitude scattered is constant with respect to the transform variable $k[\sin(\theta) + \sin(\psi)]$, which means good dispersion is generated in a polar response at each wavenumber. If the surface is periodic, this will relate to grating lobes all having the same level except for the lobe in the specular reflection direction, which will be attenuated. In many ways, this is similar to the theories behind Schroeder diffusers discussed in Chapter 10.

Angus[6] first developed hybrid diffusers using maximum length sequences (MLSs). These sequences are a good starting point as they have desirable Fourier properties. There are many other bipolar sequences that have flat Fourier transforms, but MLSs are the best known. The issue of sequences is discussed in more detail in the next section.

12.3 NUMBER SEQUENCES

To gauge the quality of a number sequence for a hybrid surface, the autocorrelation function can be examined. This is because the autocorrelation function directly relates to the scattering performance of the surface (see Section 10.4.5).

12.3.1 One-dimensional MLS

The reflection coefficients for the hybrid surface are 0 (absorption) and 1 (reflection), and consequently, the number sequence used should have optimal autocorrelation properties for 0s and +1s, which means that a good *unipolar* sequence is needed. Most pseudorandom binary sequences, on the other hand, have autocorrelation properties designed with a *bipolar* sequence composed of +1s and −1s. The autocorrelation side lobe performance of a unipolar and bipolar sequence can be very different. Figure 12.6 demonstrates this for an MLS of length 7. When a sequence can be bipolar (positive and negative), the autocorrelation side lobes on either side of zero delay include cancelling effects, which enable a low side lobe energy to be created as desired. When the sequence is unipolar, no cancellation can occur and the autocorrelation side lobe levels are higher. Consequently, it would be anticipated

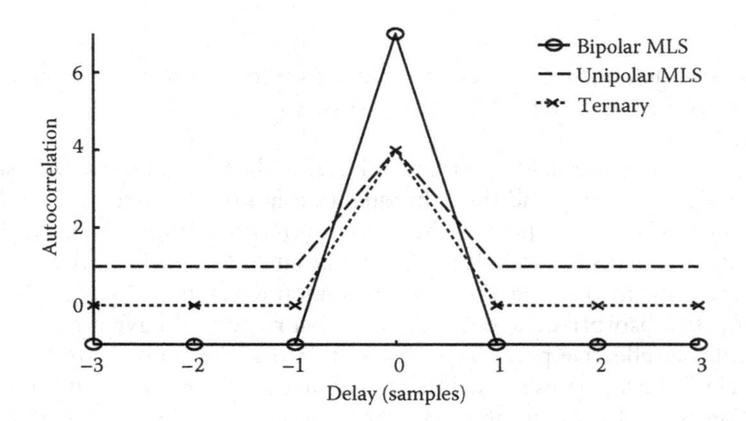

Figure 12.6 Autocorrelation function for: a bipolar (+1s and −1s) MLS, a unipolar (+1s and 0s) MLS, and a ternary sequence. The first is like a phase grating diffuser; the other two are like hybrid surfaces.

that the scattering performance would be worse for a unipolar sequence. The consequence of no cancellation is that the DC component in the power spectrum is large, as shown in Figure 12.7. This means that the energy in the specular reflection direction, when [sin(θ) + sin(ψ)] = 0, will be attenuated less for a unipolar surface in comparison to a bipolar surface.

In a bipolar MLS, the mean value of the reflection coefficients is close to 0, and consequently, the DC value of its power spectrum is close to 0 (see Figure 12.7). This means that suppression of the zeroth-order lobe in the polar response, when [sin(θ) + sin(ψ)] = 0, occurs (see Figure 12.8). The construction of diffusers based on bipolar MLSs was discussed in Section 10.4.1, as these are a type of Schroeder diffuser.

With a unipolar MLS, a hybrid diffuser is being constructed. The mean value of the reflection coefficients is about 0.5. The DC value is actually higher than the other spectral values, and consequently, the zeroth-order lobe is significantly greater than other lobes.

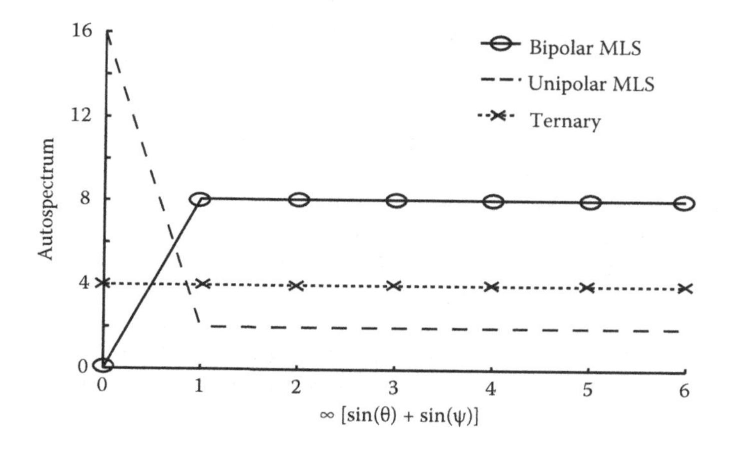

Figure 12.7 Comparison of power spectra (autospecta) for a bipolar MLS, a unipolar MLS, and a ternary sequence.

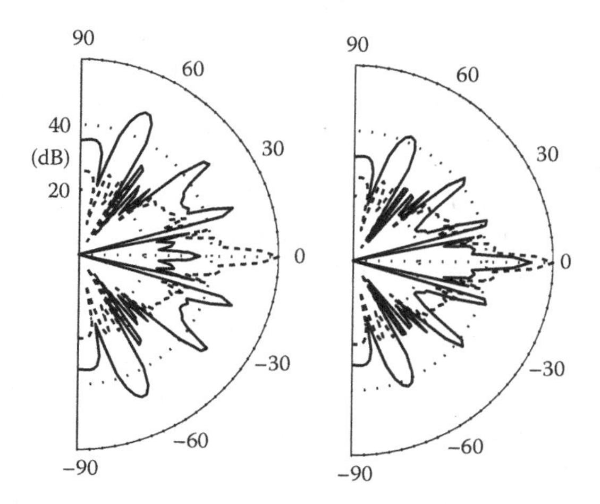

Figure 12.8 Scattering from diffusers constructed using bipolar and unipolar MLSs: ------- plane surface, shown in both figures; and ——— MLS, either bipolar (left figure) or unipolar (right figure).

Figure 12.7 illustrates this by comparing the power spectra of unipolar and bipolar MLSs. Figure 12.8 shows the polar response for a diffuser constructed from bipolar and unipolar MLSs. This is evidence that the planar hybrid surfaces will have a significant specular energy lobe, with the zeroth-order lobe being about $10\log_{10}(N)$ times larger than the other lobe energies, where N is the length of the MLS. The specular reflection lobe will be attenuated by about 6 dB compared to a plane hard surface (as expected for a surface which is 50% absorptive by surface area).

When the open area of the panel is about 50%, then an MLS is a good choice. The performance of the MLS when composed of unipolar elements is worse than when it is bipolar, as shown previously, but the MLS will still be the best possible sequence achievable (there are no better unipolar sequences, although there are some that are just as good). Problems arise if the open area of the panel needs to be reduced, as common optimal binary number sequences usually have a similar number of 0s and 1s.

12.3.2 One-dimensional optical sequences

There are a set of sequences, called optical sequences,[7] that are unipolar and have a different number of 1s and 0s. Angus[8] suggested that these could be used to overcome some of the problems associated with MLS. For instance, they can be used to form a surface whose open area is not 50%. They use an optimal sequence defined as one where the maximum of the side lobes of the autocorrelation function has the smallest possible value.

The problem with optical sequences is that the typical construction methods available result in a very low number of 1s in a long sequence. For example, a typical length 20 sequence is {0, 1, 0, 1, 0, 1, 0, 0, 0, 0, 1, 0, 1, 0, 0, 0, 0, 0, 0, 0}, which contains only five 1s and so would give a nominal open area of 25%. This occurs because the sequences were devised for very low side lobe performance, which necessitates a low occupancy of 1s, as this is the only way to achieve low side lobe energy in the autocorrelation function when no cancellation can occur. Unfortunately, for diffuser design, this makes the sequences, especially the very long sequences, not that useful. The surface will either be very reflective (if the 1s are associated with absorption) or very absorptive (if the 1s are associated with reflection).

Note that the previous sequence can also be written as {2, 4, 6, 11, 13}, where the index number gives the location of the 1s. For low-occupancy cases, this is a much more compact representation of the sequences and so will be used here.

Optical sequences are usually generated in families. These are a set of sequences that not only have good autocorrelation properties but also have low energies for the cross-correlation between family members. In an optical sequence, five parameters are used to specify their performance. ξ is the number of 1s in the sequence, N is the length, M is the family size (the number of sequences in the family), S_{xxm} is the maximum side lobe value in the autocorrelation function, and S_{xym} is the maximum value in the cross-correlation function between sequences in the same family. These parameters can be stated in an abbreviated form $\{\xi, N, M, S_{xxm}, S_{xym}\}$. For the length 20 sequence given previously, this would be expressed as $\{5, 20, M, 3, S_{xym}\}$.

There are some sequences with a reasonable occupancy, for example, {3, 7, 1, 1, 1}, but the length of these sequences is too short, as a designer should always look to maximize the repeat length to minimize periodicity effects. This means that large N sequences are required, and then the number of 1s becomes too small using the normal optical sequence generation techniques. Consequently, when N is large, three construction techniques are suggested, as outlined here. In reality, a combination of techniques might be needed.

The first technique involves starting with a sequence with too few 1s and increasing the number of 1s. Given a $\{\xi, N, M, S_{xxm}, S_{xym}\}$ family of sequences, it is possible to increase the number of 1s, but at the penalty of increasing the maximum value of the autocorrelation and cross-correlation. This is best illustrated with an example: A set of optical sequences {3, 25, 4, 1, 1} are {1, 2, 7}, {1, 3, 10}, {1, 4, 12}, {1, 5, 14}. The sequences are taken in pairs and a bit-wise OR taken between the pairs to form new sequences. So if both sequences have a 0 bit, then the new sequence has a 0 bit; otherwise, the new sequence has the bit set to 1. In the previous case there are six unique combinations, as sequences are not combined with themselves. The new sequences are {1, 2, 3, 7, 10}, {1, 2, 4, 7, 12}, {1, 2, 5, 7, 13}, {1, 3, 4, 10, 12}, {1, 3, 5, 10, 13}, {1, 4, 5, 12, 13}. These form a new family of sequences with the property $\{\xi' = 5, N' = 25, M' = 6, S'_{xxm} = 2, S'_{xm} = 3\}$, where the prime is used to denote the new sequence. The number of 1s has increased from three to five, but both the autocorrelation and cross-correlation properties have degraded. The maximum side lobe value in the autocorrelation has increased from 1 to 2.

This process, in general, produces new sequences where the number of 1s is $\xi' \geq 2\xi - S_{xym}$, where ξ and S_{xym} are the values for the original sequences. If too many 1s are generated after the OR operations, some 1s are randomly chosen to be changed to 0s. This was not necessary in the previous example. The autocorrelation and the cross-correlation of the new sequences will be $S'_{xxm} \leq 2S_{xxm} + 2S_{xym}$ and $S'_{xym} \leq \xi + 3S_{xym}$, respectively. In the previous construction example, the new sequences are considerably better than the worst case given by these upper bounds. For diffuser design, repeated application of this process is likely to be needed.

The second technique is very similar to the first, but in this case, you would take a sequence with too many 1s and reduce their number. For example, it should be possible to start from a family of MLSs and, via logical operations, reduce the number of 1s to the desired value.

The third technique involves constructing a family of optical sequences and then concatenating these together. As the family will have mutual low correlation, they should work well in a concatenated longer sequence.

12.3.3 One-dimensional ternary and quadriphase sequences

Many of the standard construction methods for general ternary sequences are inappropriate because they do not generate the right balance of –1, 0, and +1 elements for hybrid surfaces. Many sequences have very few 0 elements in them, and consequently, the surfaces would not be very absorbing. This arises because most applications of number theory want to maximize the efficiency of the sequence—efficiency in this context meaning the power carried by a signal based on the sequence. The sequence also needs to have a similar number of –1s and 1s because this maximizes the attenuation of the specular reflection.

Correlation identity derived ternary sequences[7] can have an efficiency of ≈50%, meaning that half the sequence are 0s and will therefore have a nominal absorption coefficient of ≈0.5, provided the design parameters are chosen correctly. They are formed from two MLSs of order m and length $N = 2^m - 1$; the order of the sequences must obey $m \neq 0 \bmod 4$.

First, it is necessary to find a pair of MLSs with suitable cross-covariance properties. The process is to form an MLS and then sample this sequence at a different rate to form a complementary sequence. For example, if the sample rate is $\Delta n = 2$, then every second value from the original sequence is taken. The sample rate is chosen using either $\Delta n = 2^k + 1$ or $\Delta n = 2^{2k} - 2^k - 1$. A parameter e is defined as $e = \gcd(m, k)$, where gcd() is the greatest common divisor. This must be chosen so that m/e is odd, as this gives the correct distribution of cross-covariance values.

Under these conditions, the two MLSs have a cross-covariance $S_{ab}(\tau)$, which has three values. The total number of 1s and –1s in the sequence will be given by $\approx N(1 - 2^{-e})$. This is therefore the amount of reflecting surface on the diffuser, and so at high frequency, it would be anticipated that the absorption coefficient of the surface, α, would be $\approx 1 - 2^{-e}$. If the aim is to achieve a surface with $\alpha \approx 0.5$, this means choosing $e = 1$, which means the order of the MLS, m, must be odd.

Consider an example of $N = 31 = 2^5 - 1$. e is required to be a divisor of m so that m/e is odd, and this can be achieved with $k = 1$ as this makes $e = \gcd(k, m) = 1$. A possible sample rate is $\Delta n = 3$.

The first part of the MLS used is {1, 0, 0, 0, 0, 1, 0, 0, 1, 0, 1, 1, 0, 0, 1, 1, 1, 1, 1,...}. Taking every third value then gives a second MLS: {1, 0, 0, 0, 0, 1, 1, 0, 0, 1, 0, 0, 1, 1, 1, 1, 1, 0, 1,...}. This then gives a cross-covariance (after the MLS sequences are made bipolar) where 7 occurs 10 times, –1 occurs 15 times, and –9 occurs 6 times.

The ternary sequence, c_n, is formed from this cross-covariance—a rather surprising and remarkable construction method. The sequence is $2^{-(m+e)/2}(S_{ab}(\tau) + 1)$. This sequence has an ideal autocovariance with a peak value of 2^{m-e} and out-of-phase values that are 0. Applying this to the previous pair of MLSs yields the ternary sequence: {0, 0, 1, 1, –1, 1, –1, 0, 0, 0, 1, 1, 0, 1, –1, –1, 0, 1, 0,...}.

The autocovariance indicates the advantages that might be expected from ternary sequence diffusers in comparison to unipolar binary sequence diffusers. The autocovariance functions for the ternary and unipolar binary sequences are shown in Figure 12.6. The unipolar binary sequence has constant out-of-phase values, but they are not 0. This leads to diffusers with a significant specular component in their polar pattern. Perfection can be achieved using a ternary sequence as the out-of-phase values are all 0.

The ternary sequence has a good reflection coefficient autospectrum because it is constant; an example is shown in Figure 12.7. This means that a surface based on a single ternary sequence produces more even scattering than one using an MLS. For a periodic structure where many repeats of the sequence are placed side by side, this will result in all the grating lobes having the same energy.

12.3.4 Optimized sequences

For short sequences, it is possible to use a computer search to find the best sequence. For a small sequence length, say $N \leq 20$ for a binary sequence, it is possible to do an exhaustive search of every possible combination to find the best sequences. Every possible combination of bits with the correct number of 1s are tested by first constructing the autocorrelation function for each case and then finding the sequence or sequences that have the smallest maximum value for the autocorrelation side lobes. As N increases, however, this rapidly becomes a very time-consuming process to carry out. The number of unique combinations to search is given by

$$^{N}C_r = \frac{N!}{\xi!(N-\xi!)}, \tag{12.2}$$

where ξ is the number of 1s in the sequence and ! indicates factorial. Consequently, for $N = 20$, $\xi = 10$, there are nearly 200,000 combinations, and this roughly doubles for every additional bit.

For, say, $20 < N < 48$, a numerical optimization[9] can be used to search for the best sequence. In this case, optimization algorithms are used to avoid the need to test every possible combination, but even so, this is still a slow process. For an $N = 48$ binary sequence with 24 1s and 24 0s, there are about 10^{13} unique sequence combinations to search.

Unlike the numerical optimization of Schroeder well depths discussed in Chapter 10, this is a discrete function optimization. In other words, the values of the optimization parameters, which are the locations of the 1s in a binary sequence, can only take discrete values. This means that the best algorithm for tackling this problem is a genetic algorithm as it can explicitly represent the discrete sequences as genes.

A genetic algorithm is a technique for searching for optimum configurations in engineering problems. Figure 12.9 illustrates how a typical genetic algorithm works. It attempts to mimic the process of evolution that occurs in biology. A population of individuals is randomly formed. Each individual is determined by their genes; in this case, the genes are simply the binary sequence values, indicating where hard and soft patches should be placed on the surface. Each individual has a fitness value that indicates how good they are. In this case, it is the largest energy in the autocorrelation side lobes. Over time, new populations are produced by breeding and the old populations die. Offspring are produced by pairs of parents breeding. An offspring has a gene sequence that is a composite of the sequences from the parents. A common method for doing this is multiple point cross-over. For each bit in the sequence, there is a 50% chance of the child's bit coming from parent A and a 50% chance of the bit being from parent B. Mutation is also used. This is a random procedure whereby there is a small probability of any bit in the child sequence changing during breeding. Mutation allows sequences outside the parent population to be made.

Selecting sequences to breed and be killed can be done randomly. As with conventional evolution theory, the fittest are most likely to breed and pass on their genes, and the least fit, the most likely to die. By these principles, the fitness of successive populations should improve. This process is continued until the population becomes sufficiently fit, so that the sequence produced can be classified as optimum.

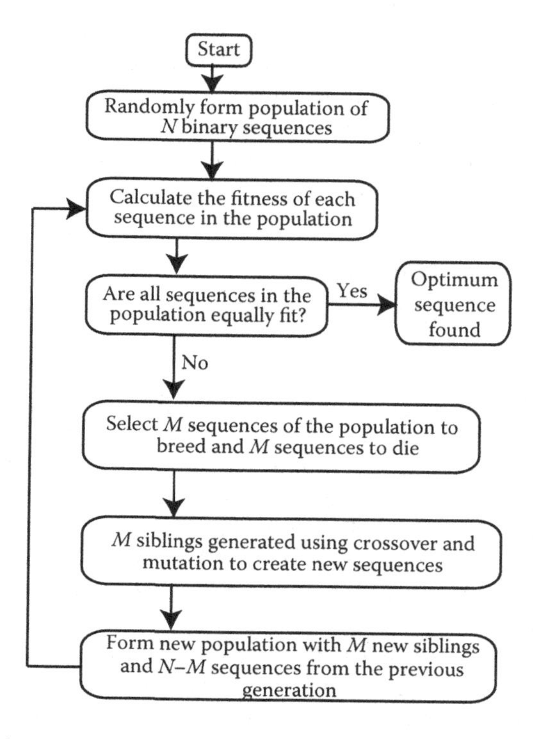

Figure 12.9 Optimization of numerical sequences using a genetic algorithm.

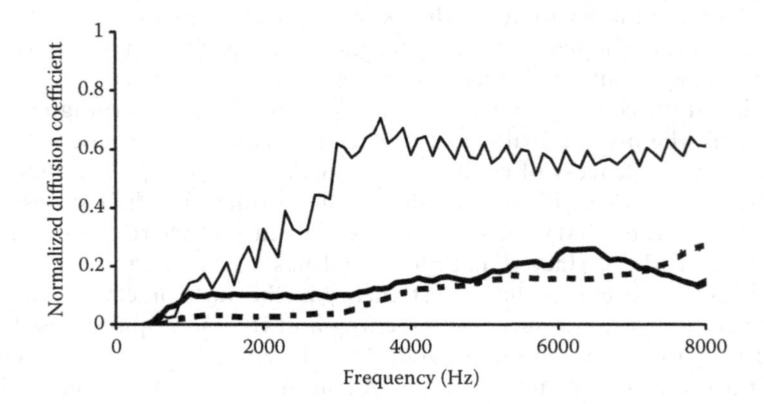

Figure 12.10 Normalized diffusion coefficient for three surfaces showing how an optimized sequence can improve on a random sequence (normal incidence): ······· hybrid surface formed using a typical random sequence, ▬▬ hybrid surface formed using an optimized sequence, and —— 25 cm fibreglass. (Original data from Cox, T.J., D'Antonio, P., *Proc. 107th Convention Audio Eng. Soc.*, preprint 5062, 1999.)

An additional advantage of using a numerical optimization is that it gives complete control over the reflectivity of the diffuser. This can be specified as a desirable characteristic; any individual not having the desired reflectivity will be scored as less fit. Consequently, the trait of undesired reflectivity will die off.

Figure 12.10 shows the diffusion coefficient for a hybrid surface formed from an optimized sequence, compared to a hybrid surface formed from a random sequence.[9] An improvement in performance is seen for most frequencies, although the improvement is not great. The main problem with this design technique is that it is impossible to find long optimal sequences because the number of possible sequences to search becomes too large. One solution to this problem is to generate a family of good sequences of relatively low N with low mutual cross-correlation and to concatenate them together to get a longer sequence. This is possible because during the optimization, both the autocorrelation and cross-correlation properties can be considered at the same time.

12.3.5 Two-dimensional sequences

A hemispherical scatterer such as that shown in Figure 12.2 requires a 2D sequence, and this can be achieved by three methods[7,10]: folding, modulation (known as Kronecker product in number theory), and periodic multiplication. Consider constructing a surface with dimensions (in terms of number of patches) of $N \times M$. Whether a sequence can be constructed depends on the values of N and M.

Schroeder[11] showed that a folding technique called the Chinese remainder theorem could be applied to phase grating diffusers based on polyphase sequences. D'Antonio[1] used the same technique for a binary hybrid diffuser. The Chinese remainder theorem folds a 1D sequence into a 2D array and yet preserves the good autocorrelation and Fourier properties. To use this method, N and M must be coprime. By *coprime*, it is meant that the only common factor for the two numbers is 1.

Consider a length 15 sequence that will be wrapped into a 3 × 5 array. The elements are sequentially labelled $a_1, a_2, a_3,..., a_{15}$. The 1D sequence is written down the diagonal of the array, and as it is periodic, every time the edge of the array is reached, the position is folded back into the base period. The process is illustrated in Figure 12.11.

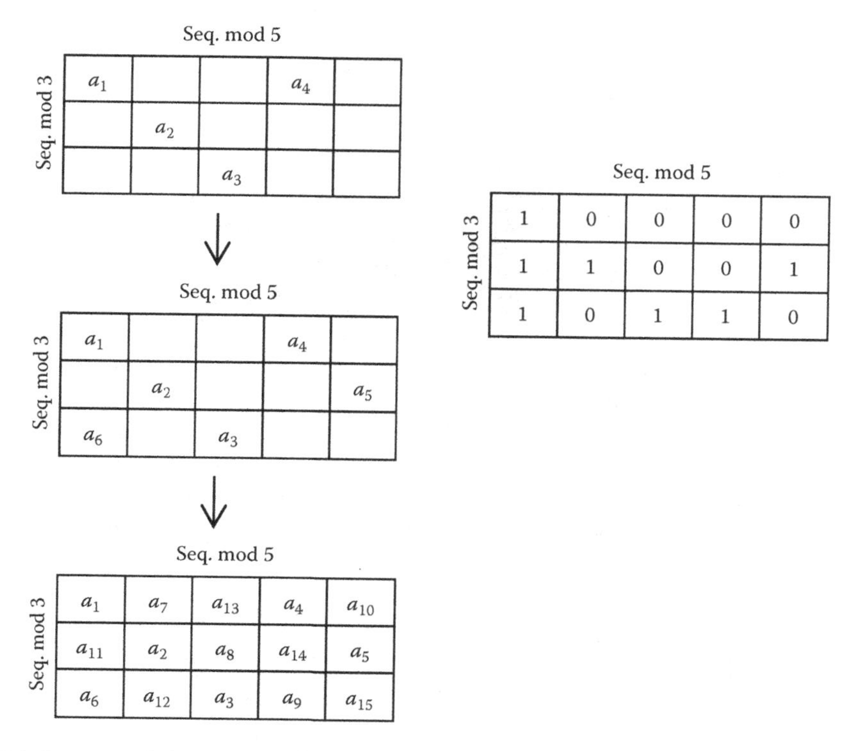

Figure 12.11 Left: some of the steps involved in folding a length 15 sequence, $a_1 \ldots a_{15}$ into a 2D array using the Chinese remainder theorem. Right: the wrapping of the MLS sequence {1, 1, 1, 0, 1, 1, 0, 0, 1, 0, 1, 0, 0, 0, 0}.

Table 12.1 shows another way of viewing this process. The coordinates (column, row) of the elements $a_1, \ldots, a_{15}$ are determined by modulo indexing. The subscript is called the index, n. The column for element a_n is determined by n modulo 5 and the row is determined by n modulo 3.

This folding technique still maintains the good autocorrelation properties of the sequence. For example, Figure 12.12 shows the autocorrelation for a unipolar MLS folded into a 3 × 5 array. The same autocorrelation properties are achieved in terms of the side lobe energy values. The Chinese remainder theorem can be applied before or after the Fourier transform, as illustrated in Figure 12.13. Consequently, the folding technique preserves the ideal Fourier properties.

Modulation is a process that allows diffusers to be arranged in a non-periodic fashion by modulating one or more base shapes with a binary sequence (see Section 10.5). Another way of viewing the outcome of this process is that it forms a single longer length sequence. A very similar process can be used to form arrays using ternary and binary sequences and arrays. Two sequences (or arrays) are modulated together to form a longer sequence (or array).

Table 12.1 The Chinese remainder theorem expressed as modulo existing

Index, *n*	0	1	2	3	4	5	6	7	8	9	10	11	12	13	14
Column coordinate = *n* modulo 5	0	1	2	3	4	0	1	2	3	4	0	1	2	3	4
Row coordinate = *n* modulo 3	0	1	2	0	1	2	0	1	2	0	1	2	0	1	2

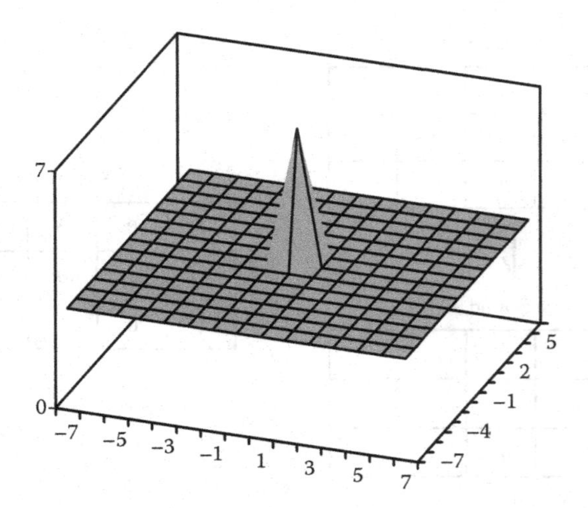

Figure 12.12 The autocorrelation function for a unipolar N = 15 MLS sequence folded using the Chinese remainder theorem.

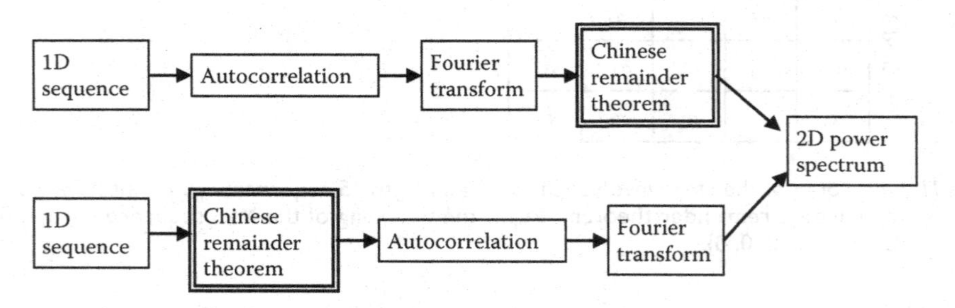

Figure 12.13 The Chinese remainder theorem can be applied before or after the Fourier transform, and consequently, it preserves good autocorrelation properties when used to fold 1D sequences into 2D arrays.

To illustrate this, consider making a ternary array by modulating a ternary sequence with a perfect aperiodic binary array. (Note that it is important to modulate the array by the sequence and not vice versa.) Consider a length 7 correlation identity derived ternary sequence, $a = \{1, 1, 0, 1, 0, 0, -1\}$; this is used to modulate the perfect aperiodic binary array, $b = \begin{Bmatrix} -1 & -1 \\ -1 & 1 \end{Bmatrix}$ to form a 2 × 14 length array, c, given by

$$c = \begin{Bmatrix} -1 & -1 & -1 & -1 & 0 & 0 & -1 & -1 & 0 & 0 & 0 & 0 & 1 & 1 \\ -1 & 1 & -1 & 1 & 0 & 0 & -1 & 1 & 0 & 0 & 0 & 0 & 1 & -1 \end{Bmatrix}.$$

This array has ideal periodic autocorrelation properties. As the binary array has no zeros, the modulated array has the same proportion of absorbent patches as the original ternary sequence; for long sequences, the proportion tends towards 50%.

There is only one known perfect aperiodic binary sequence, the one shown previously. Consequently, there are only six array sizes that can be constructed by this method with ≈50% efficiency for $NM \leq 2^{16}$.

The final design process is to use periodic multiplication. Two arrays can be multiplied together to form a larger array. Consider array 1 to be $b_{p,q}$ of size $N_b \times M_b$, which has an efficiency of E_b, and array 2 to be $c_{p,q}$ of size $N_c \times M_c$, which has an efficiency of E_c. Then the new array is a product of the periodically arranged arrays, $b_{p,q} \cdot c_{p,q}$ of size $N_b N_c \times M_b M_c$ and the efficiency will be $E_b E_c$. A necessary condition is that N_b and N_c are coprime, and so are M_b and M_c; otherwise, the repeat distance for the final arrays will be the least common multiples of N_b and N_c in one direction and M_b and M_c in the other.

For example, consider a 7 × 3 ternary array derived by folding a sequence made using a Singer difference set[7]:

$$c = \begin{Bmatrix} 1 & 0 & 1 \\ 0 & 1 & 0 \\ -1 & 1 & -1 \\ -1 & 1 & -1 \\ 1 & 0 & 1 \\ -1 & 1 & -1 \\ 1 & 0 & 1 \end{Bmatrix}.$$

This has an efficiency of 76%. Consider also a perfect aperiodic ternary array d_2:

$$d_2 = \begin{Bmatrix} 1 & 1 \\ 0 & 0 \\ 1 & -1 \end{Bmatrix},$$

which has an efficiency of 67%.

c and d_2 can be multiplied together to from a 21 × 6 array with ideal autocorrelation properties and an efficiency of 0.76 × 0.67 × 100% = 51%.

It is also possible to use a combination of these design methods. Perfect array construction methods, ones that produce zero side lobe energy in the autocorrelation, tend to generate rectangular grids of holes. But it is possible to construct optimal binary sequences on a hexagonal array pattern as well.

In the design, it is necessary to consider the balance between 0s and 1s for binary sequences and –1s, 0s, and 1s for ternary sequences. The right balance needs to be struck to achieve the desired absorption characteristic. For ternary sequences, a rough balance between the number of –1s and 1s is needed so that the specular reflected energy is sufficiently attenuated.

Once the array is formed, any periodic section can be chosen and many other manipulations can be carried out while still preserving the autocorrelation properties. Procedures that can be done on their own or in combination include the following:

- Using a cyclic shift to move the pattern around: $c_{p,q} = b_{p+u,q+v}$, where u and v are integers and the indexes $p + u$ and $q + v$ are taken modulo N and M, respectively.
- Mirror image the array: $c_{p,q} = b_{\pm p,\pm q}$
- Invert the sequence: $c_{p,q} = -b_{p,q}$
- Rotation by 90°: $c_{p,q} = b_{q,p}$
- Under sample the array: $c_{p,q} = b_{up,vq}$, provided both u, N and v, M are coprime.

These will not change the acoustic performance but may change the visual aesthetic. These processes can also help to make the array more asymmetric, which can be useful in modulation.

The main problem in forming these arrays is that there is only a limited set of array sizes. It has been shown,[12] however, that by relaxing the requirement for ideal autocorrelation enables more array sizes to be formed. For example, where there are a large number of elements in a sequence, it may be possible to truncate the sequence, losing one or two elements, and still gain good (but not ideal) autocorrelation properties. This type of truncation might then give the right sequence length for folding into an array with the desired size.

12.4 ABSORPTION

Figure 12.14 shows the random incidence absorption coefficient for a planar binary surface with and without the perforated mask. This shows that the hybrid surface is behaving like the Helmholtz absorbers discussed in Chapter 7. It is possible to predict the absorption characteristics using the transfer function matrix method. Problems arise, however, because the hole spacing is not regular and many holes are too close together for the normal assumptions used when modelling sound propagation through a perforated sheet. Nevertheless, it seems possible to at least predict the trends of the absorption.

The amount of added mass in the holes determines the increase in absorption at bass frequencies. For the BAD panel, if additional bass absorption is required, it is possible to reduce the open area (this will not be true for all geometries), as shown in Figure 12.14. Alternatively, thicker mineral wool layers can be used, or the panel can be spaced from the wall to effectively increase the backing depth and so lower the resonant frequency. The effect of changing the backing depth is shown in Figure 7.9, giving the expected trends.

The drop-off in high-frequency absorption can be explained by the open area of the panel. In a simplistic analysis, a surface with a 50% open area would be expected to have an absorption coefficient of 0.5 at high frequency. Figure 12.14 shows the effect of reducing the open area on the absorption coefficient. Boundary element modelling (BEM) predictions show that the absorption coefficient response is similar for ternary diffusers, if a little less

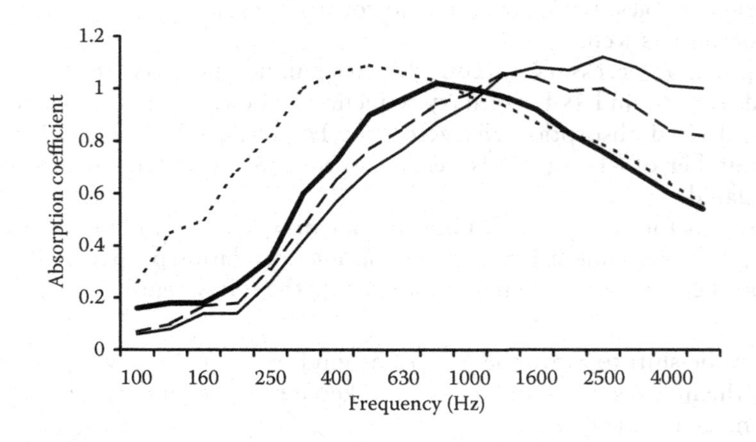

Figure 12.14 Random incidence absorption coefficients for various planar hybrid surfaces (BAD panels): ——— 25 mm fibreglass; - - - - BAD 16 mm hole, 25 mm fibreglass backing; ━━━ BAD 13 mm hole, 25 mm fibreglass backing; and ······ BAD 13 mm hole, 51 mm fibreglass backing.

smooth with respect to frequency. It is assumed that this is due to reflections from the wells providing out-of-phase reflections when compared to other parts of the surface, and therefore, waves combine to put energy into the reactive field.

12.5 ACCURACY OF THE FOURIER THEORY

The Fourier theory used so far in this chapter is approximate. The absorption coefficient measurements give information about the likely accuracy of the design principles based on this simple theory. Some of the key assumptions behind the theory are as follows:

- The absorption coefficient of the soft patches is 1 and there is no phase change on reflection.
- The Kirchhoff boundary conditions are true.

Usually, the absorption coefficient of the soft patches is not 1 at low frequency because there is insufficient depth of mineral wool to cause complete absorption. At mid and low frequencies, mutual interactions across the panel render the Kirchhoff boundary conditions inaccurate. The Kirchhoff boundary conditions assume that the pressure on the hard patches is twice the incident pressure and on the soft patches is just the incident pressure. This is true only when the patch width is large compared to wavelength. It is only at high frequency, when the wavelength becomes smaller, that the Kirchhoff boundary conditions will be reasonable. Consequently, the previous discussions are really only true for the highest frequencies of interest. To put this another way, for a 50% open area sample predicted with the simple Fourier theory, the absorption should be 0.5. Figure 12.14 shows that this is not achieved; for instance, at mid frequency, the absorption is around 1.

Despite these problems, it is still possible to learn something from the Fourier theory. The theory shows that as with all diffusers, periodicity is a problem as energy gets concentrated in the lobe directions. Consequently, the repeat distance of the diffuser should be made as large as possible. This can be done using a modulation scheme, as discussed for Schroeder diffusers in Section 10.5 and geometric surfaces in Chapter 11, or by designing the diffuser based on a large number sequence.

Consider using a modulation scheme where two different hybrid surfaces are used. One of these is denoted A, the other, B. A wall is filled by arranging the hybrid surfaces according to a pseudorandom sequence; for example, if a length 5 Barker sequence is used, the arrangement of the surfaces would be AAABA. This is a method for reducing periodicity effects.

Modulation schemes pose problems for devices based on unipolar sequences. The most successful modulation scheme for Schroeder diffusers was to modulate a surface with a phase inverted version of the surface. In a unipolar device, it is difficult to see how exact broadband phase inversion can be achieved. However, it is possible to go some way to achieving the inversion and, therefore, good performance. Consider a unipolar MLS a, an approximate inverse sequence can be found by adding 1 modulo 2, so the inverse sequence is given by $a + 1$ modulo 2. These two base shapes can then be arranged in a pseudorandom order to form a modulated array to reduce periodicity effects.

For example, consider the top three graphs in Figure 12.15. Graph (a) shows the autocorrelation properties for a periodic sequence of 13 $N = 127$ unipolar MLS sequences. As expected, there are periodicity lobes every 127 units. This is compared in graph (c) to a random sequence of length 1651 (= 13 × 127) showing no periodicity lobes. A modulated arrangement is shown in graph (b) using a hybrid surface and its inverse. The ordering of the two surfaces was determined by the length 13 pseudorandom sequence with the best unipolar aperiodic

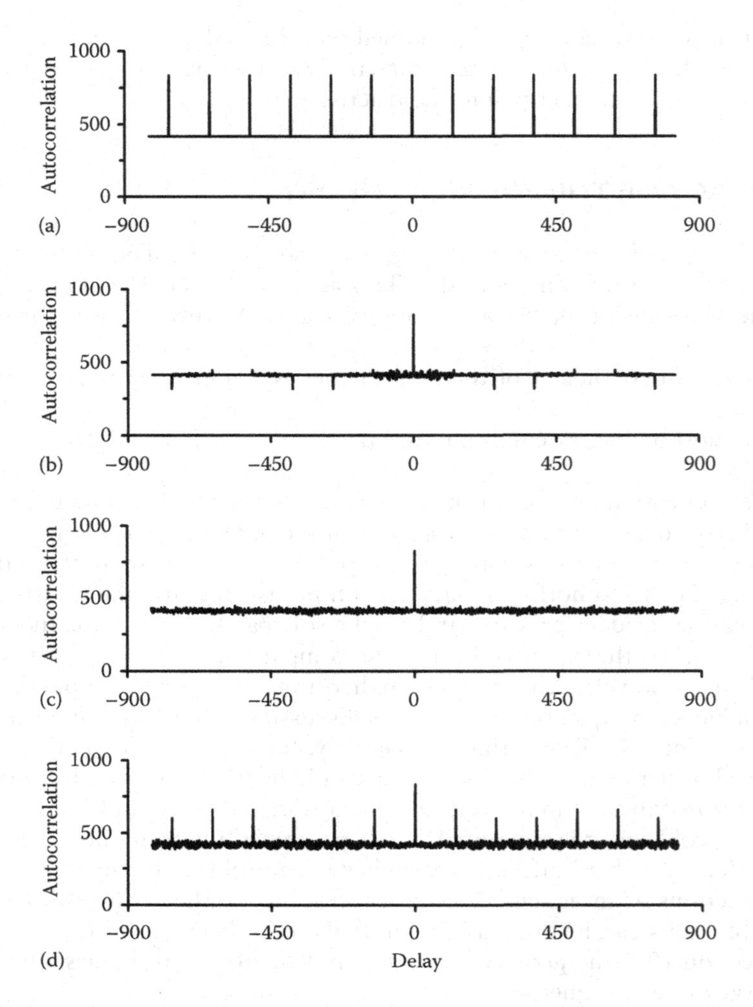

Figure 12.15 Autocorrelation properties for four sequences: (a) periodic, (b) modulated using inverse, (c) random length 1651, and (d) one base shape modulation.

autocorrelation properties. The effects of periodicity have been greatly reduced. Figure 12.16 shows the effects of the modulation in terms of a scattered level polar distribution. The modulation has removed the periodicity effects for the non-zero-order lobes. In that respect, the modulation has been successful, but the dominant characteristic of the polar response is still the zero-order specular lobe. The only way to deal with this is to introduce some possibility of phase cancellation, for example, by curving the surface or using a ternary sequence.

A ternary sequence can be more readily modulated with its inverse. For instance, if the first ternary sequence was {1, 1, 0, 1, 0, 0, −1}, then the complementary sequence used in modulation is the inverse of this: {−1, −1, 0, −1, 0, 0, 1}. The modulated arrangement is highly effective, but only over certain frequency ranges. When multiples of half wavelengths fit into the −1 well, then the reflection coefficients of the ternary sequence and its inverse are identical, and undesirable periodicity lobes result.

Another technique for modulation would be to use two sequences from the same family with mutually good cross-correlation properties, although this does not work as well as the

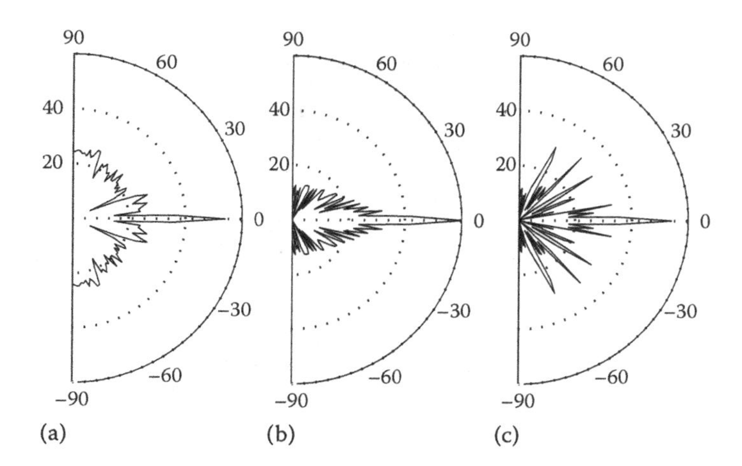

Figure 12.16 Scattering from three surfaces showing how modulation can improve the performance of hybrid surfaces—(a) modulated MLS hybrid surface, (b) planar non-absorbing reflector, and (c) periodic MLS hybrid surface.

inverse process described previously. It is also possible to modulate using number sequences of different lengths—a similar process was described for Schroeder diffusers in Section 10.5, where it was called *orthogonal modulation.*

As with Schroeder diffusers, it is possible to modulate a single asymmetrical base shape. The advantage of this technique is that it needs only one type of mask to be made. One of the masks shown in Figure 7.8 could be rotated by 90° and arranged with the original mask in a random arrangement. Brief tests on 1D sequences show this to be less successful than using two or more different base shapes. Graph (d) in Figure 12.15 shows the autocorrelation coefficient for a 1D asymmetric single base shape modulation. In this case, the sequence order is just reversed to give the second base shape. While not as good as the modulation using two base shapes (graph b), one base shape modulation still gives a better result than the periodic arrangement.

12.6 DIFFUSE REFLECTIONS

Measuring hybrid surfaces to get the scattering performance is awkward. The normal measurement techniques described in Chapter 5 have used scale models of diffusers to allow measurements far enough away from the surface. For hybrid surfaces, accurate scale models cannot be easily produced because the impedance properties of porous materials do not scale in the same way as the wavelength in air. It would be possible to empirically find a substitute material to use in the scale models, but this has not been done. Nevertheless, diffusion measurements have been carried out on full-scale samples, even though the receivers and source are too close. For example, Figure 12.17 shows the scattered polar response from a planar hybrid surface and just the porous backing material for the 1.25 kHz one-third octave band. The energy in the specular zone is attenuated by 6 dB. This is exactly as would be expected for a 50% open area panel. Some additional side lobe energy is produced compared to the flat hard surface.

Consider an analysis now based on the simple Fourier model. Figure 12.18 shows polar distributions for one example frequency. For this comparison, five surfaces were predicted: a curved hybrid surface with a binary sequence of hard and soft patches, a plane hybrid surface based on the length 31 MLS, a plane hard surface, an $N = 7$ QRD, and an optimized

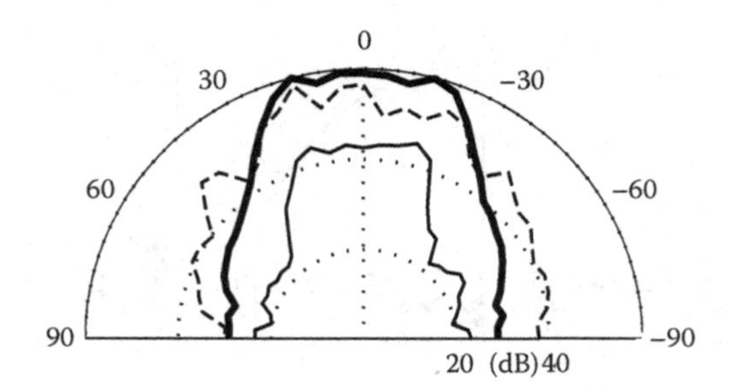

Figure 12.17 Measured scattering from a binary amplitude diffuser (BAD) compared to a hard surface and a piece of absorbent: - - - - BAD, ▬▬ non-absorbing flat reflector, and —— 25 mm fibreglass. (Data from Angus, J.A.S., D'Antonio, P., *Proc. Audio Eng. Soc.*, preprint 5061 (D-5), 1999.)

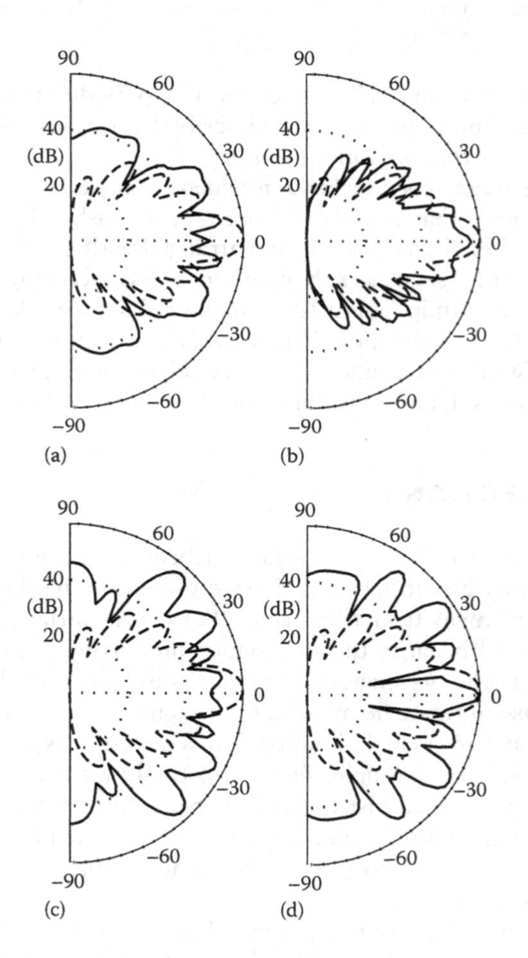

Figure 12.18 Scattering for five surfaces; in each case, the dotted line represents a planar non-absorbing surface of the same size as the diffusers. (a) Curved hybrid surface; (b) flat hybrid surface; (c) curved hard diffuser; and (d) $N = 7$ QRD.

rigid curved surface. This last surface was 30 cm deep, which is typical of the non-absorbing diffusers used in performance spaces; the hybrid curved surface is much shallower, about 7 cm deep.

Moving from a flat hard surface to a plane hybrid device increases the dispersion, as was also found in the measurements reported in Figure 12.17. However, curving the hybrid surface produces more dramatic improvements. The quality of dispersion for the curved hybrid surface is only a little worse than that of the rigid optimized curved surface, which is four times as deep. This seems to indicate that a hybrid curved surface is a good method for generating more diffuse reflections from a restricted depth, provided partial absorption is acceptable or wanted. It should be remembered that practical hybrid surfaces are often mainly absorptive up to about 2 kHz, and consequently for these surfaces, it is only at mid to high frequencies that dispersion needs to be considered.

12.6.1 Boundary element modelling

It is known that the Kirchhoff boundary conditions are not accurate at low to mid frequencies, and so the Fourier model used in the previous section is unreliable. Using BEM, it is possible to calculate the scattering over a hemisphere including interactions across the surface, provided the absorbent patches remain locally reacting. BEMs have been shown to give accurate results for hybrid surfaces before, when compared with measurements.[13] They also give accurate results for Schroeder diffusers, and consequently, it would be anticipated that the BEM will be accurate for ternary diffusers as well.

First, consider single plane predictions on a planar hybrid and a ternary device. Figure 12.19a shows the scattering for the one-third octave band centred on the design frequency; Figure 12.19b shows the scattering at an octave higher.[4] At odd multiples of the design frequency (e.g., graph a) the well in the ternary diffuser provides waves that help cancel the specular reflection and so the device offers more even scattering and a reduced specular

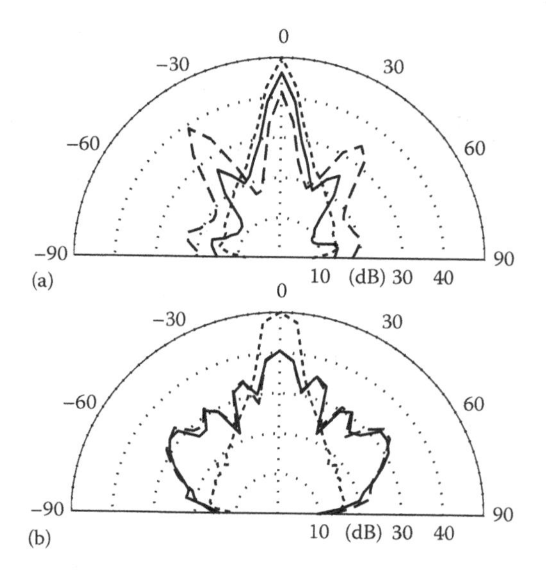

Figure 12.19 Scattering from three devices predicted using a BEM model. ——— planar binary hybrid, - - - - ternary, and ······· plane surface for the one-third octave whose centre frequency is (a) the design frequency and (b) twice the design frequency. (After Cox, T.J., Angus, J.A.S., D'Antonio, P., *J. Acoust. Soc. Am.*, 119, 310–319, 2006.)

lobe in comparison to the planar hybrid surface. At even multiples of the design frequency, however (e.g., graph b), the scattering from the planar hybrid and ternary devices is similar. At these even multiples of the design frequency, the waves propagating in the well of the ternary diffuser return in phase with reflections from the flat sections of the device, and so the wells are ineffective. To overcome these critical frequencies at even multiples of the design frequency, more well depths must be introduced. If two different well depths are used, for instance, a quadriphase diffuser results.[4]

Figure 12.20 shows the normalized diffusion coefficient for three hybrid surfaces. It shows how a ternary diffuser improves over the planar binary surface for most frequencies, except in the one-third octave band centred on 4 kHz where there is a critical frequency for the ternary device. It also shows how using an additional different well depth in a quadriphase diffuser overcomes the critical frequency.

So far, the performance of the ternary devices has been discussed only at harmonics of the design frequency. Between these frequencies, the phase of the reflection coefficient offered by the well of fixed depth is neither –1 or +1. The waves reflected from this well will be partly out-of-phase with the waves from the flat parts of the device. Consequently, the performance is still improved over the unipolar binary diffuser for these in-between frequencies.

Now consider predictions for hemispherical devices using a thin panel, periodic BEM formulation. Figure 12.21a shows the scattering from a 4 × 4 array of the BAD panel at 3 kHz, which is an MLS-based hybrid planar surface. This is compared to a flat hard surface in Figure 12.21b. The polar balloon is shown side on and illustrates the drop in specular reflection energy as was found with the simpler theories. It also shows that a strong specular component still exists from this hybrid surface. Figure 12.21c is for a perforated mask where the holes are regularly spaced. This shows that the side lobe performance is worse than the BAD panel, confirming the usefulness of a pseudorandom hole arrangement for getting more diffuse reflections. Figure 12.21d is for a perforated mask where the hole locations are determined randomly. The performance is similar to, but slightly worse than, that of the BAD panel. This shows the superiority of using pseudorandom number sequences rather than a random hole arrangement.

Figure 12.22 compares the scattering from the BAD panel to a plane surface at two contrasting frequencies. At the low frequency (Figures 12.22c and d), the mineral wool is not providing much absorption and so the BAD panel behaves similarly to the planar hard surface, although a little additional dispersion is achieved. At the higher frequency (Figures 12.22a and b), the

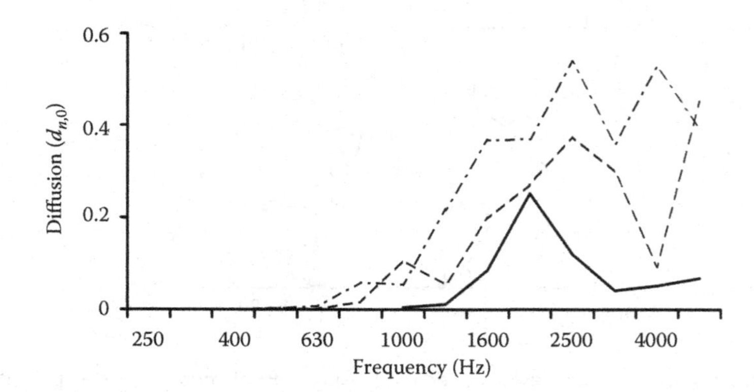

Figure 12.20 Predicted normalized diffusion coefficient for three hybrid surfaces: ——— planar binary; - - - - ternary; and – - – - – ·· quadriphase.

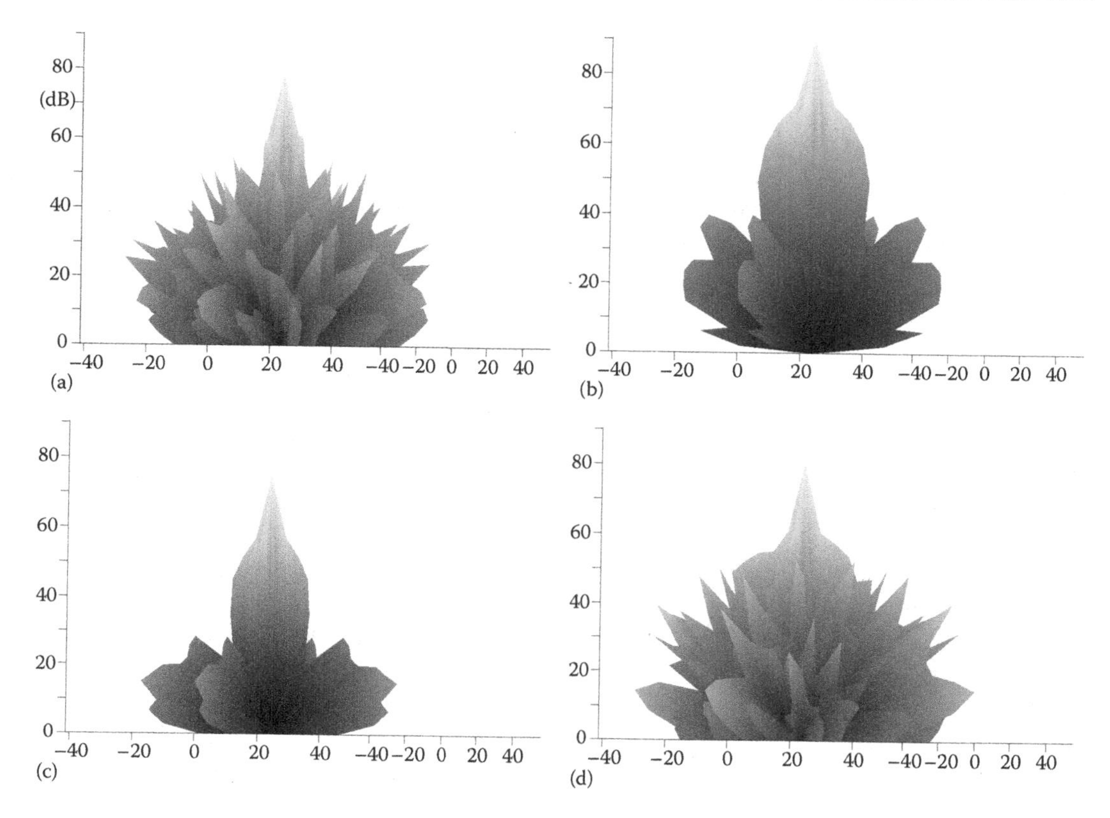

Figure 12.21 Scattering from arrays of three surfaces at 3 kHz. (a) BAD panel; (b) flat hard surface; (c) a perforated mask with regular hole spacing; (d) a random perforated mask. 3D polar balloons viewed from side.

difference between the surfaces is marked, as would be expected, with the hybrid surface generating some more dispersion.

12.6.2 Planar devices

Polar responses for flat hybrid surfaces show some dispersion, but the reflection is still dominated by specular reflection. Calculated values of the diffusion coefficient therefore tend to be small, as illustrated in Figure 12.4. Why is more diffusion not created? Figure 12.23 shows the front view for a flat hybrid surface, where the shaded patches are absorbing and the white patches are reflective. The grid spacing of the patches is w.

At low frequency, where the wavelength is much greater than the patch width, $\lambda \gg w$, due to evanescent waves in front of the surface, the device behaves as a material with the average admittance of the device. The contrast between soft and hard patches is lost. Some attenuation of the specular reflection can occur, but no dispersion, with the attenuation limited by the depth and type of porous absorbent used.

At high frequency, where the wavelength is much smaller than the patch width, $\lambda \ll w$, each reflective patch provides a specular reflection. There is no mechanism that can disperse sound to oblique angles, and so what is achieved is a specular reflection that is attenuated.

At mid frequency, $\lambda \approx w$, edge diffraction from the patches can create reflections to oblique angles and also some absorption can occur. Further understanding of the behaviour can be

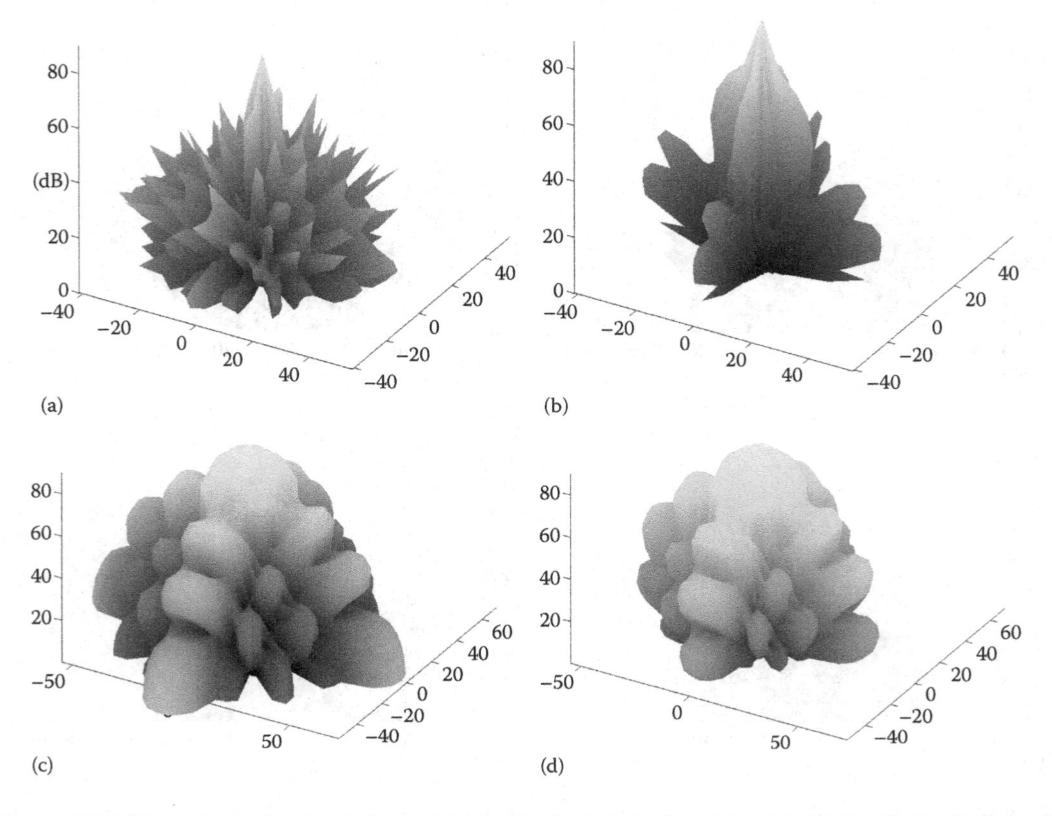

Figure 12.22 Hemispherical polar balloons. (a) 3 kHz BAD hybrid surface; (b) 3 kHz flat hard surface; (c) 400 Hz BAD hybrid surface; (d) 400 Hz flat hard surface.

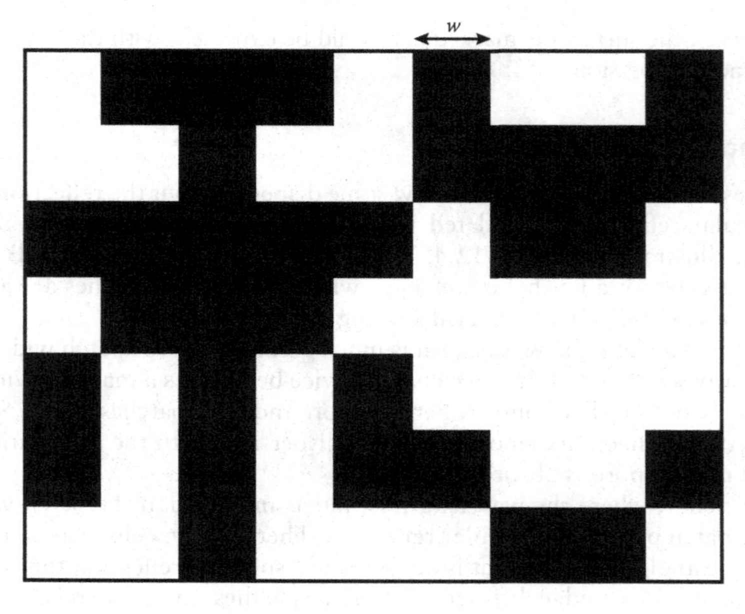

Figure 12.23 Front view of a planar hybrid surface.

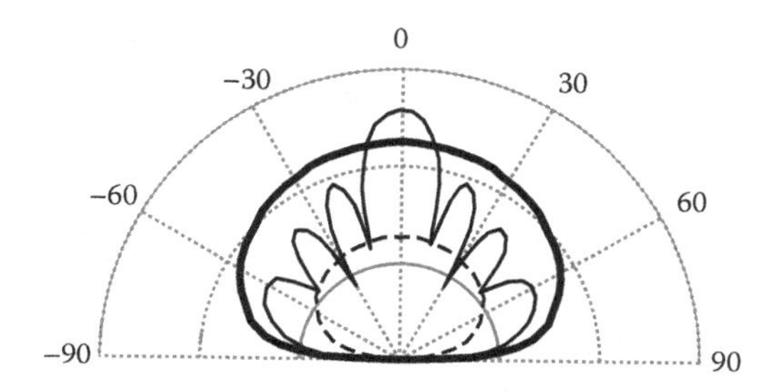

Figure 12.24 Reflected sound pressure level for a strip of width *w* calculated via a BEM model for three different wavelengths: ——— $\lambda = w/4$, ▬▬▬ $\lambda = 2w$, and ------- $\lambda = 10w$. (Data from Dadiotis, K., "Improving Phase Grating and Absorption Grating Diffusers", PhD thesis, University of Salford, 2010, p. 112.)

developed by considering how a strip of width w would behave on its own. Figure 12.24 shows the reflection from such a strip.[14] At low frequency, the case shown is $\lambda = 10w$, the sound pressure level scattered into all angles is similar but is small. For the high-frequency example, $\lambda = w/4$, the reflection from the strip is stronger, but it is concentrated in the specular reflection direction so there is a lack of dispersion. But there is a middle range of wavelengths where the reflection is both strong and spread over different angles.

The region is approximately $w \leq \lambda \leq 3w$, and this is where some dispersion is created by edge diffraction, and the surface is neither so small that the reflection is very weak or too large so that specular reflection dominates.[14] A single strip at best creates dispersion over a little more than one-and-a-half octaves, with the maximum dispersion created by edge diffraction occurring when half a wavelength fits across the width.

The areas of reflecting material in Figure 12.23 are complex shapes and have different sizes. Although this means the edge diffraction effects are spread over a wider bandwidth, at any one frequency, the edge diffraction will be weakened because only a small part of the panel has patches of the right size to create dispersion.

12.7 SUMMARY

Hybrid surfaces allow designers to control reverberation and provide diffuse reflections in a single device. To get reasonable dispersion, they need to be shaped, a few wells added, or some other way of modulating the depths is needed, such as using slats in front arranged according to a pseudorandom number sequence.

REFERENCES

1. P. D'Antonio, Planar binary amplitude diffusor, Patent No. US 5817992 (1998).
2. P. D'Antonio and T. J. Cox, Acoustical treatments with diffusive and absorptive properties and process of design, Patent No. US 6112852 (2000).
3. P. D'Antonio and T. J. Cox, *Hybrid Amplitude-Phase Grating Diffusors*, US patent 7,428,948 B2 (2008).
4. T. J. Cox, J. A. S. Angus, and P. D'Antonio, "Ternary and quadriphase sequence diffusers", *J. Acoust. Soc. Am.*, **119**(1), 310–9 (2006).
5. Z. Davidovic and B. Petrovic, "Acoustical design of control room for stereo and multichannel production and reproduction—A novel approach", *Proc. Audio Eng. Soc. 129th Convention* (2010).

6. J. A. S. Angus, "Sound diffusers using reactive absorption grating", *Proc. 98th Convention Audio Eng. Soc.*, preprint 3953 (1995).
7. P. Fan and M. Darnell, *Sequence Design for Communication Applications*, John Wiley & Sons, New York (1996).
8. J. A. S. Angus and P. D'Antonio, "Two dimensional binary amplitude diffusers", *Proc. Audio Eng. Soc.*, preprint 5061 (D-5) (1999).
9. T. J. Cox and P. D'Antonio, "Optimized planar and curved diffsorbors", *Proc. 107th Convention Audio Eng. Soc.*, preprint 5062 (1999).
10. M. F. M. Antweiler, L. Bomer, and H. D. Luke, "Perfect ternary arrays", *IEEE Trans. Inf. Theory*, **36**(3), 696–705 (1990).
11. M. R. Schroeder, *Chaos, Power, Laws: Minutes from an Infinite Paradise*, W. H. Freeman & Co., New York (1991).
12. H. D. Luke, H. D. Schotten, and H. Hadinejad-Mahram, "Binary and quadriphase sequences with optimal autocorrelation properties: A survey", *IEEE Trans. Inf. Theory*, **49**(12), 3271–82 (2003).
13. L. Xiao, T. J. Cox, and M. R. Avis, "Active diffusers: Some prototypes and 2D measurements", *J. Sound Vib.*, **285**(1–2), 321–39 (2005).
14. K. Dadiotis, "Improving Phase Grating and Absorption Grating Diffusers", PhD thesis, University of Salford (2010), p. 112.

Chapter 13

Absorbers and diffusers in rooms and geometric room acoustic models

So far, this book has discussed how to design diffusers and absorbers, but mostly in isolation of where and how they are applied. This chapter starts by presenting some of the issues arising from the application of absorbers in rooms, especially the problems of translating absorption coefficients between the impedance tube, free field, reverberation chamber, and real room applications. It then proceeds to discuss how absorption and diffuse reflections are represented in geometric room acoustic models (GRAMs) and how this affects prediction accuracy.

13.1 CONVERTING ABSORPTION COEFFICIENTS

13.1.1 From impedance tube or free field to random incidence

In Chapter 4, various methods for measuring absorption were outlined. These included impedance tube, free field, and random incidence techniques. Unfortunately, it is not easy to translate between free field or impedance tube measurements and random incidence values for a variety of reasons, and these will be discussed in the following paragraphs. Being able to do these translations is extremely useful because the free field and impedance tube experiments are done in a controlled environment, which is ideal for validating prediction models. Furthermore, impedance tube measurements can be done on small samples—ideal when developing new absorbents. To make these measured coefficients useful to practitioners, however, they need converting into random incidence values.

The translation from a set of angle-dependent, free field absorption coefficients to a random incidence value is normally carried out using Paris's formula[1]

$$\alpha_s = \int_0^{\pi/2} \alpha(\psi)\sin(2\psi)\,d\psi, \tag{13.1}$$

where α_s is the random incidence absorption coefficient and $\alpha(\psi)$ is the absorption coefficient in the free field at an incident angle ψ. This formula is derived by considering the sound incidence on a surface in a diffuse space, the $\sin(2\psi)$ term arising because of solid angle considerations.

If the surface is locally reacting, then it is possible to just measure the normal incidence surface impedance, apply the formulations given in Equations 2.21 and 2.24 to get the angle dependent absorption coefficient, and, from Equation 13.1, get the random incidence value. In theory, for a locally reacting surface, a single measurement at 55° will suffice, as this is the same as the random incidence value. Many surfaces, however, are not locally reacting, as discussed in Chapter 6.

Makita and Hidaka[2] examined the problem of translating from free field to random incidence coefficients for homogeneous and isotropic porous materials. They measured different polyurethane foams in the impedance tube and the reverberation chamber and compared the random incidence absorption coefficients, derived from the impedance tube measurement using Paris's formulation, and the reverberation chamber measurements. Discrepancies arise because

- The reverberation chamber is not completely diffuse and the intensity is not uniform, leading to Paris's formula being inaccurate and some angles of incidence being emphasized over others.
- Diffraction at the edges of the sample creates excess absorption in the reverberation chamber measurements. The impedance discontinuity at the edges of the sample causes additional sound to bend into the sample as this has a lower speed of sound.
- The mounting conditions are often different, which may affect how the frame of the absorbent vibrates.
- The assumption of local reaction, so that small sample experiments in the impedance tube can be translated to a large sample measurement in the reverberation chamber, may not be correct.

Makita and Hidaka carried out a series of reverberation chamber measurements on the foams with different sample sizes to get the absorption coefficient of an infinite array using extrapolation. Figure 13.1 shows the inferred absorption coefficient for an infinite sample, compared to the 12 m^2 sample, which is the largest recommended size in ISO 354.[3] The absorption coefficients are shown for one typical sample of the six foams tested. The average difference in the absorption coefficient between the finite and infinite samples for all six foams is also shown. Figure 13.1 illustrates that great differences in the absorption coefficient can be obtained due to edge effects, even if the edges of the samples are covered with a reflective frame, as might be done for one mounting condition in ISO 354.

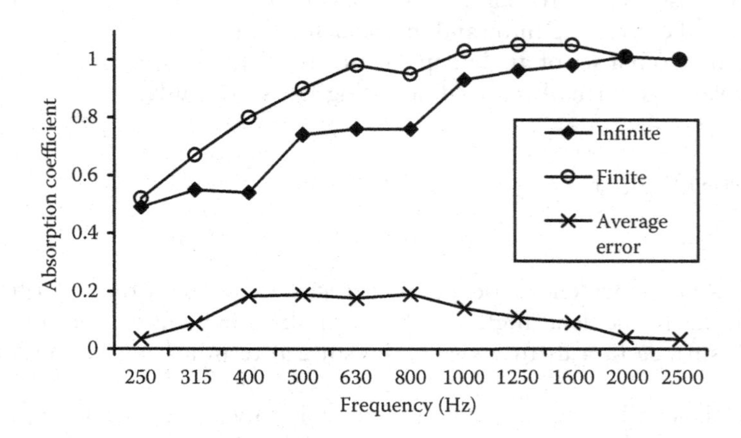

Figure 13.1 Comparison of measured random incidence absorption coefficients for finite and infinite samples for one material. Also shown is the average error between the finite and infinite sample absorption coefficients for six materials. (Data from Makita, Y., Hidaka, T., *Acustica*, 66, 214–220, 1988.)

Bartel[4] examined the extrapolation from finite to infinite area absorption. The paper also contains a useful and comprehensive literature review of previous work in this area. The simplest model for absorption coefficient extrapolation assumes a linear relationship. This can be represented by

$$\alpha_\infty = \alpha_s + mE, \tag{13.2}$$

where α_∞ is the absorption coefficient for an infinite sized sample; α_s is the absorption coefficient for a finite sized sample, for example, a 12 m^2 sample in a reverberation chamber measurement; m is a constant; and E is the ratio of the specimen perimeter to the specimen area.

The sensitivity of the absorption coefficient to the perimeter-to-area ratio, E, varies with material type and frequency. Figure 13.2 shows a plot of the absorption coefficient for different perimeter-to-area ratios, measured in the reverberation chamber, illustrating that the edge effects can be very significant. This is with the edges of the sample covered with a hard, rigid strip. Figure 13.3 shows how the value of the constant m varies with frequency and material type. Bartel also investigated the effect of sample shape. Sample shape affected the absorption coefficient only by at most 3%, so the size of the perimeter-to-area ratio is much more important than the shape of the sample's edge.

Deriving the constant m requires a series of measurements on different sample sizes in the reverberation chamber. Consequently, Equation 13.2 does not help get the random incidence absorption coefficients from impedance tube measurements because impedance tube experiments do not give values for m. Bartel[4] reports, however, a formulation attributed to Northwood[5] that allows the perimeter effect to be predicted and so allows the translation from impedance tube to random incidence values. The normalized admittance β of the sample is measured in the impedance tube and is given by

$$\beta = \frac{\rho c}{z} = g - jb, \tag{13.3}$$

where z is the surface impedance, ρ is the density of air, g is the real part of the admittance, b is minus one times the imaginary part of the admittance, and c is the speed of sound in air.

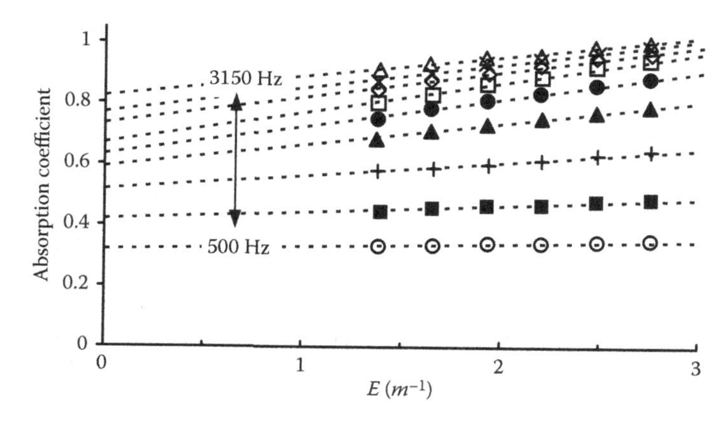

Figure 13.2 Random incidence absorption coefficient as a function of sample perimeter divided by sample area. Shown for different one-third octave bands. ○ 500; ■ 630; + 800; ▲ 1000; ● 1250; □ 1600; ◇ 2000; × 2500; and △ 3250 Hz. (Data from Bartel, T.W., *J. Acoust. Soc. Am.*, 69, 1065–1074, 1981.)

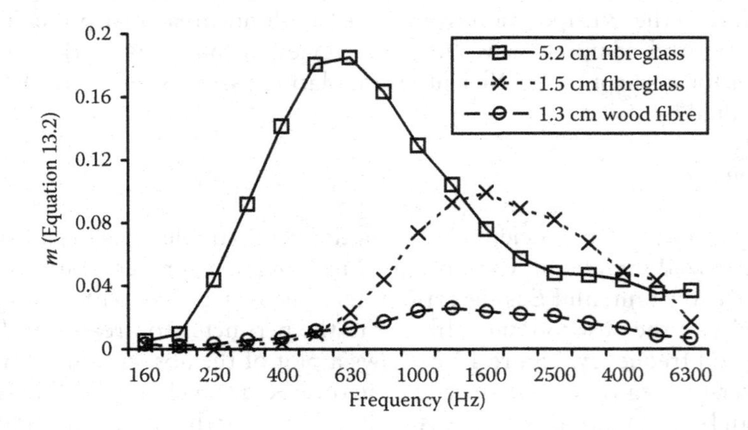

Figure 13.3 Variation of constant *m* in Equation 13.2, which measures the perimeter effect for random incidence absorption coefficient. Results for three different materials are given as indicated. (Data from Bartel, T.W., *J. Acoust. Soc. Am.*, 69, 1065–1074, 1981.)

The average absorption coefficient for a finite rectangular sample is then given by

$$\alpha = \frac{16g}{\pi}\int_0^{\pi/2}\int_0^{\pi/2}\frac{\sin^3(\varphi)\,d\varphi d\theta}{|\beta|^2(A^2+B^2)+2(gA+bB)\sin(\varphi)+\sin^2\varphi}, \tag{13.4}$$

where

$$A = \int_0^{ka\sin\varphi}\cos(x\sin(\theta))J_0(x)\,dx - \frac{1}{ka\sin\varphi}\int_0^{ka\sin\varphi}x\cos(x\sin(\theta))J_0(x)\,dx, \tag{13.5}$$

and

$$B = \int_0^{ka\sin\varphi}\cos(x\sin(\theta))N_0(x)\,dx - \frac{1}{ka\sin\varphi}\int_0^{ka\sin\varphi}x\cos(x\sin(\theta))N_0(x)\,dx, \tag{13.6}$$

where $a = 2/E$ and J_0 and N_0 are the Bessel functions of the first and second kind, respectively. These formulations can also be used to gain random incidence predictions of absorbers based on calculations of the surface impedance using the equations in Chapters 6 through 8.

Makita et al.[6] presented a revised Paris's formula that accounts for inaccuracies in the cosine law formulation. The formulations derived are complex but enable the effects of the boundary layer on the absorption to be accounted for. The boundary layer effect is not normally as big as the perimeter effect but can still be significant.

If current efforts to improve the reproducibility of reverberation chamber measurements are successful (see Section 4.3), then the translation from impedance tube and free field measurements should become easier.

13.1.2 From the reverberation chamber to real rooms

The problem of extrapolating from reverberation chamber measurements to whole room design has already been discussed in Section 4.3.1 in relation to audience seating. But seating is not the only surface to suffer from problems. Even when the inaccuracies in the extrapolated absorption coefficient are numerically small, the resulting reverberation time in a space can be wrong because the inaccuracy in the total absorption might be large. The problems centre on two issues, edges and non-diffuseness.

1. *Edges*: The sample size tested in the reverberation chamber is often smaller than that applied in the final application. Consequently, the effects of edges are often greater in the reverberation chamber than when the absorbent is used to treat a room. This can lead to significantly different effective absorption coefficients. The solution could be to determine the absorption of the edges, as was done for seating in Section 4.3.1. Alternatively, it is possible to test different sample sizes in the reverberation chamber, and from there extrapolate to the sample size used in the real room. The formulations given in the previous section for the influence of sample perimeter can also be used to extrapolate the absorption of large areas of absorbent from the smaller reverberation chamber samples.
2. *Non-diffuseness*: The acoustic conditions in the reverberation chamber and the real room may be very different. Reverberation chambers usually have test samples on only one surface, the floor, whereas the real room might have absorption more evenly distributed around the space. This often leads to the random incidence absorption coefficient measured in the reverberation chamber not matching the effective absorption coefficient in the real room. The most common scenario is that both the reverberation chamber and real room are non-diffuse, but they are non-diffuse in different ways.

The effect of the diffuseness of the space is difficult to account for. If the non-diffuseness is generated in a simple manner in a room with simple geometry, for instance, all the absorption on one surface of a cuboid room with low scattering, then there are reverberation time formulae that can deal with these cases. (Gomperts[7] carried out a comprehensive review of reverberation time formulations.)

One solution is to use a GRAM to fully model the sound distribution in the reverberation chamber and also the room where the absorber is to be applied, rather than assume that both are diffuse fields. This is an appealing solution, but as shall be discussed in the next section, the application of absorption coefficients in computer models is not necessarily straightforward and also depends on both the values of the scattering coefficients and the scattering distribution used. This also requires the modeller to have access to the reverberation times with and without the sample in the reverberation chamber, and these are almost never available.

An ISO working group is examining how to improve the reverberation chamber method (see Section 4.3). If better laboratory measurement methods can be devised, this should help reduce the errors in extrapolating from the reverberation chamber to real rooms.

13.2 ABSORPTION IN GRAMs

GRAMs are becoming a core tool for practitioners and researchers designing or investigating the propagation of sound within a space. The correct modelling of absorption is not always straightforward, however.

GRAMs calculate the sound propagation within a space using techniques such as[8] ray tracing, beam or cone tracing, image source, acoustic radiance transfer, or hybrid approaches (hybrid approaches are combinations of two or more methods). They are high-frequency approximations to the true sound propagation and do not properly deal with the wave nature of sound.

For readers unfamiliar with GRAMs, it is necessary to describe how the sound might be modelled, but for brevity, only one type of modelling will be described. Here, the simplest form of ray tracing will be considered, as this is the easiest to describe. In this method, the sound energy is modelled as rays that propagate around the room like rays of light. When the ray hits a surface in the room, it is reflected from the surface, and if no scattering is considered, the angle of reflection equals the angle of incidence. There is a spherical receiver in a room, and every time a ray passes through this, the reflection contributes to the energy impulse response (or echogram). This process is illustrated in Figure 13.4.

Every time a ray reflects from the surface, the energy of the ray is decreased by a factor of $1 - \alpha$, where α is the absorption coefficient of the surface. One problem is that absorption coefficient tables published in the literature often contain values greater than 1. How should these be translated for use in a GRAM, where values greater than 1 would create negative energy in the model? Normally, the value needs to be truncated to 1 (or a little below 1). Furthermore, many GRAMs are now producing auralizations of the sound field within the space to allow clients to hear the effect of design changes. To get a natural rendition of the sound field, a wide audio bandwidth is required, yet absorption coefficients are normally only available for a restricted bandwidth of 100 Hz to 5 kHz[9] (and it is unlikely that the GRAM will be valid at 100 Hz anyway). These and other issues are discussed below.

Although a GRAM has knowledge of the angle of incidence that a ray strikes a surface, it is usual for a random incidence absorption coefficient to be applied to every reflection. If local reaction is assumed, and the normal incidence impedance of surfaces are known, it would be possible to predict the absorption coefficient as a function of incident angle and use these in computer models instead. Whether a room has a *mixing* or *non-mixing* geometry is important. Mixing geometries are one where the reverberant field tends towards a stochastic process, and in this context, it means that every reflection path involves every room surface equally. For mixing geometries with some scattering present, this is probably not necessary, and using a random-incidence absorption coefficient should be sufficient. But for non-mixing geometries or coupled spaces, angle-dependent absorption coefficients may improve predictions.[10]

One cause for the absorption coefficients to be greater than 1 in reverberation chamber measurements is that Sabine's formulation[11] becomes inaccurate when the absorption is high ($\alpha > 0.2$–0.4). According to Sabine's formula, a room constructed from completely

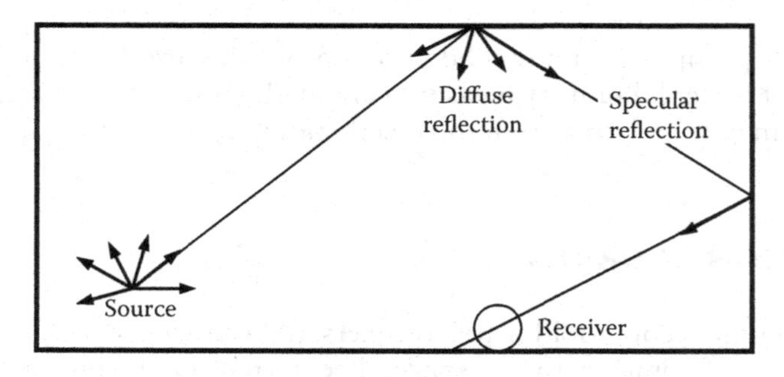

Figure 13.4 Picture of ray tracing in a room.

absorbing walls still has a reverberation time greater than 0. Some have suggested that alternative reverberation time formulations, such as Eyring and Millington, should be used to calculate the absorption coefficients from the reverberation time measurements. But that requires modellers to have access to the reverberation times measured in the reverberation chambers and the geometry of the laboratory to allow them to rederive the absorption coefficients. Using more exacting reverberation time equations may help predictions, but it does not prevent measured absorption coefficients often exceeding 1, due to edge effects and the non-diffuse nature of the reverberation chamber (see Section 13.1.2).

An alternative solution to the problem of mapping from the reverberation chamber to the real room prediction is empirical fitting. As a surface treatment is used over time, an understanding of how the absorption coefficient varies between the reverberation chamber and the GRAM of the real room is developed. If the reverberation time was underestimated this time, next time, the modeller might put in a smaller absorption coefficient. This is obviously not very satisfactory, especially as the role of scattering needs to also be considered, but it is the reality of applying many scientific models to engineering problems. This is why experienced geometric room acoustic modellers with a good grasp of physical acoustics usually produce better predictions than novices do. This also creates problems with the verification of computer models. In some poor-quality scientific papers, the absorption coefficients for surfaces are found by fitting predictions to match measurements, and then the same measurements are used to show that the numerical method works!

So far, the issue of phase change on reflection has not been considered. The true reflection from a surface should include changes in magnitude and phase, and consequently, the pressure reflection coefficient or impedance is needed. This is problematical for many reasons. Impedance data for surfaces are not readily available. Furthermore, many building surfaces have non-local reacting behaviour, which creates problems in measuring and implementing the phenomena in GRAMs. For many predictions, it is probably unnecessary to go into this detail, which is fortunate, as it would pose some difficult problems to overcome. For smaller spaces, however, room modes created by interference cause significant problems.[12]

There have been round-robin trials of GRAMs comparing the accuracy of the different techniques.[13] One of the key findings is that the accuracy of the prediction models is highly dependent on the quality of the input data, including the absorption and scattering coefficients of surfaces. So to summarize, defining absorption coefficients (and impedance) for GRAMs appears to be a tricky problem. Fortunately, most experienced practitioners can usually produce estimations of absorption coefficients that are good enough, using previous knowledge. This is not an entirely satisfactory situation, as *best guesses* should not be part of a prediction model.

13.3 DIFFUSE REFLECTIONS IN GRAMs

GRAMs on their own cannot precisely represent sound as a wave. They inherently assume that the wavelength of sound is much shorter than any dimension of any object within the room; this enables sound to be modelled as particles, rays, beams, or as coming from image sources. Unfortunately, the wavelength of sound is relatively large at low to mid frequencies, and consequently, GRAMs that do not attempt to predict the effects of surface and edge scattering are liable to produce poorer results. In real life, some sound will inevitably be scattered into angles other than the specular reflection direction upon reflection from most surfaces.

The problems of not modelling diffuse reflections in GRAMs are well established. For example, Hodgson[14] used a ray tracing model that took diffuse reflections into account by redistributing diffuse energy according to Lambert's law.[1] Hodgson concluded that in simple empty rooms, the effects of diffuse surface reflections are negligible in proportionate rooms, while in disproportionate rooms, the effects can be considerable. To take another example, in the first round-robin study of GRAMs,[15] three prediction models were found to perform significantly better than others. For the 1 kHz octave band, these three prediction models produced results approximately within one subjective difference limen, while the less successful computer models produced predictions inaccurate by many difference limen. What differentiated the three best models from the others was the inclusion of methods to model scattering.

If the overall room shape and sizes and orientations of surfaces are such that they will cause reflections to be well mixed for purely geometrical reasons, a diffuse field may be created even if no rough or scattering surfaces are used. In this context, mixing means that the reflection paths involve all the surfaces of the room. For mixing room shapes, the reverberation time can be well predicted even without diffuse reflection modelling, even if predictions of other parameters such as clarity may suffer.[16] However, as it is difficult to know in advance if a room shape is mixing, it is best to always include diffuse reflection modelling in a GRAM, as otherwise poor estimations of acoustic parameters may result.

The most obvious error created by a lack of diffuse reflection modelling is an overestimation of reverberation time.[14,17] This is especially true in enclosed spaces where absorption is concentrated on one surface, such as in concert halls, or when the room shape is highly disproportionate, such as in large factories with low ceilings. In some halls, the choice of scattering coefficients has a greater impact on the estimated reverberation times than the uncertainty in the absorption coefficients.[18] Torres and Kleiner[19] found that changes in the scattering coefficient in GRAMs are audible in auralizations and that the diffuse reflections should be modelled with frequency dependence. A round robin on GRAMs[13] found that the biggest errors with the models were consistently at low frequencies, presumably because of their inability to model diffraction and interference effects correctly because these are most prominent at low frequencies.

There are many different methods for incorporating diffuse reflections into a GRAM. Dalenbäck et al.[18] and Savioja and Svensson[8] gave a comprehensive survey of the techniques. Many of the methods are similar or just variants. Consequently, a few of the most important and commonly used techniques are outlined here. While there are many possible diffuse reflection methods, it is not known which modelling technique, if any, is intrinsically more accurate. It is known that a diffuse reflection model is needed for accurate predictions, however. One key determining factor is, however, the computing time. Accurate diffuse reflection models are relatively simple for low orders of reflections but often become increasingly computationally expensive as the reflection order increases. However, as reflection order increases, predictions often become less sensitive to inaccuracies in reflection models.

13.3.1 Ray redirection

This method is a ray scattering process as suggested by Kuttruff.[1] A wall is considered to absorb a proportion of the energy given by the absorption coefficient α. The proportion of reflected energy diffused is given by the scattering coefficient s. This means that $(1 - \alpha)s$ is distributed according to Lambert's law or another polar distribution, and the remaining reflected energy $(1 - \alpha)(1 - s)$ is reflected in a specular manner. The direction of the diffuse reflection is determined by two random numbers. The angle of azimuth is chosen by a random number in the interval $\{-\pi, \pi\}$, and the elevation angle is given by the inverse cosine of the square root of a random number that is chosen from the interval $\{0, 1\}$. A variant on this technique is vector-based scattering. This produces only one reflection that is the vector sum of the specular and diffuse reflection.

Concentrating all the diffusely reflecting energy for one ray into one direction is a simplification, but within the reverberant sound field, there are a large number of reflections and rays to average out random errors in the response. For the early sound field, instead of giving many weak reflections from diffusing surfaces spread over time, a receiver gets fewer stronger reflections. Even for the first-order reflection, however, there will usually be multiple rays reflecting from different parts of a surface, so the situation is not as bad as might appear at first.

Generating multiple rays from a diffuser rather than redirecting one ray would improve predictions in some cases, but this becomes very computationally expensive as the number of rays rapidly increases.

13.3.2 Transition order using particle tracing

The reflection model is separated by a user-defined transition order.[20] Reflections with orders lower than the transition order are purely specular. After the transition order, sound rays are treated as energy packets similar to a normal ray tracing method. At each subsequent wall reflection after the transition order, a secondary impulse source is created at the reflection point, which radiates into the room as an elemental area source. The energy is then regrouped into a ray and traced forward in a direction given by a random process, in which either a purely specular, or a diffuse direction, will be chosen depending on whether the value of a random number generated by the program is greater or smaller than the surface's scattering coefficient. If the reflection is diffuse, then its direction is determined by a second random process based on Lambert's diffusion law.

The choice of transition order is dependent on the hall shape rather than size.[21] In rectangular rooms, a transition order of 0 in low-frequency bands and 1 to 3 in the high-frequency bands is found to be appropriate. In fan-shaped halls, where correct modelling of the specular reflections is important to account for the influence of the hall geometry, a higher transition order is also required in the low-frequency bands. Generally, the choice of transition order should be based on the importance of the specular components in the early reflections in defining the acoustic character of the hall. An order of 1 or higher should be used only when the sound field is significantly affected by the specular components, such as at high frequencies or with strong reflecting surfaces. In real halls, where the sound field is expected to be more diffuse, lower transition orders should be used.

The main problem with this method of diffuse reflection modelling is that the concept of a transition order is not physically satisfactory, since diffuse reflections should occur even at the very first reflection, rather than suddenly being switched on at a transition order.

13.3.3 Diffuse energy decays with the reverberation time of the hall

Upon each reflection, the scattering coefficient is used to define the fraction of energy diffusely scattered into non-specular angles, while the remaining energy is reflected into the specular reflection angle. The diffuse energy is ambient energy spread throughout the room volume. This energy is assumed to decay exponentially, with the decay constant being determined by Eyring's formulation. A visibility check is used to ensure that the right surfaces contribute to the received sound.[22]

The problem with this is that it assumes that the Eyring's formulation is correct, which is not necessarily true, especially if the space is non-diffuse. Some have used an iterative procedure to gradually improve the estimated reverberation time, but this slows computation as it requires multiple passes through the algorithm.

13.3.4 Radiosity and radiant exchange

The diffuse part of a reflection is stored in memory while the specular ray tracing is continued. Later, the stored diffuse energy is emitted and all subsequent surface reflections are assumed to be diffuse. These methods use a stochastic radiative exchange process to propagate sound between surfaces. This radiosity can be modelled by integral equations but more commonly uses simple heat exchange formulations. The radiant exchange takes place at a time interval given by the mean free time between reflections.

13.3.5 Early sound field wave model

Another suggestion is to use wave-based models, such as a time domain Fourier approach, to model the early sound field before resorting to a ray tracing with randomized ray redirection for the later sound field. This has the advantage of potentially being more accurate in modelling the early sound field while allowing ray tracing to take over where it is computationally more efficient and sufficiently accurate in many cases.

13.3.6 Edge scattering for small surfaces

Scattering coefficients are often used to account for the scattering caused by both the surface roughness and also the limited size of surfaces and edge diffraction. It is possible to deal with surface roughness and the effects of edges separately.[23–25] This can produce a more physically correct and robust model of scattering.

13.3.7 Distributing the diffuse energy

The models discussed previously have many common features, one of which is that the diffusely reflected energy from a surface is modelled as radiating from the surface with a particular spatial distribution. In most current models, Lambert's law is used to determine this distribution. Another possibility is to disperse the sound reflected from a surface using a probability distribution based on the scattered pressure polar response measured or predicted in the free field. Problems would arise because the polar response would be from a surface of significant finite extent, whereas many GRAMs would require the correct dispersion from a point on the surface. This needs consideration in designing the model, and a method to reverse engineer the point reflectivity function from the finite-sized polar response is needed. This can be done using Farina's method for characterizing diffusers, which was discussed in Section 5.4.1. The situation is more complex with beam tracing as the beam may interact with part of a surface, and so the reflectivity function would neither be the point reflectivity function nor the finite-sized polar response. There is a further problem in dealing with situations where only part of the surface is illuminated, as might happen when objects cast shadows on surfaces, or with directional sources. Polar responses are usually measured or predicted using complete illumination by omnidirectional sources.

The most common dispersion law for GRAMs is Lambert's law, also referred to as the cosine law. It states that the intensity scattered from a surface follows a cosine distribution with respect to the incident and reflected sound angle to the wall. This is illustrated in Figure 13.5 for a normal incident source. Stated in terms of equations for a ray tracing case,[1]

$$I_r = \frac{I_0 ds \cos(\psi)\cos(\theta)}{\pi r^2}, \tag{13.7}$$

where I_r is the reflected intensity at the surface, I_0 is the incident intensity at the surface, θ the angle of the receiver to the surface normal, ψ is the angle of the source to the surface normal, ds is the area of the surface being considered, and r is the receiver radius.

This formulation is a simple statement of solid angle projections. For instance, the solid angle is 0 for sources and receivers close to grazing. If the surface is partly absorbing, then the intensity should be attenuated by the corresponding intensity reflection coefficient, but to simplify matters for now, the surface will be assumed non-absorbing in this discussion.

True rooms are not purely specular and nor are they purely diffuse following Lambert's law. For the reverberant field, however, the evidence is that the sound field more closely matches a diffuse case than a specular one,[1] especially at mid to high frequency.

Figure 13.6 shows the scattering distributions for a surface with $s = 0.01$, 0.5, and 0.99, where the diffuse reflection is modelled using Lambert's law. Also included for comparison in Figure 13.7 are the polar distributions for a periodic diffuser, a single cylinder, a plane surface, and a surface with small random roughness. Each surface was 2 m wide and the source and receivers are in the far field. The scattered distribution, using the Lambert model, does not match many of the real surfaces very well for many cases. The closest is the random rough surface, which approximately matches the $s = 0.99$ case. Consequently, GRAMs are not producing scattering distributions close to real diffuser scattering. However, this is probably not that important for the reverberant sound field, where there are a large number of reflections over which the errors in any one diffuse reflection tend to average out, or at least mask any inaccuracies. It is more of a problem, however, for the early sound field, when precise modelling of first-order reflections is needed.

Lambert's law is a natural choice for GRAMs because it is the asymptotic high-frequency case, which is intrinsically the frequency range where the models are most correct. Furthermore, it

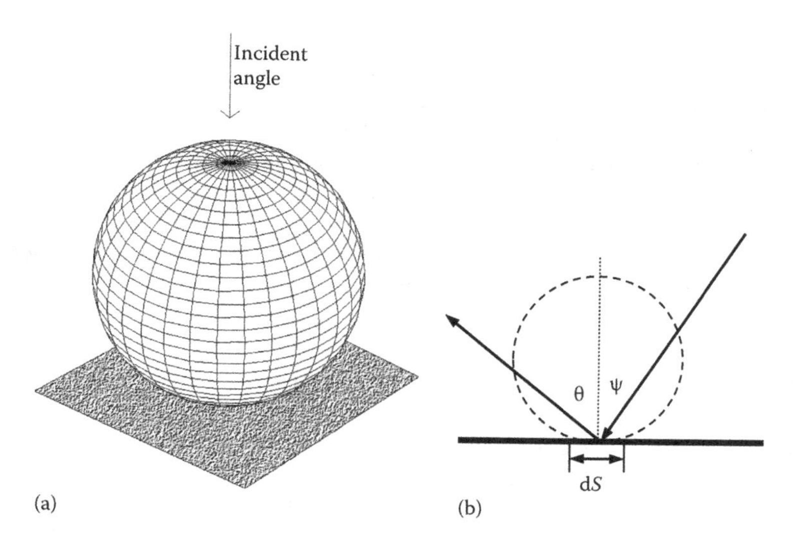

Figure 13.5 Two views of the Lambert distribution of intensity. (a) 3D view; (b) cross-section.

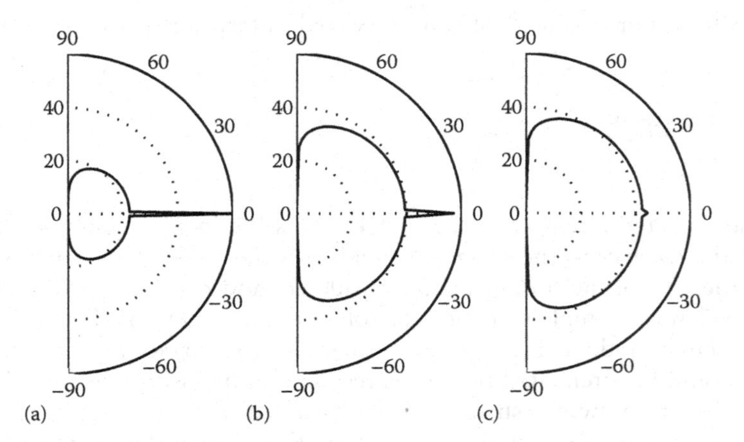

Figure 13.6 Typical far field energy distributions in a GRAM for different values of scattering coefficient s and a normal incident source. (a) s = 0.01; (b) s = 0.5; (c) s = 0.99.

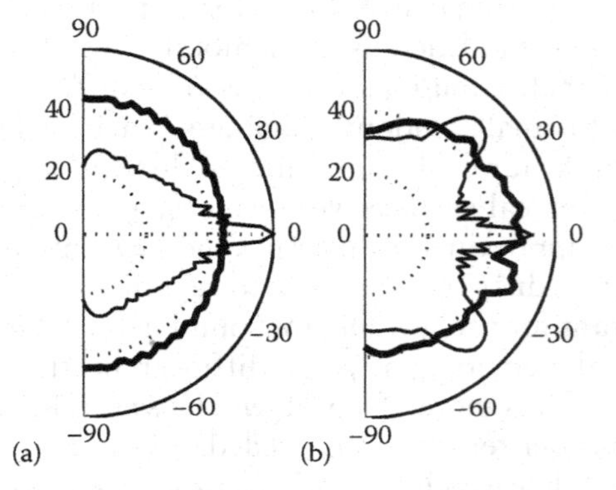

Figure 13.7 Some typical far field one-third octave polar response patterns for different surfaces. (a) ——— plane surface and ▬▬▬ single semicircle; (b) ▬▬▬ random rough surface and ——— periodic diffuser.

deals with scattering from a point, which fits with the philosophy of most models. Lambert's law deals with high-frequency incoherent scattering, but in acoustics, the wavelength is often comparable to surface roughness. Indeed, diffusers such as Schroeder diffusers would not function as designed if the scattering were incoherent. It is true to say, however, that at high frequencies, the scattering from surfaces will approximate Lambert's law. At the key frequencies for room design, and for first-order reflections, some advocate that it is better to approximate the scattering from specialist diffusers according to a uniform energy distribution rather than Lambert's law in GRAMs, but others disagree.

There is a further problem with Lambert's law with single-plane diffusers. Single-plane diffusers produce dispersion in one plane and are very common in spaces; examples include pipe work, balcony fronts, 1D Schroeder diffusers, and columns. Yet most current GRAMs disperse reflections according to Lambert's law, scattering the sound in all directions over a hemisphere. (Dalenbäck gives a method for adjusting the scattering coefficient values to produce anisotropic scattering.[26]) Using hemispherical scattering for devices that disperse in one plane only has the potential to cause prediction errors, again particularly acute for the early sound field or some coupled spaces. It might be thought that this prediction inaccuracy is less important for the later sound field, and indeed, the averaging effect of multiple numerous reflections probably makes the model less sensitive to incorrect modelling of the reflections. In recent years, however, attention has been drawn to the importance of late lateral energy in auditoria[27] and its role in spatial impression. Consequently, for a correct auralization of large music spaces, it may be important to correctly model the spatial distribution of late sound, and correct modelling of anisotropic scatterers is probably needed to achieve this.

13.3.8 Scattering coefficients

In Chapter 5, methods for measuring and analyzing the reflections from surfaces were given, including characterization using a scattering coefficient. The scattering coefficient is intended to be used as an input to GRAMs. The success of the new scattering coefficient, however, appears to be mixed, with some suggesting that it works well within GRAMs, and others reporting problems. This is unsurprising given that the required scattering coefficient is known to be dependent on the reflection modelling technique used in the GRAM.

In the past, there has not been a defined way to gain scattering coefficients for surfaces, so researchers have adopted an empirical approach to investigations. They have examined what scattering coefficients are needed to gain accurate predictions in rooms by a trial-and-error process. The results from these investigations are presented here as they give some guidelines as to what scattering coefficients might be used. One problem with getting the correct scattering coefficient using this approach, however, is the interrelationship between the absorption and scattering coefficients used in the GRAM and the predicted acoustic parameters. The reverberation time predicted in a space will depend on both the absorption and scattering coefficients. The effect is most marked for disproportionate spaces or ones where the absorption is unevenly distributed. It is therefore not correct to determine the scattering coefficient by simply adjusting the reverberation time prediction until it matches measurement, as there is usually considerable uncertainty as to what the absorption coefficient of surfaces should be, and this also affects the reverberation time.

Prediction models use various approaches to model diffuse reflections, and consequently, different prediction models will require different values of the scattering coefficient, even for modelling the same wall under the same room conditions. Lam[22] investigated the scattering coefficients required for three different diffuse modelling algorithms, using scale models with largely smooth walls. In simple proportionate rooms, where the room dimensions are comparable, the predictions were similar whatever the method used to model diffuse reflections and the scattering coefficient applied. In a highly disproportionate room, however, the scattering coefficient required to gain accurate results varied with the algorithm used to model the diffuse reflections. The required scattering coefficient varied between 0.25 and 1 for the three GRAMs considered. It was also found that different scattering coefficients were required to give accurate predictions of different acoustic parameters within the same prediction model.

Lam further investigated the effect of scattering coefficients in a concert hall. The trend with scattering coefficients is that going from a zero coefficient to a low value, say $s = 0.1$ or 0.2, can make a large difference to predicted acoustic parameters. Increases of the scattering coefficient beyond these values created a much smaller effect.[22] Kang[28] found a similar result for city squares. He found that changing the scattering coefficient from 0 to 0.2 had a significant effect on the sound level and reverberation parameters, but increasing the scattering coefficient beyond 0.2 had much less effect. Consequently, the sensitivity of acoustic parameters to scattering coefficients is non-linear.

Figure 13.8 shows the early lateral energy fraction (ELEF) for different scattering coefficients in a real concert hall. Measurements are compared with various predictions. The

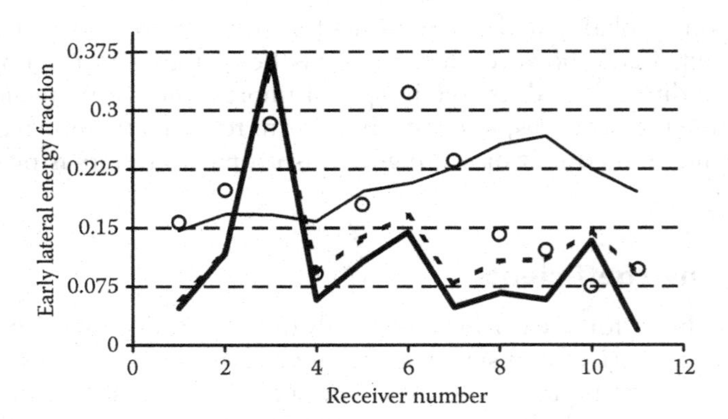

Figure 13.8 Variation of ELEF with different scattering coefficients. The grid lines in the y direction are spaced at the subjective difference limen. ○ measured; ━━ s = 0; ········· s = 0.1 for walls, 0.7 for seats; and ——— s = 1. (After Lam, Y.W., *J. Acoust. Soc. Am.*, 100, 2193–2203, 1996.)

difference limen for ELEF is about 0.075,[29] so differences bigger than this value are significant. ELEF is chosen because it is probably the parameter most affected by the diffuse reflection modelling algorithm, as it depends purely on the directional characteristics of the early sound field. This shows that the predictions are sensitive to getting the correct scattering coefficients.

There will be cases, however, where accurate predictions cannot be achieved whatever the values of the scattering coefficients. This will be particularly noticeable for acoustic parameters that are very sensitive to the early sound field, such as clarity and ELEF.[30] The early sound field contains only a few reflections, and consequently, inaccuracies in the modelling of the sound field are most apparent. These cases show that the diffraction effects present in real-life reflections are far more complicated than the simple scattering assumed in GRAMs. It is only by summing over a large number of reflections that the simple scattering assumptions can be regarded as valid approximations. Fortunately, in many situations, there are enough reflections that inaccuracies tend to average out, but in some cases, such as coupled rooms and under deep balconies, this will not be true. Certainly, a GRAM should not be used to evaluate first-order reflection paths without understanding the limitations of the predictions, unless the surfaces are simple, large, and planar or generate diffuse hemispherical dispersion.

Researchers who have investigated what values of scattering coefficients are required in GRAMs to gain reasonable predictions have found that it varies between different models, especially for rooms that are not mixing. One thing that all the studies agree on is that surfaces should never be modelled with a scattering coefficient of 0, even large flat surfaces. Lowest recommended values tend to be between 0.05 and 0.2. There is a greater risk associated with underestimating scattering coefficients than with overestimating them. The suggested coefficient values for seating again illustrate how different reflection models can require different values. Lam[22] set the scattering coefficient to 0.7 for seating, whereas Dalenbäck[31] recommends 0.4–0.7 for 125 Hz–4 kHz on audience areas and Zeng et al.[23] recommend 0.6–0.7. For generally rough surfaces, Dalenbäck suggests high values of 0.8 where the roughness is of the order of, or higher than, the wavelength, and gradually lower below. For example, if the roughness scale is 0.3 m, set 0.8 for 1–4 kHz, 0.6 for 500 Hz, 0.3 for 250 Hz, and 0.15 for 125 Hz. Zeng et al.[23] recommend mid-frequency values of between 0.3 and 0.8 in large halls, where the GRAM has greatly simplified the geometry, and between 0.02 and 0.05 where the major details of the hall are explicitly modelled and the GRAM deals with finite-sized panel effects.

Scattering coefficients have also been used when modelling streets. Onaga and Rindel[32] estimated values for the sum of the absorption and scattering coefficients for building façades ranging from 0.1 to 0.25. Ismail and Oldham[33] found scattering coefficient values ranging from 0.09 to 0.13. Although these coefficients are relatively small, the evidence is that for late sound some way from the source, scattering is the dominant mechanism in street canyons.

Empirically determined parameters from old literature have to be used with caution, however, because commercial GRAMs may have changed over the decades and now use different reflection models.

As an additional approach, Dalenbäck suggests testing for the sensitivity to scattering coefficients. This is done by calculating acoustic parameters with an initial reasonable guess of scattering coefficients, and with diffuse modelling switched off. The modeller can then examine if the resulting acoustic parameters differ substantially. If the parameters vary greatly, then the scattering coefficients have to be more carefully estimated and it might be wise to include in the room design some options for final fine tuning of the reverberation time or use a complementary technique such as scale model measurement.

Appendix C gives tables of scattering coefficients for a variety of surfaces. These are correlation scattering coefficients, calculated using a method outlined in Section 5.3.6. These were calculated using BEM and represent single-plane surfaces, such as arrays of cylinders. Appendix D includes published scattering coefficients measured according to ISO 17497-1. The use of the table values within a GRAM will take a little interpretation, however, as the scattering coefficient required in GRAMs varies between different diffuse reflection modelling algorithms.

One deficiency in the Table C.3 data is the raised values for random incidence scattering coefficients at low frequencies. This arises because the scattering from the edge and the rear of the test sample is different from a flat surface; a pragmatic fix would be to set the scattering coefficients to 0 where the depth is less than a quarter of a wavelength. Furthermore, the 2D predictions in Table C.3 are for the plane of maximum scattering, whereas most diffuse reflection algorithms will interpret these values as being for hemispherically scattering devices and distribute the scattered energy according to Lambert's law. So, although the predictions were produced using single-plane scatterers, they will probably be better matched to hemispherical scatterers in many GRAMs. The values in Table C.3 for semicylinders will probably better match the required values for hemispheres, the Table C.3 values for 1D Schroeder diffusers will probably better match the required values for 2D Schroeder diffusers, and so on. When modelling a single-plane device (cylinder, 1D Schroeder diffuser), it may be necessary to reduce the scattering coefficient table values if the GRAM distributes energy according to Lambert's law using a single scattering coefficient.

13.4 SUMMARY

This chapter has presented some of the problems associated with going from isolated predictions or measurements of surface properties to whole room predictions. It has considered the problems associated with the use of absorption coefficients in simple statistical models, as well as the role of absorption and scattering coefficients in GRAMs. There are many gaps in understanding and knowledge, and many problems still remain and need further research. The next and final chapter will look at how active technology can provide absorption and diffuse reflections.

REFERENCES

1. H. Kuttruff, *Room Acoustics*, 5th edn, Spon Press, Oxon, UK (2009).
2. Y. Makita and T. Hidaka, "Comparison between reverberant and random incident sound absorption coefficients of a homogeneous and isotropic sound absorbing porous material—Experimental examination of the validity of the revised cos θ law", *Acustica*, **66**, 214–20 (1988).
3. BS EN ISO 354:2003, "Acoustics—Measurement of sound absorption in a reverberation room".
4. T. W. Bartel, "Effect of absorber geometry on apparent absorption coefficients as measured in a reverberation chamber", *J. Acoust. Soc. Am.*, **69**(4), 1065–74 (1981).
5. T. D. Northwood, "Absorption of diffuse sound by a strip or rectangular patch of absorptive materials", *J. Acoust. Soc. Am.*, **35**, 1173–7 (1963).
6. Y. Makita, T. Hidaka, J. Kaku, and J. Yoshimura, "Re-examination of the set of numerical values of the functions which the revised cos θ law and the revised Paris' formula comprise", *Acustica*, research note, **74**, 159–63 (1991).
7. M. C. Gomperts, "Do the classical reverberation time formulas still have the right to existence?", *Acustica*, **16**, 255–68 (1965).
8. L. Savioja and U. P. Svensson, "Overview of geometrical room acoustic modeling techniques", *J. Acoust. Soc. Am.*, **138**(2) 708–30 (2015).
9. M. Kleiner, B.-I. Dalenbäck, and P. Svensson, "Auralization—An overview", *J. Audio Eng. Soc.*, **41**(11), 861–75 (1993).
10. L. Nijs, G. Jansens, G. Vermeir, and M. van der Voorden, "Absorbing surfaces in ray-tracing programs for coupled spaces", *Appl. Acoust.*, **63**, 611–26 (2002).
11. W. C. Sabine, *Collected Papers on Acoustics*, Harvard University Press, Cambridge, MA (1922).
12. I. Bork, "Report on the 3rd round robin on room acoustical computer simulation—Part II: Calculations", *Acta Acust. Acust.*, **91**(4), 753–63 (2005).
13. I. Bork, "A comparison of room simulation software—The 2nd round robin on room acoustical computer simulation", *Acustica*, **86**, 943–56 (2000).
14. M. Hodgson, "Evidence of diffuse surface reflections in rooms", *J. Acoust. Soc. Am.*, **89**, 765–71 (1991).
15. M. Vorländer, "International round robin on room acoustical computer simulations", *Proc. 15th ICA*, **II**, 689–92 (1995).
16. B.-I. Dalenbäck, "The importance of diffuse reflection in computerized room acoustic prediction and auralization", *Proc. IoA(UK)*, **17**, 24–34 (1995).
17. Y. W. Lam, "On the parameters controlling diffusion calculation in a hybrid computer model for room acoustics prediction", *Proc. IoA(UK)*, **16**(2), 537–44 (1994).
18. B-I. Dalenbäck, M. Kleiner, and P. Svensson, "A macroscopic view of diffuse reflection", *J. Audio Eng. Soc.*, **42**, 793–807 (1994).
19. R. R. Torres and M. Kleiner, "Audibility of 'diffusion' in room acoustics auralization: An initial investigation", *Acust. Acta Acust.*, **86**, 919–27 (2000).
20. G. Naylor, "Treatment of early and late reflections in a hybrid computer model for room acoustics", *Proc. 124th Acoust. Soc. Am. Meeting*, Paper 3aAA2 (1992).
21. Y. W. Lam, "The dependence of diffusion parameters in a room acoustics prediction model on auditorium sizes and shapes", *J. Acoust. Soc. Am.*, **100**, 2193–203 (1996).
22. Y. W. Lam, "A comparison of three diffuse reflection modelling methods used in room acoustics computer models", *J. Acoust. Soc. Am.*, **100**(4), 2181–92 (1996).
23. X. Zeng, C. L. Christensen, and J. H. Rindel, "Practical methods to define scattering coefficients in a room acoustics computer model", *Appl. Acoust.*, **67**, 771–86 (2006).
24. B.-I. Dalenbäck, *CATT Acoustic User's Manual V8*, Gothenburg, Sweden (2005).
25. A. Farina, "A new method for measurement of the scattering coefficient and the diffusion coefficient of panels", *Acustica*, **86**, 928–42 (2000).
26. B.-I. Dalenbäck, "Modeling 1D-diffusers—The missing link", http://www.catt.se/Lambert-1D-CATT.pdf, accessed 2 September 2015.

27. J. S. Bradley, R. D. Reich, and S. G. Norcross, "On the combined effects of early- and late-arriving sound on spatial impression in concert halls", *J. Acoust. Soc. Am.*, **108**(2), 651–61 (2000).
28. J. Kang, "Numerical modeling of the sound fields in urban squares", *J. Acoust. Soc. Am.*, **117**(6), 3695–706 (2005).
29. T. J. Cox, W. J. Davies, and Y. W. Lam, "The sensitivity of listeners to early sound field changes in auditoria", *Acustica*, **79**(1), 27–41 (1993).
30. ISO 3382-1:2009, "Acoustics—Measurement of room acoustic parameters—Part 1: Performance spaces".
31. RPG, http://www.rpginc.com/research/reverb01.htm, accessed 11 April 2008.
32. H. Onaga and J. H. Rindel, "Acoustic characteristics of urban streets in relation to scattering caused by building facades", *Appl. Acoust.*, **68**, 310–25 (2007).
33. M. R. Ismail and D. J. Oldham, "A scale model investigation of sound reflection from building facades", *Appl. Acoust.*, **66**(2), 123–47 (2005).

Chapter 14

Active absorbers

The absorbers and diffusers discussed in previous chapters have difficulty altering low-frequency sound. Low frequencies have long wavelengths, which means treatments have to be large to perturb or absorb the wavefronts. Active control technologies offer the possibility of bass absorption or diffuse reflections from relatively shallow surfaces, as well as a capability for variable acoustics. An example application for active absorption is the control of modes in small rooms. The cost and difficulties of implementation are considerable, however, and this is one reason why this technology has not been more widely applied.

Active absorption has much in common with active noise control; indeed, in many ways, it is the same concept, just reorganized by a slightly different philosophy. Olson and May carried out pioneering experiments, and they suggested an active noise control method based on interference.[1] In their method, an electroacoustic feedback loop was used to drive the acoustic pressure to 0 near an error microphone placed close to a secondary loudspeaker. This is illustrated in Figure 14.1. More sophisticated systems alter the surface impedance of the control loudspeaker towards a desired target value. They may be configured as feed-forward or feedback devices and are often constructed around single channel, filtered-x least mean square (LMS) adaptive filter algorithms. A more effective method places resistive material in front of the control surface (loudspeaker) to gain energy dissipation. The active system then maximizes the particle velocity through the material. This concept was also first suggested by Olson and May.

14.1 SOME PRINCIPLES OF ACTIVE CONTROL

In this section, some basic principles of adaptive filtering and active control are outlined. The particular form of a control system is dictated by the physics of the environment in which it operates and the control task to which it is set. However, broad classifications of control systems exist that are useful in differentiating between certain very different approaches. These classifications distinguish feedforward from feedback control systems, which may or may not be adaptive to changes in their operating environment.

Consider the system in Figure 14.2. The signal s is corrupted by the addition of the noise signal n at the first summing node, generating the observable signal d. At the second summing node, a signal y is subtracted from d. The result of this subtraction is the error signal, e.

- If $y = n$, then the noise corruption on the signal s is removed, $e = s$; this is the ideal.
- If y is a reasonable approximation of n, then some of the noise contamination is removed, $e \approx s$; this is more realistic of what happens with active control systems.
- If y is largely uncorrelated with n, then the second summing node represents an additional source of noise, further corrupting the signal s in e; this is to be avoided.

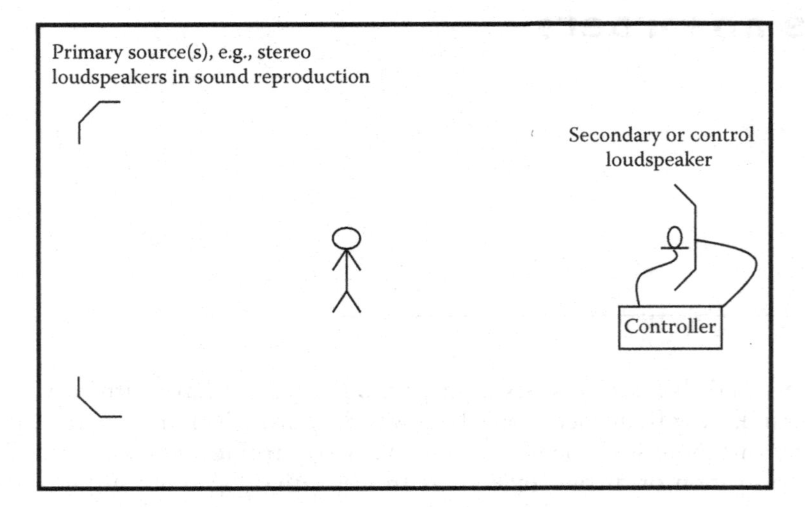

Figure 14.1 Schematic of active absorption in a small room. In this case, a microphone close to the secondary loudspeaker is used as an error signal for the controller to minimize.

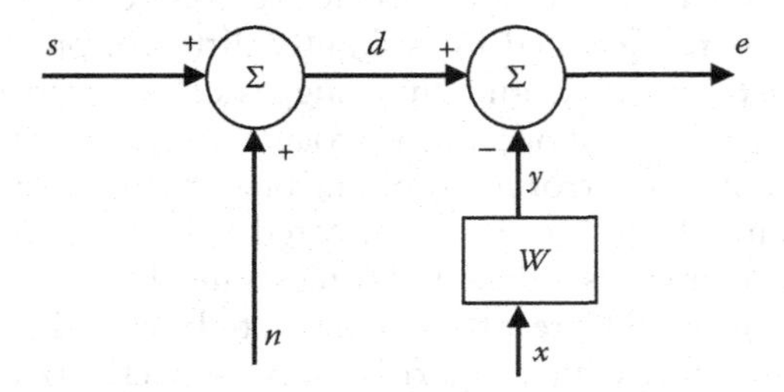

Figure 14.2 A basic active noise control system to remove noise *n* from signal *s*.

The cancelling signal y is derived by filtering operations—through the filter block W, which is an adaptive filter, i.e., a filter that can change its coefficients to achieve the required control—on the reference signal, x. The optimal configuration of the adaptive cancelling filter, W, is the inverse of the filter relating n to x. Then, the noise added at the first summing node is perfectly cancelled at the second summing node. Such perfect performance is never achieved in practice for several reasons, most important among which are the imperfect implementation of the cancelling filter and the imperfect correlation between the noise n and the reference x.

Consider the problem of imperfect correlation. The attenuation of the noise component in d is a function of the coherence between the noise and reference signal. The attenuation increases as the coherence increases, and useful levels of noise attenuation can be achieved only with high coherence between the reference and the noise signal; this can pose problems in electroacoustic applications.

Ideally, an analytical solution for the necessary filter W would be derived; however, computing the coefficients of the filter W is usually a non-trivial problem. Fortunately, a computationally efficient iterative approach to the identification of the necessary filter coefficient exists; this is an adaptive filter running under the LMS algorithm.

The technique uses an iterative search process to find the filter W that minimizes the error e. The LMS algorithm discovered by Widrow and Hoff[2] has been found to be robustly stable in many practical applications. It is also a clear, simple, and computationally efficient

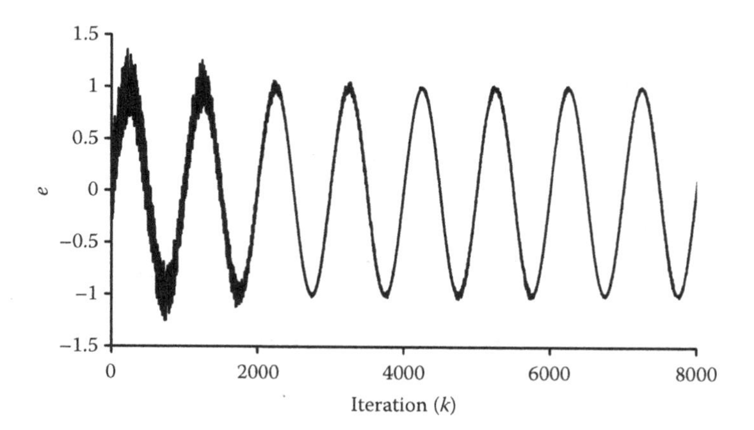

Figure 14.3 Removing noise from a sine wave using active control.

approach for identifying *W*. There are other techniques for solving minimization problems, but the LMS approach forms the basis of most adaptive noise cancelling systems.

The weights of the filter *W* are updated using the following equation. The coefficients of the adaptive filter at the $k + 1$ iteration are given by

$$W_{k+1} = W_k + 2\alpha e_k x_k, \tag{14.1}$$

where subscript k refers to iteration number, W_k is the vector of adaptive filter coefficients at iteration k, e_k is the error, α is the update rate, and x_k is input to the adaptive filter.

The performance of the LMS algorithm is illustrated here by an example. A simulation of a discrete time implementation of Figure 14.1 was coded, in which

$$n_k = 0.5x_k + 0.2x_{k-1}. \tag{14.2}$$

A length 2 adaptive filter *W* was updated using the LMS algorithm (Equation 14.1) with s as a simple sinusoid and x a random signal.

The error signal is shown in Figure 14.3. The initial noise is seen to be quickly cancelled, leaving a sinusoidal wave, the signal s. The decay of the noise follows a roughly exponential form, which is due to the convergence behaviour of the LMS algorithm approximating the first-order convergence of a steepest descent algorithm.

The convergence of the two coefficients of the adaptive filter *W* is shown in Figure 14.4. The weights are seen to approach the optimal values implied by Equation 14.2. This is very similar to the system shown in Figure 14.2 except that the white noise signal is fed direct to x and then filtered to get the signal n.

Having studied some fundamentals of noise cancelling, it is now possible to consider the practicalities of active impedance systems.

14.2 AN EXAMPLE ACTIVE IMPEDANCE SYSTEM AND A GENERAL OVERVIEW

Figure 14.5 shows a possible feedforward controller for an active impedance system.[3–6] A signal generator is driving the *primary source* in the top left of the diagram. The role of the

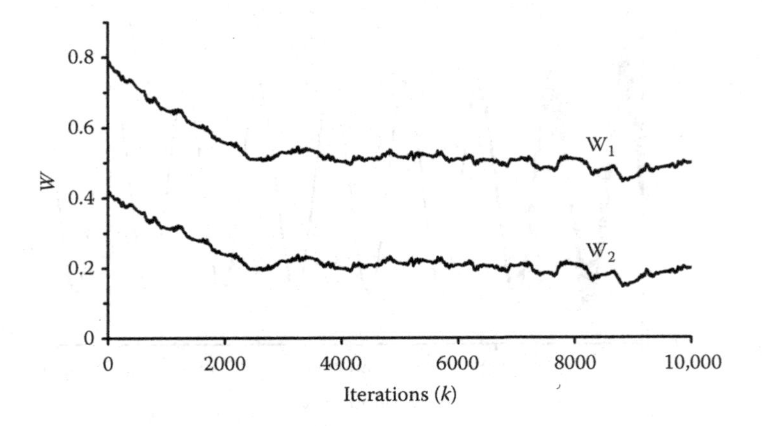

Figure 14.4 Filter coefficients for filter W during training.

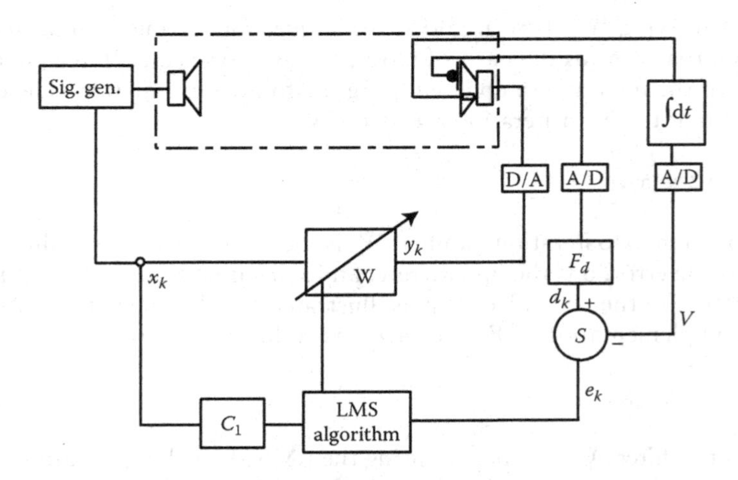

Figure 14.5 A feedforward active impedance control system.

primary source is to generate acoustic waves for the controller to operate on. In the diagram shown, this is constrained within a pipe (shown dotted), but it could be within other spaces. The sound from the primary source then propagates to the control surface (*secondary loudspeaker*) shown top right.

The control surface is instrumented to sense pressure and velocity. Consequently, the impedance at the surface of the controller is known. Using an LMS algorithm, it is possible to alter how the control loudspeaker moves to force the surface impedance to be some desired value.

The velocity, v_k, at the control surface is sensed by integrating the output from a miniature accelerometer mounted on the cone surface. An alternative technique would use two closely spaced microphones.[7,8] The pressure is measured using a surface microphone. This pressure is then passed through the filter F_d. F_d is the desired admittance, and consequently, the signal d_k is the desired velocity. The desired and actual velocities are subtracted to give an error signal, e_k. If this error was 0, then the surface admittance is as desired. If the error is not 0, then the LMS algorithm is used to change the weights of the adaptive filter W, to reduce the error. Consequently, there is an adaptation time over which the error gradually converges to a small value, preferably 0. Setting the correct value for the update rate given in Equation 14.1 is crucial to achieving training—too large a value and the system never converges, too small a value and the convergence is very slow. There can be problems with instability during training.

The input to the adaptive filter is a signal x_k, which must correlate with the primary source signal, otherwise the control surface just adds additional noise. This signal can be derived from two places, forming either a feedback or feedforward system. In a feedforward case, the signal x_k is an electronic feed from the primary signal source. This is the case shown in Figure 14.5. The great advantage of feedforward is that it forms a stable system and no unstable feedback can occur. The disadvantage is that an electronic feed from the source signal is required, which means the active surface could be used with electroacoustic sound reproduction systems (i.e., loudspeakers), but not sound production systems such as acoustic musical instruments or speech.

In the feedback case, a microphone picks up the signal from the primary source, as illustrated in Figure 14.6. This can actually be the miniature microphone on the surface of the control loudspeaker. With this system, however, there is potential for instability, as a loop is formed that will become unstable if the gain exceeds 1. The solution to this problem is to insert a feedback compensation filter, F_1, that is designed to cancel the feedback path. The feedback cancellation is awkward, however, and if not entirely successful, the system will become unstable. Alternatively, highly directional loudspeakers and microphones can be used to steer energy from the control source away from the microphone connected to the reference input, but performance is then frequency dependent.

The usual system is to train the coefficients of the adaptive filter and, once the error is sufficiently low, to fix the coefficients. In this example, adaptation is used purely as an efficient method for obtaining the filter coefficients W, which may not be analytically derived.

In order that the impedance converges to the correct value, the measurements must provide a true and accurate measurement of the actual ratio of the pressure to the particle velocity. Any error in these measurements will result in convergence to a value other than that desired by the user. For instance, transduction will introduce uneven frequency responses onto the signals. C_1 is called the *plant model*, and its role is to compensate for the frequency responses of the transducers and other components. The filter that models the plant is that referred to in the phrase *filtered-x*.

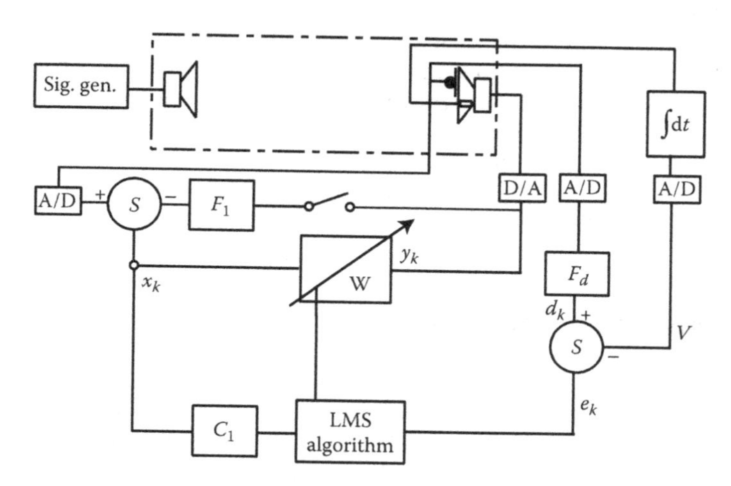

Figure 14.6 A feedback active impedance control system.

The design of the plant model presents significant problems. It may be sufficient to measure the frequency response of the plant off-line with a noise or impulsive test signal and fit this with a finite impulse response filter providing a reasonable estimate of the actual plant response. The accuracy of the plant model appears to determine whether or not convergence of the filter will be achieved and over what timescale adaptation may take place without the risk of instability.[9]

There is some tolerance to plant model errors, which is fortunate since, sometimes, the object of employing an adaptive active control system is to enable a controller to track changes in its environment during operation. Any such changes will introduce errors between the assumed and the current plant response. Where gross run-time alteration of the plant response is anticipated, a run-time measurement of the plant that continually updates filter C_1 may be employed. This has been attempted using maximum length sequence signals at very low levels presented simultaneously with program material.[10]

In fact, while people may refer to the active absorption systems as adaptive, this is a rather misleading name in many cases. The system might be adaptive during the training of the system, but it is most often used with the adaptation turned off, as to do otherwise risks instability in operation. But without adaptation, the system is vulnerable to changes in the physical acoustics, such as temperature changes and room occupancy.

A theoretical analysis of the significance of transduction errors for active impedance control in a 1D waveguide is presented by Darlington et al.,[11] along with measured results derived from the intentional perturbation of pressure and velocity control signals. It is concluded that the transducers and associated signal-conditioning circuits should be calibrated to within 1 dB magnitude error and 5° phase error in order to achieve an absorption coefficient greater than 0.95. This analysis is helpful in that it identifies the significance of transduction errors, but a discussion of the measurement method itself and its relation to a theoretically modelled ratio of pressure and velocity at the surface of a loudspeaker cone is not attempted. This relationship is important in two ways.

Physical measurements of the impedance at the loudspeaker cone depend on two factors—the correct transduction of cone velocity and a suitable measurement of the pressure at the cone surface. Velocity measurement can be done via a two-microphone method or an accelerometer, but the position of the accelerometer is shown to be crucial. Nicholson and Darlington[3] report that at frequencies as low as 150 Hz, significant differences appear in the magnitude and phase of the velocity between accelerometers mounted at different points on the cone, as the local mass load encourages the onset of non-pistonic motion of the cone. Accelerometer locations where the dust cap meets the cone are most suitable. Nicholson also investigated microphone locations immediately adjacent to the control source cone and found that a frame mounting 5 mm from the dust cap was best. It is important that the microphone does not directly pick up the effects of the cone vibration (the velocity), as otherwise, the system becomes unstable.

Having given some sense of how an adaptive system might work in principle, the following sections detail the application of these types of controllers.

14.3 ACTIVE ABSORPTION IN DUCTS

When the system described in Section 14.2 is constrained to 1D plane waves, the controller is very successful. This would be the case for low-frequency control within ducts. Figure 14.7 shows the modes in a duct with the controller turned off (so the termination is the control loudspeaker, which is not being driven, with the termination impedance being dictated by

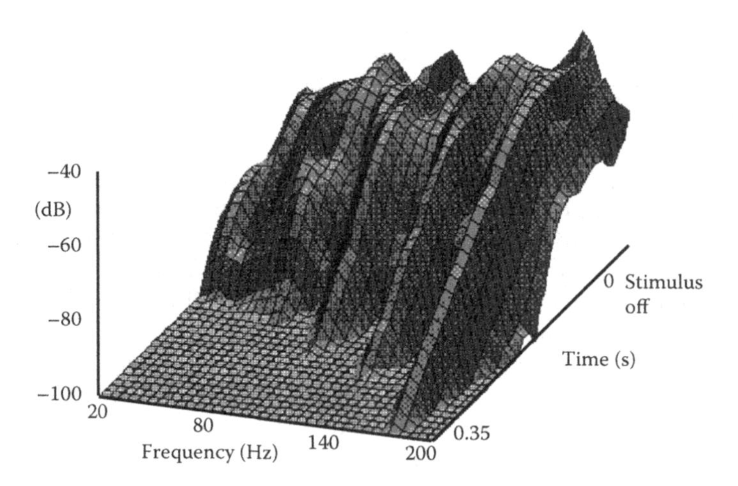

Figure 14.7 Waterfall plot of decay of modes in a 4 m duct, controller off. (After Avis, M.R., "The Active Control of Low Frequency Room Modes", PhD thesis, University of Salford, UK, 2000.)

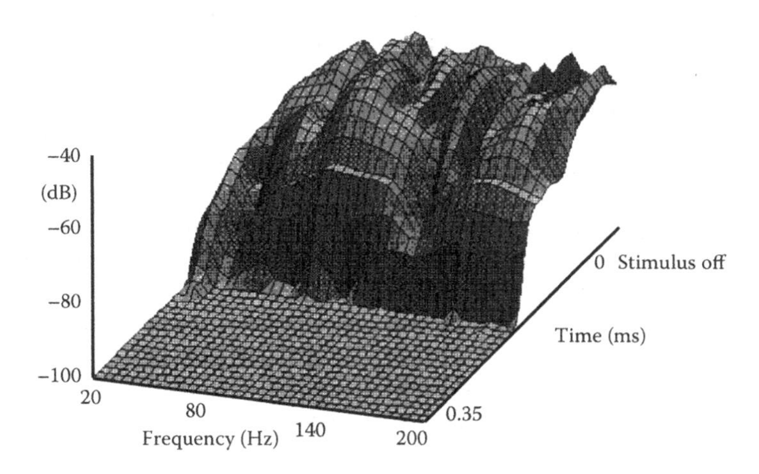

Figure 14.8 Waterfall plot of decay of modes in a 4 m duct, controller on. (After Avis, M.R., "The Active Control of Low Frequency Room Modes", PhD thesis, University of Salford, UK, 2000.)

the mechanical characteristics of the loudspeaker). The plot shows the steady-state response ($t = 0$) and the resulting decay when the primary source is turned off. Figure 14.8 shows the same situation but with the controller in operation. The ability of the controller to damp the modes and therefore make them decay faster is evident. In this case, the target impedance for the active surface was the characteristic impedance of air for plane waves.

14.4 ACTIVE ABSORPTION IN THREE DIMENSIONS

It is possible to train the active absorber in a duct to a characteristic impedance, turn the adaptation off, and then use the system within a room. Unfortunately, in this case, only small reductions in pressure are obtained. The controller surface does achieve a high absorption coefficient. Furstoss et al.[12] report absorption coefficients of about 0.9, but only close to the loudspeaker cone. Consequently, while a high absorption coefficient is achieved, the total absorption added to the room is rather small and so there is little effect on most spaces.

In a 3D environment, the relationship between the control source's surface impedance and the modal behaviour of the room is not simple. The sound field in the room is not plane, although it can be considered to result from the sum of normal modes that individually are plane waves.[13] There exist three orthogonal coordinate axes for particle velocity rather than the single axis within the duct, and the velocity of the controlling driver may lie in the plane of one coordinate or perhaps none of the three. The meaning of a characteristic impedance is therefore no longer clear, and hence, it follows that the solution for the duct is unlikely to result in optimal control in the room. Consequently, a different target function is required. For instance, it might be possible to train the system to minimize the pressure at one or more points in a room. This is then a traditional active control system, and more on these can be found in Nelson and Elliott.[14]

Alternatively, it is possible to consider the relationship between surface pressure and velocity in terms of the power radiated by the source. It can be shown that in certain circumstances, the power radiated becomes negative, corresponding to absorption of energy by the source. When a pistonic sound source radiates acoustic power at low frequency, the power radiated is proportional to pv^*, where v is the velocity, p is the total pressure, and * indicates a complex conjugate. If the velocity of the source can be controlled to minimize the power radiated, which is equivalent to maximizing the in-going intensity, the source is then absorbing acoustic power. This has, however, rather simplified the situation as the pressure across the cone is not constant, and the total pressure at the cone will contain direct and reflected components from the primary and secondary sources. The risk with maximizing the in-going intensity is that the controller will achieve this by maximizing the pressure, and so the sound pressure levels within the room will actually increase. For this reason, this control target is rather problematical.

Another problem with this system is that there is no energy dissipation. The active absorption is generated by superposition or interference. In effect, the active control system works by changing the radiation impedance of the primary sources rather than by absorbing energy from waves in the room. Consequently, what these active absorbers achieve is a reduction in radiated power.[15] Really achieving absorption requires a proper energy dissipation mechanism, and this can be achieved through hybrid designs, discussed in Section 14.5. In the following two sections, however, some experimental results from modal control using adaptive and static techniques are presented.

14.4.1 Low-frequency modal control—example results

Consider the system described in the previous section. This system will be used to try and deal with low-frequency modes that are present in a room. Figures 14.9 and 14.10 compare the pressure distribution of the primary axial mode in a room with the controller on and off. In this example, the controller reduces the steady-state pressure in the mode by about 6 dB. In this case, a single 8-inch loudspeaker is capable of almost halving the decay time of the first axial mode.

These systems work for single modes well isolated in frequency. As soon as modes become degenerate, the active controller has problems. If many modes need to be controlled, many control loudspeakers need to be used. There is probably a need for one control loudspeaker per mode. Consequently, a full control system is going to be expensive to implement.

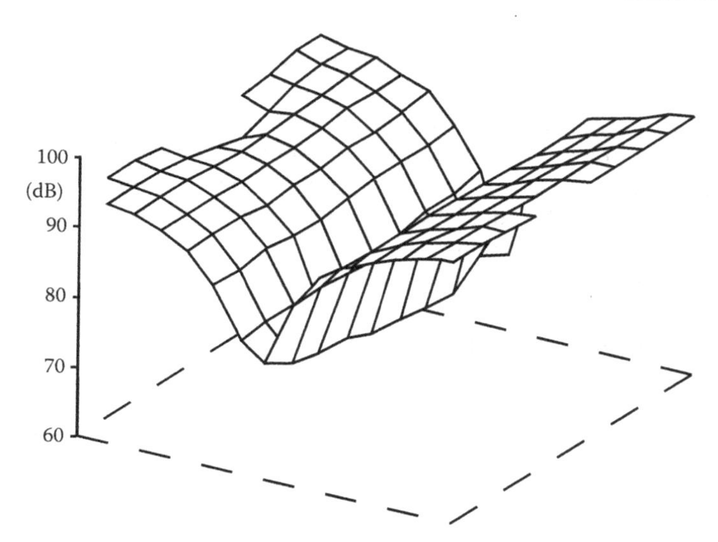

Figure 14.9 Distribution of pressures in a small room for the main axial mode (about 44 Hz); controller off. (After Avis, M.R., "The Active Control of Low Frequency Room Modes", PhD thesis, University of Salford, UK, 2000.)

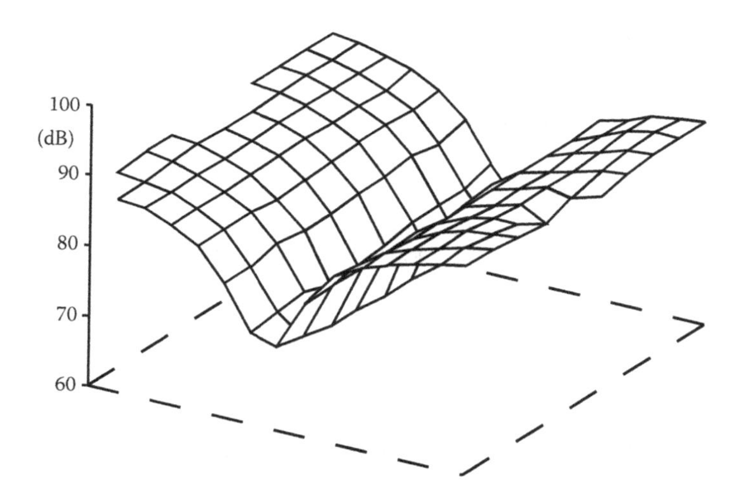

Figure 14.10 Distribution of pressures in a small room for the main axial mode (about 44 Hz); controller on. (After Avis, M.R., "The Active Control of Low Frequency Room Modes", PhD thesis, University of Salford, UK, 2000.)

14.4.2 Low-frequency modal control—alternative control regime

Adaptive systems incur significant cost, in terms of both hardware and constraints on operation due to stability and convergence issues. This has motivated several authors to look for other static control regimes for modal control.[16–19] Discussed here is a short description of one of these. Avis[17] examined an analytical modal decomposition to derive a control filter that acts to reduce the modal quality factor by relocating system poles. The aim was to go further than conventional steady-state equalization, since the detection of bass modes is more related to time than frequency domain artefacts.[20] Additionally, this has potential for controlled equalization across the whole sound field.

A sound field in a room can be expressed as a modal decomposition.[21] This implies that the sound field may be considered as the sum of a large number of second-order functions; these functions can be implemented as infinite impulse response (IIR) biquad filters. The coefficients of these filters are determined by fitting responses to measurements in the physical sound field. Figures 14.11 and 14.12 show an example of the fitting of magnitude and phase for two modes in a small room.

A secondary source is used to radiate pressures, which combine with the natural sound field of the room to generate modes with smaller Q factors, i.e., ones that decay faster. Figures 14.13 and 14.14 show a typical result. The controller is formulated such that the poles of the controlled sound field are relocated further away from the unit circle than the

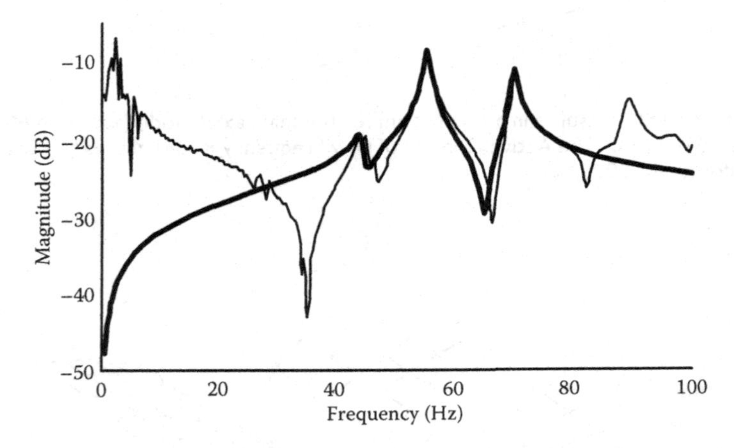

Figure 14.11 Example fitting of measured magnitude of —— modal response to ▬▬ biquad model. (After Avis, M.R., *Proc. Audio Eng. Soc., 21st Conference*, St Petersburg, 2002.)

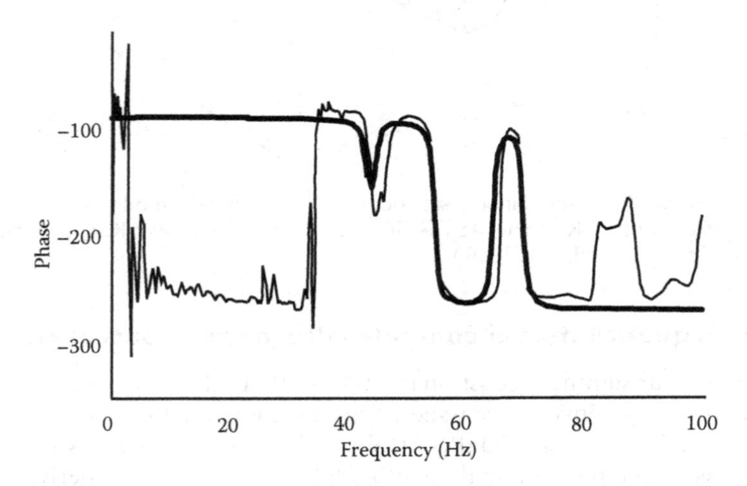

Figure 14.12 Example fitting of measured phase of —— modal response to ▬▬ biquad model. (After Avis, M.R., *Proc. Audio Eng. Soc., 21st Conference*, St Petersburg, 2002.)

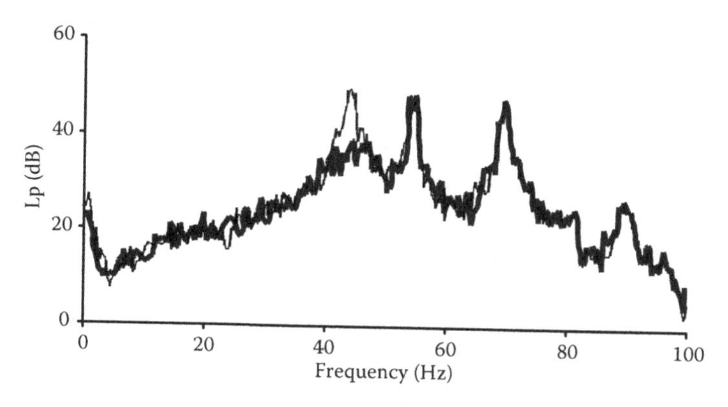

Figure 14.13 Effect of biquad controller designed for a single mode at 44 kHz: ——— controller off and ━━━ controller on. (After Avis, M.R., *Proc. Audio Eng. Soc., 21st Conference*, St Petersburg, 2002.)

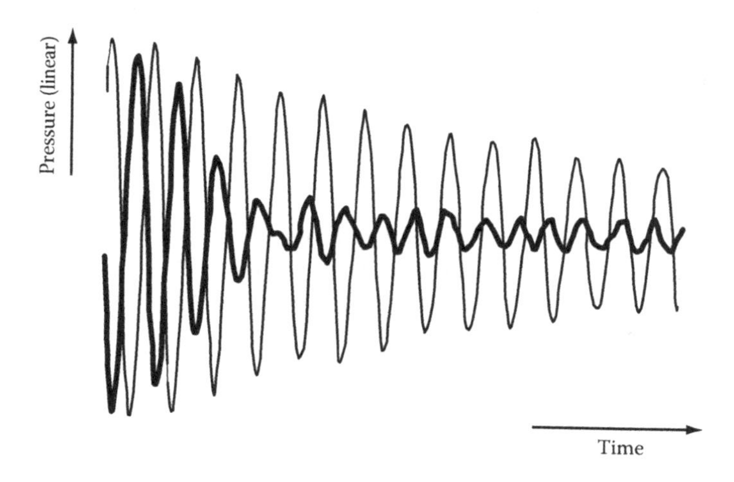

Figure 14.14 Effect of biquad controller designed for a single mode at 44 kHz: ——— controller off and ━━━ controller on. (After Avis, M.R., *Proc. Audio Eng. Soc., 21st Conference*, St Petersburg, 2002.)

uncontrolled case. The controller works well at the measurement point used to fit the IIR filters but operates less effectively at locations remote from that point.

This system can be used to control multiple modes simultaneously. Because the control technique mimics additional damping, the time, frequency, and spatial aspects of the modal nature of the sound field are all addressed simultaneously and in sympathy. The effectiveness of control is again limited to situations where the modes are widely spaced and not degenerate. The sensitivity of this control regime to changes in room conditions is unknown. Presumably, it would be necessary to regularly recalibrate the system for the damping to remain efficient.

14.5 HYBRID ACTIVE–PASSIVE ABSORPTION

The previous adaptive systems have not had explicit dissipation mechanisms included; they have worked by a process of interference or superposition. It makes sense to try to include some form of real resistance as better performance can be achieved. Consequently, this is a hybrid approach involving the combination of absorbent material with an active controller. Olson and May[1] considered the possibility of using their secondary loudspeaker to absorb sound by placing it behind acoustically resistant cloth and to use the active controller to maximize the dissipation of energy in the cloth. A concise summary of the development of the hybrid approach is given by Smith et al.[22]

Furstoss et al.[12] picked up the hybrid concept in the 1990s and made it into a useable device. It is mostly their work that is reported here. A piece of resistive material is placed in front of the active element, and the absorber is made efficient by creating a virtual quarter wavelength resonator behind (as though the resistive material is a quarter of a wavelength from a rigid wall). A typical set-up is shown in Figure 14.15. In the example shown, the surface of the control loudspeaker is instrumented to measure velocity and pressure, and this is used as inputs to an active controller, which drives the control loudspeaker. The controller is tasked with setting the appropriate backing impedance condition. The active control system avoids the need for a large air gap as would be required for a passive resonant absorber at low frequency. Furthermore, it can produce broadband efficiency rather than the limited bandwidth achieved by the passive quarter wave resonant absorbers.

At low frequency, the pressure drop across the resistive material can be given by the flow resistivity and particle velocity:

$$\frac{p_2 - p_1}{v} = \sigma d, \tag{14.3}$$

where p_2 and p_1 are the pressures at the front and the rear, respectively, of the resistive material, v is the particle velocity, σ is the flow resistivity, and d is the material thickness.

If the active element renders the backing pressure to be 0 (as would be the case with a quarter wavelength tube), the impedance of the layer is

$$z = \frac{p_2}{v} = \sigma d. \tag{14.4}$$

This is the flow resistance of the resistive material, which should be set to the characteristic impedance of air to maximize absorption.

Therefore, an alternative set-up is to place a microphone at the rear of the porous material, and a controller is then tasked with minimizing the pressure at the microphone. Figure 14.16 shows the results from such an arrangement. High absorption across a relatively wide

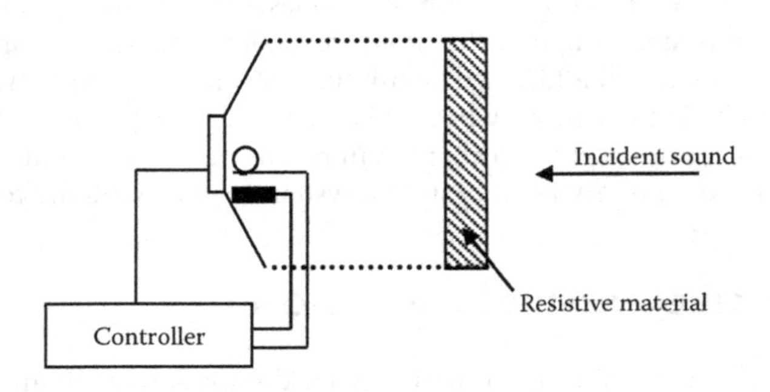

Figure 14.15 A hybrid active–passive absorber.

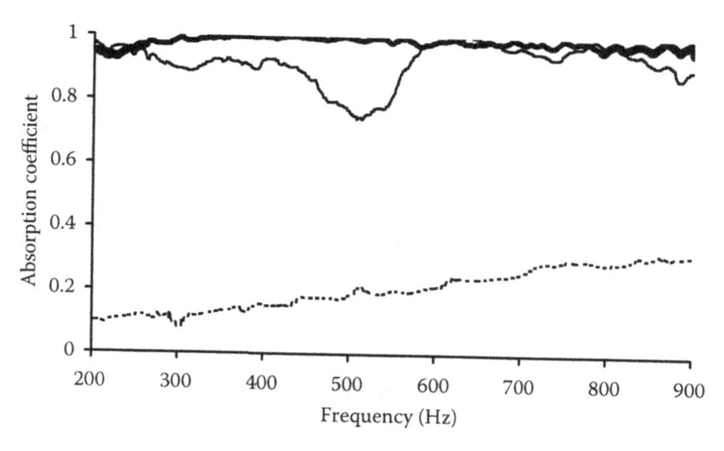

Figure 14.16 Absorption coefficient for a resistive material with and without an active controller backing minimizing the pressure at the rear of the material. The distance *r* refers to the distance of the microphone from the centre of the control loudspeaker: .... material with rigid backing; ——— hybrid absorber, $r = 10$ cm; and ▬▬ hybrid absorber, $r = 0$ cm. (After Furstoss, M., Thenail, D., Galland, M.A., *J. Sound Vib.*, 203, 219–236, 1997.)

frequency range is achieved. Absorption is not as high for oblique incident sound, averaging around 0.6–0.7 for an angle of incidence, $\psi = 60°$, because for that case, the optimal backing pressure for maximum absorption is no longer 0 (see below). When used in an array of active absorbers, good performance is achieved, although transduction problems limit the useful frequency range to 1 octave around 280 Hz.

While this regime works for low frequencies, this anechoic termination becomes less successful as the frequency increases. Furstoss et al.[12] showed that a better termination criterion is obtained by considering the optimal backing impedance more completely. Consider a porous layer between two fluids as shown in Figure 14.17. The impedance at the front face can be found using the transfer matrix approach described in Chapters 2 and 6. The impedance at the back face is

$$z_b = \frac{z_c k}{k_x} \cdot \frac{-jz_f \cot(k_x d) + z_c k/k_x}{z_f - jz_c(k/k_x)\cot(k_x d)}, \tag{14.5}$$

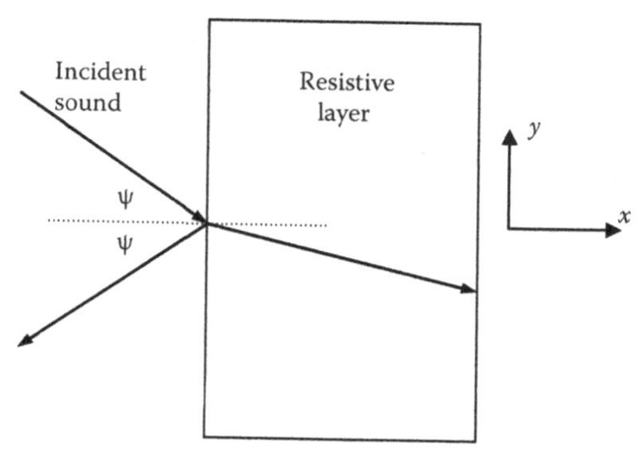

Figure 14.17 Geometry under consideration when determining optimal backing impedance for hybrid active–passive absorption.

where z_f is the impedance on the front face and k_x is the component of the wavenumber in the porous layer in the x direction. The wavenumber k and characteristic impedance z_c in the porous medium can be found using the porous absorber models given in Section 6.5. By considering Equation 14.5, the optimal backing impedance for maximum absorption can be found for a particular angle of incidence by setting $z_f = \rho_0 c_0/\cos(\psi)$.

Figure 14.18 shows the optimal backing impedance for a particular situation, where the porous material is offering a resistance close to $\rho_0 c_0$. At low frequency, the optimal backing impedance is 0, similar to a zero-pressure condition, as indicated before, but as the frequency increases, the optimal backing impedance also changes. It will also change with the porous material's resistance and the angle of incidence. Consequently, minimizing the backing pressure does not necessarily produce optimal absorption, although in the case shown, it will be fairly effective below 1 kHz. This impedance matching approach requires pressure and velocity transducers on the active control surface.

Smith et al.[22] compared the impedance matching exemplified by Equation 14.5 and pressure release control conditions. They found that the impedance matching approach was superior, requiring less control effort and achieving higher absorption coefficients. Absorption coefficients ranged from 0.8 to 1 from 100 Hz to 1 kHz.

The active controller can also be placed behind a microperforated sheet to gain wideband absorption. The principles are similar to that outlined previously for more conventional resistive material. Cobo and Cuesta[23] achieved an absorption coefficient of around 0.7 for a frequency range of 200 to 900 Hz from a full-scale system in an anechoic chamber measurement.

An alternative approach to hybrid absorption was developed by Guigou and Fuller.[24] They used a smart foam design that integrated a lightweight distributed piezoelectric PDVF actuator (the active component) between individual layers of sound absorbing foam (the passive component) such that the control can efficiently operate over a broad range of frequencies. The foam provides absorption passively at high frequencies, and the active element in collaboration with the foam provides absorption at low frequencies. In this case, the active surface is being used to reduce the radiated power from a vibrating surface.

Lissek et al.[25] developed another approach that utilizes the sensing of pressure and velocity on a control surface but avoids the complication of active noise control. Their *shunt loudspeaker* absorber uses a fixed, passive circuit connected to the loudspeakers circuit that then facilitates absorption of sound waves.

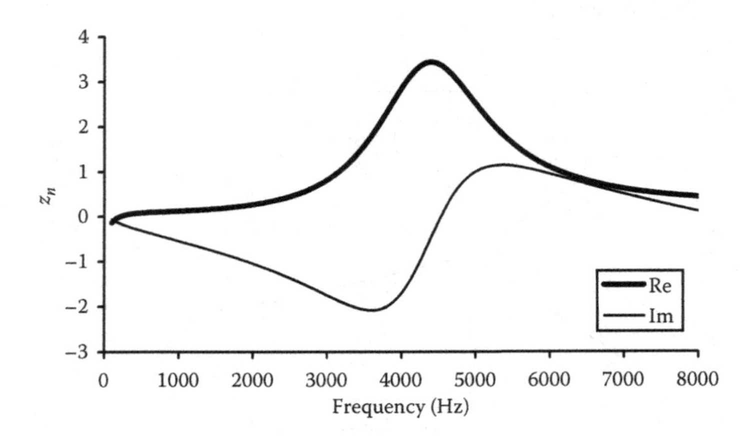

Figure 14.18 Optimal normalized backing impedance for a hybrid active–passive absorber.

14.6 SUMMARY

This chapter has discussed the use of active elements to achieve improved absorption. It is also possible to use similar methods to create diffusers that outperform passive devices.[26–28] It is hard to envisage any application where active diffusers might be used in practice, however. The main advantage of active control is that it overcomes the requirement for large passive surfaces at low frequencies when sound wavelengths are long. Unfortunately, the cost and practical difficulties associated with this technology have meant that its use is not widespread.

REFERENCES

1. H. F. Olson and E. G. May, "Electronic sound absorber", *J. Acoust. Soc. Am.*, **25**, 1130–6 (1953).
2. B. Widrow and S. D. Stearns, *Adaptive Signal Processing*, Prentice Hall, Englewood Cliffs (1985).
3. G. C. Nicholson and P. Darlington, "Active control of acoustic absorption, reflection and transmission", *Proc. IoA(UK)*, **15**(3), 403–9 (1993).
4. G. C. Nicholson, "The Active Control of Impedance", PhD thesis, University of Salford, UK (1995).
5. M. R. Avis, "The Active Control of Low Frequency Room Modes", PhD thesis, University of Salford, UK (2000).
6. G. C. Nicholson and P. Darlington, "Smart surfaces for building acoustics", *Proc. IoA(UK)*, **13**(8), 155–64 (1991).
7. F. Orduna-Bustamante and P. A. Nelson, "An adaptive controller for the active absorption of sound", *J. Acoust. Soc. Am.*, **91**, 2740–7 (1992).
8. J. Y. Chung and D. A. Blaser, "Transfer function method of measuring in-duct acoustic properties", *J. Acoust. Soc. Am.*, **68**, 907–21 (1980).
9. C. C. Boucher, S. J. Elliot, and P. A. Nelson, "The effects of modelling errors on the performance and stability of active noise control systems", *Proc. Recent Advances in Active Control of Sound and Vibration*, 291–301 (1991).
10. L. J. Ericksson and M. C. Allie, "Use of random noise for on-line transducer modeling in an adaptive active attenuation system", *J. Acoust. Soc. Am.*, **85**(2), 797–802 (1989).
11. P. Darlington, G. C. Nicholson, and S. E. Mercy, "Input transduction errors in active acoustic absorbers", *Acta Acust.*, **3**, 345–9 (1995).
12. M. Furstoss, D. Thenail, and M. A. Galland, "Surface impedance control for sound absorption: Direct and hybrid passive/active strategies", *J. Sound Vib.*, **203**(2), 219–36 (1997).
13. H. Kuttruff, *Room Acoustics*, 5th edn, Spon Press, Oxon, UK (2009).
14. P. A. Nelson and S. J. Elliott, *Active Control of Sound*, Academic Press, London (1993).
15. R. D. Ford, "Where does the power go?", *Proc. 11th ICA*, **8**, 277–80 (1983).
16. P. Herzog, A. Soto-Nicola, and F. Guery, "Passive and active control of the low-frequency modes in a small room", *Proc. 98th Convention Audio Eng. Soc.*, preprint 3951(D1) (1995).
17. M. R. Avis, "Q-factor modification for low-frequency room modes", *Proc. Audio Eng. Soc., 21st Conference*, St Petersburg (2002).
18. J. Mourjopoulos, "Digital equalisation of room acoustics", *J. Audio Eng. Soc.*, **42**(11), 884–900 (1994).
19. Y. Haneda, S. Makino, and Y. Kaneda, "Multiple-point equalisation of room transfer functions by using common acoustical poles", *IEEE Transactions on Speech and Audio Processing*, **5**(4), 325–33 (1997).
20. M. Wankling, B. Fazenda, and W. J. Davies, "The assessment of low-frequency room acoustic parameters using descriptive analysis", *J. Audio Eng. Soc.*, **60**, 325–37 (2012).

21. P. M. Morse and K. U. Ingard, *Theoretical Acoustics*, Princeton University Press, Princeton, NJ, 555–72 (1968).
22. J. P. Smith, B. D. Johnson, and R. A. Burdisso, "A broadband passive–active sound absorption system", *J. Acoust. Soc. Am.*, **106**(5), 2646–52 (1999).
23. P. Cobo and M. Cuesta, "Hybrid passive-active absorption of a microperforated panel in free field conditions", *J. Acoust. Soc. Am.*, **121**(6), EL251–5 (2007).
24. C. Guigou and C. R. Fuller, "Adaptive feedforward and feedback methods for active/passive sound radiation control using smart foam", *J. Acoust. Soc. Am.*, **104**(1), 226–31 (1998).
25. H. Lissek, R. Boulandet, and R. Fleury, "Electroacoustic absorbers: Bridging the gap between shunt loudspeakers and active sound absorption", *J. Acoust. Soc. Am.*, **129**, 2968–78 (2011).
26. T. J. Cox, M. R. Avis, and L. J. Xiao, "Maximum length sequence and Bessel diffusers using active technologies", *J. Sound Vibr.*, **289**, 807–29 (2006).
27. M. R. Avis, L. J. Xiao, and T. J. Cox, "Stability and sensitivity analyses for diffusers with single and multiple active elements", *J. Audio Eng. Soc.*, **53**(11), 1047–60 (2005).
28. L. J. Xiao, T. J. Cox, and M. R. Avis, "Active diffusers: Some prototypes and 2D measurements", *J. Sound Vib.*, **285**(1–2), 321–39 (2005).

Appendix A: Table of absorption coefficients

Material	Frequency (Hz)					
	125	*250*	*500*	*1000*	*2000*	*4000*
Curtains or drapes						
Light velour 0.338 kg/m² hung straight in contact with wall[1]	0.04	0.05	0.11	0.18	0.30	0.35
Medium velour 0.475 kg/m², hung straight[1]	0.05	0.07	0.13	0.22	0.32	0.35
Medium velour 0.475 kg/m², draped to half area[1]	0.07	0.31	0.49	0.75	0.70	0.60
Heavy velour, 0.61 kg/m² hung straight[1]	0.05	0.12	0.35	0.48	0.38	0.36
Heavy velour, 0.61 kg/m² draped to half area[1]	0.14	0.35	0.55	0.77	0.70	0.60
Variation with draping						
Hung straight[2]	0.04	0.16	0.19	0.17	0.20	0.25
Draped to half area[2]	0.15	0.25	0.30	0.28	0.35	0.40
Draped to 40% of area[2]	0.19	0.31	0.35	0.34	0.44	0.50
Curtains in folds against wall[3]	0.05	0.15	0.35	0.40	0.50	0.50
Cotton curtains, 0.475 kg/m²						
Draped to 7/8 area[4,5]	0.03	0.12	0.15	0.27	0.37	0.42
Draped to 3/4 area[4,5]	0.04	0.23	0.40	0.57	0.53	0.40
Draped to 1/2 area[4,5]	0.07	0.37	0.49	0.81	0.65	0.54
Carpet						
Carpet heavy, on concrete[2]	0.02	0.06	0.14	0.37	0.60	0.65
Heavy carpet (same as line above) on foam rubber or 1.35 kg/m² hair felt[2]	0.08	0.24	0.57	0.69	0.71	0.73
Heavy carpet (same as 2 lines above) with latex backing on foam rubber or 1.35 kg/m² hair felt[2]	0.08	0.27	0.39	0.34	0.48	0.63
Haircord on felt[6]	0.10	0.15	0.25	0.30	0.30	0.30
Pile and thick felt[6]	0.07	0.25	0.50	0.50	0.60	0.65
No underlay (pad), woven wool loop, 1.2 kg/m² 2.4 mm pile height[2]	0.10	0.16	0.11	0.30	0.50	0.47
No underlay (pad), woven wool loop, 1.4 kg/m² 6.4 mm pile height[2]	0.15	0.17	0.12	0.32	0.52	0.57
No underlay (pad) woven wool loop, 2.3 kg/m² 9.5 mm pile height[2]	0.17	0.18	0.21	0.50	0.63	0.83
Loop pile tufted carpet,[2] 1.4 kg/m², hair underlay 1.4 kg/m²	0.03	0.25	0.55	0.70	0.62	0.84
Loop pile tufted carpet,[2] 1.4 kg/m², hair underlay 3.0 kg/m²	0.10	0.40	0.62	0.70	0.63	0.88

(Continued)

Material	Frequency (Hz)					
	125	*250*	*500*	*1000*	*2000*	*4000*
Loop pile tufted carpet,[2] 1.4 kg/m^2, hair and jute underlay 3 kg/m^2	0.20	0.50	0.68	0.72	0.65	0.90
Loop pile tufted carpet, 1.4 kg/m^2, no underlay[2]	0.04	0.08	0.17	0.33	0.59	0.75
Loop pile tufted carpet, 0.7 kg/m^2, 1.4 kg/m^2 hair underlay pad[2]	0.10	0.19	0.35	0.79	0.69	0.79
16 mm wool pile with underlay[1]	0.20	0.25	0.35	0.40	0.50	0.75
9.5 mm wool pile no underlay on concrete[1]	0.09	0.08	0.21	0.26	0.27	0.37
Cord carpet[3]	0.05	0.05	0.10	0.20	0.45	0.65
Thin (6 mm) carpet on underlay[7]	0.03	0.09	0.20	0.54	0.70	0.72
6 mm pile carpet bonded to closed-cell foam underlay[7]	0.03	0.09	0.25	0.31	0.33	0.44
Thick (9 mm) carpet on underlay[2]	0.08	0.08	0.30	0.60	0.75	0.80
Needle felt 5 mm stuck to concrete[8,9]	0.01	0.02	0.05	0.15	0.30	0.40
Thin carpet cemented to concrete[10]	0.02	0.04	0.08	0.2	0.35	0.4
Other floors						
Wood block/lino/rubber flooring[6]	0.02	0.04	0.05	0.05	0.1	0.05
Parquet fixed with asphalt, on concrete[1]	0.04	0.04	0.07	0.06	0.06	0.07
Wood on solid floor[1]	0.04	0.04	0.03	0.03	0.03	0.02
Floors, wood[2]	0.15	0.11	0.10	0.07	0.06	0.07
Wood platform, large airspace below[1]	0.40	0.30	0.20	0.17	0.15	0.10
Floor boards on joist floor[6]	0.15	0.20	0.10	0.10	0.10	0.10
Floors, concrete or terrazzo[2,11]	0.01	0.01	0.015	0.02	0.02	0.02
Concrete floor[10]	0.01	0.02	0.02	0.02	0.02	0.02
Linoleum or vinyl stuck to concrete[9,12]	0.02	0.02	0.03	0.04	0.04	0.05
Linoleum, asphalt tile, or cork tile on concrete[2,5,13]	0.02	0.03	0.03	0.03	0.03	0.02
Layer of rubber, cork, linoleum and underlay, or vinyl and underlay, stuck to concrete[9,14]	0.02	0.02	0.04	0.05	0.05	0.10
Cork, lino or rubber tile on solid floor[1]	0.04	0.03	0.04	0.04	0.03	0.02
25 mm cork on solid backing	0.05	0.1	0.2	0.55	0.6	0.55
Slate[1]	0.01	0.01	0.01	0.02	0.02	0.02
Theatre seating, unoccupied						
Beranek's values[15]	0.19	0.37	0.56	0.67	0.61	0.59
Average of nine modern seating designs, 0.9 m row spacing[16]	0.34	0.46	0.64	0.71	0.77	0.85
One seat type, 0.8 m row spacing[16]	0.29	0.39	0.61	0.74	0.83	0.88
Same seat as line above, 0.9 m row spacing[16]	0.25	0.35	0.58	0.70	0.78	0.84
Same seat as two lines above, 1 m row spacing[16]	0.23	0.34	0.52	0.65	0.73	0.75
Upholstered seating[6]	0.45	0.60	0.73	0.80	0.75	0.64
Upholstered seating, well upholstered[17]	0.44	0.60	0.77	0.89	0.82	0.70
Upholstered seating, leather covered[17]	0.40	0.50	0.58	0.61	0.58	0.50
Seating, occupied						
Occupied theatre seating average from References 1 and 16	0.41	0.58	0.80	0.90	0.92	0.89
Audience on timber seats (1/m^2)[2]	0.16	0.24	0.56	0.69	0.81	0.78
Audience on timber seats (2/m^2)[2]	0.24	0.4	0.78	0.98	0.96	0.87
Orchestra with instruments (1.5 m^2/person)[2]	0.27	0.53	0.67	0.93	0.87	0.8
Wooden pews (100% occupancy)[17]	0.57	0.61	0.75	0.86	0.91	0.86

(Continued)

Material	Frequency (Hz)					
	125	*250*	*500*	*1000*	*2000*	*4000*
Wooden chairs (100% occupancy)[17]	0.60	0.74	0.88	0.96	0.93	0.85
Wooden pews (75% occupancy)[17]	0.46	0.56	0.65	0.75	0.72	0.65
Standing audience						
2.7 people/m² (see Reference 18)	0.23	0.46	0.95	1.21	1.20	1.14
Miscellaneous						
Water surface in swimming pool[19]	0.01	0.01	0.01	0.01	0.02	0.02
Water surface in swimming pool[2]	0.008	0.008	0.013	0.015	0.02	0.025
Marble or glazed tile[2]	0.01	0.01	0.01	0.01	0.02	0.02
Solid wooden door[9,14]	0.14	0.10	0.06	0.08	0.10	0.10
Ventilation grille[8,9]	0.60	0.60	0.60	0.60	0.60	0.60
Egg boxes[20]	0.01	0.07	0.43	0.62	0.51	0.70
Anechoic chamber wall (wedges)	0.997	0.997	0.997	0.997	0.997	0.997
Wood						
Plywood panelling, 1 cm thick[2,11]	0.28	0.22	0.17	0.09	0.1	0.11
22 mm chipboard, 50 mm cavity filled with mineral wool[9,14]	0.12	0.04	0.06	0.05	0.05	0.05
3–4 mm plywood sheets, >75 mm cavity with 25–50 mm mineral wool[8,9]	0.50	0.30	0.10	0.05	0.05	0.05
Plywood/hardwood, air space[6]	0.32	0.43	0.12	0.07	0.07	0.11
6 mm wood fibreboard on laths, cavity >100 mm deep[9,14]	0.30	0.20	0.20	0.10	0.05	0.05
Fibreboard, solid backing[6]	0.05	0.1	0.15	0.25	0.3	0.3
Fibreboard, 25 mm air space[6]	0.3	0.3	0.3	0.3	0.3	0.3
9.5–12.7 mm wood panelling, 5–10 cm air space behind[1]	0.30	0.25	0.20	0.17	0.15	0.10
Wood, 50 mm thick	0.01	0.05	0.05	0.04	0.04	0.04
Concrete						
Rough concrete[21]	0.02	0.03	0.03	0.03	0.04	0.07
Smooth unpainted concrete[9,14]	0.01	0.01	0.02	0.02	0.02	0.05
Smooth concrete, painted or glazed[9,14]	0.01	0.01	0.01	0.02	0.02	0.02
Concrete block, coarse[2]	0.36	0.44	0.31	0.29	0.39	0.25
Concrete block, painted[2,5,13]	0.10	0.05	0.06	0.07	0.09	0.08
Porous concrete blocks without surface finish,[9] 400–800 kg/m³	0.05	0.05	0.05	0.08	0.14	0.20
Clinker concrete, no surface finish,[8,9] 800 kg/m³	0.10	0.20	0.40	0.60	0.50	0.60
Bricks and blocks						
Brick, unglazed[2]	0.03	0.03	0.03	0.04	0.05	0.07
Brickwork, plain painted[6]	0.05	0.04	0.02	0.04	0.05	0.05
Smooth brickwork with flush pointing, painted[19]	0.01	0.01	0.02	0.02	0.02	0.02
Brick, unglazed, painted[2]	0.01	0.01	0.02	0.02	0.02	0.03
Smooth brickwork with flush pointing[9,14]	0.02	0.03	0.03	0.04	0.05	0.07
Smooth brickwork, 10 mm deep pointing, pit sand mortar[8,9]	0.08	0.09	0.12	0.16	0.22	0.24
Breeze block[6]	0.2	0.3	0.6	0.6	0.5	0.5
Plaster						
Lime cement plaster[14]	0.02	0.02	0.03	0.04	0.05	0.05
Glaze plaster[9,14]	0.01	0.01	0.01	0.02	0.02	0.02

(Continued)

Material	Frequency (Hz)					
	125	250	500	1000	2000	4000
Painted plaster surface[8,9]	0.02	0.02	0.02	0.02	0.02	0.02
Plaster with wallpaper on backing paper[9,14]	0.02	0.03	0.04	0.05	0.07	0.08
Plaster, gypsum, or lime, rough finish on lath[10,22]	0.02	0.03	0.04	0.05	0.04	0.03
Plaster, gypsum, or lime, smooth finish on lath[2]	0.14	0.1	0.06	0.04	0.04	0.03
Plaster, gypsum, or lime, smooth finish on lath[10,22]	0.02	0.02	0.03	0.04	0.04	0.03
Plaster, on laths/studs, air space[6]	0.3	0.1	0.1	0.05	0.04	0.05
Plaster, gypsum, or lime, smooth finish on tile or brick[2]	0.013	0.015	0.02	0.03	0.04	0.05
Plaster, lime, or gypsum on solid backing[6]	0.03	0.03	0.02	0.03	0.04	0.05
Acoustics plaster[6]	0.30	0.35	0.5	0.7	0.7	0.7
Acoustics plaster, 40 mm thick[23]	0.31	0.55	0.84	0.78	0.71	0.54
Acoustics plaster, 68 mm thick[23]	0.47	0.74	0.76	0.65	0.62	0.49
Plasterboard						
Gypsum board, 1.27 cm nailed to studs with 4.1 m c-t-c[2]	0.29	0.1	0.05	0.04	0.07	0.09
Plasterboard on frame, 9.5 mm boards, 10 cm empty cavity[9,24]	0.11	0.13	0.05	0.03	0.02	0.03
Plasterboard on frame, 9.5 mm boards, 10 cm cavity filled with mineral wool[9,24]	0.28	0.14	0.09	0.06	0.05	0.05
Plasterboard on frame, 13 mm boards, 10 cm empty cavity[9,24]	0.08	0.11	0.05	0.03	0.02	0.03
Plasterboard on frame, 13 mm boards, 10 cm cavity filled with mineral wool[9,24]	0.30	0.12	0.08	0.06	0.06	0.05
2 × 13 mm plasterboard on steel frame, 5 cm mineral wool in cavity, surface painted[9,12]	0.15	0.10	0.06	0.04	0.04	0.05
Glazing						
Glass, ordinary window glass[2,11]	0.35	0.25	0.18	0.12	0.07	0.04
Single pane of glass,[6] 3–4 mm	0.2	0.15	0.1	0.07	0.05	0.05
Single pane of glass,[6] >4 mm	0.1	0.07	0.04	0.03	0.02	0.02
Single pane of glass, 3 mm[9,24]	0.08	0.04	0.03	0.03	0.02	0.02
Double glazing, 2–3 mm glass, 1 cm gap[8,9]	0.10	0.07	0.05	0.03	0.02	0.02
Double glazing, 2–3 mm glass, >3 cm gap[9,24]	0.15	0.05	0.03	0.03	0.02	0.02
Glass, large panes, heavy glass[2,5,13]	0.18	0.06	0.04	0.03	0.02	0.02
Wools and foam						
25 mm fibreglass, rigid backing[25]	0.08	0.25	0.45	0.75	0.75	0.65
2.54 cm fibreglass,[2] 24 to 48 kg/m³	0.08	0.25	0.65	0.85	0.8	0.75
2.5 cm fibreglass, 2.5 cm airspace[2]	0.15	0.55	0.8	0.9	0.85	0.8
5 cm fibreglass, rigid backing[25]	0.21	0.50	0.75	0.90	0.85	0.80
7.5 cm fibreglass, rigid backing[25]	0.35	0.65	0.80	0.90	0.85	0.80
10 cm fibreglass, rigid backing[25]	0.45	0.90	0.95	1.00	0.95	0.85
5 cm mineral wool (40 kg/m³), glued to wall, untreated surface[8,9]	0.15	0.70	0.60	0.60	0.85	0.90
5 cm mineral wool (40 kg/m³), glued to wall, surface sprayed with thin plastic solution[8,9]	0.15	0.70	0.60	0.60	0.75	0.75
5 cm mineral wool (70 kg/m³) 30 cm in front of wall[8,9]	0.70	0.45	0.65	0.60	0.75	0.65

(Continued)

Material	*Frequency (Hz)*					
	125	*250*	*500*	*1000*	*2000*	*4000*
5 cm wood-wool set in mortar[8,9]	0.08	0.17	0.35	0.45	0.65	0.65
5.1 cm fibreglass, panels with plastic sheet wrapping and perforated metal facing[2]	0.33	0.79	0.99	0.91	0.76	0.64
5.1 cm fibreglass,[2] 24–48 kg/m^3	0.17	0.55	0.8	0.9	0.85	0.8
Acoustic tile, 1.27 cm thick[5]	0.07	0.21	0.66	0.75	0.62	0.49
Acoustic tile, 1.9 cm thick[5]	0.09	0.28	0.78	0.84	0.73	0.64
Polyurethane foam, 2.5 cm thick	0.16	0.25	0.45	0.84	0.97	0.87
Thermafleece, sheep wool absorbent 100 mm thick[26]	0.47	0.86	1.00	0.94	0.96	1.02
Ballast						
Ballast or other crushed stone, 3.18 cm, 15.2 deep[2]	0.19	0.23	0.43	0.37	0.58	0.62
Ballast or other crushed stone, 3.18 cm, 30.5 cm deep[2]	0.27	0.58	0.48	0.54	0.73	0.63
Ballast or other crushed stone, 3.18 cm, 45.7 cm deep[2]	0.41	0.53	0.64	0.84	0.91	0.63
Ballast or other crushed stone, 0.64 cm, 15.2 cm deep[2,11]	0.22	0.64	0.7	0.79	0.88	0.72
Microperforated absorber						
4 cm cavity[23]	0.08	0.27	0.70	0.35	0.11	0.04
40 cm cavity[23]	0.64	0.56	0.41	0.28	0.13	0.06
Diffusers						
Hybrid absorber-diffuser (BAD panel mounted on 2.5 cm fibreglass)[23]	0.17	0.40	0.86	1.00	0.84	0.61
2D $N = 7$ QRD, design freq. = 500 Hz[23]	0.14	0.12	0.14	0.20	0.09	0.12
2D $N = 7$ QRD as line above, with cloth covering[23]	0.16	0.17	0.28	0.41	0.26	0.3
1D $N = 7$ QRD, design freq. = 500 Hz[23]	0.11	0.1	0.07	0.08	0.06	0.06
1D $N = 7$ QRD as line above, with cloth covering[23]	0.13	0.14	0.2	0.24	0.20	0.23
Green wall systems						
Data from Azkorra et al.[27]	0.46	0.42	0.36	0.38	0.45	0.50
Data from Wong et al.[28] (100% greenery)	0.09	0.23	0.43	0.46	0.50	0.48
Data from Yang et al.[29]	0.61	0.62	0.69	0.68	0.68	0.72
Top soil with different percentage of vegetative cover[29]						
0%, bare soil	0.27	0.57	0.76	0.90	0.89	0.84
20%	0.34	0.63	0.79	0.92	0.89	0.81
40%	0.39	0.68	0.83	0.95	0.90	0.83
60%	0.45	0.72	0.85	0.97	0.90	0.80
80%	0.46	0.73	0.85	0.97	0.89	0.72
100%, completely covered	0.49	0.75	0.89	0.98	0.91	0.73
Top soil with different moisture content[29]						
12.5%	0.26	0.55	0.73	0.89	0.85	0.67
17%	0.25	0.51	0.69	0.81	0.78	0.57
20.4%	0.23	0.45	0.57	0.64	0.58	0.38
23.8%	0.20	0.36	0.43	0.46	0.38	0.20
25.4%	0.14	0.28	0.33	0.34	0.29	0.12
34.1%	0.07	0.22	0.23	0.24	0.20	0.06

REFERENCES

1. L. L. Beranek, *Acoustics*, McGraw-Hill, New York (1954).
2. C. M. Harris (Ed.), *Handbook of Noise Control*, 2nd edn, McGraw-Hill, New York (1991).
3. D. Templeton (Ed.), *Acoustics in the Built Environment*, 2nd edn, Architectural Press, Oxford, UK (1997).
4. V. S. Mankovsky, *Acoustics of Studio and Auditoria*, Focal Press, New York (1971).
5. F. Alton Everest, *Master Handbook of Acoustics*, 4th edn, McGraw-Hill, New York (2001).
6. A. Fry (Ed.), *Noise Control in Building Services*, Pergamon Press, Oxford, UK (1987).
7. P. H. Parkin, H.R. Humphreys, and J. R. Cowell, *Acoustics, Noise and Buildings*, Faber and Faber, London (1979).
8. J. Kristensen, "Sound Absorption Coefficients—Measurement, evaluation, application", Note 45, Statens Byggeforskningsinstitut, Hrsholm (1984) (in Danish).
9. C. Lynge, *ODEON Room Acoustics Program, User Manual* DTU, Denmark (2001).
10. L. L. Beranek and T. Hidaka, "Sound absorption in concert halls by seats, occupied and unoccupied, and by the hall's interior surfaces", *J. Acoust. Soc. Am.*, **104**(6), 3169–77 (1998).
11. Physikalisch-Technische Bundesanstalt, http://www.ptb.de/en/index.html, accessed 2003.
12. J. Petersen, "Rumakustik", SBI-anvisning 137, Statens Byggeforskningsinstitut, Hrsholm (1983).
13. R. W. Young, "Sabine reverberation and sound power calculations", *J. Acoust. Soc. Am.*, **31**, 912–21 (1959).
14. H. W. Bobran, *Handbuch der bauphysik*, Verlag Ulstein, Berlin (1973).
15. L. L. Beranek, "Audience and chair absorption in large halls. II", *J. Acoust. Soc. Am.*, **45**, 13–9 (1969).
16. W. J. Davies, R. J. Orlowski, and Y. W. Lam, "Measuring auditorium seat absorption", *J. Acoust. Soc. Am.*, **96**, 879–88 (1994).
17. D. A. Bies and C. H. Hansen, *Engineering Noise Control: Theory and Practice*, 2nd edn, E&FN Spon, London (1996).
18. N. W. Adelman-Larsen, E. R. Thompson, and A. C. Gade, "Suitable reverberation times for halls for rock and pop music", *J. Acoust. Soc. Am.*, **127**(1), 247–55 (2010).
19. V. O. Knudsen and C. M. Harris, *Acoustical Designing in Architecture*, John Wiley, New York (1953).
20. Riverbank Acoustical Laboratories of IIT Research Institute, Report, http://www.acoustics first.com/docs/egg.pdf, accessed 1 April 2008.
21. ISO/TR 11690-3, "Acoustics—Recommended practice for design of low-noise workplaces containing machinery—Part 3: Sound propagation and noise predictions in workrooms" (1997).
22. D. Davis and C. Davis, *Sound System Engineering*, Focal Press, Indianapolis, IN (1997).
23. RPG Diffusor Systems Inc., http://www.rpginc.com, accessed 2003.
24. W. Fasold and H. Winkler, *Bauphysikalische entwurfslehre, band 4: Bauakustik*, VEB Verlag für Bauwesen, Berlin (1976).
25. L. E. Kinsler, A. R. Frey, A. B. Coppens, and J. V. Sanders, *Fundamentals of acoustics*, 4th edn, John Wiley & Sons (2000).
26. Greenshop, http://www.greenshop.co.uk/documents/Thermafleece/Thermafleece_acoustic.pdf, accessed 1 April 2008.
27. Z. Azkorra, G. Pérez, J. Coma, L. F. Cabeza, S. Bures, J. E. Álvaro, A. Erkoreka, and M. Urrestarazu, "Evaluation of green walls as a passive acoustic insulation system for buildings", *Appl. Acoust.*, **89**, 46–56 (2015).
28. N. H. Wong, A. Y. K. Tan, P. Y. Tan, K. Chiang, and N. C. Wong, "Acoustics evaluation of vertical greenery systems for building walls", *Build. Environ.*, **45**(2), 411–20 (2010).
29. H. S. Yang, J. Kang, and C. Cheal, "Random-incidence absorption and scattering coefficients of vegetation", *Acta Acust. Acust.*, **99**(3), 379–88 (2013).

Appendix B: Normalized diffusion coefficient table

See Section 5.2.5 for how this was constructed.

Surface	Angle of incidence (°)	Frequency (Hz) 100	125	160	200	250	315	400	500	630	800	1000	1250	1600	2000	2500	3150	4000	5000
1. Effect of changing diffuser periodicity and width. Semicylinder(s) non-absorbing surfaces, radius 0.3 m (1 cm flat section between each period)																			
1 period, 0.61 cm wide	0	0.02	0.00	0.00	0.00	0.65	0.96	0.92	0.96	0.97	0.90	0.93	0.95	0.94	0.95	0.97	0.98	0.98	0.98
	57	0.06	0.07	0.16	0.38	0.59	0.18	0.43	0.37	0.55	0.53	0.64	0.70	0.74	0.77	0.80	0.82	0.85	0.86
	Random	0.06	0.07	0.11	0.28	0.56	0.50	0.56	0.66	0.80	0.76	0.77	0.80	0.82	0.82	0.84	0.85	0.86	0.87
2 periods, 1.22 m wide	0	0.16	0.18	0.15	0.06	0.02	0.17	0.60	0.62	0.71	0.43	0.50	0.72	0.65	0.77	0.77	0.73	0.77	0.80
	57	0.15	0.12	0.18	0.53	0.40	0.50	0.30	0.26	0.66	0.69	0.66	0.69	0.72	0.73	0.76	0.76	0.80	0.83
	Random	0.23	0.16	0.09	0.26	0.32	0.38	0.38	0.33	0.64	0.55	0.62	0.69	0.71	0.73	0.74	0.74	0.78	0.79
4 cylinders, 2.44 m wide	0	0.02	0.00	0.01	0.05	0.03	0.04	0.13	0.58	0.32	0.16	0.24	0.38	0.31	0.48	0.44	0.41	0.43	0.54
	57	0.04	0.18	0.21	0.19	0.29	0.34	0.12	0.02	0.49	0.31	0.33	0.49	0.51	0.53	0.64	0.66	0.64	0.66
	Random	0.00	0.05	0.03	0.07	0.09	0.18	0.19	0.19	0.38	0.29	0.34	0.45	0.47	0.49	0.54	0.59	0.61	0.62
6 periods, 3.66 m wide	0	0.00	0.02	0.02	0.01	0.02	0.02	0.05	0.21	0.22	0.10	0.19	0.21	0.26	0.39	0.32	0.38	0.49	0.72
	57	0.07	0.16	0.16	0.10	0.18	0.32	0.09	0.03	0.40	0.18	0.25	0.43	0.42	0.46	0.59	0.58	0.60	0.66
	Random	0.00	0.00	0.00	0.00	0.07	0.14	0.14	0.12	0.28	0.19	0.26	0.36	0.40	0.42	0.48	0.55	0.60	0.65
12 periods, 7.32 m wide	0	0.00	0.00	0.01	0.01	0.01	0.01	0.01	0.04	0.14	0.08	0.17	0.48	0.35	0.63	0.57	0.72	0.66	0.81
	57	0.06	0.06	0.06	0.03	0.04	0.32	0.13	0.07	0.30	0.09	0.21	0.40	0.42	0.49	0.55	0.58	0.68	0.67
	Random	0.00	0.00	0.00	0.00	0.03	0.10	0.10	0.09	0.20	0.12	0.22	0.38	0.42	0.46	0.56	0.64	0.66	0.70
2. Effect of surface depth, six semiellipses, non-absorbing, each width 0.6 m, total with 3.66 m (1 cm flat section between semiellipses)																			
1 cm deep	0	0.02	0.02	0.01	0.00	0.00	0.00	0.00	0.00	0.00	0.00	0.00	0.00	0.00	0.00	0.00	0.01	0.02	0.03
	57	0.01	0.03	0.02	0.00	0.00	0.00	0.00	0.00	0.00	0.00	0.00	0.00	0.00	0.00	0.00	0.01	0.02	0.03
	Random	0.01	0.02	0.02	0.00	0.01	0.01	0.00	0.00	0.00	0.00	0.00	0.00	0.00	0.00	0.01	0.01	0.01	0.02
2 cm deep	0	0.02	0.02	0.01	0.00	0.00	0.00	0.00	0.00	0.00	0.00	0.00	0.01	0.01	0.01	0.02	0.04	0.08	0.13
	57	0.01	0.03	0.02	0.00	0.00	0.00	0.00	0.00	0.00	0.00	0.00	0.00	0.01	0.01	0.01	0.03	0.08	0.10
	Random	0.01	0.03	0.02	0.01	0.02	0.03	0.02	0.01	0.02	0.01	0.01	0.01	0.01	0.02	0.03	0.04	0.06	0.08

(Continued)

Surface	Angle of incidence (°)	Frequency (Hz) 100	125	160	200	250	315	400	500	630	800	1000	1250	1600	2000	2500	3150	4000	5000
5 cm deep	0	0.01	0.01	0.01	0.00	0.00	0.00	0.00	0.01	0.03	0.03	0.04	0.07	0.10	0.14	0.24	0.24	0.34	0.36
	57	0.00	0.03	0.02	0.00	0.00	0.05	0.04	0.03	0.03	0.03	0.03	0.04	0.05	0.06	0.10	0.12	0.15	0.15
	Random	0.01	0.03	0.03	0.02	0.05	0.06	0.04	0.03	0.04	0.04	0.03	0.05	0.08	0.12	0.14	0.15	0.20	0.25
10 cm deep	0	0.01	0.01	0.01	0.00	0.00	0.00	0.00	0.05	0.26	0.22	0.19	0.32	0.26	0.32	0.57	0.50	0.70	0.73
	57	0.00	0.03	0.02	0.00	0.02	0.22	0.15	0.12	0.14	0.14	0.16	0.17	0.19	0.24	0.31	0.40	0.41	0.40
	Random	0.02	0.06	0.07	0.06	0.06	0.10	0.07	0.05	0.11	0.15	0.19	0.23	0.25	0.29	0.39	0.40	0.47	0.51
20 cm deep	0	0.00	0.01	0.01	0.00	0.01	0.01	0.02	0.28	0.41	0.18	0.02	0.16	0.28	0.27	0.57	0.58	0.58	0.54
	57	0.00	0.03	0.03	0.04	0.09	0.36	0.16	0.03	0.22	0.27	0.34	0.31	0.29	0.38	0.47	0.53	0.54	0.59
	Random	0.00	0.00	0.00	0.01	0.06	0.11	0.11	0.11	0.20	0.25	0.23	0.31	0.32	0.37	0.45	0.50	0.56	0.61
30 cm deep (semicylinders)	0	0.00	0.02	0.02	0.01	0.02	0.02	0.05	0.21	0.22	0.10	0.19	0.21	0.26	0.39	0.32	0.38	0.49	0.72
	57	0.07	0.16	0.16	0.10	0.18	0.32	0.09	0.03	0.40	0.18	0.25	0.43	0.42	0.46	0.59	0.58	0.60	0.66
	Random	0.00	0.00	0.00	0.00	0.07	0.14	0.14	0.12	0.28	0.19	0.26	0.36	0.40	0.42	0.48	0.55	0.60	0.65
3. Triangles, non-absorbing, 3.66 m wide (0.01 cm flat section between each period)																			
15 periods, 60° angle	0	0.00	0.00	0.00	0.00	0.00	0.00	0.00	0.00	0.00	0.00	0.00	0.00	0.05	0.01	0.00	0.01	0.01	0.01
	57	0.07	0.09	0.07	0.05	0.09	0.06	0.06	0.11	0.07	0.16	0.05	0.04	0.23	0.17	0.15	0.23	0.24	0.23
	Random	0.00	0.00	0.00	0.00	0.00	0.00	0.00	0.06	0.05	0.08	0.05	0.05	0.13	0.11	0.15	0.15	0.14	0.14
9 periods, 45° angle	0	0.00	0.00	0.00	0.00	0.00	0.00	0.00	0.00	0.01	0.03	0.03	0.00	0.02	0.02	0.01	0.01	0.01	0.01
	57	0.00	0.03	0.03	0.01	0.00	0.02	0.26	0.19	0.05	0.13	0.26	0.15	0.22	0.20	0.15	0.07	0.06	0.07
	Random	0.00	0.00	0.00	0.00	0.00	0.00	0.12	0.11	0.07	0.13	0.16	0.14	0.18	0.20	0.16	0.13	0.15	0.10
6 periods, 30° angle	0	0.00	0.00	0.00	0.00	0.00	0.00	0.00	0.07	0.39	0.21	0.22	0.35	0.40	0.33	0.36	0.33	0.22	0.17
	57	0.00	0.01	0.02	0.01	0.08	0.32	0.15	0.09	0.21	0.18	0.16	0.20	0.15	0.17	0.16	0.14	0.09	0.05
	Random	0.00	0.00	0.00	0.00	0.04	0.09	0.08	0.09	0.17	0.18	0.19	0.22	0.24	0.24	0.23	0.20	0.16	0.12
3 periods, 18° angle	0	0.01	0.01	0.01	0.01	0.07	0.23	0.33	0.28	0.35	0.32	0.30	0.30	0.28	0.23	0.19	0.15	0.12	0.10
	57	0.02	0.15	0.34	0.29	0.19	0.19	0.16	0.18	0.16	0.16	0.16	0.17	0.19	0.17	0.13	0.10	0.09	0.02
	Random	0.05	0.08	0.11	0.10	0.08	0.10	0.15	0.19	0.20	0.20	0.21	0.21	0.20	0.17	0.15	0.12	0.10	0.07

(Continued)

Surface	Angle of incidence (°)	Frequency (Hz)																	
		100	*125*	*160*	*200*	*250*	*315*	*400*	*500*	*630*	*800*	*1000*	*1250*	*1600*	*2000*	*2500*	*3150*	*4000*	*5000*
4. Semiellipses mounted on 3.63 m wide flat baffle, non-absorbing, each semiellipse 0.6 m wide, 0.2 m deep																			
One semiellipse in middle of baffle	0	0.00	0.00	0.03	0.09	0.11	0.10	0.17	0.14	0.08	0.02	0.02	0.04	0.02	0.03	0.03	0.01	0.01	0.04
	57	0.11	0.27	0.23	0.29	0.31	0.29	0.26	0.26	0.22	0.16	0.11	0.08	0.06	0.09	0.15	0.15	0.07	0.06
	Random	0.08	0.13	0.15	0.19	0.22	0.22	0.22	0.24	0.21	0.17	0.15	0.14	0.16	0.16	0.17	0.13	0.11	0.10
Three semiellipses with 0.6 m flat section between	0	0.01	0.02	0.01	0.02	0.12	0.50	0.33	0.28	0.30	0.10	0.12	0.16	0.17	0.10	0.08	0.16	0.17	0.25
	57	0.05	0.31	0.43	0.28	0.23	0.38	0.23	0.33	0.31	0.40	0.52	0.42	0.20	0.22	0.44	0.41	0.27	0.67
	Random	0.05	0.13	0.14	0.14	0.17	0.27	0.24	0.31	0.31	0.29	0.25	0.33	0.33	0.32	0.34	0.37	0.34	0.35
5. Optimized curved surfaces, 3.6 m wide, modulated arrays																			
3 periods, 30 cm deep	0	0.09	0.10	0.06	0.13	0.54	0.49	0.28	0.39	0.53	0.28	0.35	0.52	0.63	0.42	0.73	0.74	0.31	0.79
	57	0.21	0.51	0.50	0.33	0.20	0.23	0.12	0.16	0.31	0.60	0.59	0.52	0.59	0.58	0.56	0.57	0.55	0.47
	Random	0.16	0.16	0.19	0.20	0.20	0.23	0.18	0.27	0.38	0.44	0.48	0.54	0.59	0.54	0.55	0.59	0.56	0.59
6 periods, 20 cm deep	0	0.06	0.04	0.02	0.01	0.00	0.01	0.02	0.28	0.52	0.35	0.35	0.29	0.18	0.38	0.15	0.31	0.33	0.54
	57	0.08	0.06	0.05	0.00	0.08	0.38	0.21	0.11	0.31	0.44	0.53	0.60	0.51	0.64	0.57	0.54	0.52	0.58
	Random	0.09	0.04	0.03	0.00	0.05	0.13	0.16	0.18	0.29	0.42	0.39	0.45	0.47	0.56	0.50	0.53	0.56	0.56
6 periods, 10 cm deep	0	0.04	0.02	0.01	0.02	0.01	0.01	0.03	0.04	0.05	0.05	0.05	0.41	0.56	0.30	0.64	0.29	0.63	0.63
	57	0.07	0.10	0.09	0.09	0.10	0.08	0.11	0.10	0.41	0.44	0.49	0.47	0.52	0.46	0.46	0.36	0.42	0.41
	Random	0.07	0.10	0.11	0.10	0.10	0.11	0.13	0.10	0.20	0.27	0.33	0.44	0.47	0.44	0.47	0.46	0.48	0.51
6 periods, 5 cm deep	0	0.06	0.03	0.02	0.01	0.00	0.00	0.00	0.01	0.00	0.00	0.01	0.10	0.16	0.29	0.45	0.56	0.51	0.64
	57	0.07	0.06	0.05	0.01	0.01	0.01	0.00	0.01	0.12	0.15	0.12	0.09	0.10	0.13	0.18	0.29	0.29	0.33
	Random	0.07	0.05	0.04	0.02	0.04	0.04	0.03	0.05	0.10	0.12	0.09	0.10	0.13	0.21	0.30	0.36	0.39	0.45

(Continued)

Surface	Angle of incidence (°)	Frequency (Hz)																	
		100	125	160	200	250	315	400	500	630	800	1000	1250	1600	2000	2500	3150	4000	5000
6. Hybrid surfaces, 3.6 m wide, modulated arrays																			
3 periods, flat hybrid surface	0	0.07	0.02	0.02	0.01	0.00	0.00	0.00	0.00	0.00	0.00	0.00	0.00	0.01	0.01	0.01	0.01	0.01	0.02
	57	0.13	0.08	0.06	0.02	0.01	0.01	0.01	0.01	0.00	0.01	0.02	0.03	0.06	0.09	0.08	0.14	0.29	0.36
	Random	0.15	0.11	0.09	0.06	0.04	0.04	0.02	0.01	0.01	0.01	0.01	0.03	0.07	0.10	0.11	0.13	0.13	0.16
3 periods, curved, 2.5 cm deep	0	0.06	0.02	0.02	0.00	0.00	0.00	0.00	0.01	0.01	0.01	0.01	0.03	0.04	0.05	0.06	0.15	0.22	0.28
	57	0.12	0.08	0.04	0.01	0.01	0.04	0.03	0.04	0.05	0.05	0.08	0.11	0.19	0.21	0.19	0.28	0.29	0.49
	Random	0.15	0.10	0.08	0.05	0.04	0.07	0.05	0.05	0.07	0.06	0.06	0.10	0.15	0.16	0.19	0.24	0.28	0.32
3 periods, curved, 7.5 cm deep	0	0.05	0.01	0.02	0.01	0.00	0.00	0.01	0.03	0.05	0.08	0.24	0.39	0.52	0.52	0.24	0.32	0.67	0.56
	57	0.11	0.07	0.03	0.01	0.01	0.11	0.09	0.14	0.39	0.30	0.33	0.44	0.43	0.41	0.40	0.40	0.42	0.40
	Random	0.15	0.10	0.06	0.04	0.04	0.10	0.09	0.11	0.21	0.20	0.24	0.36	0.42	0.40	0.33	0.40	0.49	0.47
7. Schroeder diffusers, 3.6 m wide																			
N = 7 QRD, 6 periods, 0.2 m deep	0	0.07	0.01	0.02	0.00	0.01	0.01	0.01	0.07	0.16	0.21	0.12	0.10	0.07	0.23	0.39	0.04	0.19	0.27
	57	0.13	0.14	0.11	0.05	0.11	0.02	0.10	0.24	0.37	0.28	0.32	0.12	0.31	0.45	0.43	0.32	0.37	0.60
	Random	0.07	0.04	0.00	0.00	0.00	0.00	0.04	0.22	0.25	0.22	0.23	0.09	0.23	0.35	0.36	0.23	0.25	0.42
Optimized profiled diffuser, modulated array, 6 periods, 8 wells/period, 0.17 m deep	0	0.07	0.02	0.01	0.00	0.06	0.07	0.07	0.30	0.58	0.40	0.41	0.49	0.59	0.38	0.33	0.38	0.30	0.33
	57	0.12	0.11	0.16	0.10	0.08	0.14	0.32	0.33	0.49	0.53	0.39	0.46	0.57	0.48	0.57	0.54	0.58	0.55
	Random	0.08	0.05	0.06	0.00	0.00	0.07	0.18	0.27	0.43	0.47	0.37	0.41	0.51	0.43	0.43	0.48	0.42	0.46
N = 7 PRD, 6 periods, 6 wells/period, 0.2 m deep	0	0.07	0.01	0.02	0.00	0.01	0.02	0.04	0.03	0.06	0.33	0.13	0.14	0.20	0.27	0.10	0.35	0.21	0.06
	57	0.15	0.15	0.12	0.08	0.07	0.24	0.31	0.25	0.21	0.26	0.15	0.20	0.34	0.34	0.19	0.27	0.32	0.19
	Random	0.09	0.04	0.00	0.00	0.00	0.10	0.15	0.15	0.19	0.19	0.17	0.20	0.22	0.32	0.22	0.29	0.32	0.25
Diffractal, 1 period, three orders of size, N = 7 (largest order only 6 wells), 0.5 m deep	0	0.06	0.25	0.10	0.27	0.30	0.36	0.56	0.31	0.39	0.40	0.42	0.52	0.56	0.49	0.49	0.40	0.47	0.54
	57	0.40	0.35	0.43	0.46	0.44	0.56	0.77	0.60	0.66	0.67	0.56	0.43	0.62	0.67	0.66	0.56	0.16	0.69
	Random	0.19	0.20	0.29	0.35	0.36	0.44	0.61	0.45	0.51	0.51	0.49	0.40	0.54	0.53	0.50	0.47	0.35	0.53
Optimized diffuser, modulated array, 6 periods, 12 wells/period, 0.17 m deep	0	0.06	0.02	0.02	0.00	0.04	0.05	0.20	0.61	0.58	0.51	0.58	0.32	0.31	0.49	0.37	0.30	0.42	0.53
	57	0.09	0.10	0.12	0.07	0.24	0.53	0.44	0.74	0.54	0.43	0.66	0.52	0.60	0.52	0.48	0.41	0.48	0.72
	Random	0.04	0.02	0.00	0.00	0.06	0.27	0.34	0.58	0.50	0.42	0.56	0.40	0.44	0.48	0.42	0.40	0.45	0.53

Appendix C: Correlation scattering coefficient tables

See Section 5.3.7 for how these were constructed.

Table C.1 3D BEM predictions normal incidence

Surface	Frequency (Hz)												
	250	*315*	*400*	*500*	*630*	*800*	*1000*	*1250*	*1600*	*2000*	*2500*	*3150*	*4000*
Sinusoidal cross-section													
h = 2 cm, L = 20 cm	0.00	0.00	0.00	0.00	0.00	0.00	0.00	0.00	0.02	0.31	0.30	0.54	0.65
h = 4 cm, L = 20 cm	0.00	0.00	0.00	0.00	0.00	0.01	0.01	0.01	0.11	0.90	0.97	0.93	0.95
h = 6 cm, L = 20 cm	0.00	0.00	0.00	0.00	0.01	0.03	0.01	0.04	0.20	0.89	0.56	0.19	1.00
h = 8 cm, L = 20 cm	0.00	0.00	0.00	0.01	0.02	0.03	0.03	0.08	0.15	0.40	0.06	0.24	0.28
h = 10 cm, L = 20 cm	0.00	0.01	0.01	0.03	0.06	0.04	0.05	0.10	0.09	0.05	0.25	0.44	0.32
h/L = 20	0	0	0	0	0	0.01	0.01	0.01	0.11	0.9	0.97	0.93	0.95
Triangular cross-section													
h = 2 cm, L = 20 cm	0.00	0.00	0.00	0.00	0.00	0.00	0.00	0.00	0.01	0.22	0.21	0.39	0.47
h = 4 cm, L = 20 cm	0.00	0.00	0.00	0.00	0.00	0.01	0.00	0.01	0.07	0.75	0.79	0.93	0.99
h = 6 cm, L = 20 cm	0.00	0.00	0.00	0.00	0.01	0.03	0.01	0.02	0.16	1.00	0.88	0.47	1.00
h = 8 cm, L = 20 cm	0.00	0.00	0.01	0.02	0.02	0.03	0.02	0.05	0.18	0.67	0.26	0.13	0.57
h = 10 cm, L = 20 cm	0.01	0.01	0.02	0.03	0.05	0.03	0.03	0.08	0.14	0.23	0.06	0.27	0.23
h/L = 25	0	0	0	0.01	0	0.03	0.01	0.01	0.11	0.95	0.99	0.8	1

(*Continued*)

Table C.1 (Continued) 3D BEM predictions normal incidence

Surface	*Frequency (Hz)*												
	250	*315*	*400*	*500*	*630*	*800*	*1000*	*1250*	*1600*	*2000*	*2500*	*3150*	*4000*
Rectangular cross-section battens													
h = 2 cm, L = 20 cm, w = 10 cm	0.00	0.00	0.00	0.00	0.00	0.00	0.00	0.01	0.14	0.54	0.58	0.86	0.99
h = 4 cm, L = 20 cm, w = 10 cm	0.00	0.00	0.00	0.01	0.01	0.03	0.03	0.09	0.29	0.90	0.83	0.48	0.03
h = 6 cm, L = 20 cm, w = 10 cm	0.00	0.00	0.01	0.03	0.04	0.05	0.12	0.17	0.10	0.56	0.10	0.10	0.93
h = 8 cm, L = 20 cm, w = 10 cm	0.01	0.01	0.04	0.08	0.04	0.11	0.17	0.10	0.05	0.06	0.20	1.00	0.12
h = 10 cm, L = 20 cm, w = 10 cm	0.05	0.04	0.13	0.06	0.08	0.14	0.11	0.05	0.02	0.30	0.98	0.22	0.71
h/L = 15	0	0	0	0	0	0.04	0.01	0.03	0.44	0.83	0.96	0.94	0.44

Source: Lee, H., Sakuma, T., *Appl. Acoust.*, 88, 129–136, 2015.[1]

Table C.2 3D BEM predictions random incidence

Surface	Frequency (Hz)												
	250	*315*	*400*	*500*	*630*	*800*	*1000*	*1250*	*1600*	*2000*	*2500*	*3150*	*4000*
Sinusoidal cross-section													
h = 2 cm, L = 20 cm	0.00	0.00	0.00	0.00	0.00	0.01	0.07	0.12	0.16	0.20	0.25	0.36	0.49
h = 4 cm, L = 20 cm	0.00	0.00	0.01	0.01	0.01	0.03	0.16	0.32	0.47	0.57	0.68	0.81	0.87
h = 6 cm, L = 20 cm	0.01	0.01	0.01	0.02	0.03	0.09	0.21	0.45	0.69	0.78	0.81	0.84	0.82
h = 8 cm, L = 20 cm	0.01	0.02	0.03	0.05	0.08	0.12	0.23	0.49	0.74	0.71	0.72	0.78	0.80
h = 10 cm, L = 20 cm	0.02	0.04	0.05	0.09	0.13	0.15	0.23	0.46	0.59	0.48	0.65	0.70	0.78
h/L = 35	0.09	0.08	0.09	0.11	0.05	0.1	0.22	0.48	0.75	0.79	0.79	0.81	0.82
Triangular cross-section													
h = 2 cm, L = 20 cm	0.00	0.00	0.00	0.00	0.00	0.01	0.06	0.09	0.12	0.15	0.18	0.27	0.37
h = 4 cm, L = 20 cm	0.00	0.00	0.01	0.01	0.01	0.02	0.15	0.26	0.38	0.47	0.56	0.71	0.82
h = 6 cm, L = 20 cm	0.01	0.01	0.01	0.02	0.03	0.07	0.20	0.41	0.61	0.71	0.79	0.87	0.89
h = 8 cm, L = 20 cm	0.02	0.02	0.03	0.04	0.06	0.11	0.23	0.48	0.74	0.76	0.75	0.84	0.87
h = 10 cm, L = 20 cm	0.04	0.03	0.05	0.08	0.12	0.14	0.24	0.49	0.73	0.60	0.64	0.76	0.82
h/L = 40	0.02	0.02	0.03	0.04	0.06	0.11	0.23	0.48	0.74	0.76	0.75	0.84	0.87

(Continued)

Table C.2 (Continued) 3D BEM predictions random incidence

Surface	Frequency (Hz)												
	250	*315*	*400*	*500*	*630*	*800*	*1000*	*1250*	*1600*	*2000*	*2500*	*3150*	*4000*
Rectangular cross-section battens													
h = 2 cm, L = 20 cm, w = 10 cm	0.00	0.00	0.00	0.00	0.00	0.01	0.08	0.19	0.32	0.42	0.47	0.72	0.82
h = 4 cm, L = 20 cm, w = 10 cm	0.01	0.01	0.02	0.02	0.04	0.10	0.24	0.47	0.73	0.82	0.74	0.65	0.48
h = 6 cm, L = 20 cm, w = 10 cm	0.02	0.03	0.04	0.07	0.12	0.16	0.29	0.51	0.70	0.68	0.40	0.24	0.56
h = 8 cm, L = 20 cm, w = 10 cm	0.03	0.04	0.07	0.13	0.15	0.19	0.29	0.44	0.46	0.39	0.31	0.63	0.60
h = 10 cm, L = 20 cm, w = 10 cm	0.05	0.08	0.14	0.18	0.19	0.21	0.25	0.32	0.22	0.37	0.68	0.64	0.47
h/L = 20	0.01	0.01	0.02	0.02	0.04	0.1	0.24	0.47	0.73	0.82	0.74	0.65	0.48
h = 4 cm, L = 20 cm, $w \to$ 0 cm (extremely thin)	0.01	0.01	0.01	0.02	0.03	0.06	0.15	0.14	0.12	0.23	0.21	0.24	0.22
h = 4 cm, L = 20 cm, w = 5 cm	0.01	0.01	0.01	0.02	0.04	0.09	0.22	0.40	0.41	0.41	0.60	0.55	0.41
h = 4 cm, L = 20 cm, w = 10 cm	0.01	0.01	0.02	0.02	0.04	0.10	0.24	0.47	0.73	0.82	0.74	0.65	0.48
h = 4 cm, L = 20 cm, w = 15 cm	0.01	0.01	0.01	0.02	0.03	0.06	0.20	0.40	0.69	0.70	0.60	0.46	0.36

Source: Lee, H., Sakuma, T., *Appl. Acoust.*, 88, 129–136, 2015.[1]

Table C.3 2D BEM predictions

Surface	*Angle of incidence (°)*	*Frequency (Hz)*																	
		100	*125*	*160*	*200*	*250*	*315*	*400*	*500*	*630*	*800*	*1000*	*1250*	*1600*	*2000*	*2500*	*3150*	*4000*	*5000*
Plane surfaces, non-absorbing, any size	All/any	0.00	0.00	0.00	0.00	0.00	0.00	0.00	0.00	0.00	0.00	0.00	0.00	0.00	0.00	0.00	0.00	0.00	0.00
1. Effect of changing diffuser width and periodicity, semicylinder(s) non-absorbing surfaces, radius 0.3 m (1 cm flat section between each period)																			
1 period, 0.61 m wide	0	0.06	0.09	0.13	0.20	0.29	0.38	0.43	0.45	0.59	0.78	0.82	0.80	0.92	0.89	0.92	0.94	0.95	0.96
	56.9	0.27	0.24	0.23	0.26	0.32	0.45	0.69	0.79	0.80	0.92	0.88	0.91	0.90	0.90	0.91	0.92	0.94	0.96
	Random	0.24	0.21	0.21	0.25	0.32	0.43	0.62	0.73	0.74	0.82	0.86	0.88	0.90	0.91	0.94	0.95	0.96	0.96
2 periods, 1.22 m wide	0	0.04	0.05	0.05	0.04	0.02	0.11	0.65	0.87	0.64	0.56	0.87	0.79	0.90	0.89	0.92	0.93	0.94	0.91
	56.9	0.13	0.12	0.18	0.40	0.43	0.71	0.85	0.93	0.80	0.71	0.81	0.87	0.90	0.89	0.93	0.92	0.94	0.96
	Random	0.16	0.16	0.21	0.32	0.34	0.52	0.80	0.93	0.76	0.67	0.88	0.87	0.91	0.91	0.93	0.94	0.95	0.95
4 periods, 2.44 m wide	0	0.01	0.00	0.02	0.06	0.02	0.08	0.35	0.74	0.55	0.49	0.87	0.77	0.87	0.84	0.80	0.86	0.83	0.88
	56.9	0.14	0.19	0.24	0.34	0.36	0.82	0.87	0.96	0.76	0.52	0.77	0.86	0.90	0.88	0.92	0.92	0.95	0.98
	Random	0.21	0.22	0.24	0.29	0.30	0.53	0.78	0.93	0.71	0.56	0.87	0.86	0.90	0.88	0.90	0.93	0.94	0.94
6 periods, 3.66 m wide	0	0.00	0.01	0.01	0.04	0.03	0.05	0.25	0.55	0.53	0.45	0.86	0.72	0.80	0.81	0.79	0.85	0.82	0.95
	56.9	0.15	0.20	0.23	0.26	0.24	0.84	0.89	0.97	0.74	0.48	0.75	0.85	0.89	0.87	0.94	0.95	0.97	0.98
	Random	0.22	0.23	0.24	0.26	0.26	0.52	0.76	0.91	0.68	0.51	0.86	0.85	0.89	0.88	0.90	0.94	0.94	0.95
12 periods, 7.32 m wide	0	0.00	0.00	0.01	0.02	0.02	0.03	0.13	0.33	0.48	0.40	0.86	0.76	0.84	0.87	0.83	0.81	0.87	0.90
	56.9	0.16	0.15	0.10	0.08	0.12	0.81	0.89	0.97	0.71	0.40	0.74	0.88	0.93	0.90	0.94	0.91	0.93	0.94
	Random	0.23	0.22	0.21	0.20	0.19	0.48	0.73	0.87	0.64	0.46	0.86	0.86	0.89	0.87	0.89	0.92	0.92	0.91

(Continued)

Table C.3 (Continued) 2D BEM predictions

Surface	*Angle of incidence (°)*	*Frequency (Hz)* 100	125	160	200	250	315	400	500	630	800	1000	1250	1600	2000	2500	3150	4000	5000
2. Effect of surface depth, 6 semiellipses, non-absorbing, each width 0.6 m, total width 3.66 m (1 cm flat section between semiellipses)																			
1 cm deep	0	0.00	0.00	0.00	0.00	0.00	0.00	0.00	0.00	0.01	0.01	0.01	0.02	0.02	0.04	0.04	0.10	0.14	0.37
	56.9	0.00	0.00	0.00	0.00	0.00	0.01	0.00	0.00	0.01	0.01	0.01	0.01	0.01	0.02	0.03	0.07	0.21	0.20
	Random	0.00	0.00	0.00	0.00	0.00	0.02	0.01	0.01	0.02	0.01	0.02	0.02	0.03	0.04	0.06	0.10	0.13	0.20
2 cm deep	0	0.00	0.00	0.00	0.00	0.00	0.00	0.00	0.01	0.03	0.04	0.05	0.07	0.09	0.15	0.17	0.36	0.47	0.80
	56.9	0.00	0.00	0.00	0.00	0.00	0.02	0.02	0.02	0.02	0.03	0.03	0.04	0.05	0.08	0.13	0.27	0.53	0.54
	Random	0.00	0.00	0.00	0.00	0.01	0.05	0.04	0.04	0.06	0.05	0.07	0.08	0.10	0.15	0.22	0.33	0.41	0.53
5 cm deep	0	0.00	0.00	0.00	0.00	0.00	0.00	0.01	0.06	0.20	0.23	0.33	0.40	0.51	0.67	0.77	0.86	0.84	0.92
	56.9	0.00	0.00	0.00	0.00	0.01	0.12	0.11	0.11	0.15	0.16	0.19	0.24	0.29	0.39	0.54	0.85	0.93	0.94
	Random	0.00	0.00	0.01	0.01	0.06	0.16	0.16	0.18	0.22	0.25	0.31	0.38	0.47	0.59	0.71	0.80	0.82	0.85
10 cm deep	0	0.00	0.00	0.00	0.00	0.00	0.01	0.03	0.26	0.64	0.71	0.85	0.82	0.73	0.80	0.88	0.86	0.93	0.96
	56.9	0.00	0.00	0.00	0.01	0.05	0.38	0.35	0.37	0.46	0.51	0.56	0.65	0.72	0.81	0.92	0.93	0.94	0.95
	Random	0.02	0.04	0.06	0.09	0.15	0.31	0.36	0.44	0.54	0.64	0.73	0.78	0.80	0.84	0.89	0.91	0.93	0.94
20 cm deep	0	0.00	0.00	0.00	0.01	0.01	0.03	0.12	0.66	0.91	0.61	0.26	0.78	0.86	0.76	0.91	0.88	0.87	0.86
	56.9	0.03	0.04	0.06	0.09	0.19	0.75	0.76	0.88	0.87	0.81	0.70	0.72	0.85	0.92	0.91	0.96	0.97	0.98
	Random	0.14	0.16	0.18	0.20	0.25	0.47	0.64	0.87	0.90	0.76	0.62	0.80	0.90	0.87	0.92	0.93	0.94	0.95
30 cm deep (semicylinder)	0	0.00	0.01	0.01	0.04	0.03	0.05	0.25	0.55	0.53	0.45	0.86	0.72	0.80	0.81	0.79	0.85	0.82	0.95
	56.9	0.15	0.20	0.23	0.26	0.24	0.84	0.89	0.97	0.74	0.48	0.75	0.85	0.89	0.87	0.94	0.95	0.97	0.98
	Random	0.22	0.23	0.24	0.26	0.26	0.52	0.76	0.91	0.68	0.51	0.86	0.85	0.89	0.88	0.90	0.94	0.94	0.95
3. Triangles, non-absorbing, 3.66 m wide (0.01 cm flat section between each period)																			
15 periods, 60°	0	0.01	0.01	0.00	0.00	0.00	0.01	0.02	0.03	0.04	0.06	0.05	0.06	0.37	0.12	0.06	0.12	0.13	0.17
	56.9	0.49	0.60	0.68	0.71	0.58	0.19	0.18	0.33	0.25	0.45	0.23	0.27	0.68	0.65	0.95	0.92	0.93	0.98
	Random	0.34	0.35	0.35	0.34	0.32	0.28	0.24	0.30	0.27	0.32	0.24	0.34	0.50	0.53	0.67	0.69	0.72	0.75

(Continued)

Table C.3 (Continued) 2D BEM predictions

Surface	Angle of incidence (°)	Frequency (Hz) 100	125	160	200	250	315	400	500	630	800	1000	1250	1600	2000	2500	3150	4000	5000
9 periods, 45°	0	0.00	0.00	0.00	0.00	0.01	0.04	0.03	0.07	0.16	0.29	0.30	0.17	0.23	0.22	0.22	0.19	0.23	0.28
	56.9	0.03	0.04	0.06	0.07	0.08	0.20	0.54	0.95	0.99	0.75	0.93	0.97	0.99	0.99	0.99	0.99	1.00	1.00
	Random	0.15	0.17	0.19	0.21	0.23	0.28	0.37	0.59	0.80	0.79	0.69	0.82	0.83	0.87	0.89	0.90	0.90	0.91
6 periods, 30° 3.66 m wide	0	0.00	0.00	0.00	0.00	0.01	0.02	0.06	0.36	0.96	0.98	0.81	0.99	0.89	0.97	0.97	0.99	1.00	0.99
	56.9	0.01	0.02	0.02	0.04	0.18	0.81	0.76	0.72	0.70	0.83	0.90	0.94	0.99	1.00	0.99	0.99	1.00	1.00
	Random	0.11	0.13	0.14	0.17	0.26	0.47	0.61	0.72	0.81	0.90	0.93	0.95	0.95	0.97	0.97	0.98	0.98	0.98
3 periods, 18°	0	0.00	0.01	0.01	0.02	0.17	0.49	0.66	0.78	0.90	0.99	0.98	0.95	0.99	0.99	0.99	0.99	0.99	0.99
	56.9	0.01	0.10	0.37	0.42	0.37	0.39	0.43	0.51	0.59	0.71	0.82	0.94	0.98	0.94	0.94	0.98	0.97	0.98
	Random	0.07	0.17	0.31	0.38	0.41	0.48	0.56	0.67	0.78	0.88	0.93	0.94	0.97	0.98	0.99	0.99	0.99	0.99
4. Semiellipses mounted on 3.63 m wide flat baffle, non-absorbing, each semiellipse 0.6 m wide, 0.2 m deep																			
1 semiellipse in the middle of baffle	0	0.01	0.02	0.03	0.09	0.20	0.22	0.36	0.40	0.35	0.22	0.19	0.27	0.20	0.19	0.14	0.18	0.21	0.30
	56.9	0.09	0.17	0.24	0.29	0.37	0.39	0.43	0.48	0.52	0.47	0.44	0.43	0.37	0.48	0.67	0.74	0.77	0.60
	Random	0.15	0.22	0.28	0.32	0.38	0.42	0.47	0.51	0.53	0.48	0.44	0.46	0.48	0.49	0.50	0.53	0.52	0.54
3 semiellipses with 0.6 m flat sections between	0	0.00	0.01	0.03	0.05	0.26	0.72	0.88	0.89	0.70	0.46	0.53	0.65	0.63	0.58	0.51	0.66	0.68	0.76
	56.9	0.03	0.17	0.52	0.62	0.56	0.66	0.78	0.80	0.90	0.95	0.94	0.72	0.54	0.72	0.90	0.80	0.91	0.91
	Random	0.13	0.22	0.38	0.50	0.57	0.67	0.79	0.85	0.85	0.76	0.69	0.75	0.77	0.75	0.79	0.85	0.81	0.78
5. Optimized curved surfaces, 3.6 m wide, modulated arrays																			
3 periods, 30 cm deep	0	0.00	0.03	0.06	0.16	0.59	0.95	0.99	0.95	0.88	0.71	0.88	0.89	0.92	0.79	0.92	0.93	0.82	0.95
	56.9	0.05	0.31	0.74	0.76	0.70	0.68	0.85	0.88	0.92	0.95	0.94	0.89	0.94	0.96	0.95	0.99	0.97	0.99
	Random	0.15	0.30	0.50	0.64	0.74	0.79	0.90	0.94	0.89	0.86	0.91	0.93	0.95	0.94	0.94	0.95	0.96	0.97
6 periods, 20 cm deep	0	0.00	0.00	0.00	0.01	0.02	0.04	0.14	0.61	0.81	0.94	0.80	0.78	0.68	0.84	0.60	0.84	0.78	0.86
	56.9	0.01	0.02	0.02	0.04	0.23	0.73	0.79	0.88	0.98	0.86	0.78	0.92	0.94	0.94	0.98	0.98	0.96	0.96
	Random	0.09	0.11	0.14	0.16	0.27	0.48	0.68	0.85	0.94	0.89	0.80	0.86	0.90	0.93	0.89	0.94	0.95	0.94

(Continued)

Table C.3 (Continued) 2D BEM predictions

Surface	Angle of incidence (°)	Frequency (Hz)																	
		100	*125*	*160*	*200*	*250*	*315*	*400*	*500*	*630*	*800*	*1000*	*1250*	*1600*	*2000*	*2500*	*3150*	*4000*	*5000*
6 periods, 10 cm deep	0	0.00	0.00	0.00	0.01	0.01	0.02	0.07	0.15	0.23	0.30	0.38	0.82	0.86	0.80	0.89	0.73	0.90	0.89
	56.9	0.00	0.01	0.01	0.03	0.11	0.15	0.20	0.24	0.67	0.71	0.77	0.81	0.82	0.89	0.96	0.97	1.00	0.99
	Random	0.01	0.03	0.04	0.09	0.15	0.20	0.25	0.31	0.50	0.62	0.73	0.84	0.89	0.92	0.95	0.94	0.95	0.96
6 periods, 5 cm deep	0	0.00	0.00	0.00	0.00	0.00	0.00	0.01	0.02	0.03	0.04	0.13	0.49	0.62	0.80	0.90	0.93	0.90	0.91
	56.9	0.00	0.00	0.00	0.01	0.01	0.02	0.02	0.04	0.31	0.39	0.42	0.42	0.43	0.52	0.69	0.81	0.93	0.95
	Random	0.00	0.01	0.01	0.02	0.05	0.06	0.06	0.12	0.29	0.36	0.43	0.50	0.58	0.70	0.82	0.88	0.91	0.92
7. Schroeder diffusers, 3.6 m wide																			
N = 7 QRD, 6 periods, 0.2 m deep	0	0.00	0.01	0.02	0.05	0.08	0.07	0.04	0.35	0.51	0.67	0.57	0.52	0.44	0.73	0.85	0.39	0.71	0.86
	56.9	0.04	0.06	0.12	0.26	0.41	0.17	0.16	0.71	0.75	0.84	0.71	0.51	0.74	0.91	0.83	0.79	0.67	0.93
	Random	0.12	0.15	0.19	0.27	0.36	0.30	0.25	0.69	0.78	0.72	0.69	0.51	0.66	0.86	0.82	0.69	0.70	0.88
Optimized diffuser, modulated array, 6 periods, 8 wells/ period, 0.17 m deep	0	0.00	0.01	0.02	0.04	0.20	0.21	0.27	0.65	0.78	0.80	0.85	0.86	0.95	0.92	0.95	0.82	0.88	0.96
	56.9	0.03	0.07	0.11	0.22	0.40	0.43	0.55	0.58	0.76	0.90	0.74	0.89	0.95	0.96	0.94	0.96	0.96	0.97
	Random	0.09	0.14	0.18	0.25	0.36	0.48	0.55	0.68	0.82	0.89	0.83	0.87	0.93	0.94	0.93	0.90	0.91	0.94
N = 7 PRD, 6 periods, 6 wells/ period, 0.2 m deep	0	0.00	0.01	0.02	0.05	0.04	0.14	0.19	0.14	0.32	0.81	0.65	0.69	0.88	0.88	0.97	0.98	0.97	0.42
	56.9	0.04	0.09	0.18	0.38	0.41	0.47	0.66	0.59	0.61	0.83	0.96	0.93	0.88	0.94	0.92	0.89	0.95	0.96
	Random	0.11	0.15	0.20	0.29	0.34	0.47	0.52	0.52	0.62	0.88	0.85	0.82	0.88	0.92	0.92	0.90	0.93	0.75
Diffractal, 1 period, 3 orders of size, N = 7, 0.5 m deep	0	0.16	0.45	0.28	0.39	0.70	0.85	0.87	0.83	0.73	0.79	0.89	0.86	0.96	0.96	0.89	0.84	0.91	0.90
	56.9	0.30	0.74	0.76	0.63	0.67	0.64	0.84	0.70	0.77	0.87	0.92	0.77	0.89	0.93	0.92	0.95	0.89	0.96
	Random	0.33	0.56	0.66	0.61	0.72	0.80	0.88	0.80	0.78	0.85	0.91	0.80	0.91	0.92	0.91	0.91	0.90	0.91
Optimized diffuser, 6 period modulated array, 12 wells/ period, 0.17 m deep	0	0.00	0.01	0.01	0.03	0.14	0.20	0.50	0.78	0.82	0.84	0.92	0.81	0.89	0.87	0.84	0.92	0.92	0.94
	56.9	0.04	0.08	0.13	0.24	0.50	0.70	0.67	0.90	0.92	0.92	0.95	0.88	0.91	0.96	0.97	0.99	0.98	0.98
	Random	0.14	0.17	0.21	0.27	0.40	0.61	0.69	0.90	0.90	0.91	0.92	0.87	0.87	0.90	0.91	0.93	0.94	0.94

REFERENCE

1. H. Lee and T. Sakuma, "Numerical characterization of acoustic scattering coefficients of one-dimensional periodic surfaces", *Appl. Acoust.*, **88**, 129–36 (2015).

Appendix D: Random incidence scattering coefficient table

Measured at full or model scale according to ISO 17497-1.

Surface	Frequency (Hz)																	
	100	*125*	*160*	*200*	*250*	*315*	*400*	*500*	*630*	*800*	*1000*	*1250*	*1600*	*2000*	*2500*	*3150*	*4000*	*5000*
Sinusoidal 1D corrugation [1]																		
h = 5.1 cm, L = 17.7 cm	0.02	0.02	0.03	0.03	0.04	0.04	0.05	0.05	0.1	0.13	0.22	0.42	0.59	0.72	0.8	0.85	0.87	0.89
Periodic 1D battens																		
$h = w$ = 10 cm, L = 2h [2]	0.00	0.01	0.01	0.03	0.05	0.08	0.16	0.24	0.28	0.25	0.34	0.37	0.25	0.28	0.6	0.65	0.67	–
$h = w$ = 5 cm, L = 2h [3]	0.01	0.00	0.00	0.01	0.02	0.02	0.00	0.06	0.16	0.25	0.25	0.31	0.36	0.52	0.44	0.31	0.32	0.61
$h = w$ = 10 cm, L = 2h [3]	0.00	0.01	0.09	0.05	0.02	0.14	0.21	0.32	0.44	0.47	0.44	0.36	0.26	0.36	0.52	0.61	0.5	0.63
$h = w$ = 22 cm, L = 2h [3]	0.01	0.08	0.12	0.23	0.42	0.42	0.54	0.5	0.29	0.35	0.66	0.68	0.49	0.56	0.61	0.48	0.6	0.57
Parallel V-shaped grooves cut into rubber sheet [4]																		
Percentage of baseplate surface area that is grooves, P (%) h = 21 cm, w = 24 cm, N grooves																		
N = 3, P = 20%	0.10	0.02	0.00	0.04	0.06	0.02	0.00	0.25	0.29	0.51	0.38	0.55	0.45	0.22	0.41	0.52	0.53	–
N = 5, P = 30%	0.12	0.07	0.04	0.04	0.04	0.04	0.03	0.18	0.28	0.57	0.74	0.67	0.44	0.57	0.58	0.60	0.73	–
N = 7, P = 44%	0.10	0.02	0.05	0.04	0.04	0.02	0.09	0.49	0.56	0.74	0.85	1.17	0.73	0.67	0.61	0.87	0.83	–
N = 9, P = 56%	0.09	0.01	0.01	0.00	0.00	0.04	0.03	0.10	0.08	0.57	0.90	0.88	0.86	0.62	0.62	0.83	0.81	–

(Continued)

Surface	Frequency (Hz)																	
	100	*125*	*160*	*200*	*250*	*315*	*400*	*500*	*630*	*800*	*1000*	*1250*	*1600*	*2000*	*2500*	*3150*	*4000*	*5000*
Periodic 2D blocks																		
Plan; Section (L, w, h)																		
h = *w* = 10 cm, *L* = 2h [2]	0.00	0.00	0.01	0.01	0.01	0.00	0.00	0.07	0.04	0.08	0.19	0.23	0.22	0.32	0.43	0.56	0.49	–
h = *w* = 20 cm, *L* = 2h [5]	0.00	0.00	0.03	0.00	0.00	0.00	0.10	0.11	0.16	0.17	0.20	0.26	0.31	0.34	0.45	0.49	0.47	0.44
Randomly arranged 2D blocks [5]																		
All blocks *h* = *w* = 20 cm, baseplate ø = 3.75 m																		
18% coverage density (50 blocks)	0.00	0.01	0.02	0.00	0.00	0.04	0.32	0.25	0.21	0.17	0.25	0.32	0.35	0.34	0.43	0.53	0.47	0.55
27% coverage density (75 blocks)	0.00	0.04	0.06	0.02	0.07	0.08	0.39	0.29	0.32	0.20	0.22	0.35	0.35	0.38	0.47	0.50	0.57	0.55
Pyramids [6]																		
Plan (b); Section (L, h)																		
h = 30.5 cm, *L* = *b* = 2h	0.00	0.00	0.00	0.00	0.00	0.02	0.05	0.08	0.12	0.18	0.22	0.22	0.39	0.41	0.55	0.69	0.86	1.00
h = 30.5 cm, *L* = *b* = 2h One in four pyramid corners raised from baseplate	0.00	0.00	0.00	0.00	0.01	0.05	0.05	0.10	0.21	0.30	0.38	0.49	0.68	0.74	0.82	0.93	0.99	1.00

(*Continued*)

Surface	Frequency (Hz)																	
	100	*125*	*160*	*200*	*250*	*315*	*400*	*500*	*630*	*800*	*1000*	*1250*	*1600*	*2000*	*2500*	*3150*	*4000*	*5000*
h = 30.5 cm, L = b = h	0.00	0.00	0.00	0.00	0.00	0.00	0.00	0.04	0.10	0.19	0.31	0.35	0.52	0.58	0.68	0.77	0.88	1.00
h = 30.5 cm, L = b = 2h One in four pyramid corners raised from baseplate	0.00	0.00	0.00	0.00	0.00	0.00	0.00	0.03	0.11	0.27	0.44	0.52	0.73	0.76	0.91	0.91	0.99	–
V-shaped grooves in 2D pattern, cut into rubber sheet [4] Percentage of baseplate surface area that is grooves, P (%) h = 21 cm, w = 24 cm, N grooves																		
N = 2, P = 14%	0.17	0.06	0.07	0.05	0.03	0.05	0.07	0.16	0.18	0.23	0.46	0.41	0.33	0.38	0.35	0.43	0.50	–
N = 6, P = 36%	0.11	0.04	0.07	0.06	0.04	0.05	0.07	0.17	0.23	0.40	0.72	0.69	0.57	0.64	0.63	0.69	0.75	–
N = 10, P = 50%	0.11	0.05	0.03	0.01	0.01	0.05	0.08	0.27	0.32	0.51	0.88	0.95	0.76	0.67	0.74	0.86	0.87	–
Wooden hemispheres covering [4] h = height, percentage of baseplate surface area covered, P (%)																		
h = 7.5 mm; P = 14%	0.00	0.00	0.01	0.00	0.02	0.00	0.00	0.03	0.00	0.00	0.01	0.04	0.09	0.20	0.21	0.24	0.25	
h = 7.5 mm; P = 28%	0.14	0.00	0.01	0.00	0.00	0.00	0.00	0.00	0.02	0.01	0.01	0.04	0.17	0.31	0.46	0.47	0.46	
h = 7.5 mm; P = 57%	0.12	0.00	0.00	0.00	0.03	0.00	0.00	0.00	0.00	0.00	0.11	0.09	0.13	0.40	0.57	0.52	0.41	
h = 7.5 mm; P = 71%	0.04	0.00	0.00	0.00	0.00	0.00	0.00	0.00	0.00	0.04	0.02	0.00	0.16	0.40	0.45	0.46	0.58	
h = 10 mm; P = 14%	0.01	0.01	0.00	0.00	0.00	0.00	0.00	0.00	0.00	0.02	0.10	0.12	0.25	0.29	0.28	0.35	0.31	
h = 10 mm; P = 28%	0.00	0.00	0.00	0.00	0.01	0.00	0.00	0.05	0.01	0.00	0.05	0.19	0.38	0.46	0.56	0.54	0.53	
h = 10 mm; P = 57%	0.05	0.00	0.00	0.00	0.01	0.00	0.00	0.00	0.00	0.02	0.07	0.14	0.41	0.52	0.74	0.67	0.51	

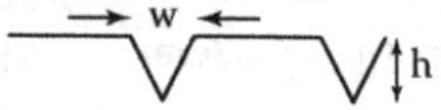

(Continued)

Surface	Frequency (Hz)																	
	100	*125*	*160*	*200*	*250*	*315*	*400*	*500*	*630*	*800*	*1000*	*1250*	*1600*	*2000*	*2500*	*3150*	*4000*	*5000*
h = 10 mm; *P* = 71%	0.05	0.00	0.00	0.00	0.01	0.00	0.00	0.00	0.00	0.03	0.06	0.14	0.34	0.57	0.54	0.66	0.53	
h = 15 mm; *P* = 14%	0.03	0.00	0.00	0.00	0.00	0.00	0.01	0.02	0.03	0.11	0.19	0.27	0.27	0.34	0.31	0.37	0.31	
h = 15 mm; *P* = 28%	0.00	0.01	0.00	0.00	0.00	0.00	0.00	0.04	0.03	0.08	0.32	0.53	0.55	0.54	0.44	0.41	0.32	
h = 15 mm; *P* = 57%	0.05	0.01	0.01	0.00	0.00	0.00	0.00	0.00	0.05	0.03	0.44	0.72	0.63	0.63	0.57	0.44	0.30	
h = 15 mm; *P* = 71%	0.03	0.01	0.02	0.00	0.00	0.00	0.00	0.00	0.00	0.11	0.33	0.48	0.62	0.79	0.56	0.54	0.57	
h = 20 mm; *P* = 14%	0.10	0.02	0.01	0.01	0.00	0.00	0.05	0.08	0.14	0.15	0.28	0.36	0.30	0.31	0.28	0.32	0.24	
h = 20 mm; *P* = 28%	0.15	0.00	0.01	0.05	0.03	0.00	0.03	0.03	0.16	0.30	0.44	0.56	0.54	0.53	0.47	0.54	0.42	
h = 20 mm; *P* = 57%	0.03	0.01	0.00	0.01	0.01	0.01	0.00	0.02	0.18	0.42	0.76	0.75	0.85	0.76	0.54	0.61	0.76	
h = 20 mm; *P* = 71%	0.09	0.00	0.00	0.00	0.00	0.00	0.00	0.03	0.11	0.27	0.58	0.81	0.70	0.71	0.67	0.77	0.68	
h = 25 mm; *P* = 14%	0.02	0.00	0.00	0.00	0.01	0.00	0.06	0.15	0.21	0.26	0.29	0.31	0.38	0.35	0.30	0.29	0.31	
h = 25 mm; *P* = 28%	0.08	0.02	0.02	0.01	0.00	0.00	0.03	0.22	0.27	0.31	0.43	0.58	0.52	0.43	0.42	0.43	0.41	
h = 25 mm; *P* = 57%	0.00	0.01	0.00	0.00	0.01	0.01	0.04	0.09	0.31	0.58	0.77	0.77	0.75	0.61	0.56	0.62	0.63	
h = 25 mm; *P* = 71%	0.05	0.00	0.00	0.00	0.01	0.01	0.04	0.06	0.23	0.47	0.73	0.78	0.83	0.75	0.65	0.70	0.77	
Vegetation with different percentage coverage of base plate																		
Box (Buxus), 20% [7]	0.01	0.01	0.01	0.00	0.00	0.03	0.02	0.00	0.02	0.02	0.00	0.00	0.00	0.00	0.01	0.01	0.10	0.06
Box (Buxus), 60% [7]	0.00	0.00	0.00	0.01	0.00	0.01	0.03	0.00	0.05	0.02	0.01	0.00	0.01	0.05	0.02	0.11	0.15	0.20
Box (Buxus), 100% [7]	0.00	0.00	0.00	0.00	0.03	0.03	0.02	0.02	0.03	0.02	0.00	0.03	0.06	0.07	0.05	0.11	0.16	0.26
Holly, 20% [7]	0.01	0.04	0.01	0.05	0.00	0.01	0.03	0.04	0.01	0.00	0.01	0.00	0.01	0	0.03	0.01	0.07	0.07
Holly, 60% [7]	0.06	0.00	0.02	0.04	0.01	0.00	0.01	0.06	0.03	0.00	0.00	0.01	0.02	0.06	0.08	0.08	0.16	0.15
Holly, 100% [7]	0.00	0.00	0.01	0.06	0.00	0.00	0.02	0.04	0.00	0.00	0.00	0.02	0.01	0.03	0.09	0.09	0.16	0.33

REFERENCES

1. M. Vorländer, J. J. Embrechts, L. De Geetere, G. Vermeir, and M. Gomes, "Case studies in measurement of random incidence scattering coefficients", *Acta Acust. Acust.*, **90**, 858–67 (2004).
2. T. Sakuma and H. Lee, "Validation of the sample rotation scheme in the measurement of random-incidence scattering coefficients", *Acta Acust. Acust.*, **99**(5), 737–50 (2013).
3. Y. J. Choi, "Effects of periodic type diffusers on classroom acoustics", *Appl. Acoust.*, **74**(5), 694–707 (2013).
4. Y. H. Kim, H. S. Jang, and J. Y. Jeon, "Characterizing diffusive surfaces using scattering and diffusion coefficients", *Appl. Acoust.*, **72**(11), 899–905 (2011).
5. L. Shtrepi, A. Astolfi, G. D'Antonio, G. Vannelli, G. Barbato, S. Mauro, and A. and Prato, "Accuracy of the random-incidence scattering coefficient measurement", *Appl. Acoust.*, **106**, 23–35 (2016).
6. A. F. Sharma and D. T. Bradley, "Assessing the effectiveness of geometrically modified pyramidal diffusers: Scattering coefficient measurements", *J. Acoust. Soc. Am.*, **134**(5), 4095 (2013).
7. H. S. Yang, J. Kang, and C. Cheal, "Random-incidence absorption and scattering coefficients of vegetation", *Acta Acust. Acust.*, **99**(3), 379–88 (2013).

Index

Note: Page numbers followed by 'f' and 't' denotes figures and tables respectively.

A

B

D

E

F

G

N

O

P

Q

R

S

T

U

V

W

X

Y

Printed in the United States
by Baker & Taylor Publisher Services